Don McAdam
University of British Columbia

Roger Winn
University College of Cape Breton

Engineering Graphics
a problem-solving approach

third edition

Toronto

Library and Archives Canada Cataloguing in Publication

McAdam, D. (Donald), 1940–
Engineering graphics: a problem-solving approach / Don McAdam, Roger Winn.—3rd ed.

Includes index.
ISBN 0-321-30819-0

1. Engineering graphics—Textbooks. 2. Engineering graphics—Problems, exercises, etc.
I. Winn, R. (Roger) II. Title.

T353.M32 2007 604.2 C2005-905846-3

ISBN 0-321-30819-0

Vice-President, Editorial Director: Michael J. Young
Editor-in-Chief: Gary Bennett
Marketing Manager: Janet Piper
Developmental Editor: Rema Celio
Production Editor: Laura Price
Copy Editor: Rohini Herbert
Proofreader: Valerie Adams
Production Coordinator: Andrea Falkenberg
Page Layout: Carol J. Anderson
Art Director: Julia Hall
Cover and Interior Design: Anthony Leung
Cover Image: Peter Christopher, Masterfile

4 5 11 10 09 08

Printed and bound in the USA.

Toronto

Brief Contents

Contents

Preface

Engineering Graphics: A Problem-Solving Approach, Third Edition, is an innovative text that provides a fresh perspective on engineering graphics. It is designed for engineering students taking a one-semester engineering graphics course.

A first-year engineering program is often full of mathematics, chemistry, and physics courses. An engineering graphics course may be the only engineering course in that year. In many instances, students have little knowledge of materials, manufacturing, and mechanics. Often, they have no background in engineering, and therefore they are not equipped to do any engineering calculations (although calculations make up only a very small part of engineering work). This is where *Engineering Graphics* fills an important gap.

The goal of this text is to engage students in critical thinking about graphics problems. It is also a quick source of information that many students use in their engineering courses and careers. However, it is not intended to train drafters in all aspects of engineering drawing.

Engineers must know how to read engineering drawings and understand the methods and conventions used in creating them. They must be able to visualize three-dimensional objects from two-dimensional drawings; explain their ideas and plan their work with sketches; apply what they know to new problems; approach projects in an organized, systematic way; and communicate their findings in an effective manner. This book is intended to develop all of these skills.

The text is divided into 10 chapters that cover the principles of projection, sketching, sectioning, dimensioning, engineering drawings, visualization, intersection and development, and presenting technical information. It also has a series of appendices that provide a wealth of useful supplementary information on such topics as geometry, vector analysis, standards, symbols, and notations.

Each chapter presents information in an easy-to-read format with plenty of illustrations. To assist students who may know little about materials, manufacturing, and machining processes, the authors use practical examples and a step-by-step approach to explain a variety of engineering drawings and calculations. In addition, all chapters in the text have a large number of graphics problems that challenge students to apply their knowledge. The problems range from basic ones that establish familiarity with the chapter content to challenging, open-ended ones.

Open-ended problems, with several possible correct solutions, are the norm in engineering, but students traditionally do not encounter them until the later years of their study. Open-ended problems can be frustrating to many students who are accustomed to one-correct-answer problems. Open-ended problems are included in this text to encourage students to think on their own and to build confidence in their own abilities. In addition, these problems help students develop a systematic approach to problem solving that will be invaluable to them in their future engineering studies and careers.

NEW IN THIS EDITION

- Expanded coverage of tolerances, including geometric tolerancing and new figures
- A chapter on engineering design
- A chapter on the basics of solid modelling
- Folding exercises included in the text's worksheets to enhance the discussion of 3-D visualization

COVERAGE

Chapter 1 of *Engineering Graphics: A Problem-Solving Approach,* Third Edition, introduces students to the principles of projection. This information forms the foundation for Chapter 2, which introduces students to sketching.

Chapters 3 and 4 on sectioning and dimensioning introduce basic techniques, standard conventions, and guidelines. Because sectioning and dimensioning are not random processes, the techniques, conventions, and guidelines presented in the chapters help students detail simple parts correctly.

Since first-year engineering courses include students who will go into different branches of engineering and because there is so much interaction among the various branches of engineering, Chapter 5 introduces students to drawings from mechanical, civil, chemical, electrical, and electronic engineering. It is not possible to cover all aspects of engineering drawings in a one-semester course; however, some of the most common ones have been included. Students who need to learn more about a particular topic in later years will have a basic background to do so after completing Chapter 5.

In Chapter 6, students learn about visualization, with particular emphasis on points, lines, and planes. This chapter also includes information on contours and 3-D representation.

While many engineering drawings are created with computer programs, such programs do not think for the user; a computer is merely a tool to help users be more productive. People who do not know anything about engineering drawings, rules, and conventions find a computer program of little use to them. Just because something is drawn using a computer does not make it correct. The computer adage "garbage in, garbage out" always applies. Users need to know what they want to do before they can use their computers effectively. To underscore this point, the last section of Chapter 6 shows students how to solve various problems using a computer.

Chapter 7 introduces students to intersection and development. This chapter includes many practical examples to help students achieve a thorough understanding of the concepts presented in the chapter.

Chapter 8 is an introduction to engineering design. It is certainly not possible to cover much on engineering design in one short chapter, and it is intended to give first-year students an introduction to the design process. Some of the sources of information are discussed and the importance of standards emphasized. Sources of standards are given in Appendix D. Teamwork is discussed, as it is so important to the success of any engineering project. A list of projects suitable for beginning students is given. These projects have been tried with good results. Students were able to design and build a mechanism to meet the objectives using simple materials that were easy to find.

Chapter 9 covers some basics of solid modelling. This chapter gives examples of common procedures in modelling software and is not based on a specific program. It is intended to give an introduction to the power of modelling programs and what they can do.

Drawings are one form of engineering communication, but technical writing is even more important. Engineers spend 60 to 70 percent of their time communicating in one form or another. They need to be able to answer questions and convey their ideas in an understandable manner. Few people will advance very far in engineering if they cannot express their ideas in a clear, concise, concrete, correct, and courteous manner (the five Cs of communication). Chapter 10 introduces students to the basics of technical writing and provides hints on how to prepare effective communications, such as business letters and engineering reports.

SUPPLEMENTS

Worksheets that include folding exercises, as well as all the figures required for the chapter problems, are shrink-wrapped with the text.

The following supplements are also available:

- *Instructor's Manual.* This manual contains solutions for most of the problems in the text.
- NEW—*Guide to Solid Edge® v18*, by Robert Fleisig. Solid Edge is a mid-range software package in everyday use in industry. This manual is a practical tool for working with the Solid Edge modelling software. Its focus is to help students learn how to use Computer Aided Design (CAD) software tools for the design of mechanical parts. At the same time, students will learn to read detail and assembly drawings, as well as improve their visualization skills.

ABOUT THE AUTHORS

This book is based on many years of teaching engineering graphics to engineering students and many years of engineering practice as professional engineers. Don McAdam is a registered professional engineer in British Columbia, and Roger Winn is a certified manufacturing engineer in Nova Scotia; between them, they have over 70 years of engineering experience in academic institutions and industry.

Don McAdam worked for a number of years in the nuclear industry after receiving his B.Sc. from the University of Alberta. He received a Ph.D. in 1975 and has been in the Mechanical Engineering Department at the University of British Columbia since then and has been in charge of the engineering graphics courses. He has been a consultant to industry and the legal profession on numerous things, including engineering drawings, and currently teaches engineering design and heat transfer.

Roger Winn has worked in heavy industrial engineering, in tool and die, and with engineering consultants. He also acts as a consultant on Geometric Tolerancing Application to manufacturing industries. Since 1987, he has taught at the University College of Cape Breton in Sydney, Nova Scotia, where he is currently chair of the engineering department and coordinator of the Bachelor of Technology in Manufacturing program. Roger has a Master's Degree in Computer Aided Manufacturing and is a senior member of the Society of Manufacturing Engineers.

ACKNOWLEDGMENTS

The publishers and authors would like to thank the following people who reviewed the third edition and/or portions of the manuscript:

Robert V. Fleisig, *McMaster University*
Gerry Hoye, *University of Alberta*
Joan Hunter, *St. Clair College*
Robert Jones, *University of Regina*
Dragan Jovanov, *Sheridan College*
Erik Luczak, *Red River College*
Emeka Oguejiofor, *St. Francis Xavier University*
Elvis Pittman, *College of the North Atlantic*
Filippo A. Salustri, *Ryerson University*
Kenneth H. Slaney, *College of the North Atlantic*

We also thank those we worked with at Pearson Education Canada for all their help with this project: Lori Will, the sponsoring editor; Kelly Torrance, the acquisitions editor; Meaghan Eley and Rema Celio, the developmental editors; and Laura Price, the production editor.

I would especially like to thank my wife, Karen, for her patience, understanding, and support throughout the creation of this book; I dedicate this book to her. (D.M.)

Many thanks to my family for their support while I was working on this book, which I dedicate to my son, Patrick, for the time missed while working on it! (R.W.)

When I was a kid growing up in Far Rockaway, I had a friend named Bernie Walker. We both had 'labs' at home, and we would do various 'experiments.' One time, we were discussing something—we must have been eleven or twelve at the time—and I said, 'But thinking is nothing but talking to yourself inside.'

'Oh yeah?' Bernie said. 'Do you know the crazy shape of a crankshaft in a car?'

'Yeah, what of it?'

'Good. Now, tell me: how did you describe it when you were talking to yourself?' So I learned from Bernie that thoughts can be visual as well as verbal.

Richard P. Feynman, Nobel Laureate, Physics, What Do You Care What Other People Think? *New York: W.W. Norton and Company Inc., 1988*

CHAPTER

1 Principles of Projection

Engineers spend only a small percentage of their time doing the calculations you spend so long in school learning to do. They spend most of their time communicating with others verbally; by listening; through the written word, e-mail, memos, letters, reports, and specifications; and through engineering drawings and sketches. An idea, or design, begins in the engineer's head but is of no use to anyone unless it can be communicated clearly to others—not only to other engineers, but to marketing, manufacturing, and, perhaps most importantly, to a sponsor or management that must finance the project. This is done with words as well as with pictures and drawings. Pictures, conceptual drawings, and artists' conceptions are used in the initial stages for presentation, but when the project is approved for construction, there can be thousands of engineering drawings and volumes of specifications.

Since engineering drawings represent three-dimensional (3-D) objects, you must be able to visualize them as 3-D when you see them represented on a two-dimensional (2-D) surface. You must make the translation from a 2-D representation on a piece of paper or a computer screen to a 3-D object in your mind. This can be easy if it is a familiar object. You will have no trouble visualizing a car after looking at a drawing of a car, but what if it is something you have never seen before? This is where understanding the techniques and conventions of engineering drawing comes in. And what if the object is something that you designed and is only a picture in your mind? How will you put that into two dimensions so that someone else can understand what you are doing? This book is designed to develop your ability to think visually and help you do these things.

Every profession has its own unique language, and engineering is no exception. *Engineering graphics* is the visual language of engineering. There are conventional ways of presenting information that every engineer must know. Once they have been learned, we do not even think about them, but they must be learned. An engineering drawing can be very intimidating. There are thick lines, thin lines, solid lines, various types of dashed lines, numbers, and text, not to mention hundreds of special symbols that relate to various branches of engineering. Even text can contain abbreviations and symbols. No one expects you to know all of the symbols and abbreviations—there are too many—but you will get to know the common ones related to your branch of engineering.

The first chapter deals with the different kinds of drawings used in engineering, the most common of which is orthographic drawings showing different views of an object. This is the kind of drawings used in manufacture and construction. You will learn the standard conventions of locating the views and how to select the main view of a drawing. You will also learn the different types of pictorial drawing that show more than one side of an object. These are still 2-D drawings but are often, wrongly, called 3-D drawings. They are used in solid modelling and sketching.

Since ideas start in the mind of the engineer, the first visual record is usually a sketch, and this is often a pictorial sketch because it shows more information. Engineers create many sketches to record their ideas (thinking sketches), to explain their ideas to others (talking sketches), and to plan their work before formal drawings are created with a computer program. Planning your work with a sketch can save you a lot of time, and preliminary sketches can be done very quickly. No one expects you to be an artist, but by using simple construction techniques, you can create understandable sketches. They will not be exact, but that is not the point. No one is going to measure them anyway. They will enable you to quickly record your ideas and explain them to others. Try explaining a safety pin without the aid of a sketch or picture. Chapter 2 deals with orthographic and pictorial sketching.

If something is going to be made, the details and dimensions must be specified, and there are recognized standards for doing this. Someone making the things you design will expect the information to be given in this standard way. If it is not, mistakes will be made that will take time, cost money, and could even take lives. Chapters 3 and 4 deal with the conventional techniques used in showing internal details (called sectioning) and dimensioning. These are not complicated, but you must know them in order to understand the information given on an engineering drawing. You, as an engineer, may not actually make an engineering drawing yourself, but you will have to read one, and if you are the designer, you will sign it and put your stamp on it—i.e., take responsibility for it. You cannot blame the person who drew it if something goes wrong. As the engineer who signs and stamps the drawing, you are responsible. You must understand everything about the drawings for which you take responsibility.

An engineering drawing is drawn full size using a computer program, but it is not printed full size, unless it is small and will fit on standard-size drawing paper. Large objects are printed at reduced size. A reduction scale is used when the drawing is printed, and the scales used are specified. Drawing scales are discussed in Chapter 4.

Organizations such as the Canadian Standards Association (CSA), the American Society of Mechanical Engineering (ASME), the American Society for Testing and Material (ASTM), and the International Organization for Standardization (ISO), to name a few, publish standards on a large number of products and materials. Standards ensure that when you specify a material or a component that meets a recognized standard, you know what you are getting. A list of standards pertaining to engineering drawings is given in Appendix D, along with Web sites of organizations that establish standards. Drawing format, layout, and drawing sizes are all standardized. This is discussed in more detail in Chapter 5, and examples of drawings from different branches of engineering are discussed.

As a first- or second-year engineering student, you will know little about strength of materials, material properties, and other things necessary to analyze an engineering problem. These things come later. You can solve graphical engineering problems, with a bit of instruction. Chapters 6 and 7 contain engineering problems that may be solved without knowing advanced analytical methods. In fact, the easiest way to solve some of them is by drawing the appropriate views, to get the information you want. Working on these problems also helps you develop spatial perception skills—skills that are valuable no matter what kind of engineering you plan on taking.

The engineering design process is discussed in Chapter 8. Engineering drawing is one of the tools used in engineering design. Drawings are used throughout the design process to develop and explain ideas. Detailed design drawings give the information necessary to build the product. Sketching is a fast way to record and explain your ideas. Pictorial drawings done with a computer show how parts and components will fit together, and what they look like when seen from different viewpoints. This technique, called solid modelling, is dealt with in Chapter 9. It allows the engineers to visualize designs quickly, make changes easily, and create new designs faster.

While engineering drawing and sketching are important ways that engineers communicate, the written word is also extremely important. You will not advance very far in engineering, or in any other profession, if you cannot communicate your ideas in writing. It is not possible to cover technical communication in one chapter, but Chapter 10 is intended to start you off on the right track. Some drawing courses will not cover this material, but it is included because of its importance to the education of an engineer. Perhaps you took engineering because you did not like English and thought you would not have to be bothered with it in engineering, but it is vital to your advancement as an engineer. The basics of technical communication are given in Chapter 10 to assist you in this vital part of engineering communication.

Let's start by looking at how objects are formed.

HOW OBJECTS ARE FORMED

Objects are formed by combining simple, basic shapes called **primitives**. A rectangular prism, shown in Figure 1.1, is an example of a basic shape. Objects can also be formed by combining surfaces to enclose the boundary of the object. Only four shapes (or surfaces)—plane, cylinder, cone, and sphere—are needed to define most parts. Ninety to 95 percent of parts can be modelled by planes, cylinders, and cones. The process is called **solid modelling** and will be covered in more detail later, but first, we must deal with some basics. We will combine shapes to illustrate how objects are formed.

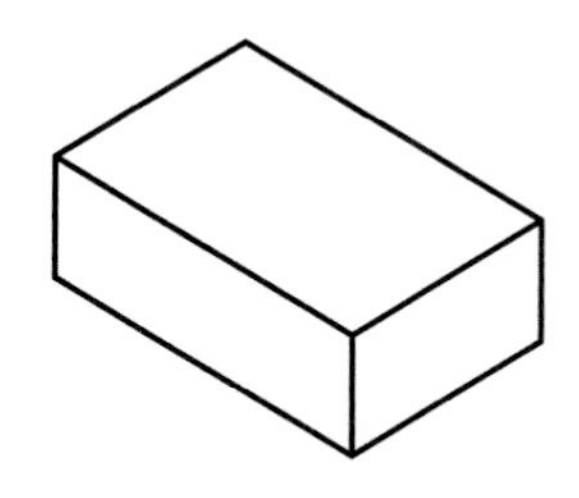

Figure 1.1 Rectangular prism

Basic shapes can be enlarged, shrunk, combined, or subtracted from other basic shapes to create new, more complicated objects. For example, if a small rectangular prism is removed from a larger rectangular prism, a new object is created (Figure 1.2). The shape that is subtracted is considered to be a negative volume.

The new object, shown in Figure 1.3, is more complex because it has more surfaces, but it is still made up of simple shapes. The usual way of doing this is to start with some solid shape and create the shape you want by removing other shapes.

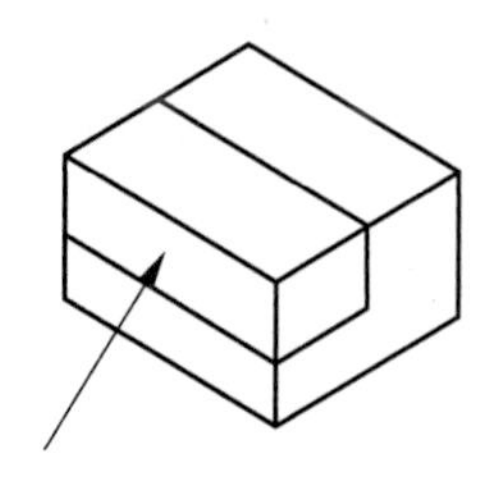

Figure 1.2 Removing one basic shape from another to create a new object

Obviously, a more complex object can be created by combining as many basic shapes as required. A cylinder, triangular prism, and rectangular prism, shown in

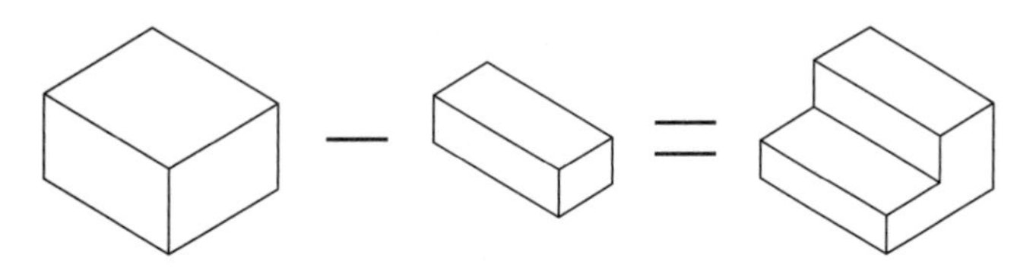

Figure 1.3 New object resulting from removing a rectangular prism

Figure 1.4, can be combined to form a simple building with a chimney, as shown in Figure 1.5. A **line of intersection** between the cylinder and the triangular prism is created where the two solids intersect.

Figure 1.6 shows how a new object can be created by removing two triangular prisms from a rectangular prism. These can be considered negative volumes. The order in which the two triangular prisms are removed does not matter.

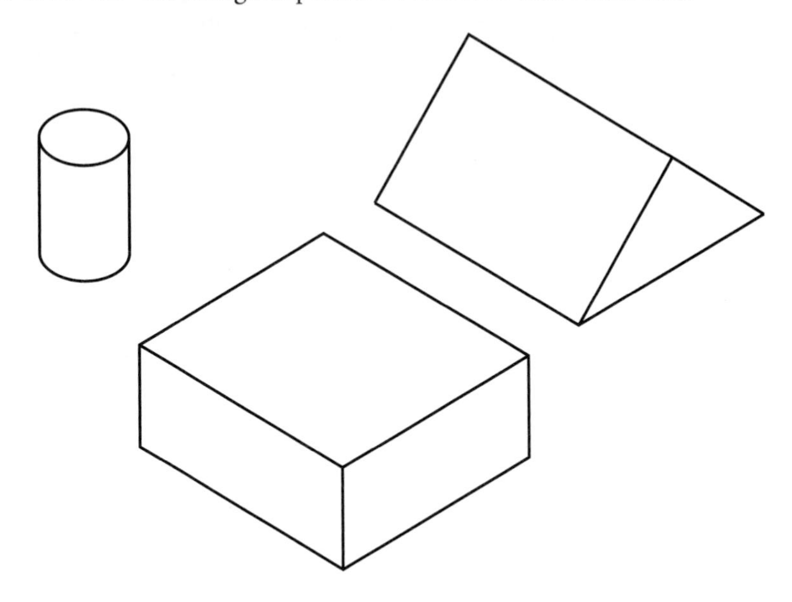

Figure 1.4 Three basic shapes: cylinder, triangular prism, and rectangular prism

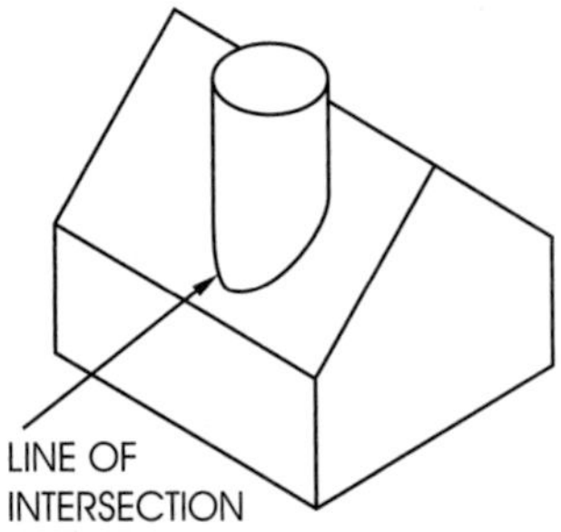

LINE OF
INTERSECTION

Figure 1.5 A building created by combining a triangular prism and a rectangular prism with a cylinder

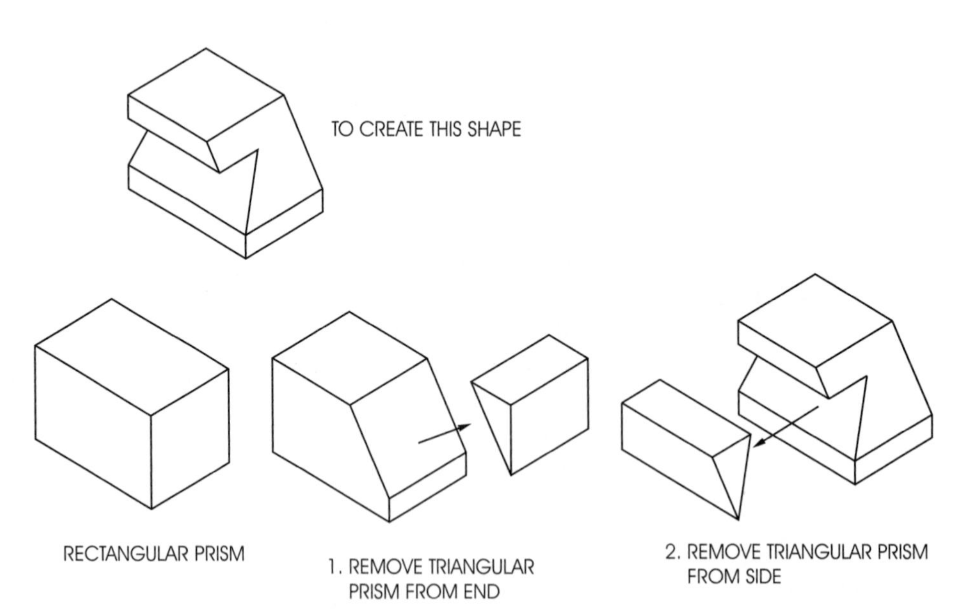

Figure 1.6 Creating a new object by removing basic shapes

You must exercise judgment when combining basic shapes; otherwise you might draw objects that cannot exist. For example, removing two small rectangular prisms from opposite corners of a larger prism can result in an object that cannot exist, as illustrated in Figure 1.7. The connection between the two parts is a line that has only one dimension. It has no strength, and the object does not stay together.

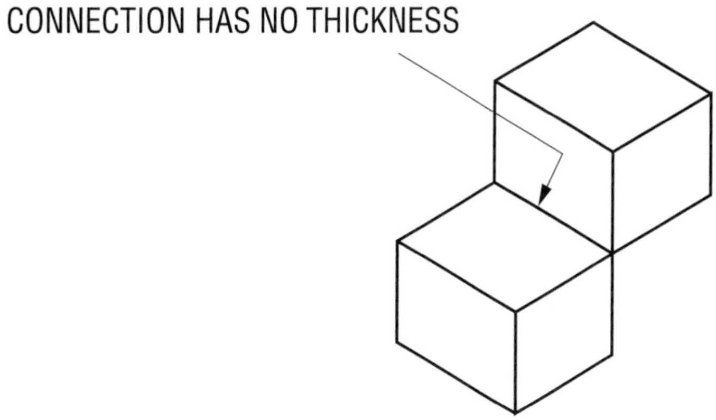

CONNECTION HAS NO THICKNESS

Figure 1.7 An impossible object

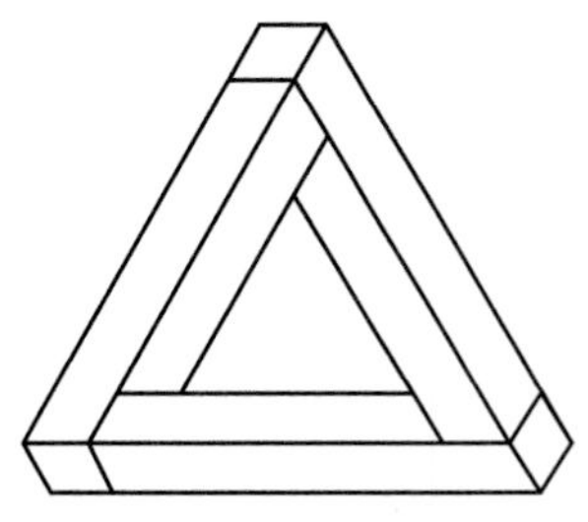

Figure 1.8 An impossible object

Similarly, it is possible to combine three square prisms to form an impossible object, which can be drawn in two dimensions but cannot be constructed in three dimensions. Figure 1.8 is an example of an impossible object constructed from three square bars. This "object" can be drawn but cannot be made.

More basic shapes are shown in Figure 1.9. A sphere and cube are not shown. A sphere is simply a circle on a drawing, and a cube is simply a special case of a rectangular prism.

Complex shapes, which require the combination of many basic shapes, can be quickly created using a computer. The solid model is a pictorial representation of an object and is the easiest way to visualize it. Pictorial representations give the illusion of three dimensions, but they are not used for fabrication. Pictorial drawings are covered later in this chapter. Engineering drawings use another method of showing 3-D objects.

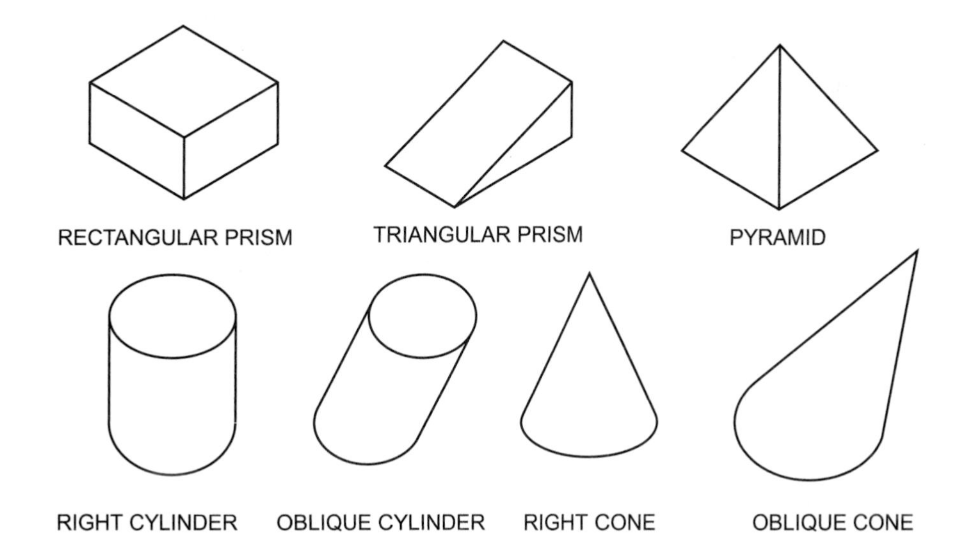

Figure 1.9 Some basic shapes

Before drawing objects, you need to understand the concept of orthographic projection.

ORTHOGRAPHIC PROJECTION

Three things must be defined before an object can be drawn: the location of the eye, or viewpoint; the object; and the picture plane on which the object is drawn. The arrangement of these three elements such that the lines of sight are perpendicular to the picture plane results in an **orthographic projection**. *Orthogonal* means "at right angles to something else."

If you put the object to be drawn behind a piece of glass and looked at it through the glass, a line drawn from your eye to a point on the object would pass through the glass at some point. This point represents a point on the object. The glass represents a picture plane on which the object could be drawn. If the viewer were located an infinite distance from the object, the line from the eye would pass through the glass picture plane at 90°, and all lines from the eye to points on the object would be parallel. This arrangement is shown in Figure 1.10.

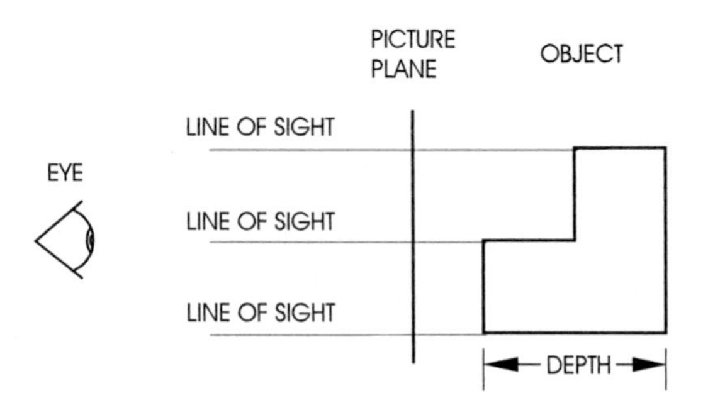

Figure 1.10 Location of object, eye, and picture plane

This type of orthographic projection, used in North America, is called **third angle projection**. A different type of orthographic projection, used in Europe and other parts of the world, is described later.

Figure 1.11 shows the points where lines of sight intersect the picture plane. These points represent different points on the object.

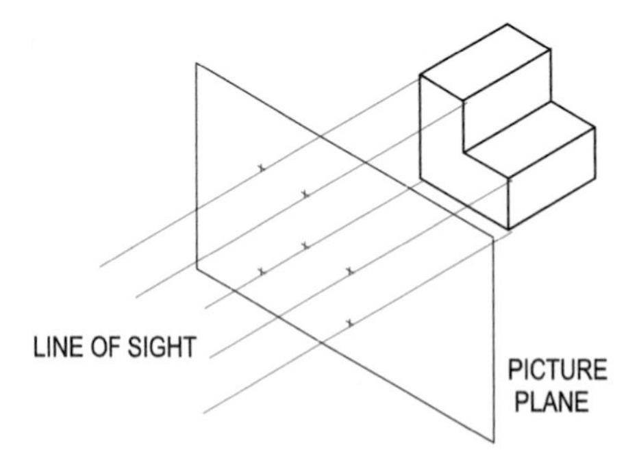

Figure 1.11 Intersection of lines of sight with picture plane correspond to points on the object

If the intersections of lines of sight with the picture plane are connected, as in Figure 1.12, the outline of the object and features that can be seen from the chosen viewpoint are seen on the picture plane.

Figure 1.13 shows a view of the object as it appears on the picture plane, which could be a computer screen or a piece of paper, and the object is behind it (inside the monitor).

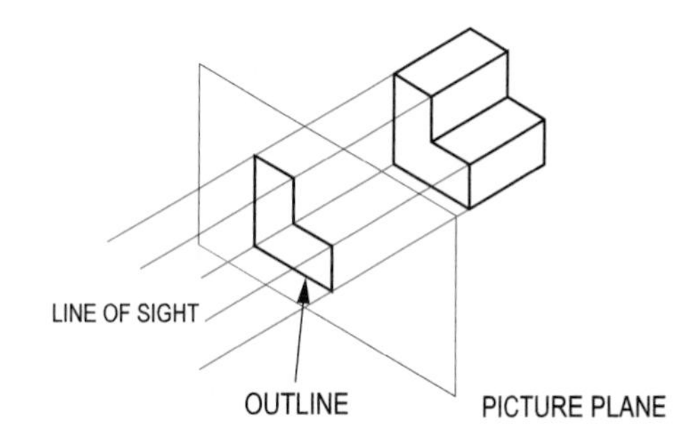

Figure 1.12 Joining points on picture plane shows object as it would appear on the picture plane

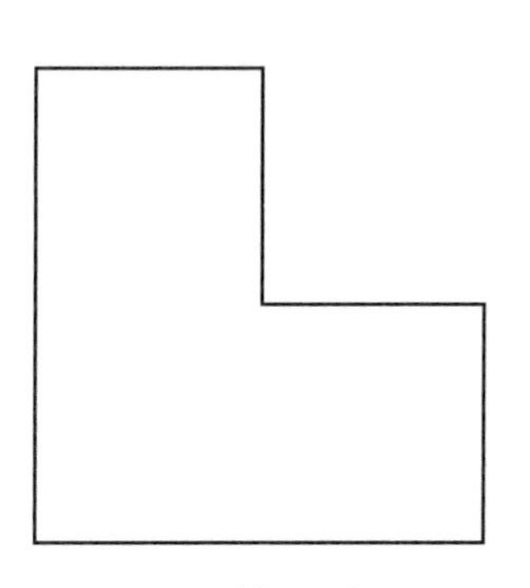

Figure 1.13 Object as it appears on the picture plane

Obviously, Figure 1.13 does not define the object entirely, so more information is required to show what it looks like. For example, there is no way to see the depth (i.e., into the paper) in Figure 1.13. We must view the object from other viewpoints before we can determine the shape. This requires more picture planes and different viewpoints.

Viewpoints and picture planes can be located anywhere, but there are six principal picture planes that are always in the same relative position. The six principal planes can be thought of as the six sides of a cube, each side representing a picture plane.

Figure 1.14 shows a cube with three sides labelled: front, top, and right side. The other three sides (left side, rear, and bottom) cannot be seen. The front, top, and right side correspond to the side of the object that is seen on that face of the cube. These three sides are called **principal planes**.

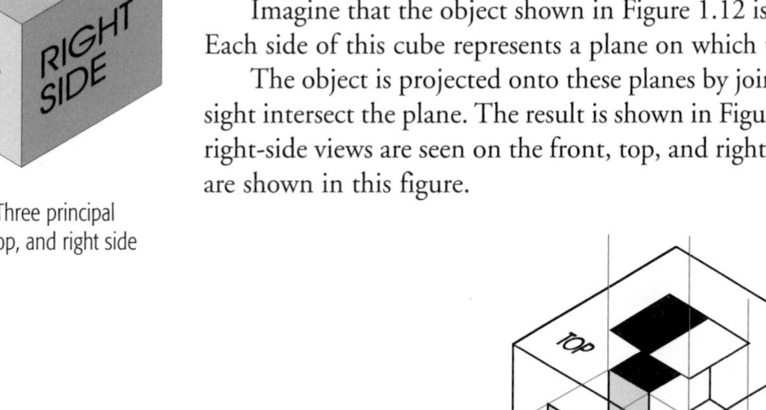

Figure 1.14 Three principal planes: front, top, and right side

Usually, only three planes are required to define an object.

Imagine that the object shown in Figure 1.12 is put in the centre of a glass cube. Each side of this cube represents a plane on which the object will be projected.

The object is projected onto these planes by joining the points where the lines of sight intersect the plane. The result is shown in Figure 1.15, where the front, top, and right-side views are seen on the front, top, and right-side planes. Not all lines of sight are shown in this figure.

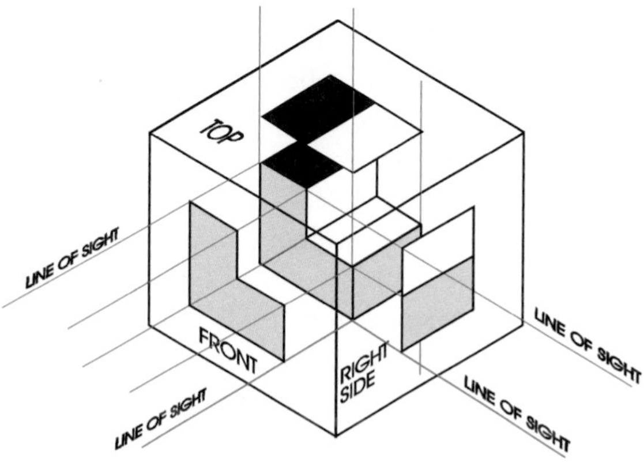

Figure 1.15 Forming the commonly used views: front, top, and right side

This shows how orthographic views are formed, but we need to represent an object on a 2-D surface, not on the sides of a glass box. To do this, the top and right-side views are folded so that they are in the frontal plane, that is, the same plane as the front view. The top is folded up and the right side is folded out so that both are in the frontal plane. This unfolding of the top and right side is shown in Figure 1.16.

The three views of the object are shown in Figure 1.17.

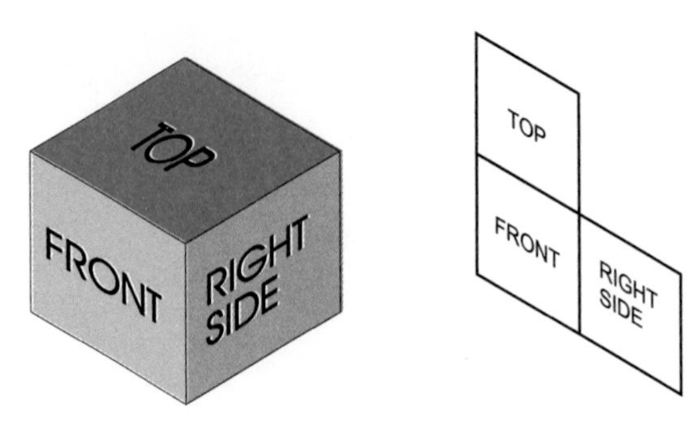

Figure 1.16 Unfolding the top and right-side views

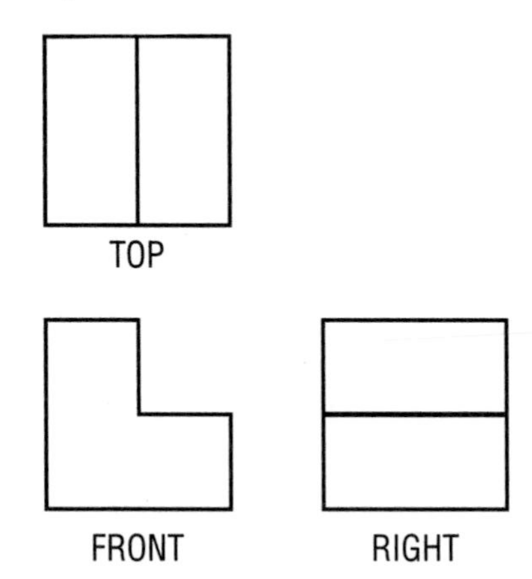

Figure 1.17 Arrangement of front, top, and right-side views

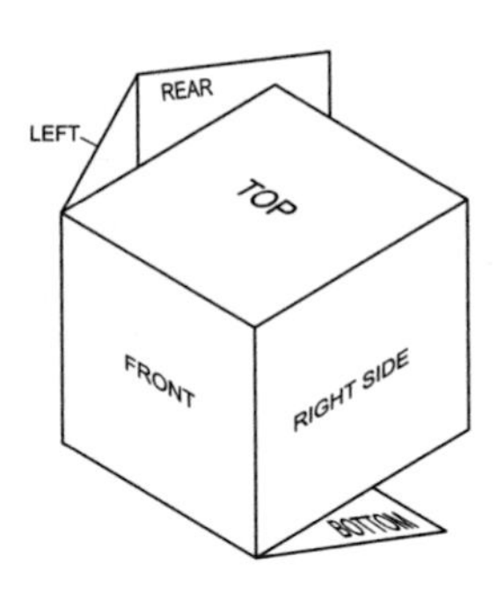

Figure 1.18 Rear, left-side, and bottom planes partially unfolded

The views are named here, but they need not be. Because these views are always in the same position relative to each other, we know what they are by their position. The length can now be seen in either the top view or the right-side view, and, of course, it must be the same in each view. Note that there is correspondence between the views. The object does not move or change in size; it is simply seen from a different viewpoint.

While there are six sides to a cube, only three have been shown. The three sides that cannot be seen in Figure 1.16 are unfolded into the frontal plane. Figure 1.18 shows the cube with the rear, left-side, and bottom faces partially unfolded so that they are visible. The top and right sides have not been unfolded in Figure 1.18. The rear view is shown folded off the left side in this figure, but it could be folded from the right side.

The result of unfolding all six sides of the cube is shown in Figure 1.19. Six views are available, and all sides of the object could be shown if required. The rear view is projected from a side view and can be shown on either the left or the right, depending on which side view is used.

All six views are shown in Figure 1.20, although it would be very unusual to require all six views to define an object. Rear and bottom views are rarely required.

Figure 1.20a shows the rear view folded off the left-side view. The left-side view shows very little. The horizontal surface cannot be seen in the left-side view, and it is shown by a dashed line that represents the edge of the horizontal surface. This is called a **hidden line**, since it is on the rear side of the object and cannot be seen in this view. (Different symbols are used to represent different types of line. All of the different types are described later in this chapter.) The edge view of the horizontal surface can be seen in the right-side view, where it is represented by a solid line. The bottom view gives the same information as does the left-side view, but this is seen better

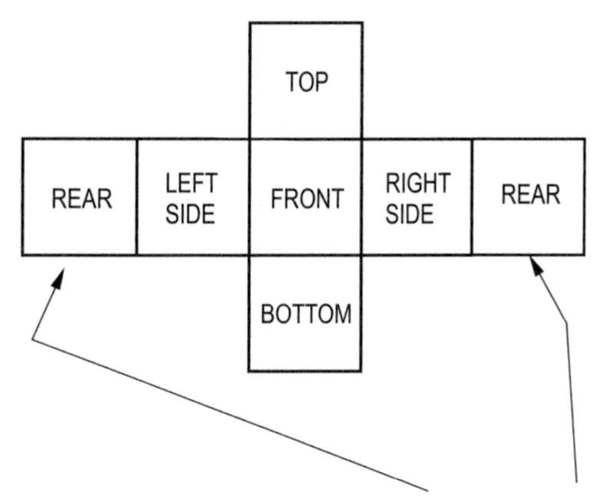

Figure 1.19 Unfolding six sides of a cube into the frontal plane

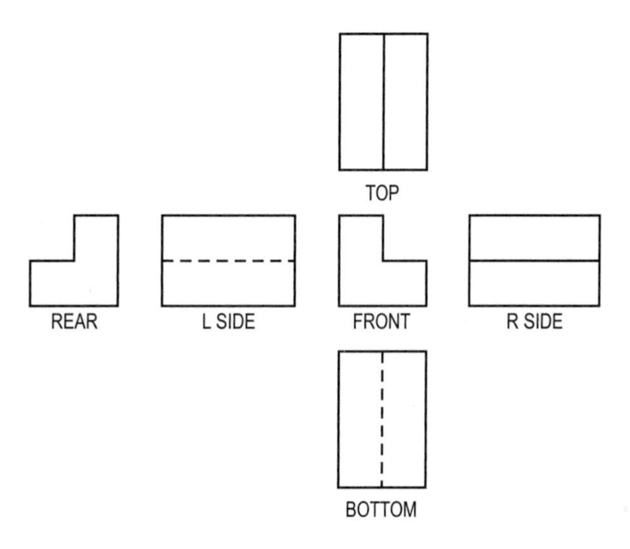

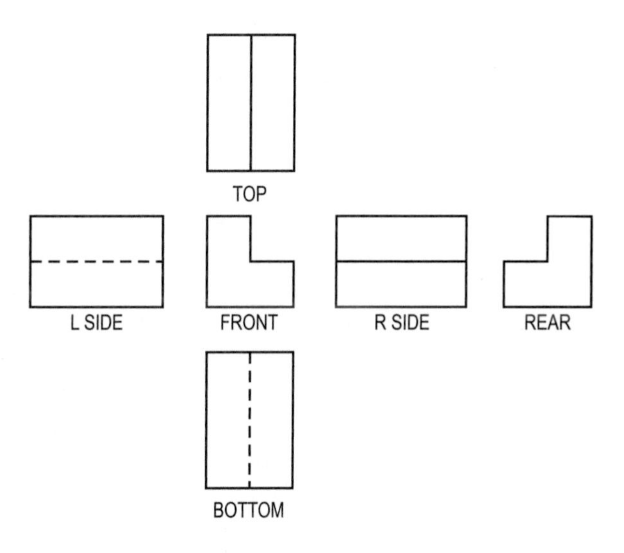

Figure 1.20 Arrangement of all six orthographic views

in the top view. The rear view is a mirror image of the front view. The arrangement shown in Figure 1.20b, where the rear view is folded off the right-side view would be the one to use. This object can be completely described with three views—front, top, and right side—and there is no need to draw the other three views.

The three most commonly used views are the front (there is always a front view), top (also called a **plan view**), and the right-side view. A left-side view would be used instead of a right-side view if a feature could only be seen in the left-side view. Some objects may require both right- and left-side views. The front and side views are often called **elevation views** because the elevation (vertical distance) can be seen in these views.

Views other than the six principal ones can be used if required. These other views, called **auxiliary views**, are used to show details that would not be seen clearly in the front, top or side views. Figure 1.21 shows an example. The true shape of the sloping face cannot be seen in the front view, and it is foreshortened in the side and top views. An auxiliary plane can be added to show the true shape of this face. This would be needed if there are details to be seen on the sloping face, for example, the location of a hole. Auxiliary views will be dealt with more fully in Chapters 5 and 6.

Now let's look at how to choose front and side views.

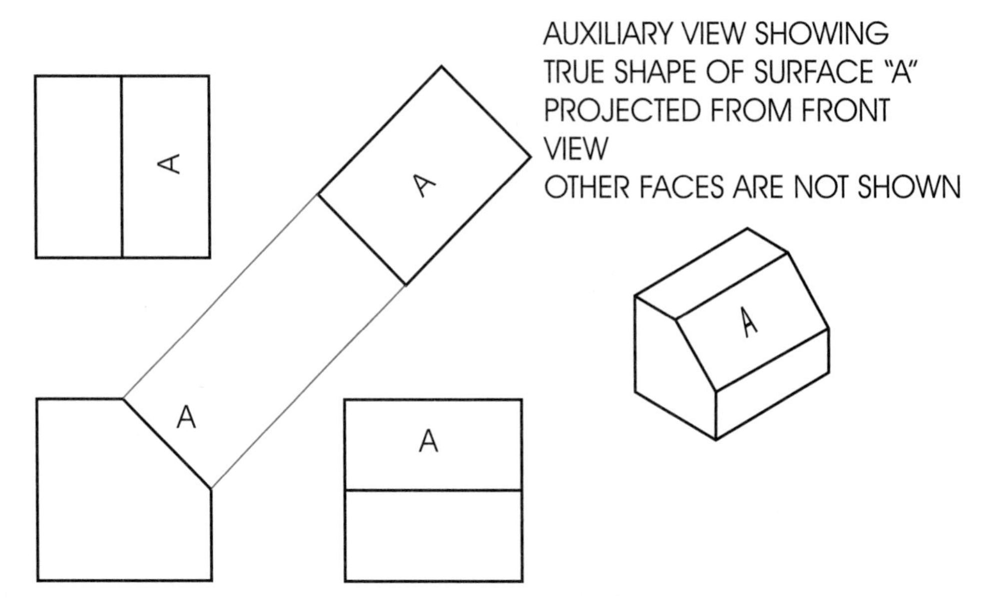

Figure 1.21 An auxiliary view is used to show details that do not show clearly on the principal views

Choosing Front and Side Views

Before drawing anything, you must decide how it should be oriented so that maximum information is given in the front view. The front view should show the shape of the object, and there should be a minimum of hidden lines. You must also think about the top and side views when deciding on the front view, since these are projected from the front view. The object should appear in the front view as it would normally be used. Sometimes this is not known, but usually it is obvious. It would not make much sense to show an automobile upside down with the tires in the air, since automobiles are not normally used in this position.

The front view of the object shown in Figure 1.22 is chosen to show the characteristic shape (an "ell" shape). The corresponding top view shows the depth. When two views are given in this orientation, the lower view is the front view. Since views always have the same locations relative to each other, there is no need to name them, unless there is a possibility of misunderstanding. If this is the case, identify at least one view.

This simple object can be defined with two views, but, to avoid misunderstanding, it is usual to draw three views. Either a left-side view or a right-side view will be added. As only one will be used, be sure to select the view that best shows details of the object. The shape of the object must be considered to determine which is the best choice in any particular case. Figure 1.23 shows how the three views would appear if a left-side view were chosen.

The middle edge of the "ell" is seen in the left-side view as a hidden line. It is better to show the right-side view, since the middle edge is visible in the right-side view. The object then appears as shown in Figure 1.24. Views selected to represent a more complex object would be selected using the same criteria.

The side view chosen depends on the object. If details are on the left side after the best front view is chosen, the left side would be drawn to minimize the number of hidden lines. Right-side views are more common; however, there is no rule that this must be the case. Sometimes, it is necessary to draw both right- and left-side views to fully define an object. If it does not matter whether a left-side view or a right-side view would define the object, a right-side view is usually used.

Another way of positioning the "ell"-shaped object is shown in Figure 1.25. The front view shows the shape; however, the corresponding top view shows one edge as a hidden line.

The front view satisfies the criterion of showing the shape, but the top view is not as illustrative as it could be. This can easily be improved by positioning the object as shown in Figure 1.24.

Another possible orientation is shown in Figure 1.26. This orientation also results in a poor front view that shows nothing of the "ell" shape. There are several ways this front view could be interpreted. The front view is the most important view, but it is only one of several views and must be considered as part of a group that defines the object in the best possible way.

Now let's look at another method of orthographic projection.

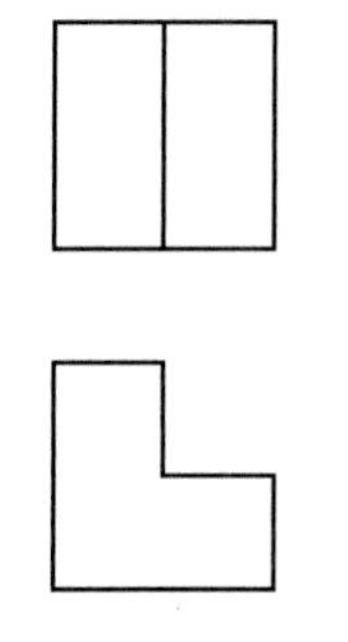

Figure 1.22 If two views are shown, they are front and top

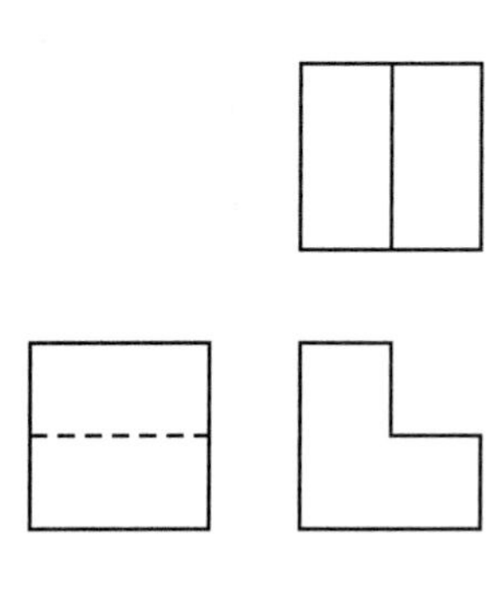

Figure 1.23 Arrangement showing front, top, and left-side views

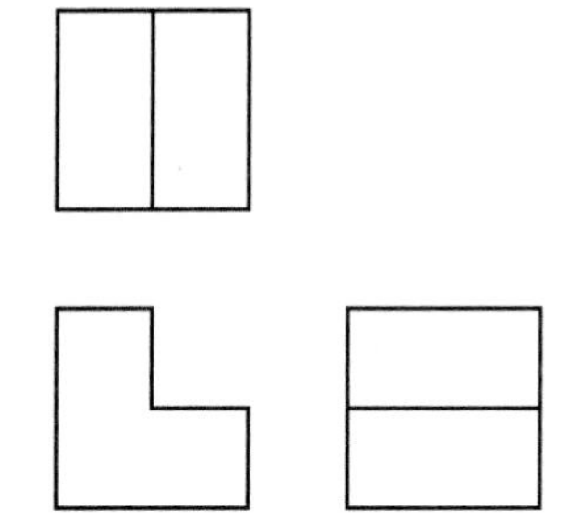

Figure 1.24 Arrangement showing the top, front, and right-side views

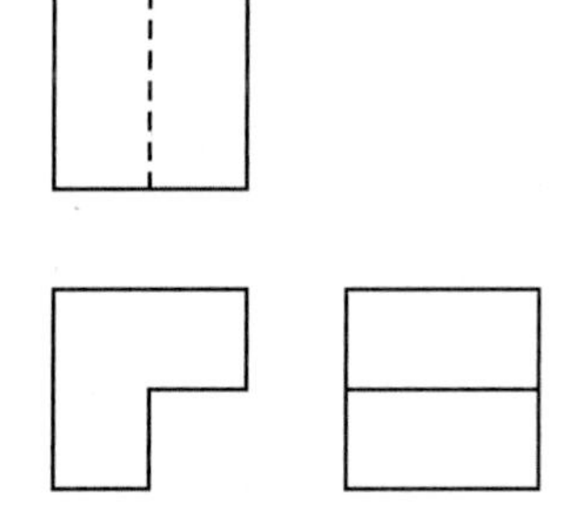

Figure 1.25 Orthographic views showing a poor choice of front view

Another Method of Orthographic Projection

A different arrangement of object, picture plane, and eye is used in Europe and other parts of the world. The picture plane is positioned to the right of the object so that the order becomes eye, object, and picture plane. This arrangement, called **first angle projection**, is illustrated in Figure 1.27. The same front view appears on the picture plane with each method of projection.

The representation on the picture plane is the same for each method of projection, but the orientation of the views is different. Figure 1.28 shows the same object drawn in third and first angle projections. A symbol (a truncated cone) is used to indicate which method of projection is used (see Figure 1.28). The symbol is usually located in the lower right corner.

Now let's move on to look at how to project between orthographic views.

Projection between Orthographic Views

We have seen how orthographic views can be created from a solid by projecting from points on the solid to projection planes, but we must also know how to project between different orthographic views. Given two views of any point, the same point in any other orthographic view can be found. The corners of one face of the solid

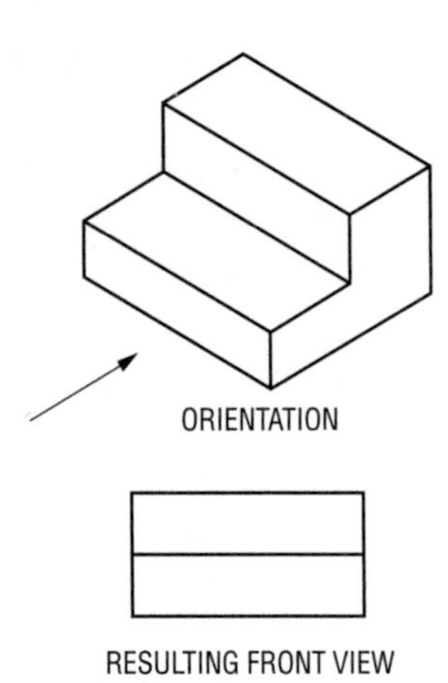

Figure 1.26 Object orientation and corresponding front view

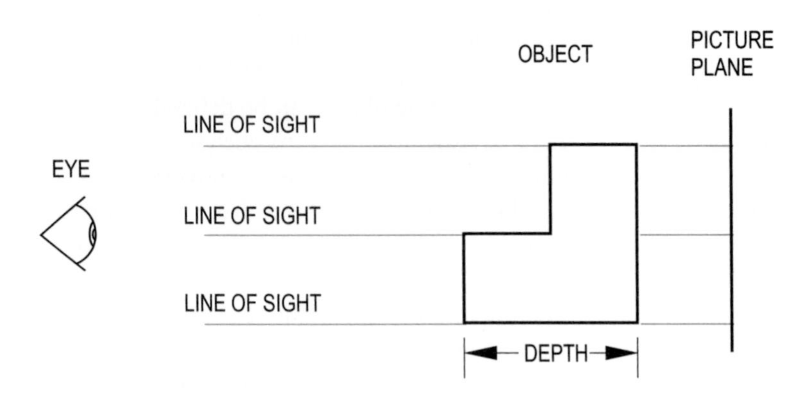

Figure 1.27 Location of object, eye and picture plane for first angle projection

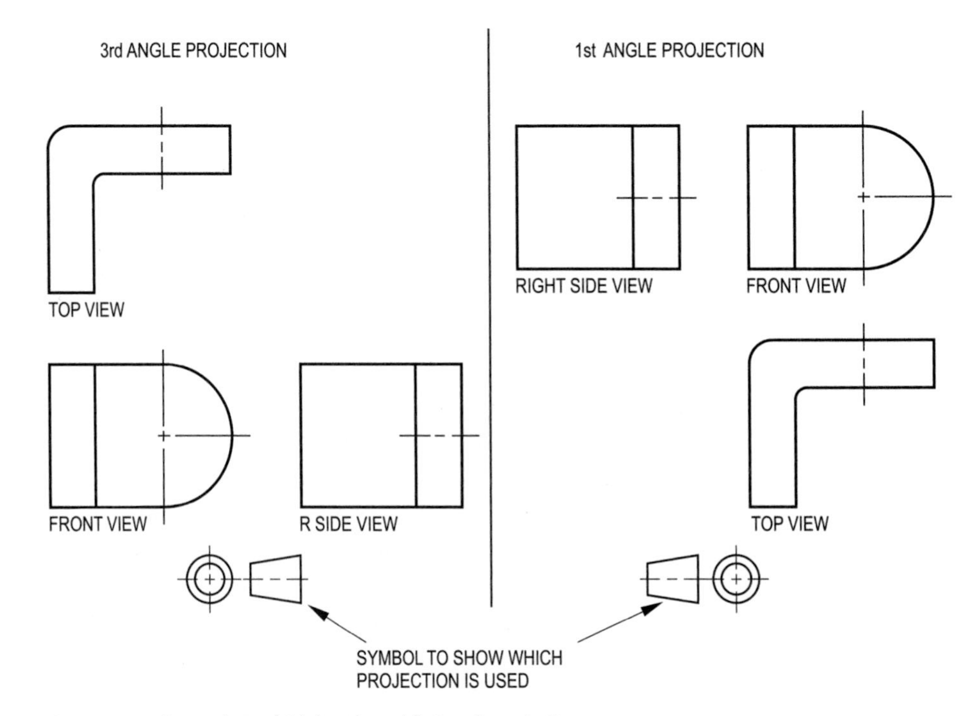

Figure 1.28 Comparison of third angle and first angle projections

shown in Figure 1.29 have been identified *abcdef* in the side view. The same points are identified in the top view. Point *a* is above *f*, and *d* is above *e*. The front view face *abcdef* can be drawn by projecting these points from the side and top views. Point *a*, in the front view, must be at the intersection of the projection line from the side view and the top view. Point *b* must be at the point where projections from the side view and top view intersect.

The complete face can be drawn by projecting all points into the front view. The points at the ends of the lines making up the surface must be joined in consecutive order. They cannot be connected in the order *abdcef*, for example. There are no lines *bd* or *ce*. The technique of labelling points is useful when determining what a more complex object looks like.

If you are wondering why the label "*i*" has not been used, it is to prevent confusion with "1" on drawings. The letter "*l*" (ell) is not used for the same reason.

A more complex object is shown in Figure 1.30. This shows orthographic views of the object created in Figure 1.6 (page 4). The front view is chosen to show the sloping end. Points in the top view are located by projecting corresponding points from

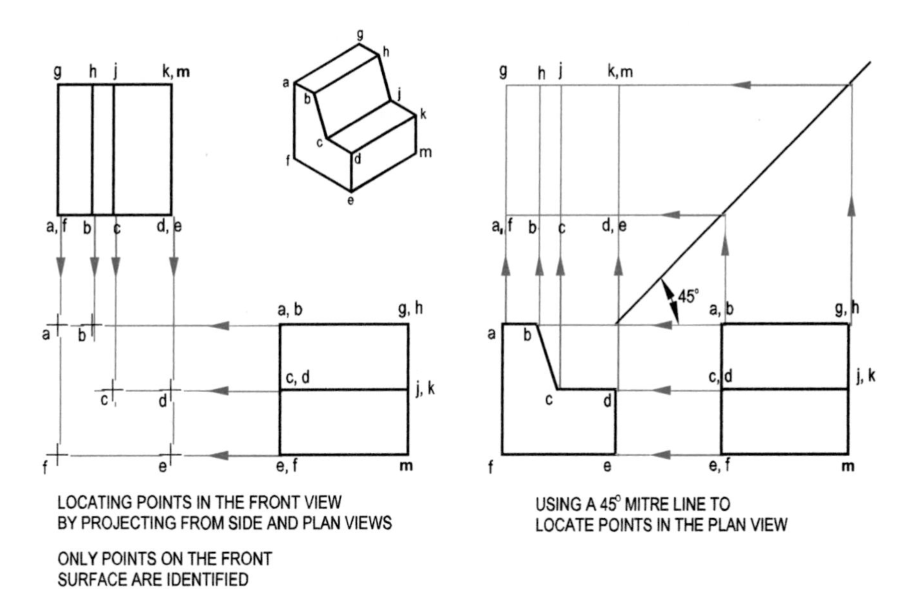

LOCATING POINTS IN THE FRONT VIEW
BY PROJECTING FROM SIDE AND PLAN VIEWS
ONLY POINTS ON THE FRONT
SURFACE ARE IDENTIFIED
USING A 45° MITRE LINE TO
LOCATE POINTS IN THE PLAN VIEW

Figure 1.29 Locating points in a third view

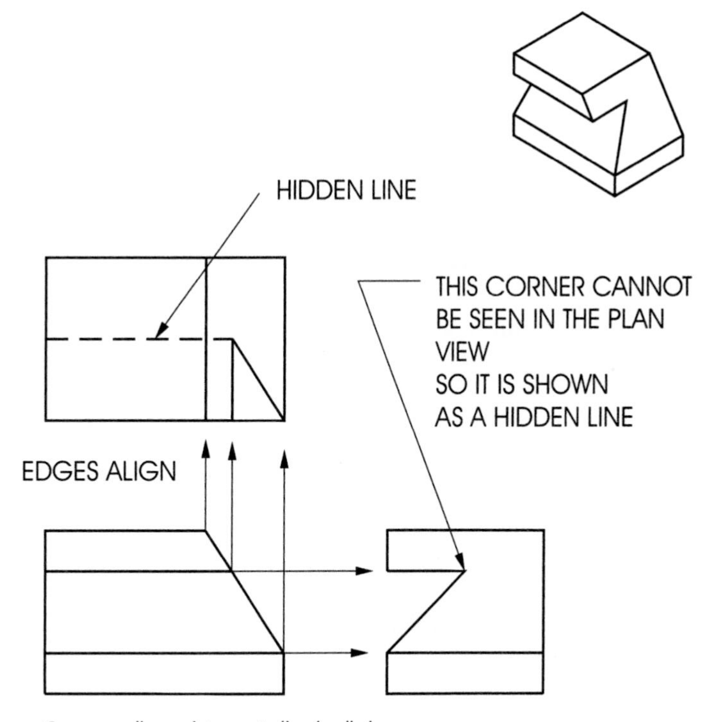

Figure 1.30 Corresponding points must align in all views

the front view. The location of the hidden line in the top view is found from the side view. The hidden line representing the rear corner of the cutout must be the same distance from the front of the object in the top view as it is in the side view.

Now let's look at how curved surfaces are shown.

Curved Surfaces

A flat, or plane, surface can be defined by straight lines around the perimeter. A curved surface can also be defined in this way, but there may be no indication that the surface is curved. For an illustration, look at Figure 1.31, which shows a rectangle. The thin,

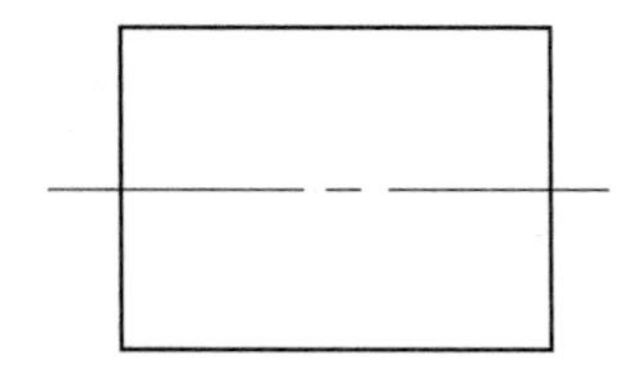

Figure 1.31 One view of a cylinder

broken line through the middle of the rectangle is a **centreline** and is not part of the object. This is a view of a cylinder, but there is little to indicate that the surface is curved. The centreline is the only clue. From this example, you will see how important it is to use more than one view to define an object with curved surfaces.

Two views of the cylinder are shown in Figure 1.32. The circular shape is seen in the side view.

One thing common to almost all curved surfaces is a centreline. The symbol for a centreline is shown in Figure 1.33.

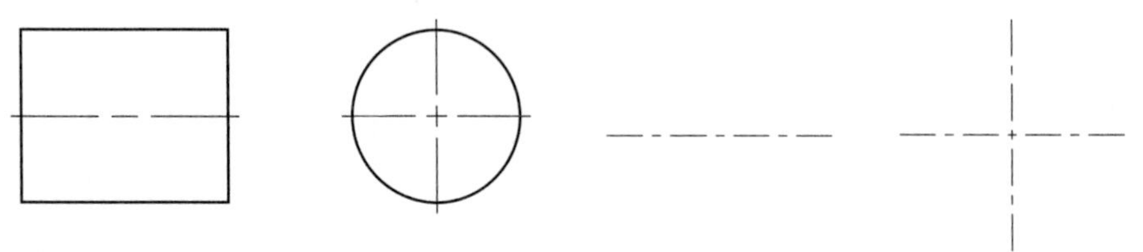

Figure 1.32 Two views of a cylinder

Figure 1.33 Centreline symbol

A centreline symbol is a sequence of long and short lines. If a centreline is used to indicate the centre of a cylinder, for example, the centre point is indicated by the intersection of the short dashes as shown in Figure 1.32. A centreline is used to indicate:

a. the centre of a circle
b. the centre of a circular arc
c. an axis of symmetry

These uses are shown in Figure 1.34. A centreline extends 2–3 mm beyond the item to which it refers and does not extend between views (see Figure 1.32). A curved surface is usually confirmed by giving two views of the object.

Now that you have a basic understanding of orthographic projection, let's move on to look at pictorial representation.

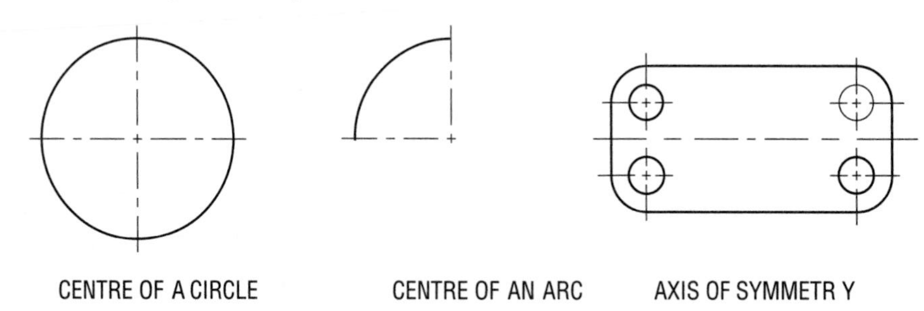

Figure 1.34 Uses of a centreline

PICTORIAL REPRESENTATION

A pictorial drawing has the illusion of three dimensions, shows more information, and is often easier to visualize, particularly by persons with no engineering background. Pictorial drawings are used as sketches to explore ideas, aid explanation, and present information. A pictorial is often used to show the relationship among different parts of an assembly and how it is put together. Pictorial drawings are not used for manufacture or construction.

Isometrics

There are several types of pictorial drawing; however, the most commonly used in engineering is the **isometric**. *Isometric* comes from a Greek word meaning "equality

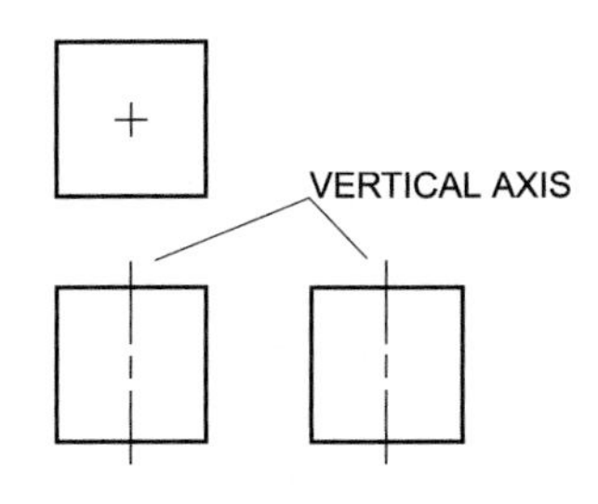

Figure 1.35 Three views of a cube

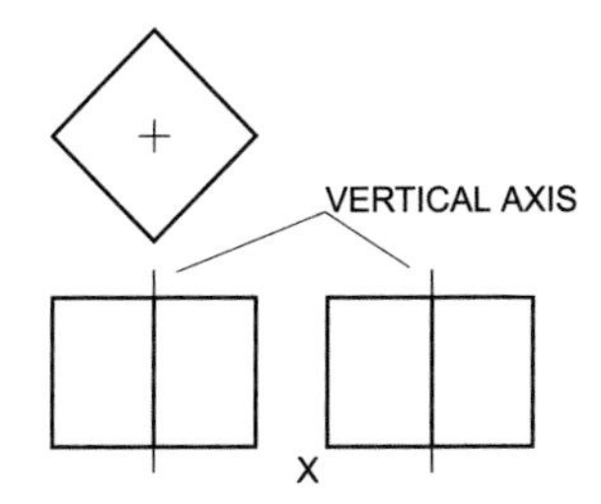

Figure 1.36 Views after rotation about vertical axis

of measure." Other types have advantages in specific applications and are discussed later. The pictorial drawings used in this book are isometrics. A cube, shown in Figure 1.35, will be used to illustrate how an isometric is created.

Figure 1.35 shows three orthographic views—top, front, and right side—of a cube. Each face appears as a square, and only one side is seen in each view. The reason for drawing a pictorial is to show more sides of the object in one view.

We have seen that different views result from different viewpoints; however, similar results can be obtained by repositioning the cube. First, the cube is rotated 45° about a vertical axis through the centre of the cube. Two sides, instead of one, can be seen in the front view (Figure 1.36); however, these sides appear shorter than they actually are. All three views of the cube after rotation are shown in Figure 1.36.

If the cube is now tipped forward about the corner "X," the top surface can be seen in the front view. Figure 1.37 shows the front view and the corresponding side and top views after tipping. Two sides and the top can now be seen in one view—the front view—and there is no need to draw the side and top views. They are shown here to illustrate how an isometric *projection* is created. It is simply an orthographic drawing with the object oriented in a particular position.

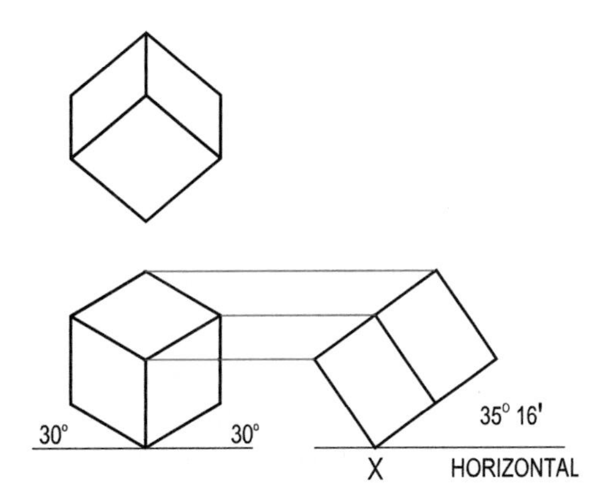

Figure 1.37 Three orthographic views after tipping about "X"

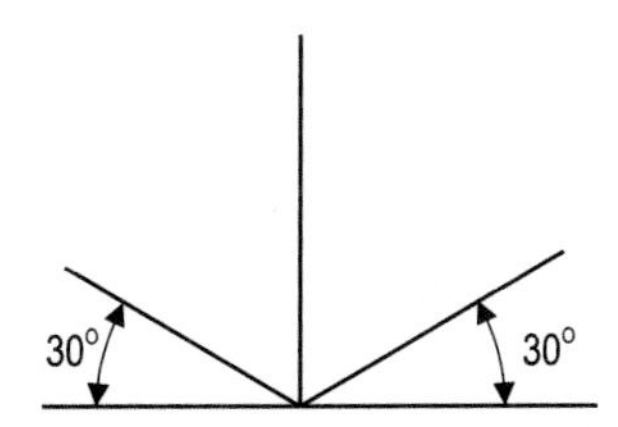

Figure 1.38 Isometric axes

The angle through which the cube is tipped is chosen so that all sides are equal length and make an angle of 30° to the horizontal. The sides are shortened to 82 percent of actual length, but this shortening is neglected in making an isometric *drawing*, and all sides are drawn actual size. This results in a small distortion, which is noticeable only in very large objects.

A set of isometric axes is shown in Figure 1.38. Computer graphics programs often have the facility to create a grid of 30° lines on the screen as an aid to creating an isometric. Special paper with a 30° grid is available for sketching.

A cube drawn as an isometric appears as shown in Figure 1.39. All sides are equal and are drawn full size. Any line parallel to an isometric axis can be measured as being true length. A line which is not parallel to an axis is not true length, as illustrated by the diagonals on the vertical sides.

All diagonals on a cube are of equal length, but it is obvious that they are not equal on the drawing.

Now let's explore some other pictorials.

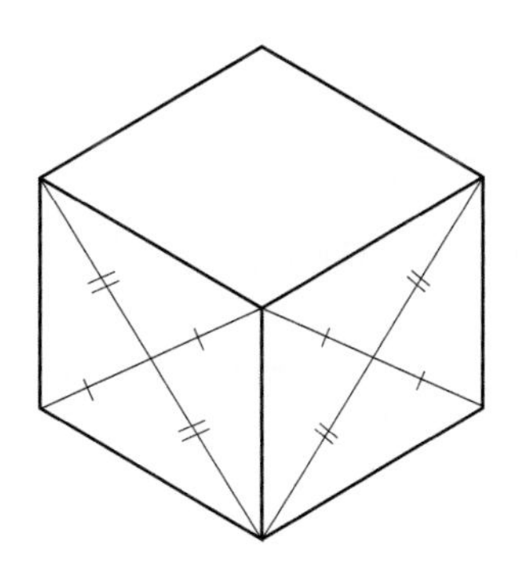

Figure 1.39 Isometric drawing of a cube

Other Pictorials

An isometric drawing is created by first placing an object in a specific position and drawing a front view as if it were full size. The position is chosen to give equal lengths on each of the axes. If different angles are used in positioning the object, a different pictorial is obtained. The object can be oriented to give the same scale on either three axes or two axes or different scales on all three axes. Three pictorial representations—**isometric**, **dimetric**, and **trimetric**—are shown in Figure 1.40. The general term for this type of drawing is an **axonometric**.

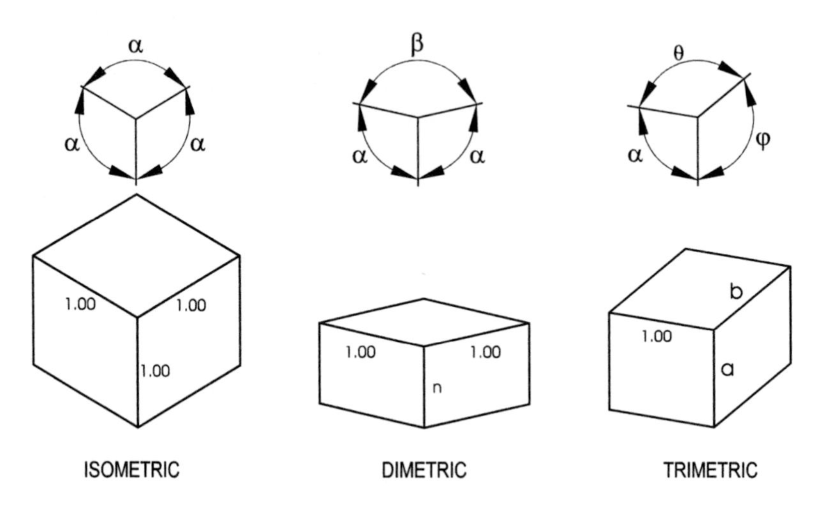

Figure 1.40 Comparison of isometric, dimetric, and trimetric projections

Angles between axes on an isometric are the same, and so the scale is the same on each axis. A dimetric drawing has equal scales on two axes, and a trimetric has a different scale on each axis. The scales must be determined. The different scales make a dimetric and a trimetric more difficult and time-consuming to draw; therefore, isometric is the most commonly used.

Pictorials are shown in solid modelling programs, and they can be rotated and viewed from different angles, thus allowing all sides to be seen. Solid modelling programs also have the capability for creating orthographic views from pictorials. Since pictorial views show more information, they are useful at the design stage and for illustration. Orthographic drawings are used for construction, and the ability to create them from pictorials used in the design stage saves time.

Oblique Projection

There are times when an object has features that may be difficult or time-consuming to represent with an isometric pictorial. In such instances, an **oblique projection** may be quicker and easier to draw. An orthographic projection results if lines of sight from the eye to the object pass through the picture plane at 90°. If lines of sight pass through the picture plane at an angle other than 90°, an oblique projection is created. Figure 1.41 shows a comparison of the arrangements for orthographic and oblique projections. Lines of sight are parallel in both types of projection, but the angle with the picture plane is different.

The front view is always drawn full size in an oblique drawing. Features in the front view are seen in their true shapes, and the circular features are easier to draw.

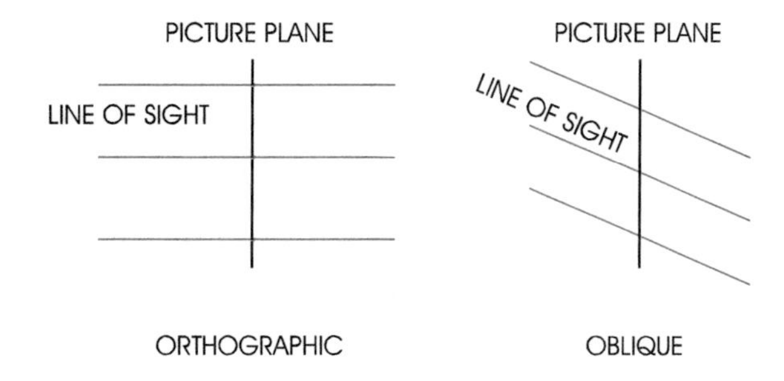

Figure 1.41 Comparison of orthographic and oblique projections

The disadvantage is that the side in the direction of the receding axis can appear distorted.

Although lines of sight can theoretically pass through the picture plane at any angle, only a few angles are used in practice. The most commonly used angles are 30°, 45°, and 60°.

Two common oblique projections are called **cavalier projection** and **cabinet projection**, but these are rarely seen today. These are similar, but distortion on the receding axis is reduced with cabinet projection. A cavalier drawing is drawn full size on all axes. If the object is long in the direction of the receding axis, there is distortion, and the object does "not look right." A cabinet drawing is drawn half size on the receding axis to compensate for this. Figure 1.42 compares cavalier and cabinet drawings.

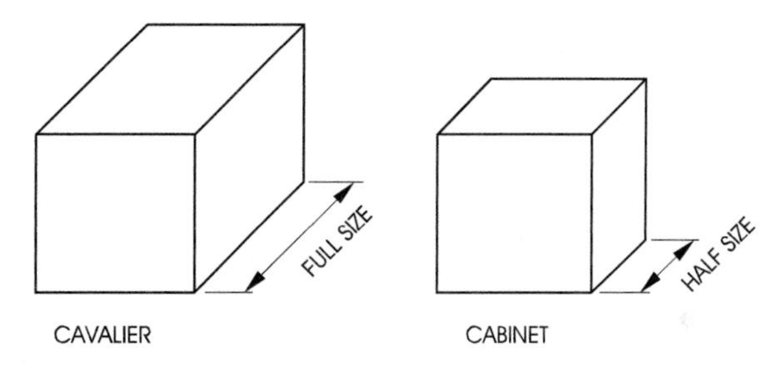

Figure 1.42 Comparison of cavalier and cabinet drawings

Figure 1.43 shows both cavalier and cabinet drawings of the same object. Distortion on the receding axis is more obvious with the cavalier drawing, and so the cabinet drawing is best for this object.

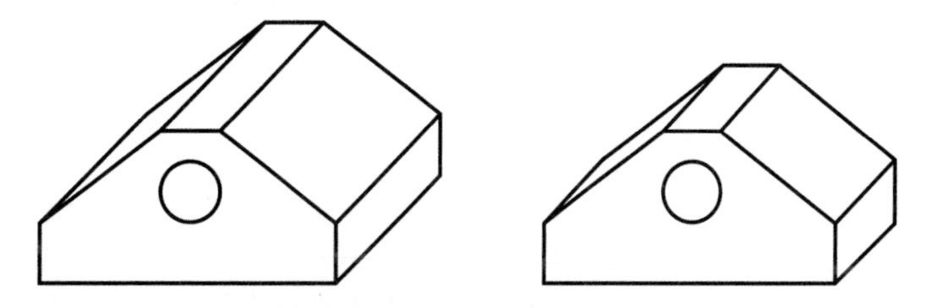

Figure 1.43 Cavalier and cabinet drawings of the same object

Perspective Drawings

A **perspective drawing** represents what the eye sees. You are all familiar with the view looking along railway tracks: The tracks seem to come together. Perspective is not used in engineering except to present an "artist's impression." Because objects appear

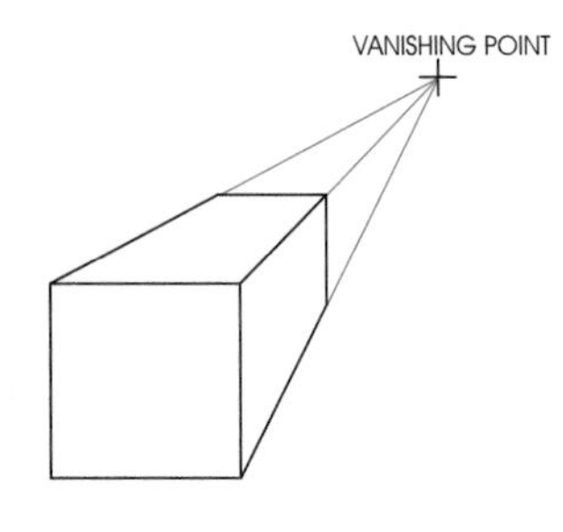

Figure 1.44 A one-point perspective drawing has one vanishing point

smaller the farther they are from the viewer, there is no distortion, and the drawing "looks right." Figure 1.44 is an example of a one-point perspective drawing.

It is referred to as a **one-point perspective drawing** because it has one vanishing point. You can attain more realistic representations if you use more vanishing points. Figure 1.45 shows a two-point perspective drawing. A three-point perspective drawing would have three vanishing points.

Computer drawing programs can be used to create perspective drawings when needed.

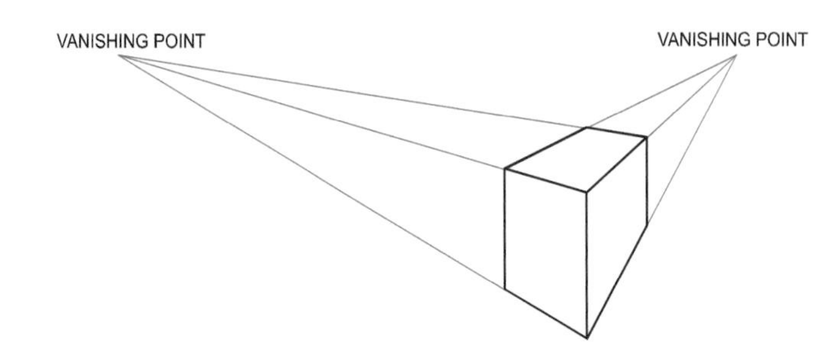

Figure 1.45 Two-point perspective drawing has two vanishing points

STANDARDS AND CONVENTIONS USED IN ENGINEERING

The purpose of an engineering drawing is to convey information, and there are standard ways of doing this to minimize the possibility of misunderstanding. Engineering drawing standards are specified in *CSA B78.1 Technical Drawing, General Principles.*

Engineering Lettering

Lettering on engineering drawings should be single stroke Gothic, and should be upper case, except for some standard international (SI) symbols, for which lower case is required. The acceptable style is shown in Figure 1.46. SI symbols are given in Appendix I.

Since most engineering drawings are done using a computer, engineering lettering is only a matter of specifying the correct font and size. The size varies depending on where text is used. Figure 1.46 shows relative sizes of lettering depending on where it is used. The minimum lettering size for dimensions and notes is 3.5 mm. On large size drawings, it is recommended that the largest letters be a minimum of 5 mm and a maximum of 7 mm. The largest letter size should be used for the title, smaller size (5 mm) for subtitles, and 3.5 mm for dimensions and notes. It is permissible to use smaller letters (2.5 mm) in revision tables. Engineering lettering has a certain style and is done in a systematic way. Engineering lettering can be either upright or sloping. Upright are shown in Figure 1.47 and sloping in Figure 1.48. These slope at 67.5° to the horizontal. Light guidelines sloping at 67.5° can be used to get consistent slopes on all letters. Some find sloping letters easier to do. Use only one style of letters on a drawing, either upright or sloping.

Engineering lettering standards are also important in other engineering work that may not be done with a computer (there are many of them). Engineering calculations and, for students, problem sets with drawings or sketches are examples. These must be understood by others. In the case of problem sets (or exams), you are likely to lose marks if the marker does not understand what you have written. Engineering calculations are filed for future reference. If someone wants to see what was done

TITLE AND DRAWING NUMBERS
ON LARGE DRAWINGS

A B C D E F G H I J K L M
N O P Q R S T U V W X Y Z
1 2 3 4 5 6 7 8 9 0

5 mm
SUB-TITLES OR MAIN TITLES
ON SMALL DRAWINGS

A B C D E F G H I J K L M
N O P Q R S T U V W X Y Z
1 2 3 4 5 6 7 8 9 0

3.5 mm
MINIMUM SIZE FOR
DIMENSIONS AND NOTES

A B C D E F G H I J K L M
N O P Q R S T U V W X Y Z
1 2 3 4 5 6 7 8 9 0

2.5 mm
FOR REVISION LISTS

A B C D E F G H I J K L M
N O P Q R S T U V W X Y Z
1 2 3 4 5 6 7 8 9 0

Figure 1.46 Standard engineering lettering

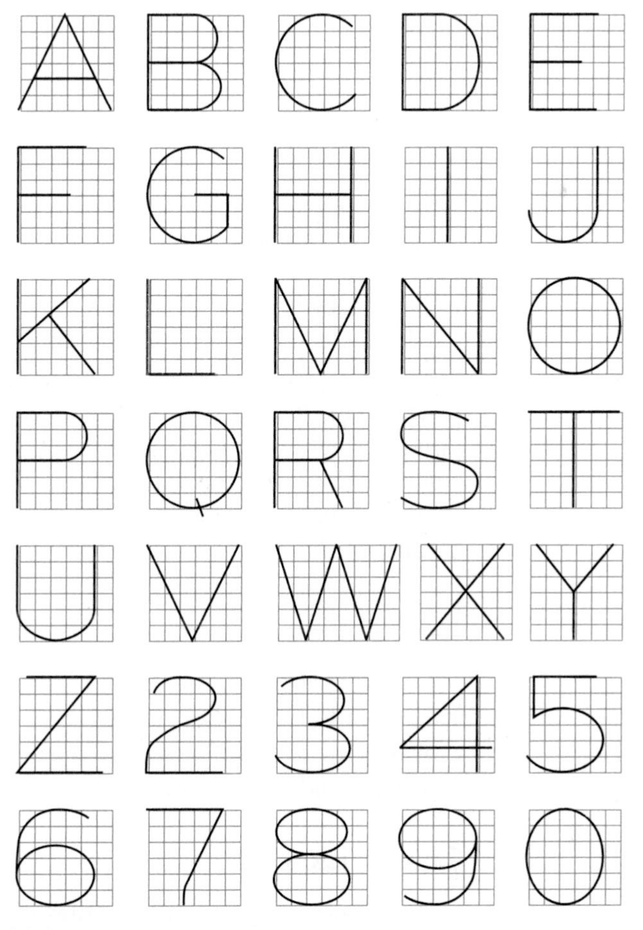

Figure 1.47 Upright lettering

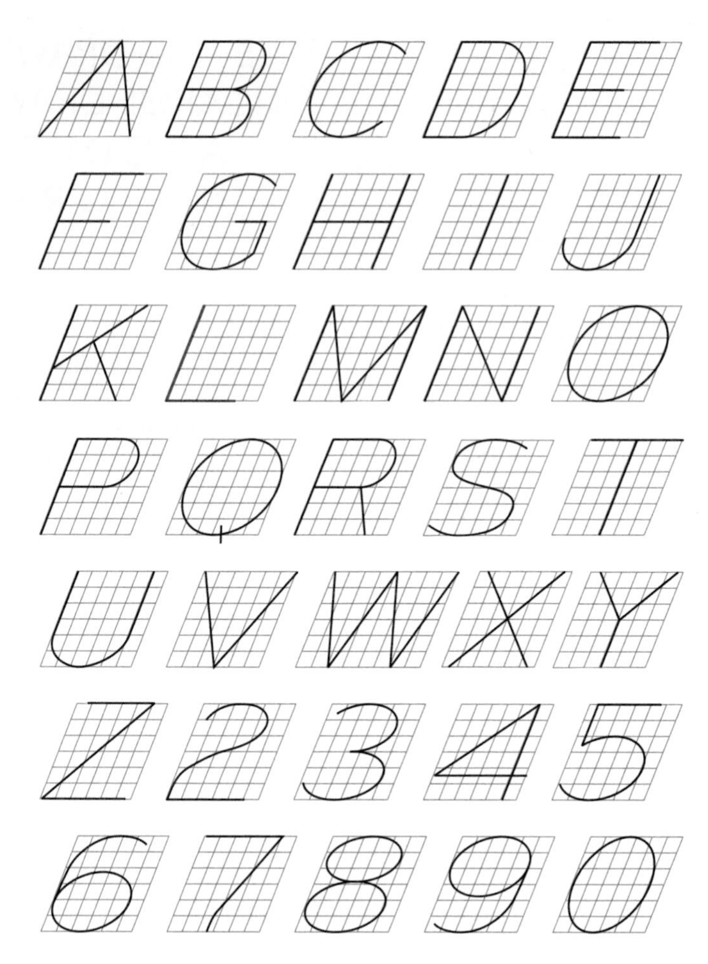

Figure 1.48 Sloping lettering

earlier, they can look at your calculations in the file. They may not be able to talk to you, and you probably would not remember anyway. Someone must understand what you did. So, similar standards apply to this sort of engineering work, which is likely to be done by hand.

Example 1.1 shows a sample of a student assignment on machine tools and processes. Lettering is done by hand and is easily readable. Drawings are done with a straight edge to illustrate what has been described in the text. This assignment was done on engineering calculation paper, with title blocks at the top giving the course, student name, and other information. This is a common format for engineering calculations, although there are variations. Some companies have their own formats for calculation sheets, and these could have the company name and logo included.

Lines

The lines on engineering drawings are also standardized, and different lines mean different things. This makes drawings easier to read and reduces the possibility of errors. Not only are there different line types, but different line thicknesses are also used. Line types and line thicknesses are standardized.

Line Thickness

Line thickness on an engineering drawing is either thick or thin. A thick line is twice the thickness of a thin line. The thickness of the line is chosen according to the size

Example 1.1

MACHINE TOOLS & PROCESSES

- FILLETS – A BUILT UP WELD AT THE INTERSECTION OF TWO SURFACES. USUALLY AT 90° ANGLES; AN INSIDE ROUNDING. THE RADII OF THE FILLET IS USUALLY ABOUT $\frac{1}{4}$ INCH. BY RELIEVING THE STRESSES IN CAST METAL, THE FILLET PROVIDES AN ADDED STRENGTH AT INSIDE CORNERS.

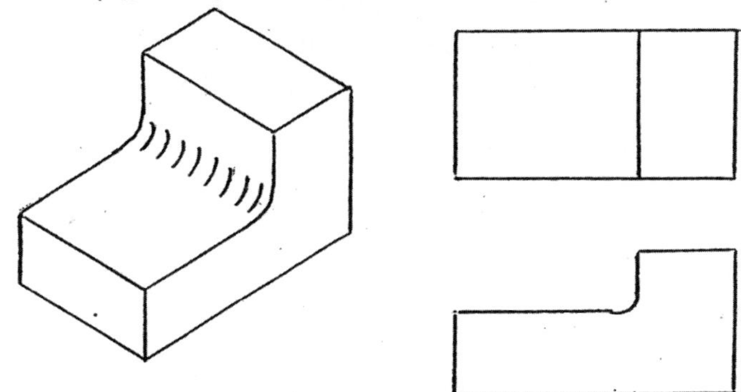

- ROUNDS – AN EXTERNAL ROUNDING ON A PART. ITS RADII IS USUALLY ABOUT $\frac{1}{4}$ INCH. ROUNDS IMPROVE APPEARANCE AND REMOVE SHARP EDGES.

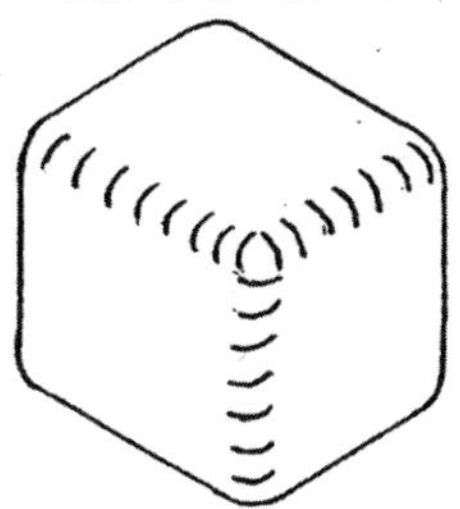

- DRILL – N. A ROTATING TOOL THAT IS INSERTED INTO A DRILLING PRESS OR TOOL FOR BORING CYLINDRICAL HOLES.

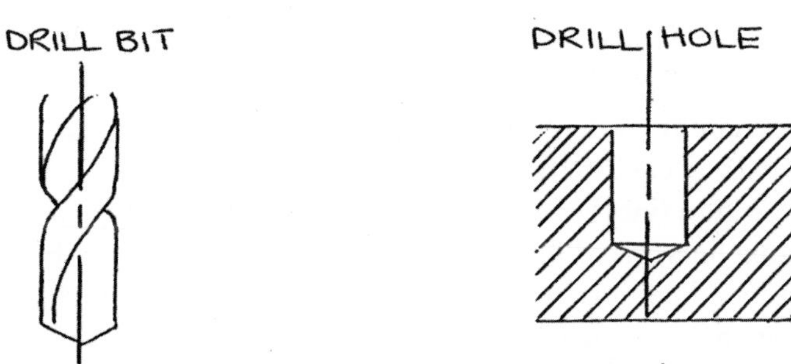

– V. BASIC METHOD OF MAKING HOLES. IT IS PERFORMED BY MOUNTING A DRILL IN THE TAIL STOCK OF THE LATHE AND ROTATING THE WORK WHILE THE BIT IS ADVANCED INTO THE PART.

(Reprinted with minor modifications with the kind permission of Veronica Zimmerman)

of the drawing, what it represents, and how it will be reproduced. The sizes that should be used are 0.25, 0.35, 0.5, 0.7, 1, 1.4, and 2 mm.

If the thickness of the outline chosen is 0.7 mm (the thick line), then the thin lines would be 0.35 mm.

Line Types

Different line types indicate different things. Line types are shown in Figure 1.49 and examples of their use in Figure 1.50.

A **visible outline** is a thick solid line used to show the outline of an object (also referred to as an **object line**).

A **hidden line** is a thin dashed line used to indicate that an edge is hidden. The line always starts and ends with a dash in contact with the lines at which they begin. An exception to this is a hidden line that forms a continuation from a visible line. In this case, there is a space between the visible line and the hidden line. Dashes must join at corners. Hidden lines may be omitted for clarity or if they are not necessary. A **centreline** is a thin line made up of long and short dashes. The long dashes are always at each end of the line. Centrelines project a short distance beyond the object to which they refer.

A **dimension line** is a thin line that shows the size of a feature. It usually has arrowheads on each end, although other line endings are sometimes used. The size is given in a break in the middle of the dimension line. The dimension is sometimes above the line, and there is no break in the line. Dimensioning is covered in Chapter 4.

An **extension line** is a thin line that extends from the end of a feature being dimensioned. A 2 to 3 mm gap is left between the outline of the object and the extension line.

A **leader** is a thin line that points from a note, or dimension, to the feature to which the note or dimension applies. There is usually an arrowhead on the end near the object, but other line endings are sometimes used depending on the feature. A leader line is a straight line. The example in Figure 1.49 shows a leader pointing from the diameter (15 mm) to the circle.

A **break line** indicates that all of the object has not been drawn. If a cylinder is very long, only a portion near each end would be drawn and this would be indicated by a break line. A break line can be thick or thin depending on how it is used. A short break (a break on a small object) would be a thick, freehand line so as not to be confused with an outline. A long break line is a thin line with "zig-zags." The number of "zig-zags" depends on the length of the break line.

A **cutting plane line** is a broken line made up of thick and thin segments. Thick sections are used at the ends of a simple section and at the ends and corners of an offset section. A cutting plane line shows where a section has been taken. Sectioning is a technique used to show interior details and is covered in Chapter 3.

Section lines are used to show interior surfaces when an object is sectioned. They are thin lines, and different types of section lines indicate different materials. The use of section lines is covered in Chapter 3. Section lines used for different materials are shown in Appendix C.

A **phantom line** is a thin line used to show some feature that is not drawn in full detail. (It would be drawn in another drawing.) Phantom lines indicate that this feature exists. It could show a mating part.

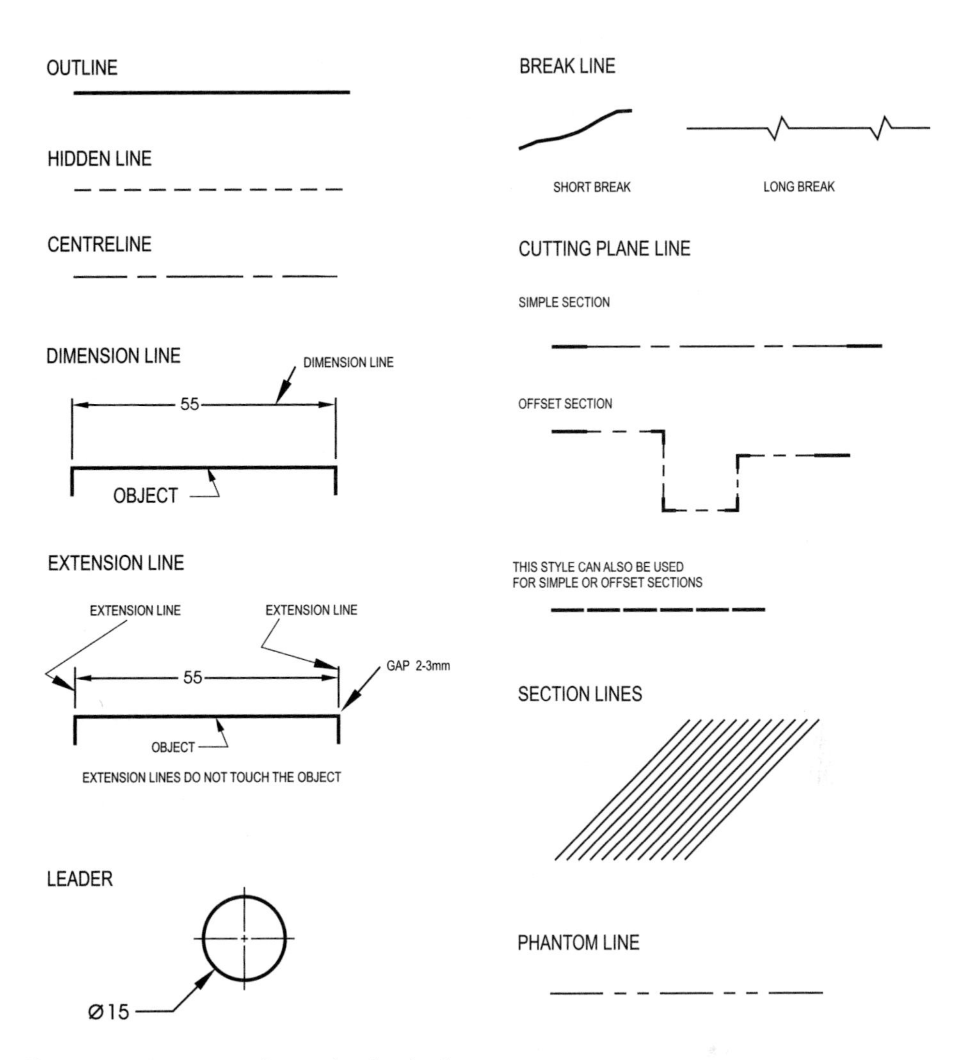

Figure 1.49 Line types used on engineering drawings

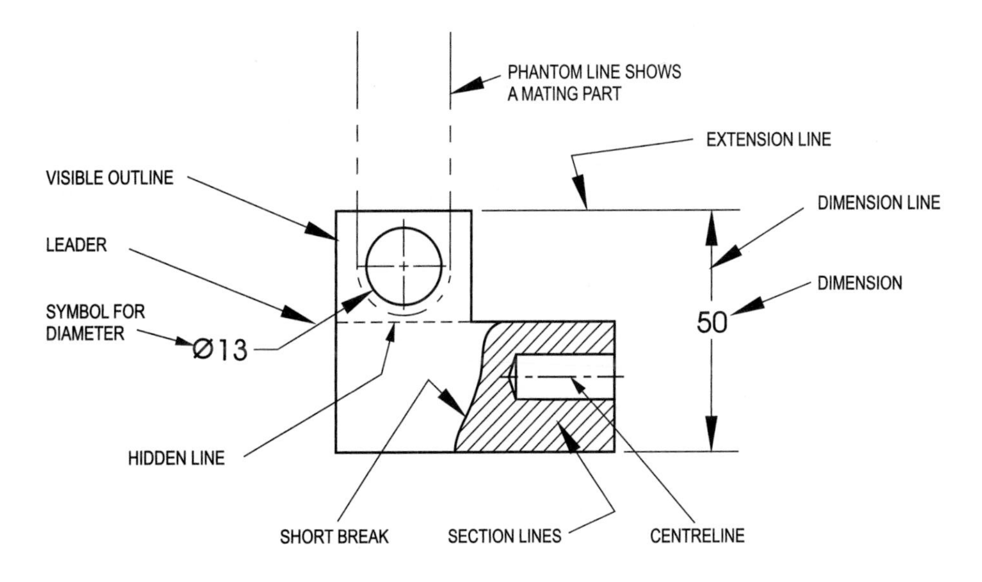

Figure 1.50 How different line types are used

Line Precedence

Often, one line on an object will appear to lie over another line. A hidden line can be behind a visible outline, or a centreline can be coincident with a visible outline. Because of this, there is a hierarchy of lines, and some line types take precedence over others. Figure 1.51 shows examples of line precedence.

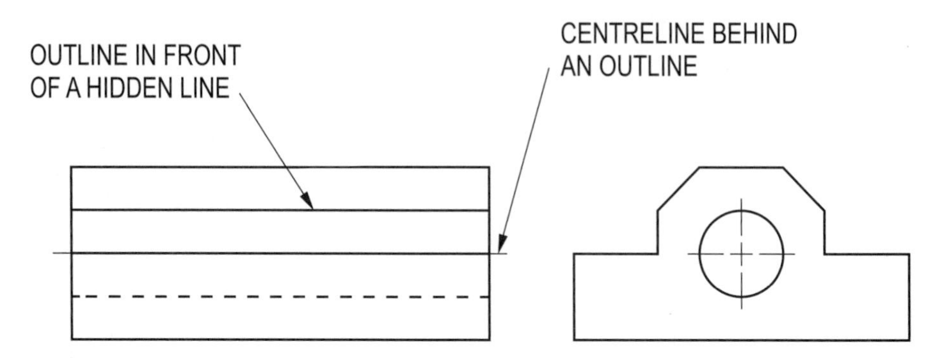

Figure 1.51 When lines are coincident, some take precedence over others

The most important lines are the visible and hidden outlines of the object, and these take precedence over other lines. Visible outlines take precedence over hidden outlines. A full line (visible line) can cover a dashed line but a dashed (hidden) line cannot cover a full line.

A centreline often appears behind a visible outline, and the outline takes precedence over the centreline. A centreline extends beyond the feature to which it refers, and so the ends can be seen.

Dimension and extension lines can be placed where they are needed and should be placed where they do not coincide with any other lines.

Other lines—break lines and phantom lines—are low-precedence lines and can be located where they do not coincide with other lines.

Problems

A. Select the best front view and the corresponding top and side views for the objects below. The sides are identified as **a**, **b**, **c**, **d**, and **e**. The views are labelled only on object one. Views are the same for all other objects.

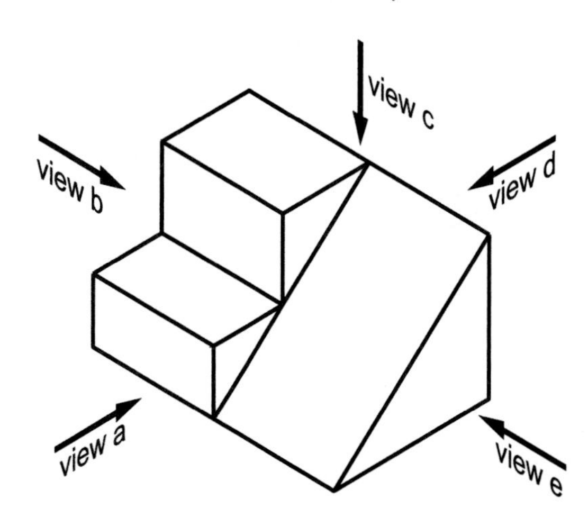

Example

If the front view is **a**, the top view is **c** and the side view is **e**.
If the front view is **e**, the top view is **c** and the side view is **a** (left side).

If the front view is **b**, the top view is **c** and the side view is **a** (right side).

If the front view is **c**, the top view is **a** and the side view is **b** (right side).
Different choices are given here to show how the name of the view depends on the viewpoint. The shape is seen best with front view **e**.

1.

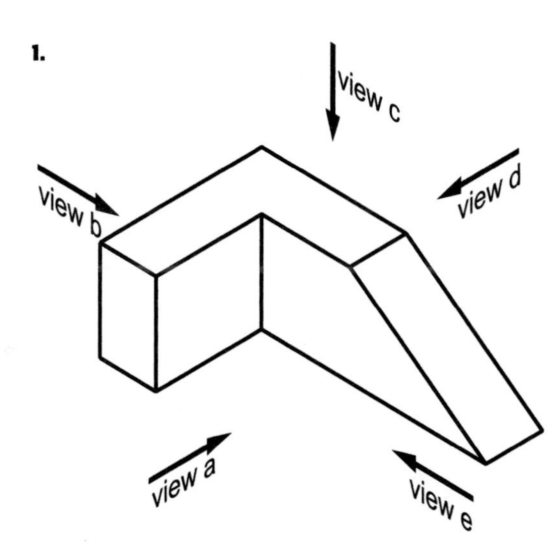

2.

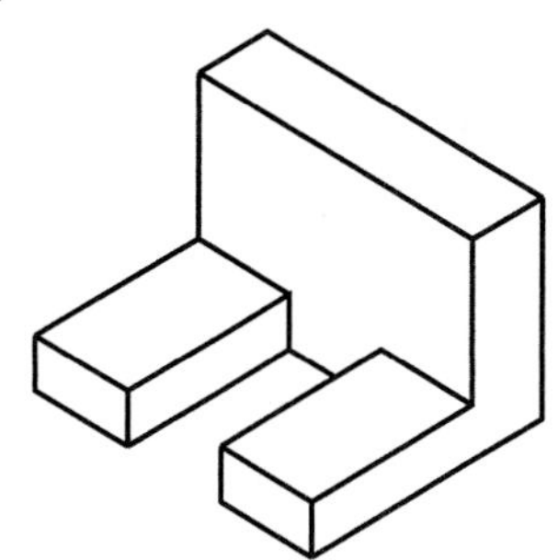

3.

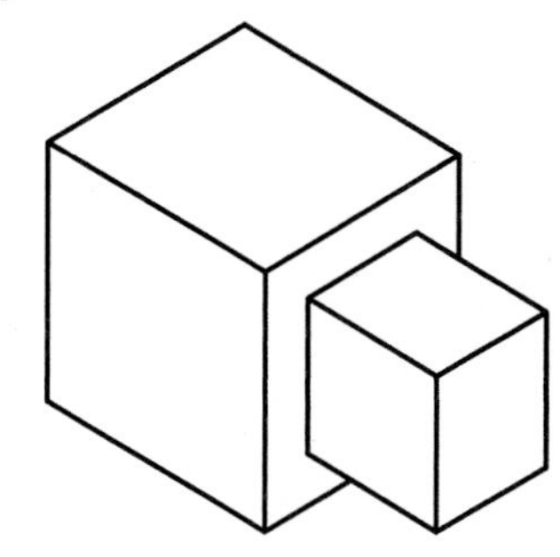

4.

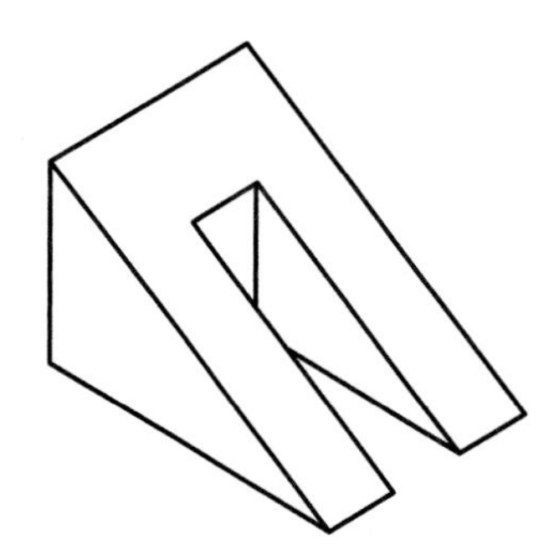

5.

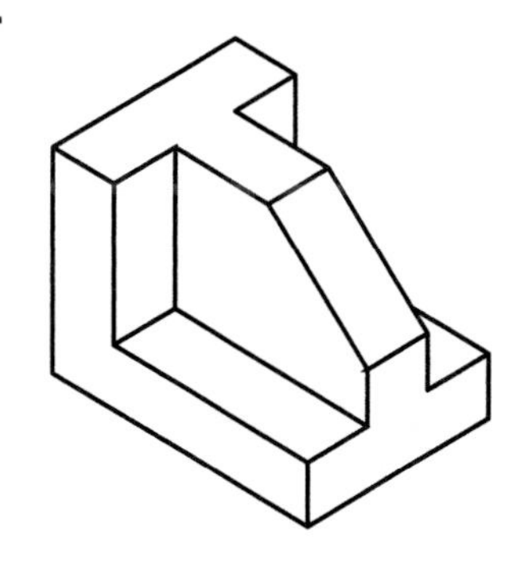

6.

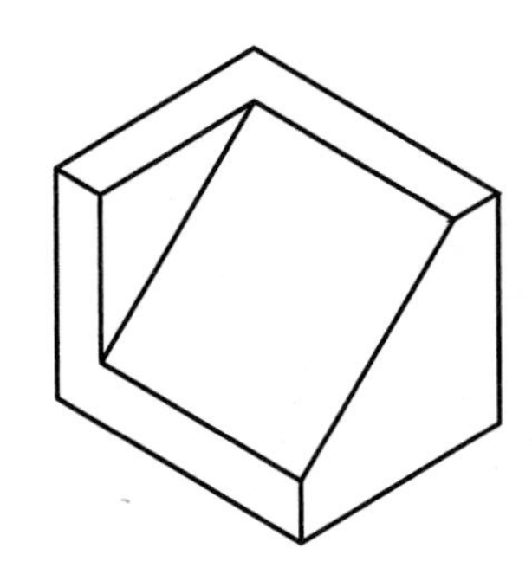

10.

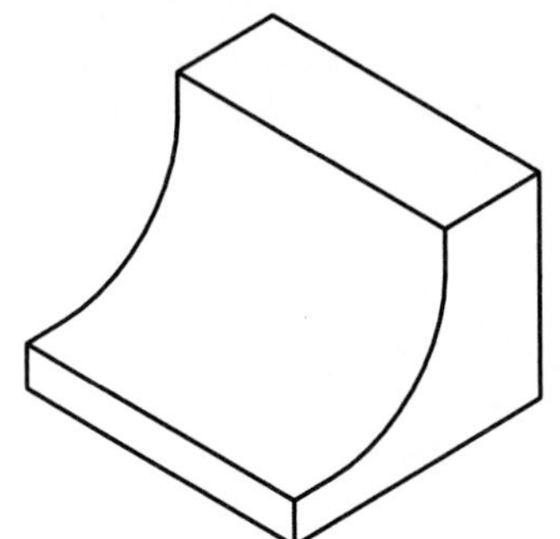

7.

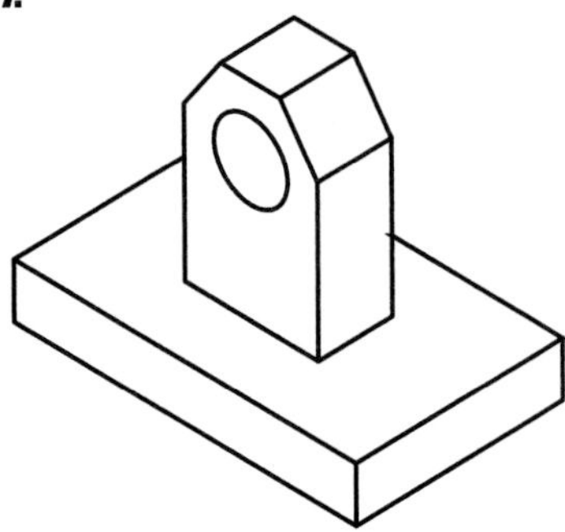

11.

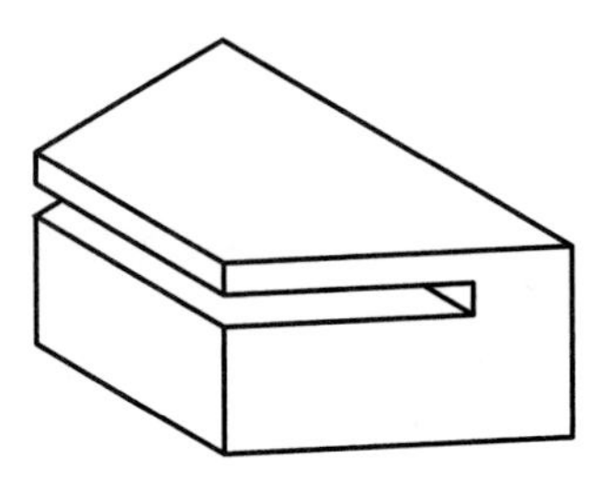

8.

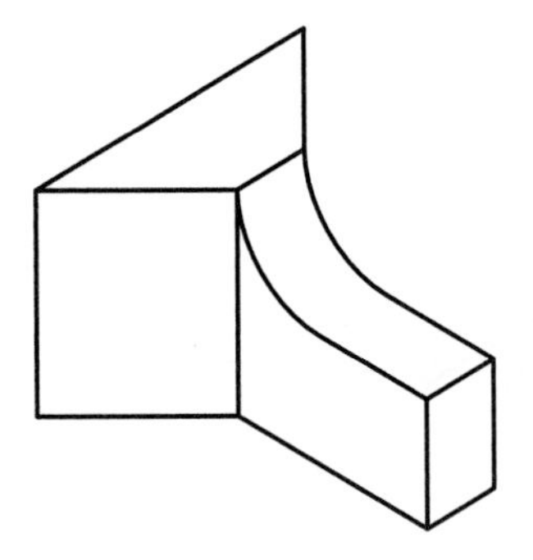

12.

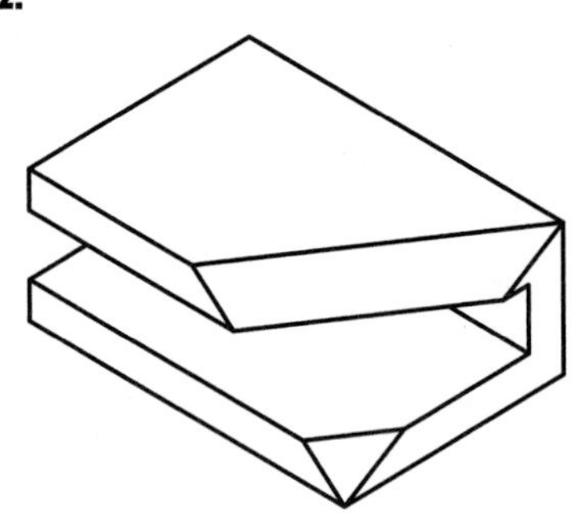

9.

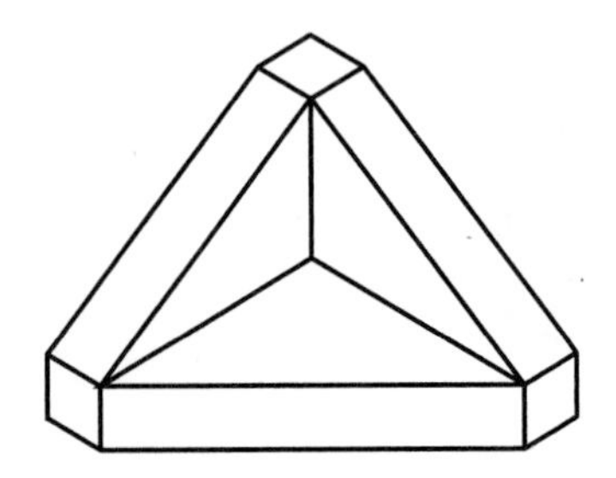

13.

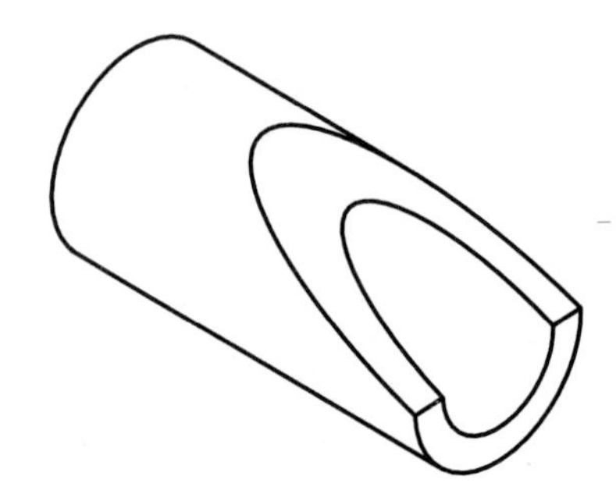

14.

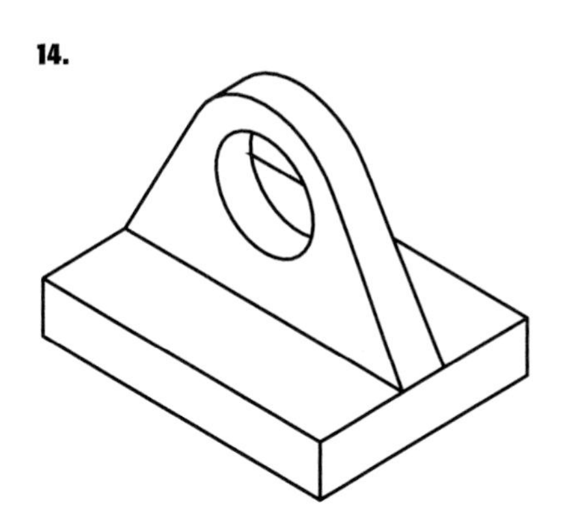

15.

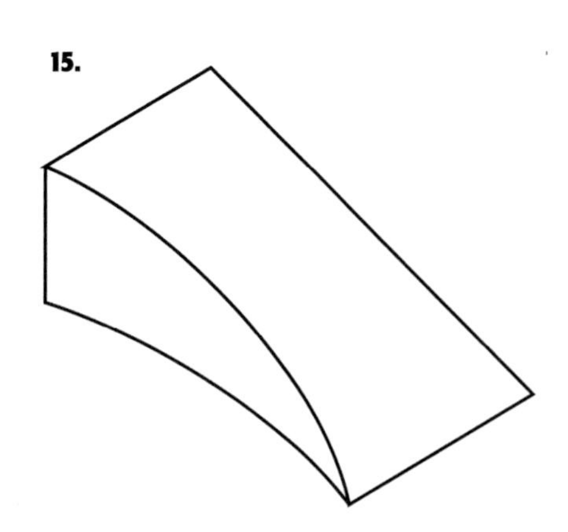

16.

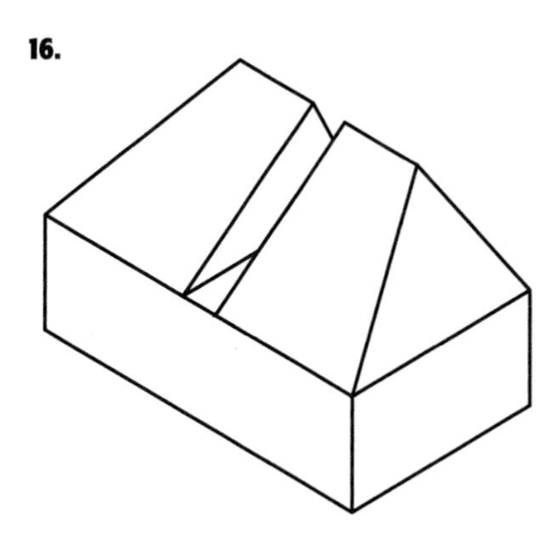

17.

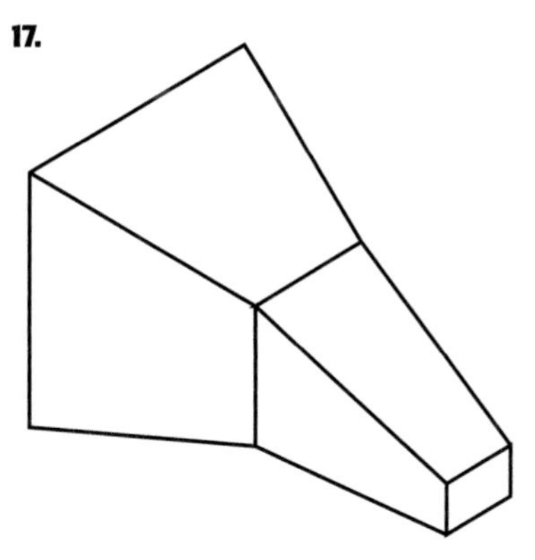

18.

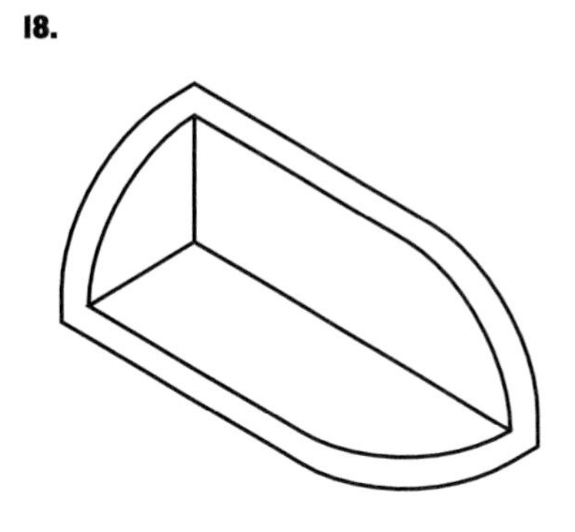

19.

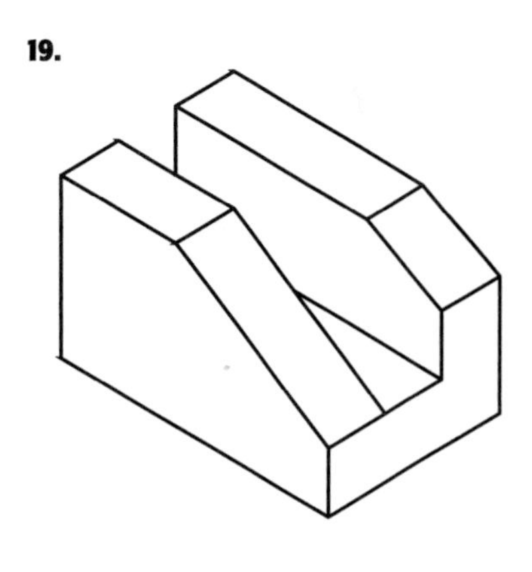

20.

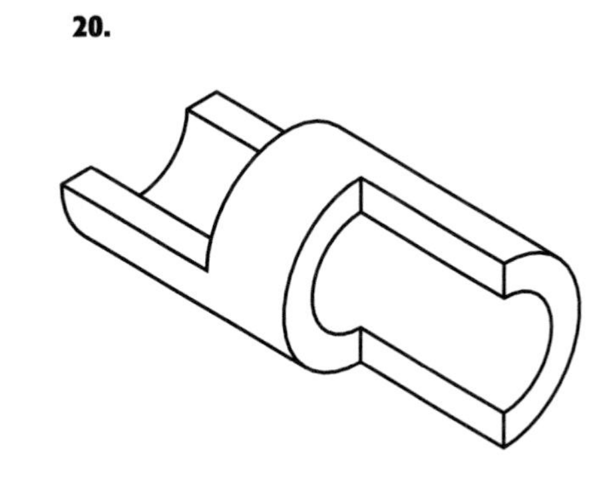

21.

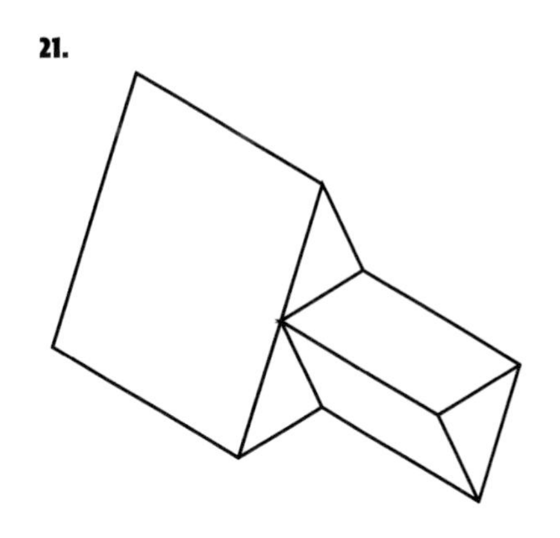

22.

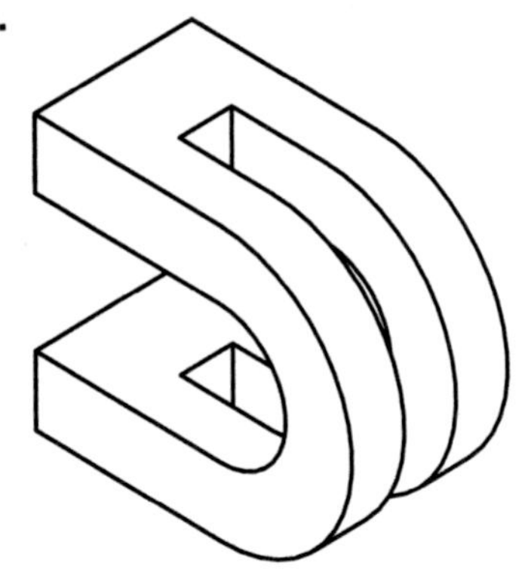

23.

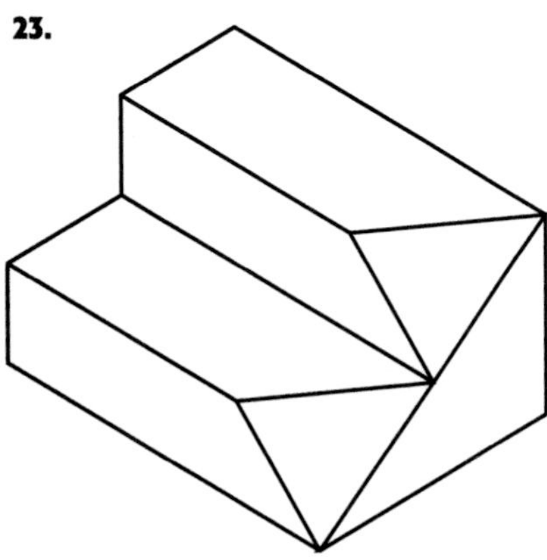

24.

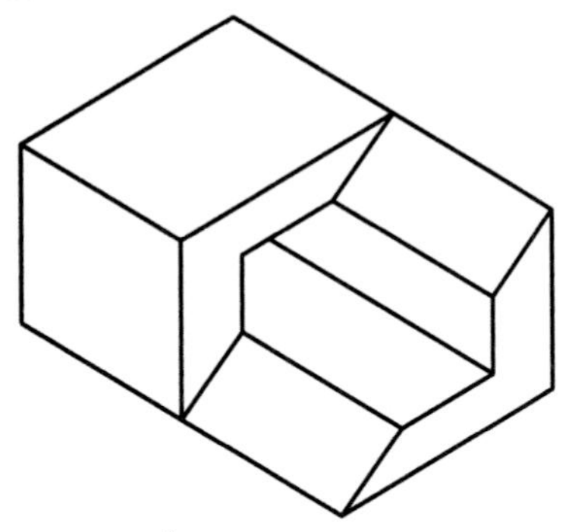

25.

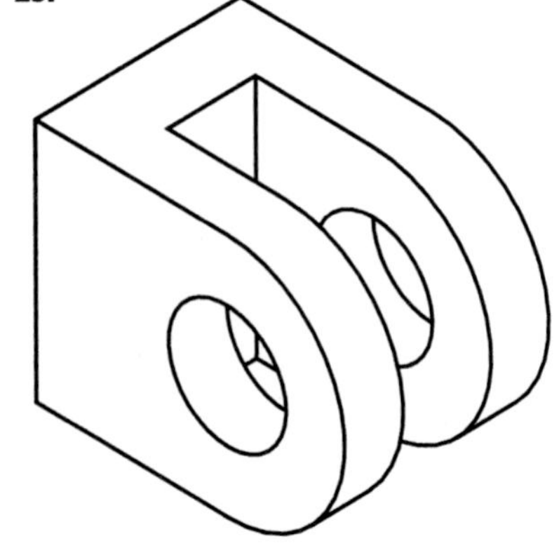

26.

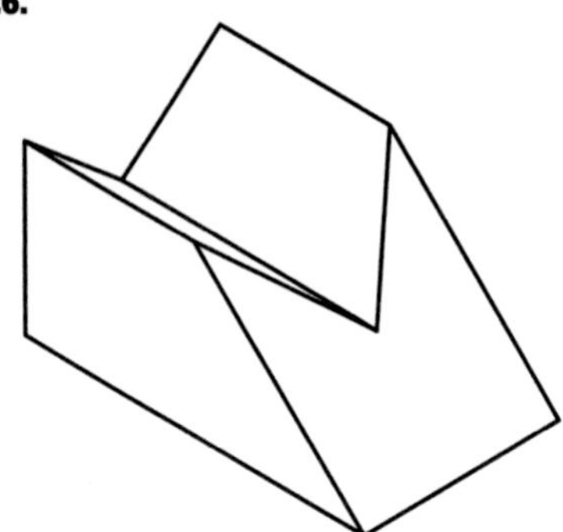

27.

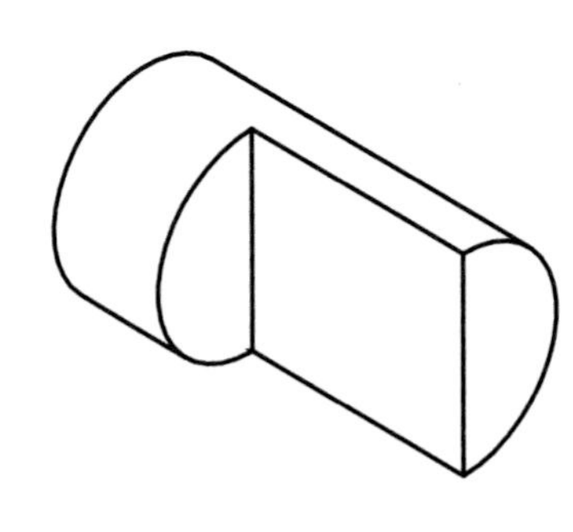

28.

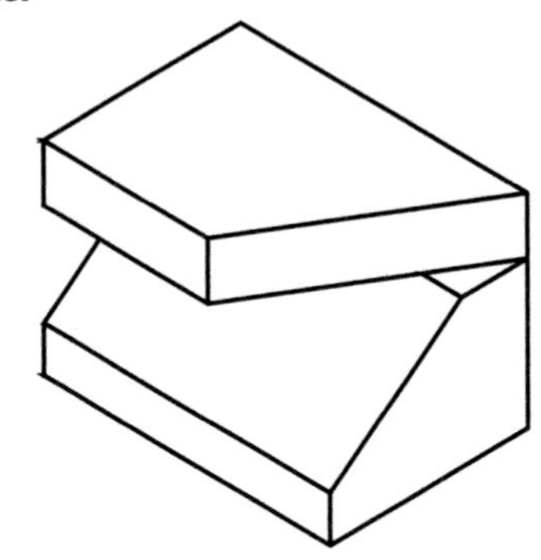

29.

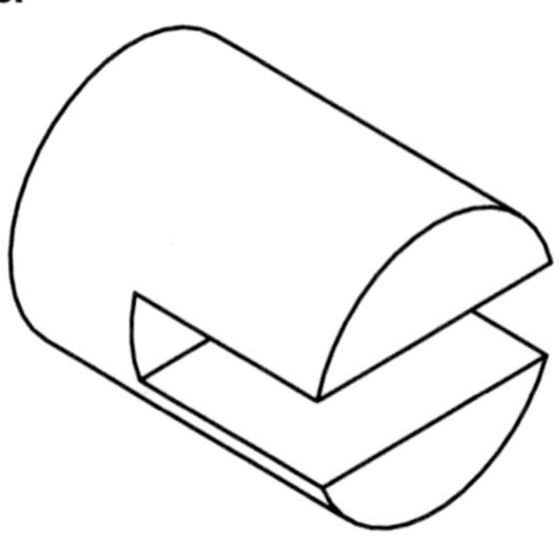

30.

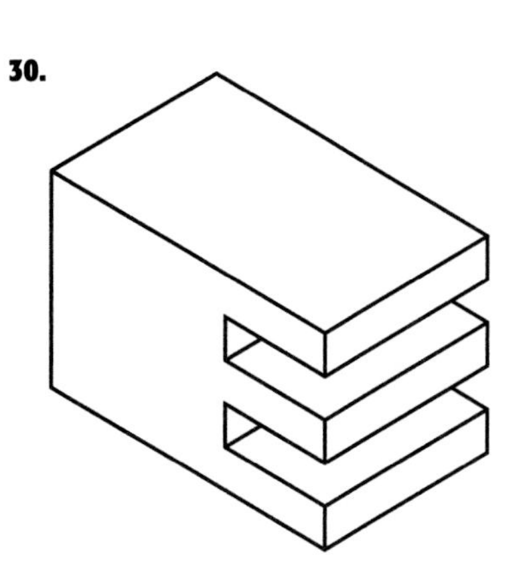

31.

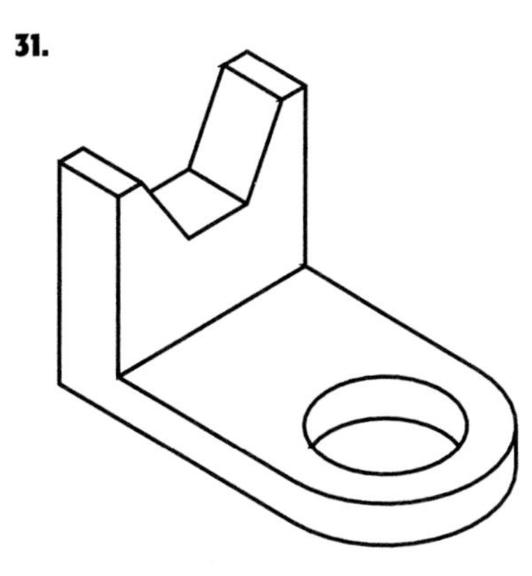

32.

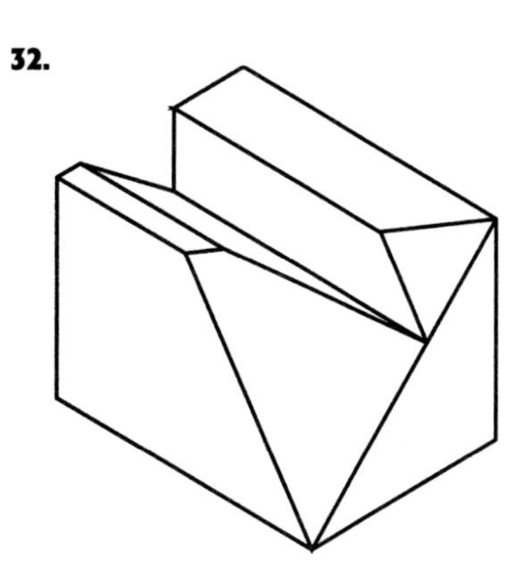

33.

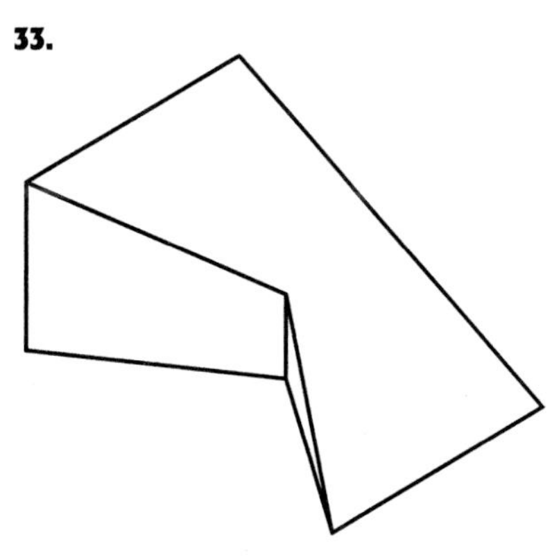

34.

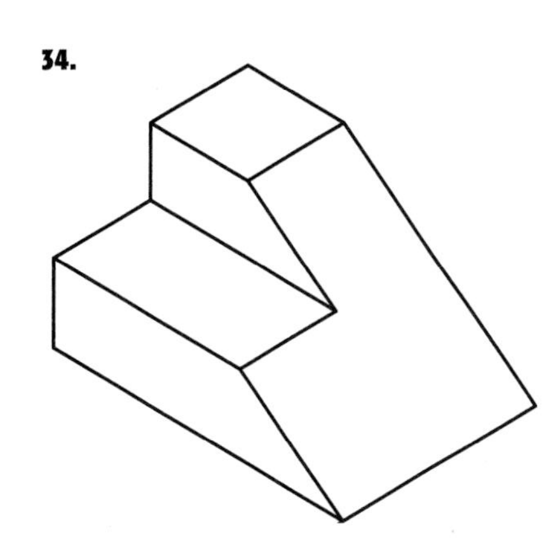

35.

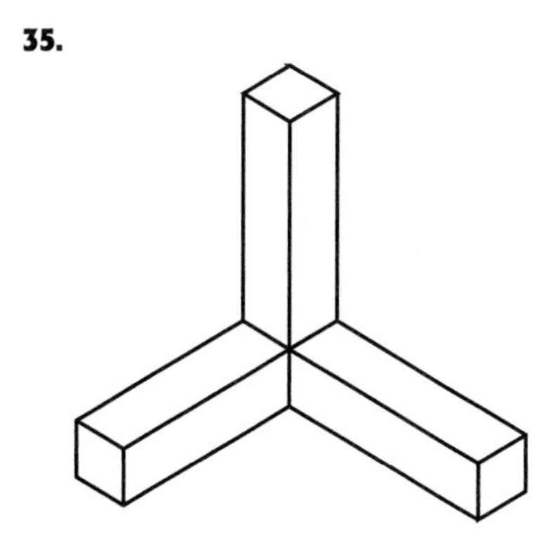

36.

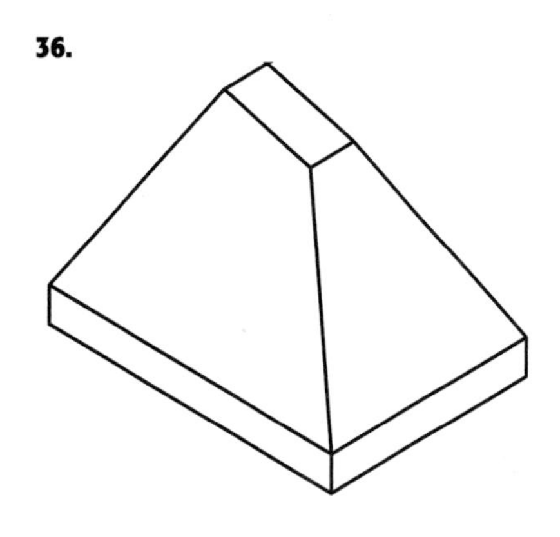

37.

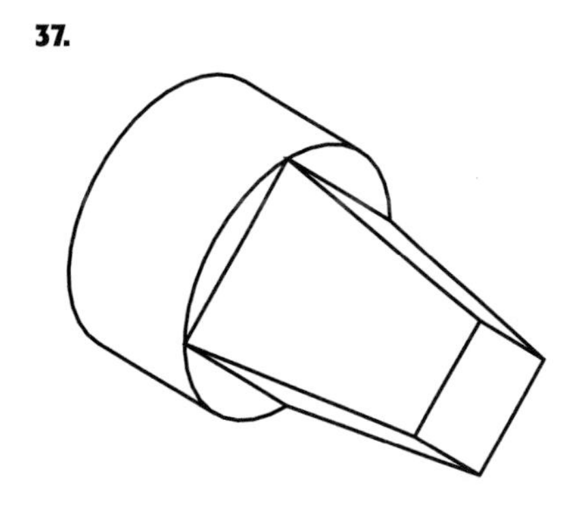

38.

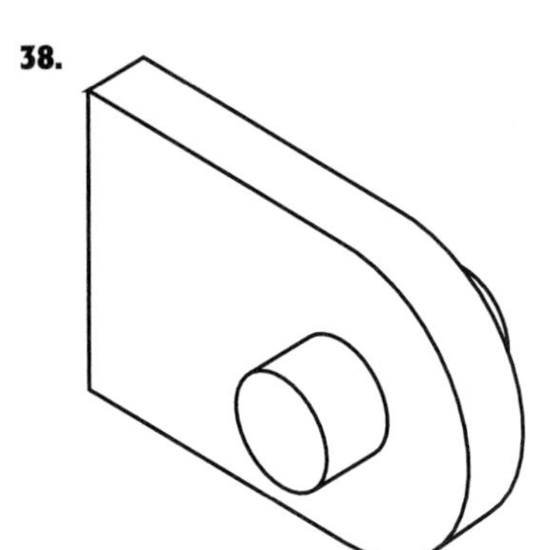

39.

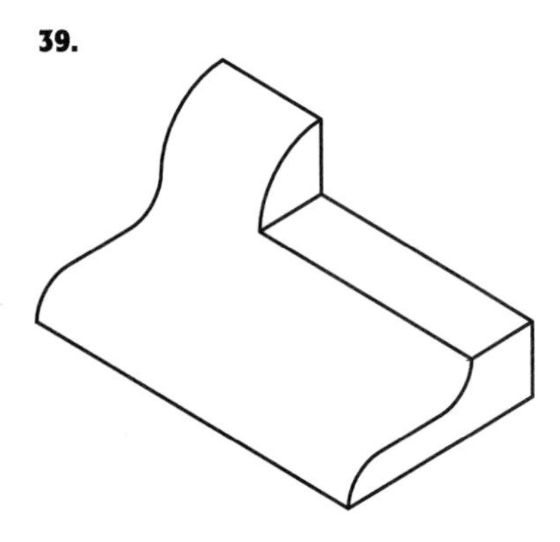

40.

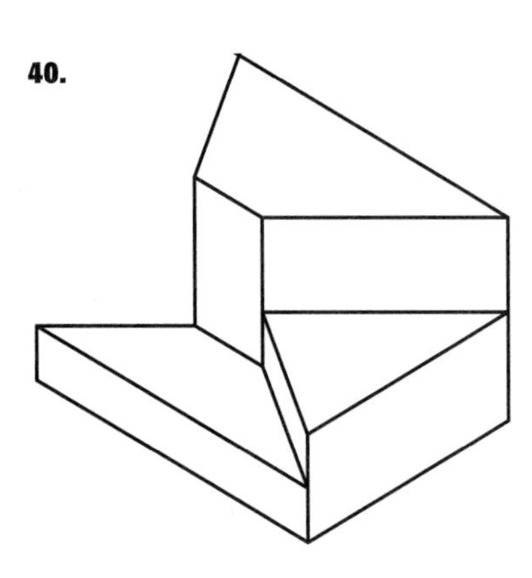

41.

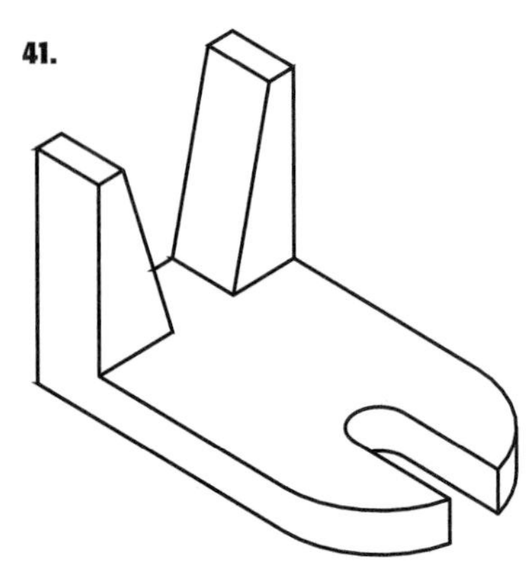

42.

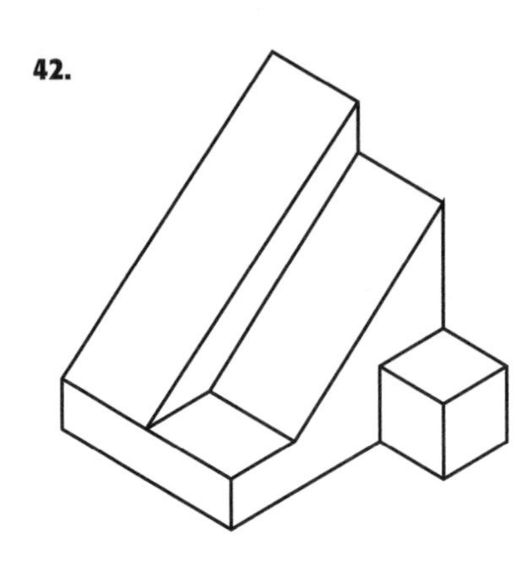

43.

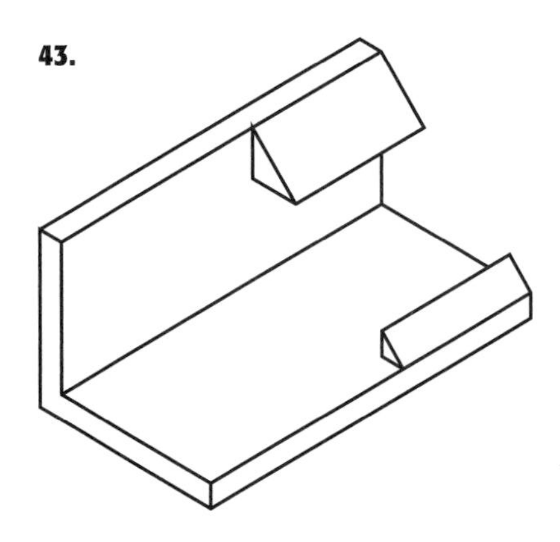

44.

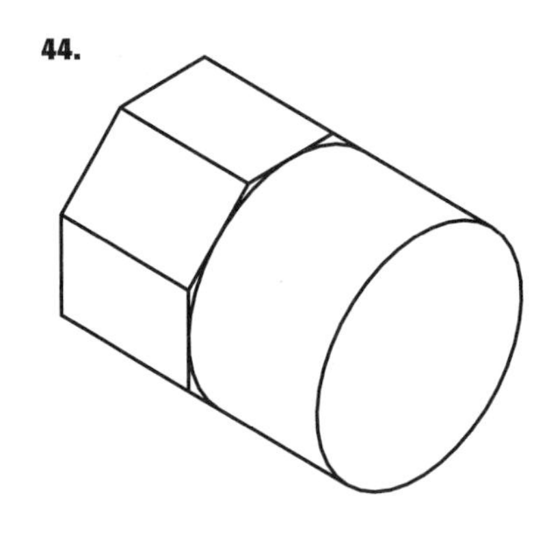

45.

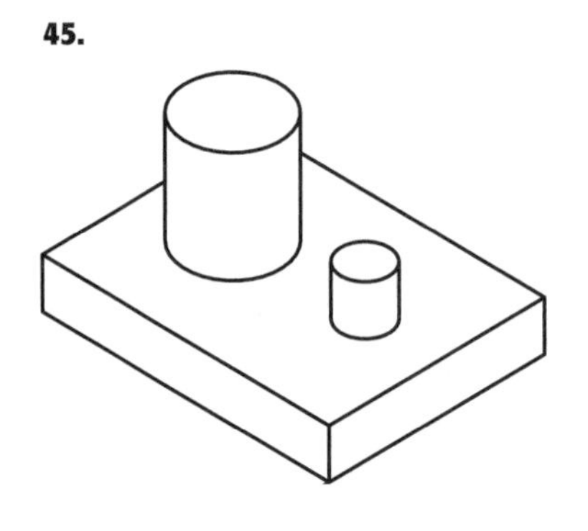

46.

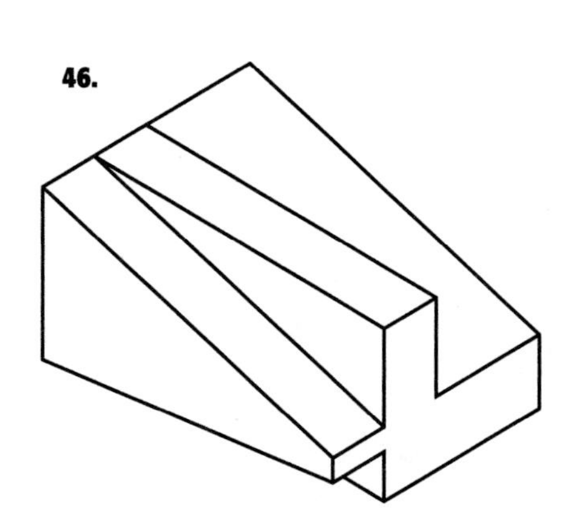

48.

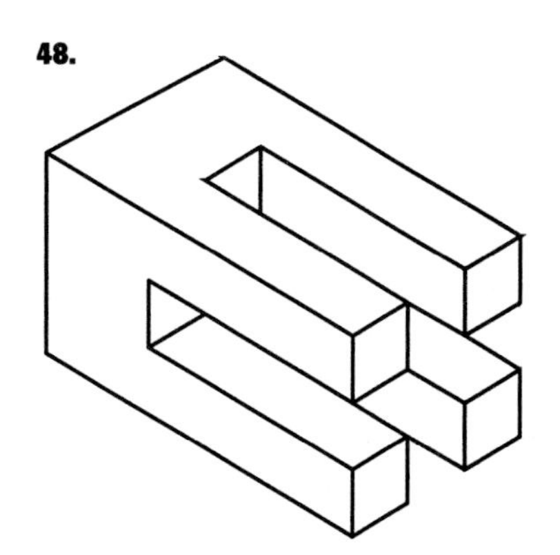

47.

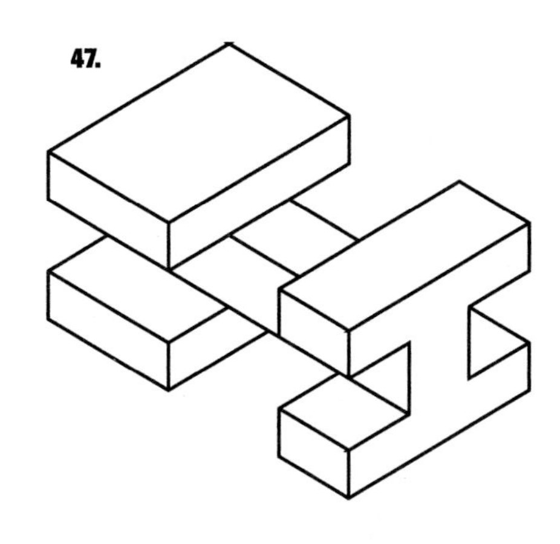

49.

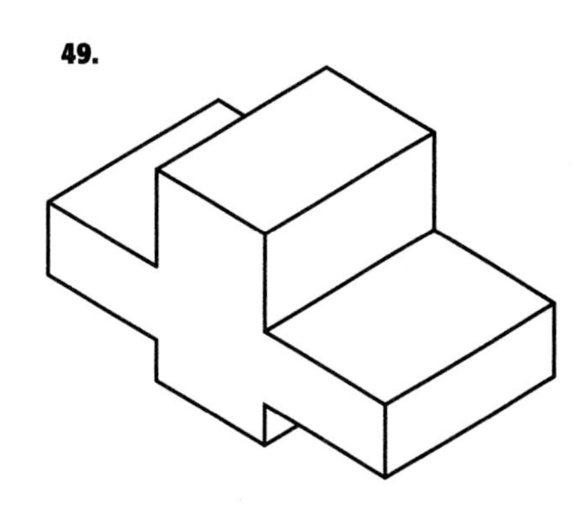

B. Identify which corners would be visible in the front, top, and right-side views of the following objects.

For example, corners FECMHG are visible in the right-side view of the object in problem 50.

50.

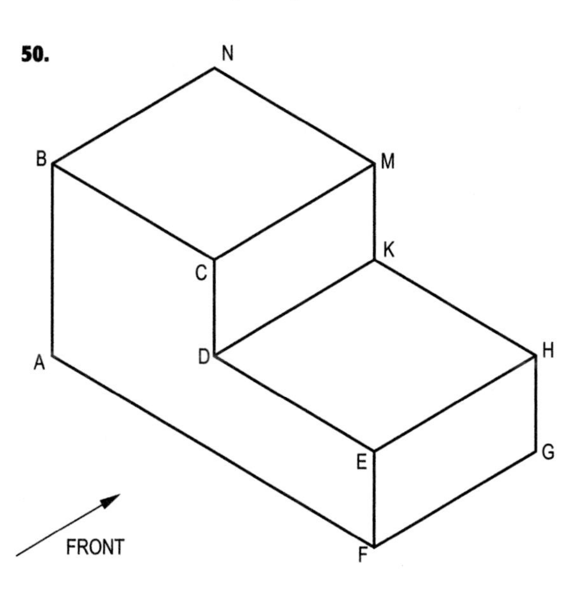

51.

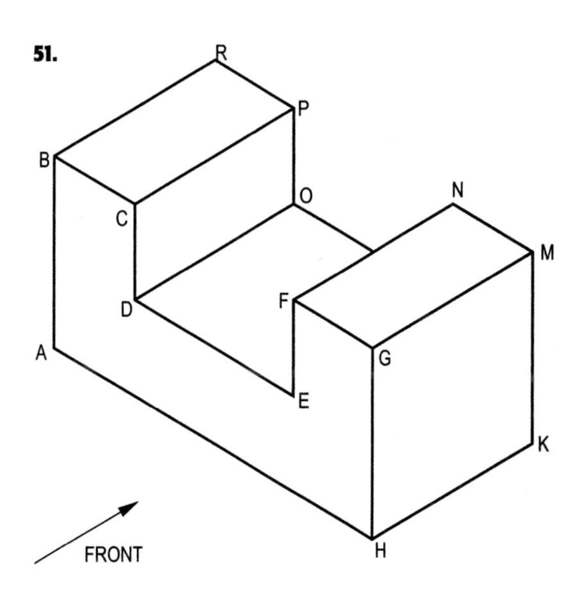

52.

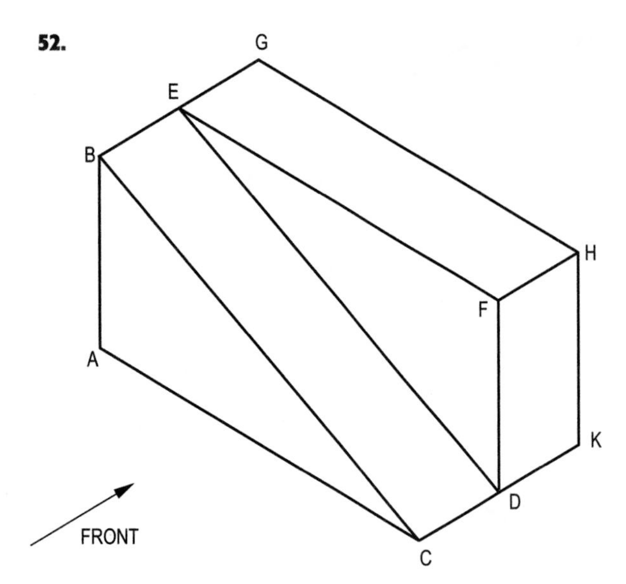
G
E
B
H
F
A
K
D
FRONT
C

53.

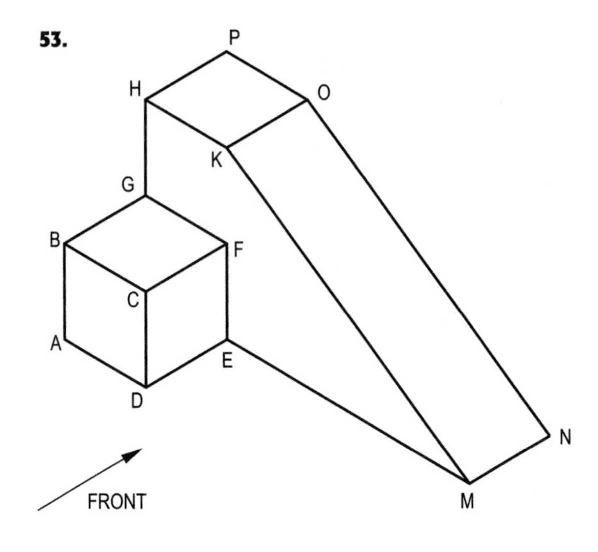
P
H
O
K
G
B
F
C
A
E
D
N
FRONT
M

54.

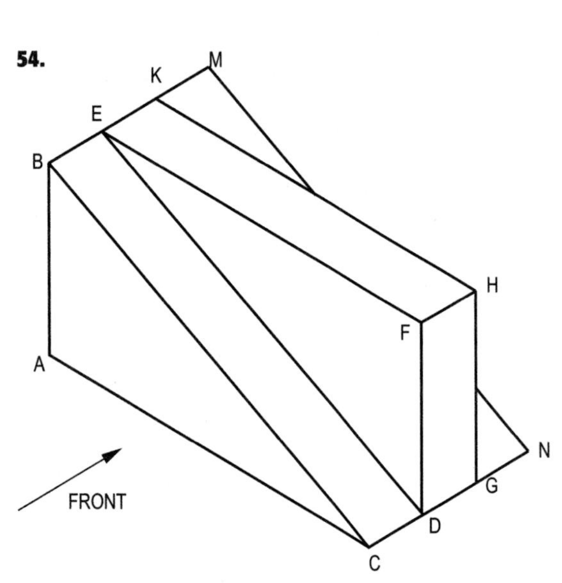
K
M
E
B
H
F
A
N
G
D
FRONT
C

55.

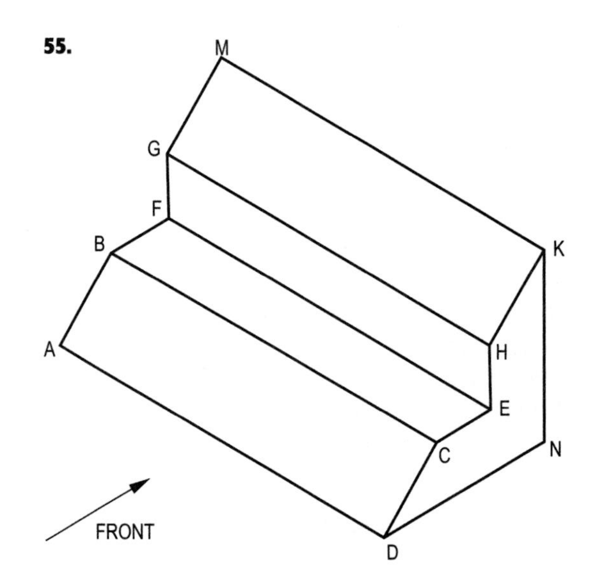
M
G
F
B
K
A
H
E
C
N
FRONT
D

C. Two views of an object are shown. Identify the correct missing view–**a**, **b**, **c**, or **d**.

56.

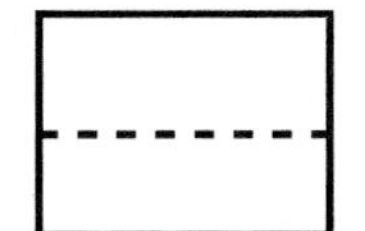

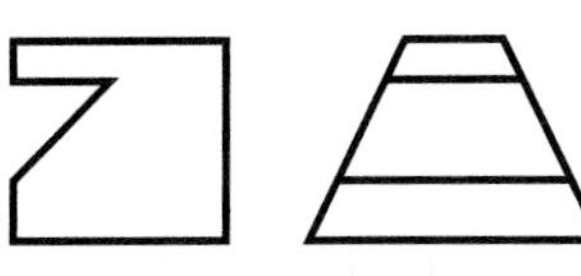
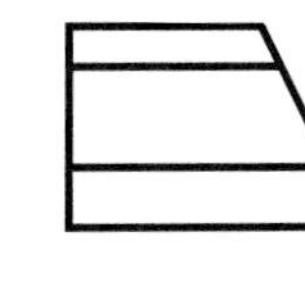
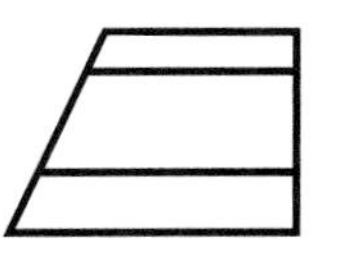
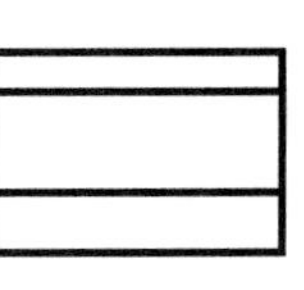

a **b** **c** **d**

57.

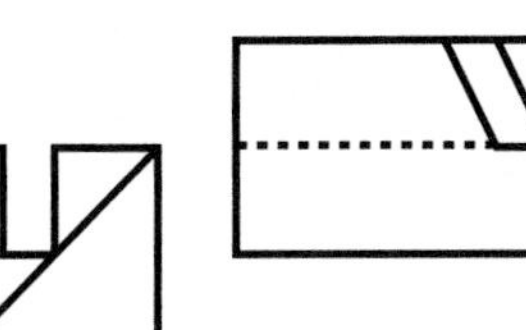
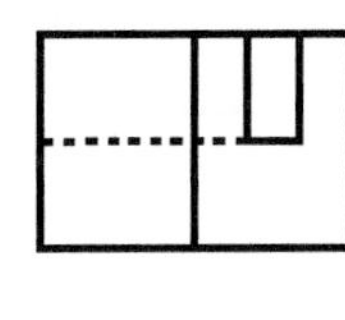
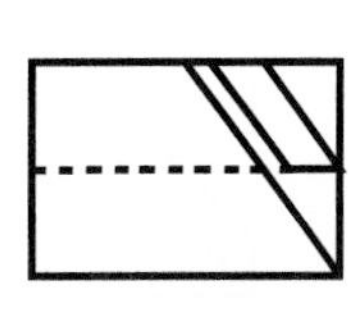
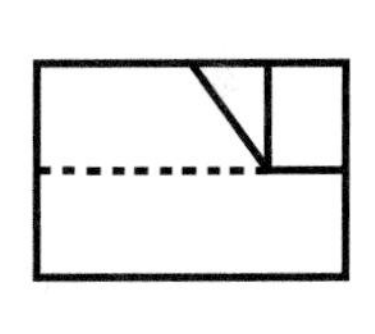

a **b** **c** **d**

58.

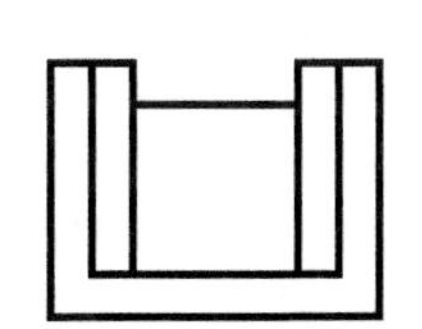
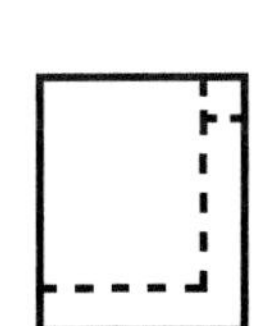
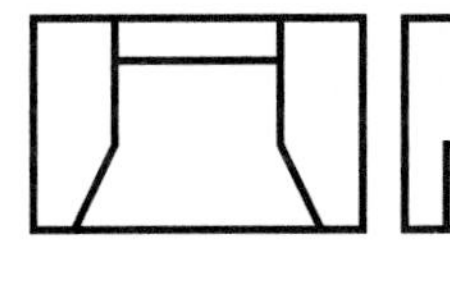
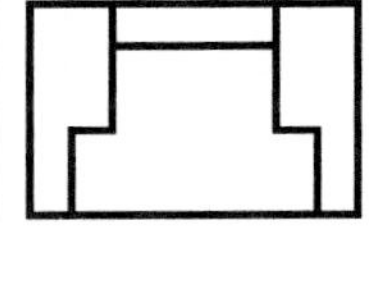
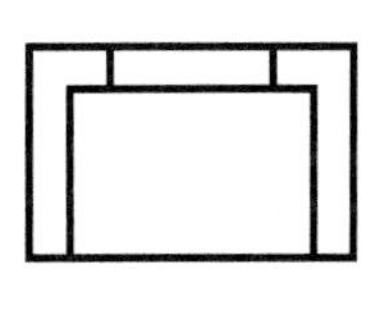
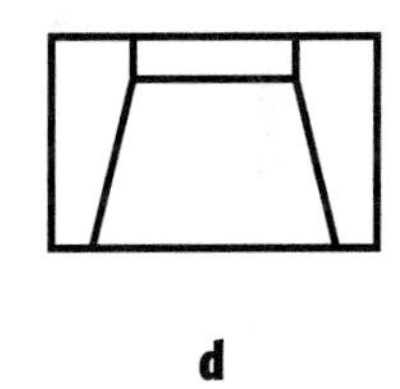

a **b** **c** **d**

59.

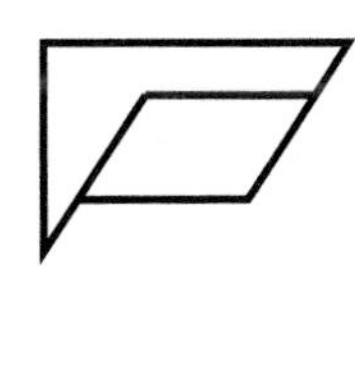

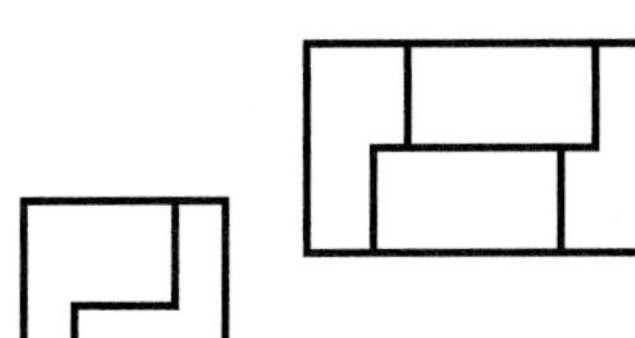
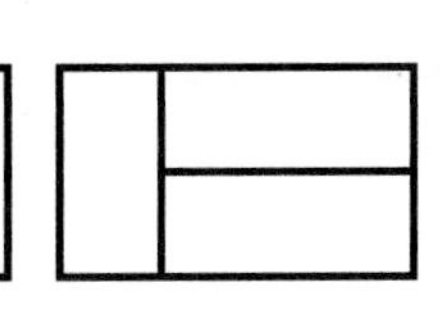
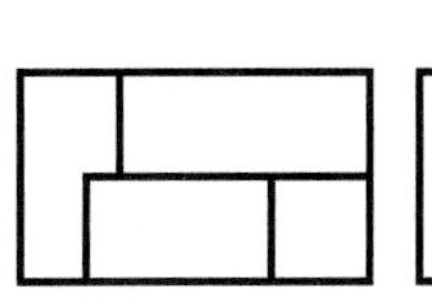
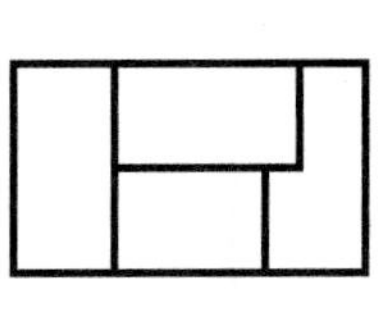

a **b** **c** **d**

60.

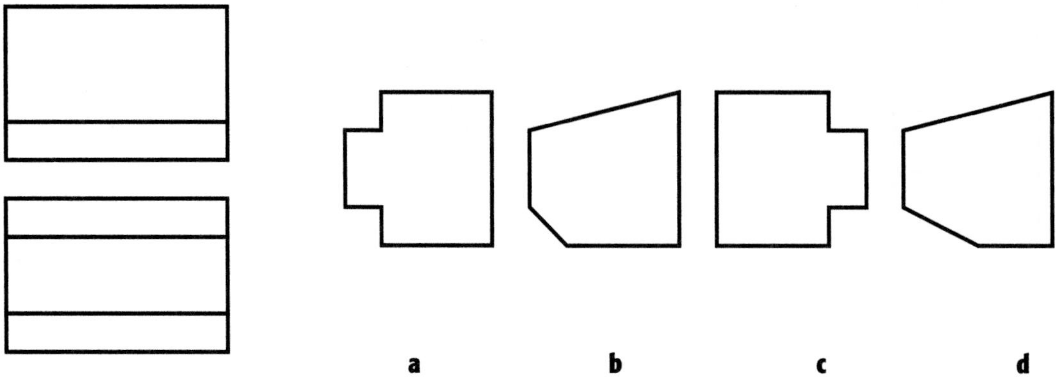

a b c d

61.

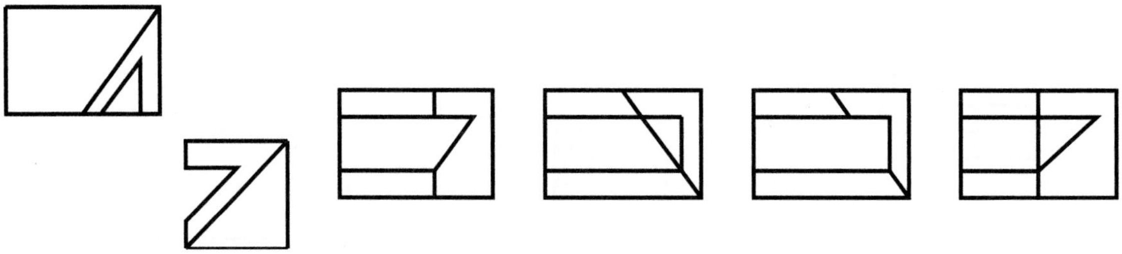

a b c d

62.

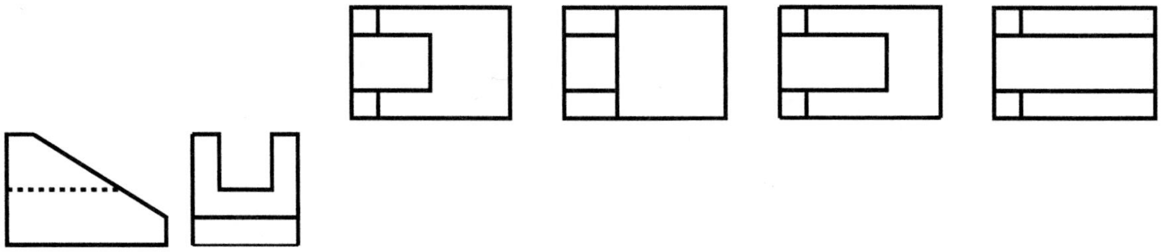

a b c d

63.

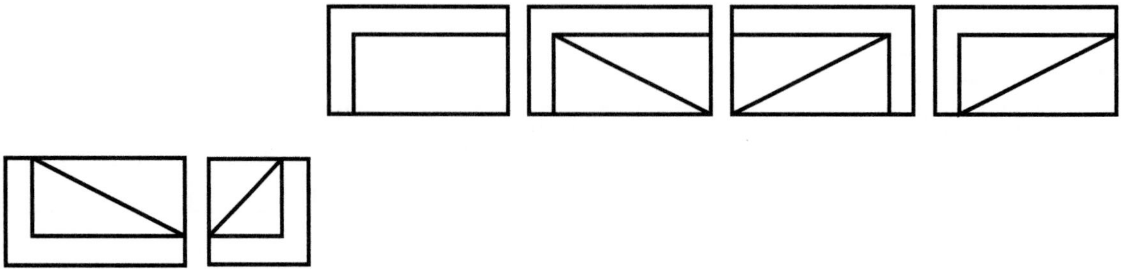

a b c d

64.

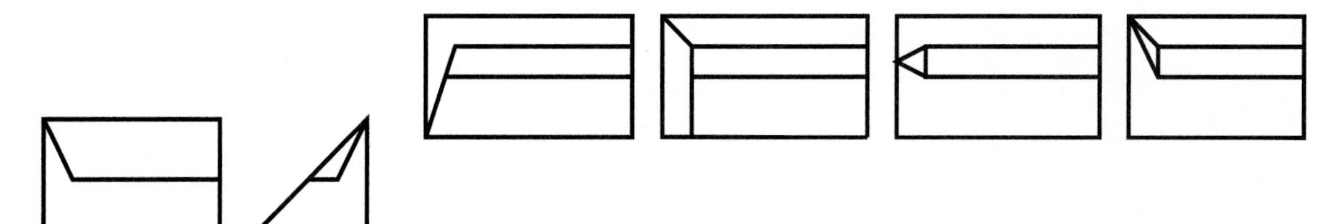

a b c d

65.

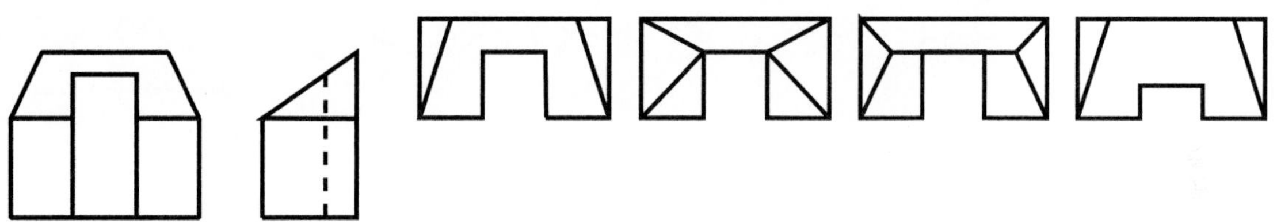

a b c d

> *An engineering drawing starts with a series of freehand sketches or roughly scaled drawings to help visualize how the object being designed will work and how the finished drawings must be made in order to show the reader exactly what the designer had in mind. Because design is a complex intellectual process, the designer uses sketches to try out new ideas, to compare alternatives, and (this is important) to capture fleeting ideas on paper.*
>
> *Eugene S. Ferguson,* Engineering in the Mind's Eye, *The MIT Press, 1992*

CHAPTER 2 Sketching

A sketch is a freehand drawing done without any aids, such as a straight edge or compass. You do not need to have any artistic talent to draw a sketch. If you do have any artistic talent, your sketches may look better than those of someone who does not, but they are still sketches. A sketch does not have to be a "work of art." Sketches do not have to be an exact representation, nor are they expected to be. A sketch can be done quickly with nothing more than pencil and paper, and it is the simplest way to record information. When you want to explain your idea to someone, sketching is often the best way to do it. Try explaining a safety pin without a picture. You can use a sketch to record an idea when it comes to you. A scrap of paper, or even a napkin, can be used to record an idea in the form of a sketch for consideration later. When you are on a job site, a sketch can be used to record information, for example, the layout of pipes in a chemical plant, or possible locations of sites for survey. The goal is to record information that you may need later. A project logbook often contains numerous sketches.

Research has shown that successful problem solvers spend more time planning and exploring a solution to a problem than actually executing the solution. Planning is an important part of the solution to any problem, and sketching is an efficient way of planning. Before you attempt to draw something using a computer, you should plan what you are going to do. It will eventually save you a considerable amount of time.

Figure 2.1 shows the several forces that act on a structural member, in this case, a beam. The goal is to design the beam, but first, we must determine the effect of the loads on it. A sketch is all that is required to show where the loads are.

The next step also involves a sketch. The beam is not moving, so something must be holding it—obviously, the reactions at the supports. These are included in another diagram, called a **freebody diagram**, which shows all the forces and reactions. Freebodies (see Figure 2.1) are discussed in more detail in Appendix B, but for now, we will simply use them to illustrate a sketch.

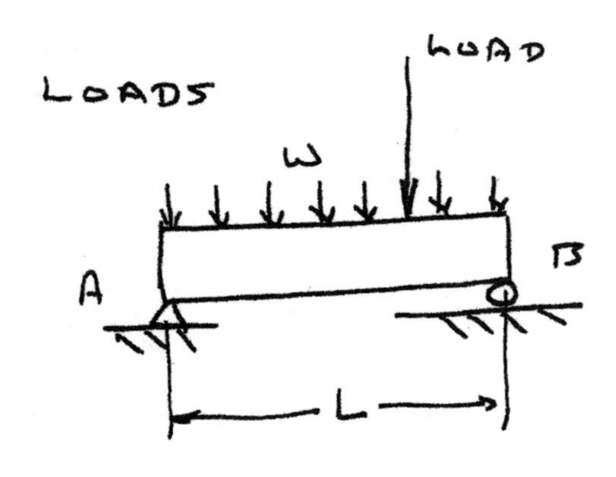

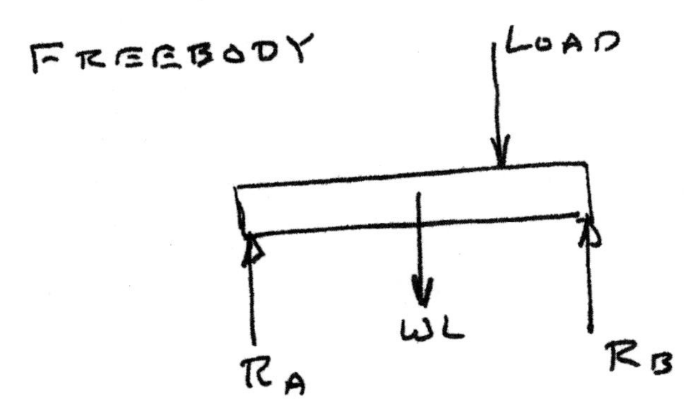

Figure 2.1 A sketch showing the loads and forces acting on a beam

Neither the length of the beam nor the length of the arrows showing loads and reactions are important in this diagram. (But the direction of the arrow is important.) There is no need to do anything other than a sketch; however, it must show the information clearly.

Figure 2.2 is a sketch to show a new arrangement for piping for a water heater. The heater is shown with a dashed line. This sketch was used to show how a bypass valve could be added to existing piping.

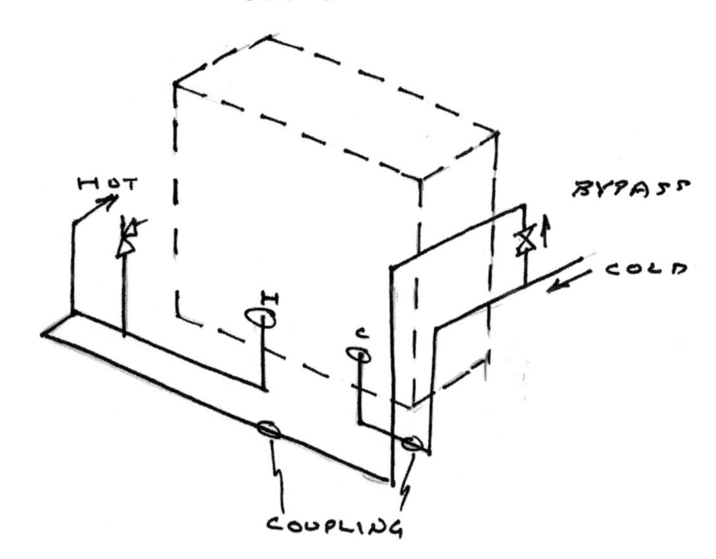

Figure 2.2 A sketch used to show changes to a piping system

Figure 2.3 shows a simple sketch that was used when planning an experiment to test different methods of restraining a domestic water heater during an earthquake. The sketch was drawn to show a proposed method of setting up the tank, the walls of the earthquake simulator representing a corner of a room.

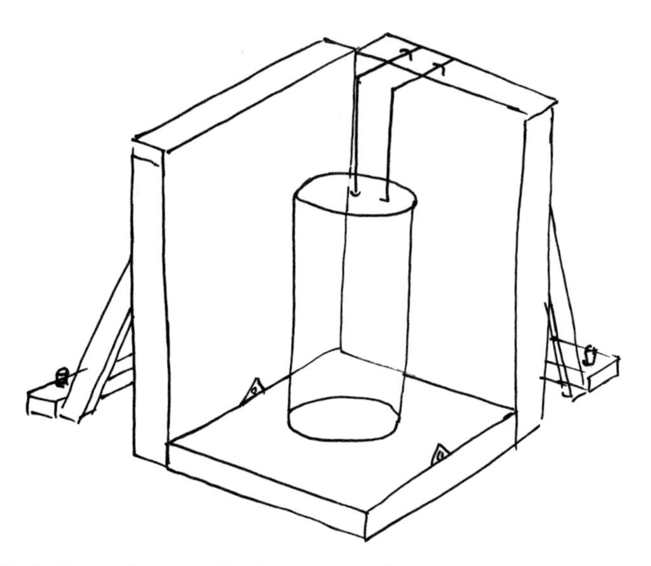

Figure 2.3 Initial sketch to explore ways of setting up an experiment

Figure 2.4 shows a more complex sketch drawn to show how the roof of a trailer could be raised. The sketch was used for discussion with the client and a manufacturer to show how this could done. Since it was not certain whether the project would go ahead, the time and expense of a drawing were not justified, and this sketch was sufficient to show the concept.

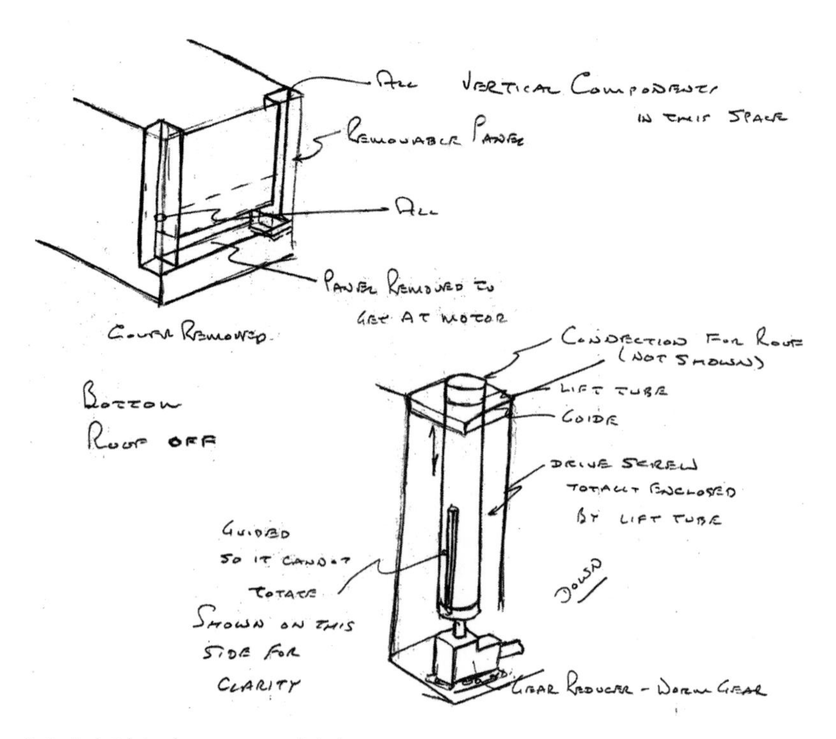

Figure 2.4 A sketch to show a proposed design

The important thing about a sketch is that it conveys the necessary information so that anyone can understand it. A pictorial, an orthographic, or both can be used. You need only learn a few techniques to produce a sketch, and all you need are a pencil and a sheet of paper.

Begin your sketch with light construction lines, some of which could be darkened at the last stage. Use a soft pencil, which allows you to draw light and dark lines by varying the pressure on the pencil. You do not need to use different pencils. An HB pencil is good for sketching.

Now let's look at some techniques for creating a sketch.

ESTIMATING PROPORTIONS

A sketch cannot show size precisely, but it should show all features in correct locations and proportions.

The simplest way to estimate proportions is "by eye." For example, in Figure 2.5 is height (H) equal to 25 or 50 percent of width (W)?

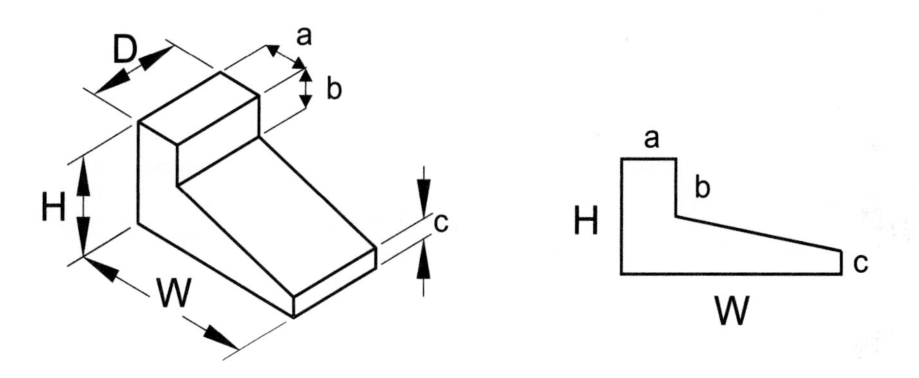

Figure 2.5 Estimating proportions by comparing one length with another

H, W, D (depth), and sides *a*, *b*, and *c* must be shown in the correct proportions. All sides will be compared with the longest side, W.

- What size is H relative to W?
- H is less than W and greater than 25 percent of W. It is 50 percent of W.
- Length *a* is less than 50 percent of W. It is taken as 25 percent of W.

It is easier to compare *b* and *c* with H, since they are parallel. Side *b* is taken as 50 percent of H and *c* as 25 percent of H.

These lengths were estimated "by eye" and are not expected to be exact, but the object had to be shown with correct proportions. Figure 2.6 shows the same object for comparison but with incorrect proportions. Neither representation gives a true picture of the object, although all sides have been shown.

Now let's move on to look at sketching straight lines.

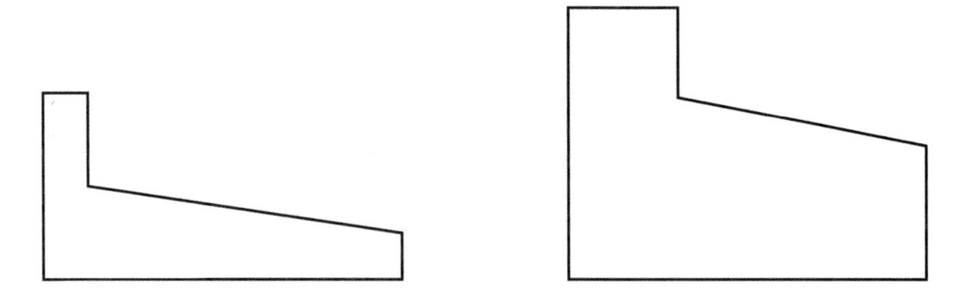

Figure 2.6 Distortion due to incorrect proportions

SKETCHING STRAIGHT LINES

A straight line is defined by a starting point and an end point or by any other point if the end point is not known. Locate the starting point, and mark it with a cross. Locate an end point on the line. An exact end point can be defined later. Start with the pencil at the starting point, and focus your eyes on the end point. Do not watch the pencil. With your eyes focused on the end point, move the pencil toward it. You will probably not hit the end point, but, with practice, you should come reasonably close. Position the paper so that your hand is in a comfortable position. You cannot do a good job if your hand and arm do not move in a comfortable manner.

Example 2.1

Sketch three orthographic views of the object shown in **Figure 2.7**.

1. Decide which elements are important before you show shape and features. An exploded view of the basic shapes that are combined to create the object is shown in **Figure 2.8**.

   There are three basic shapes: a rectangular prism and two triangular prisms. Both triangular prisms, A and B, are "removed" (negative volumes) from the rectangular prism. All surfaces are flat.

2. The front view is selected to show the sloping face in profile.

3. Since no size is specified, there is no point in measuring the isometric (see Figure 2.7); so, estimate the proportions by eye. Front, top, and side views correspond to the *x-z*, *x-y*, and *y-z* planes, respectively. Position them as specified in Chapter 1. Use light construction lines, drawn with an HB pencil, to block out sizes and locations. "Remove" prism A from the right. The construction is shown in **Figure 2.9**.

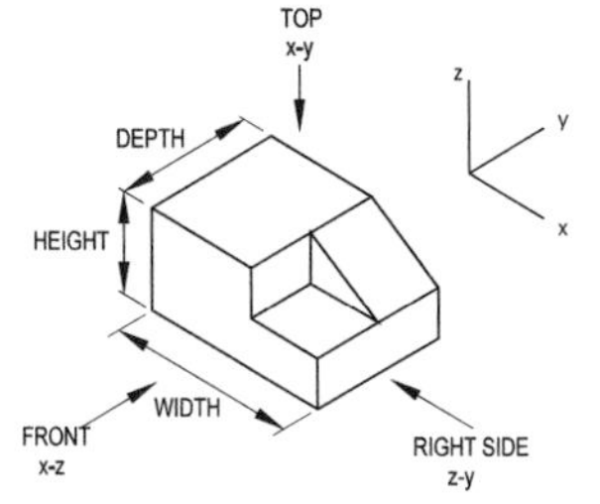

Figure 2.7 The object

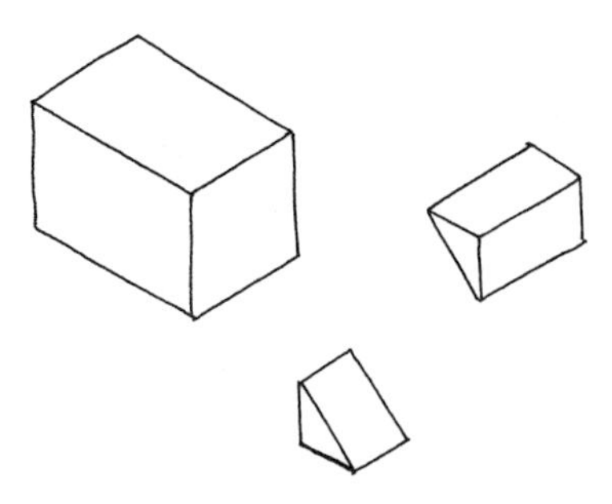

Figure 2.8 The object is broken into the basic shapes that form it

4. Remove prism B from the right end, and draw lines in all views to represent these surfaces. Views corresponding to this are shown in **Figure 2.10**.
5. All lines are shown as construction lines; there is nothing to distinguish them from outlines. The outlines will be darkened, but before this is done, check the sketch to see that it is correct. Does it represent the information given? Are the proportions of the various features correct? Do the features in one view match those in another view? If there are any errors at this stage, correct them before darkening the outlines. There is no need to erase construction lines. The finished sketch is shown in **Figure 2.11**.
6. The objective is to show the shape and the relative sizes of the features. Does this sketch satisfy this objective? If the answer is "Yes," you are finished. If not, make the appropriate changes so that the objective is met.

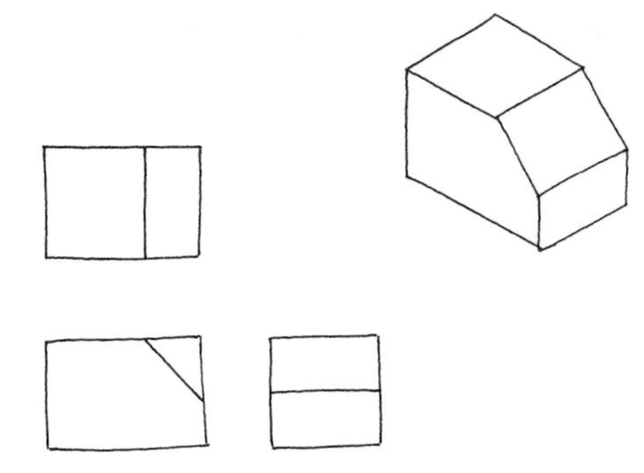

Figure 2.9 Sloping face added in all views

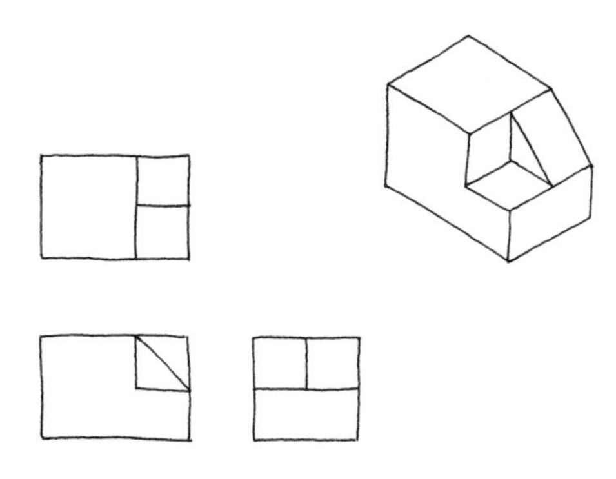

Figure 2.10 All features added as light construction lines

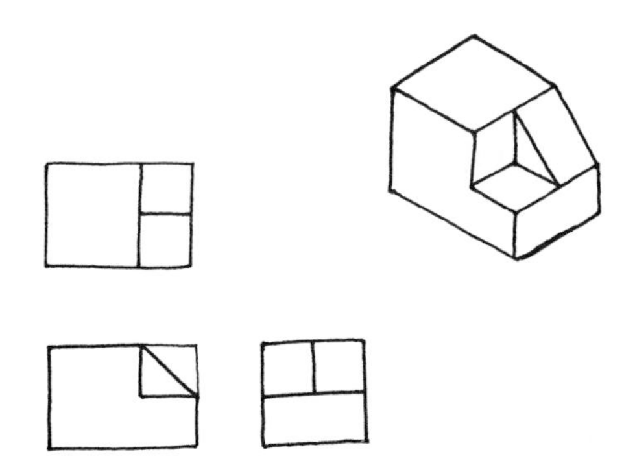

Figure 2.11 Finished sketch with outlines darkened

USING GRIDS

Sketching straight lines and estimating proportions are easier when there are lines to follow. Engineering calculation paper often has a non-reproducible grid on one side to aid sketching. Both orthographic and isometric grid papers are available (Figure 2.12). Grid papers can be easily created using vertical and horizontal lines (or lines at 30° for an isometric grid) produced with a computer drawing program and copied at whatever interval you specify.

Computer drawing programs also have grids to help locate points. An array of dots forms the grid. The user specifies the spacing between dots, which can be changed at any time. Both orthographic and isometric grids can be created. Points can be precisely located with commands that "snap" to a grid point. The grid is not part of the drawing and does not print. Locating points at grid points makes drawing much easier. Some computer programs have the provision for locating guidelines wherever you need them. You can then snap to a guideline, or the intersection of two guidelines, to locate points accurately. Guidelines are not part of the drawing and do not print.

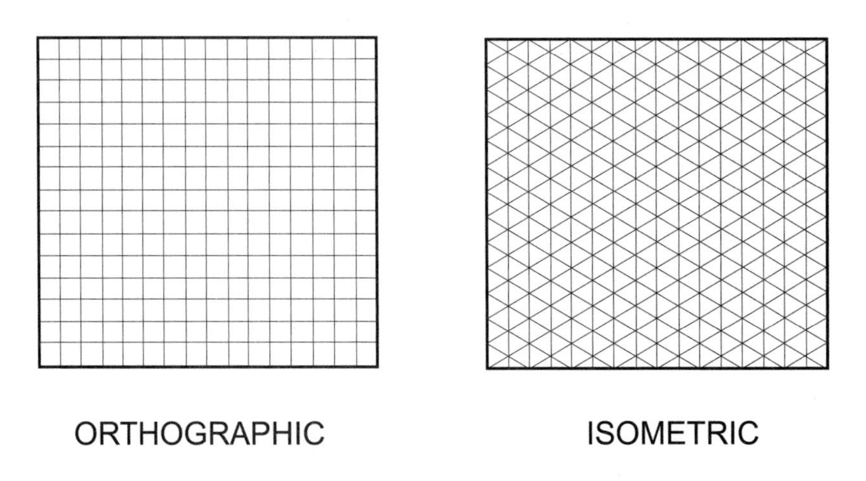

Figure 2.12 Orthographic and isometric grids

Now that you know a little more about using sketches to solve problems, estimating proportions, and sketching straight lines, try the following problems.

Problems

Sketch orthographic views of the objects below. Use two or three views to describe each object. Orient the object to give the best front view.

1.

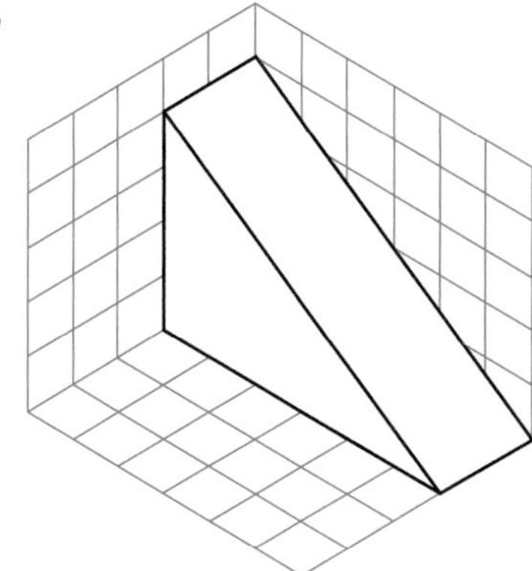

2.

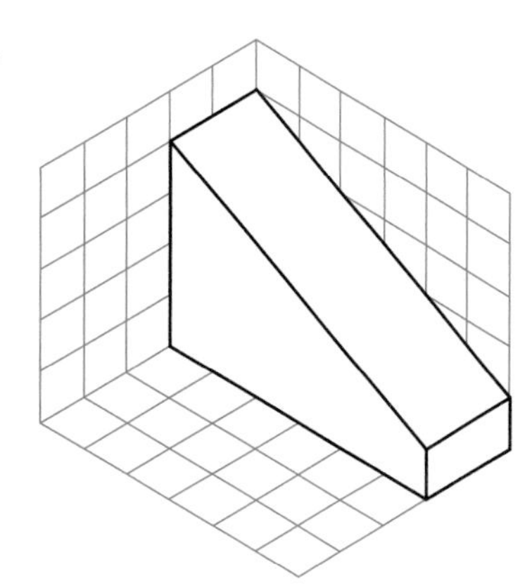

3.

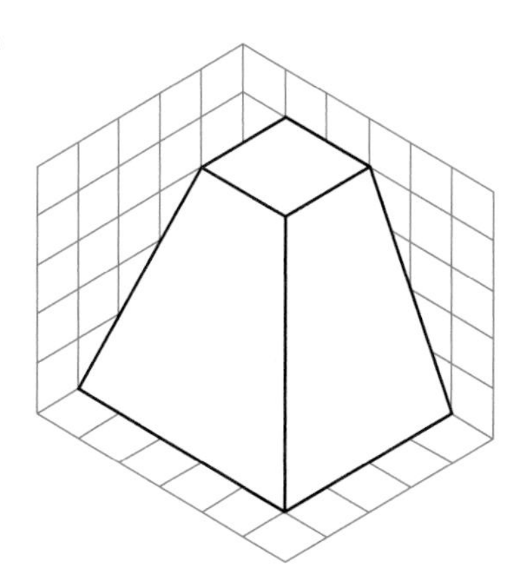

4.

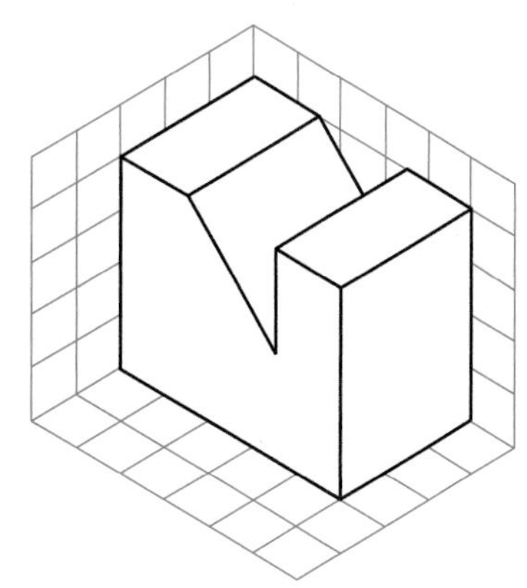

5.

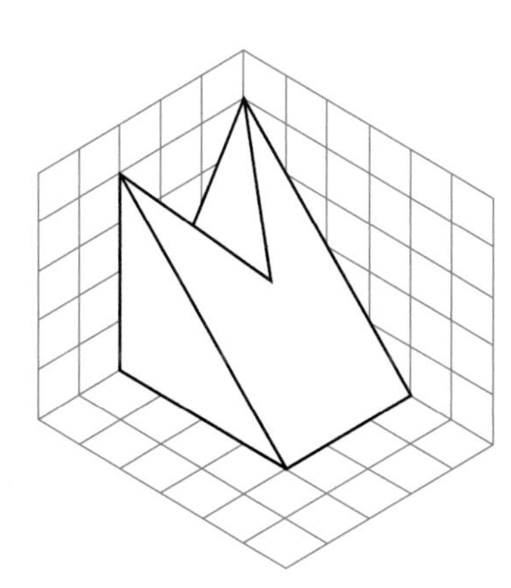

6.

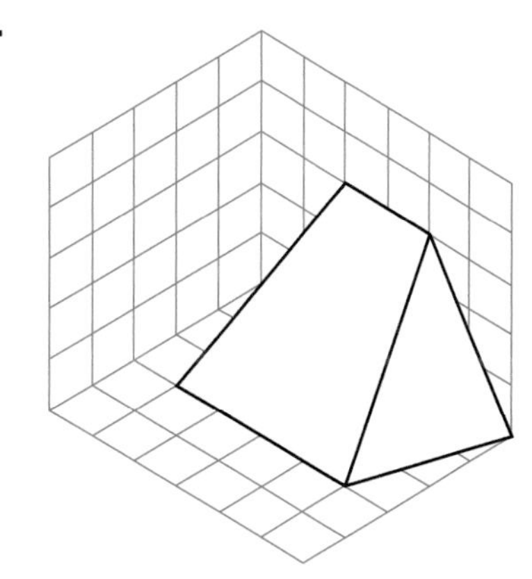

7.

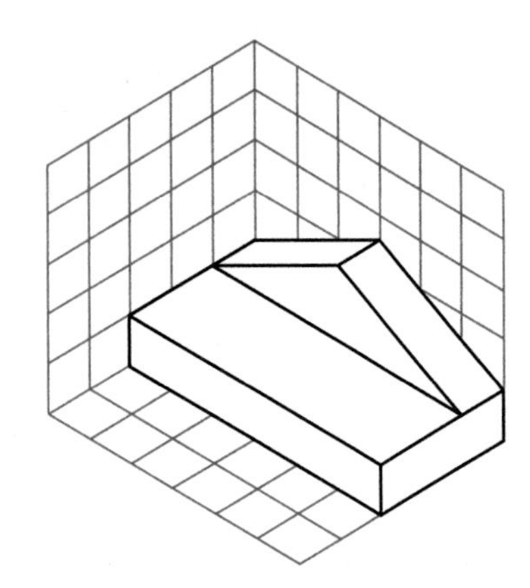

8.

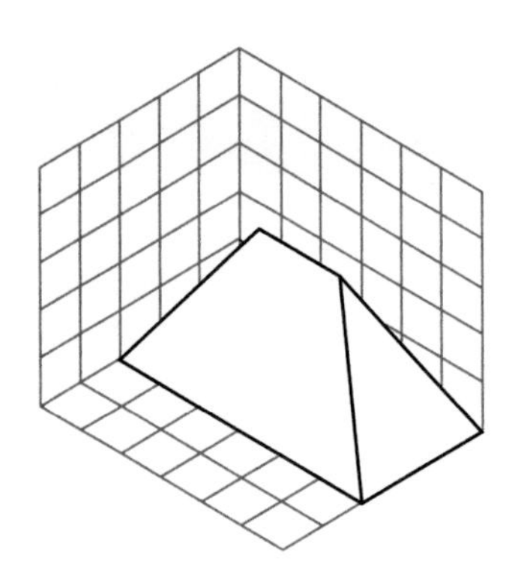

9.

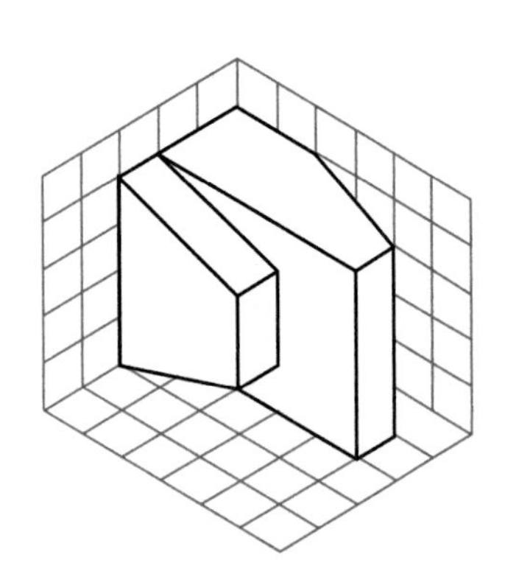

10.

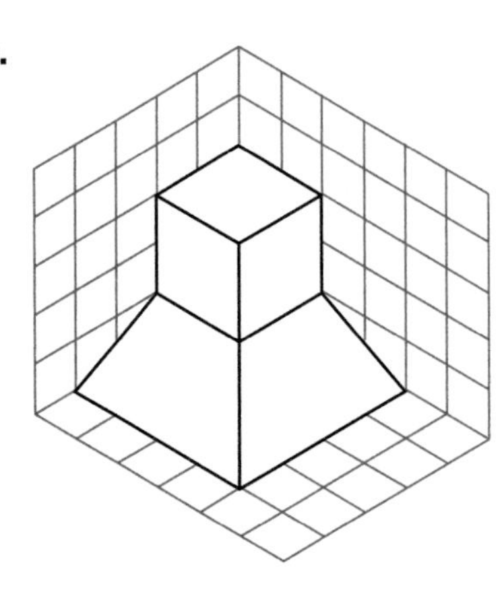

11.

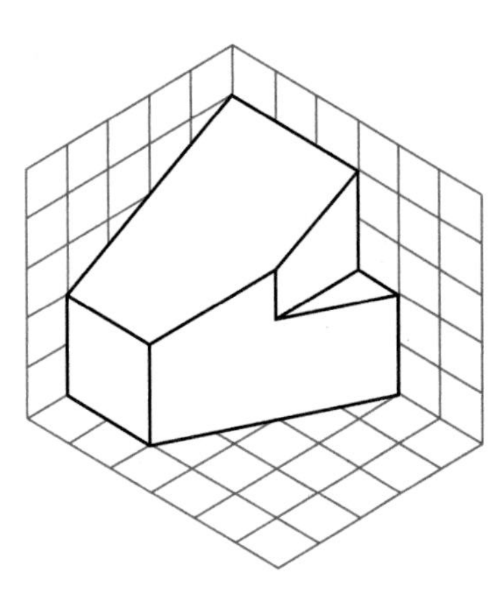

12.

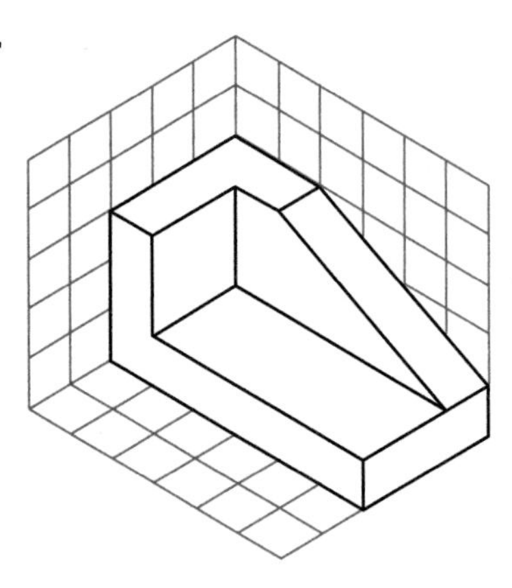

13.

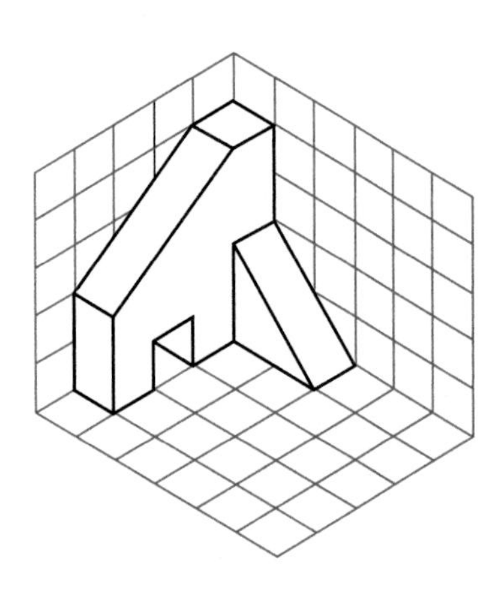

14.

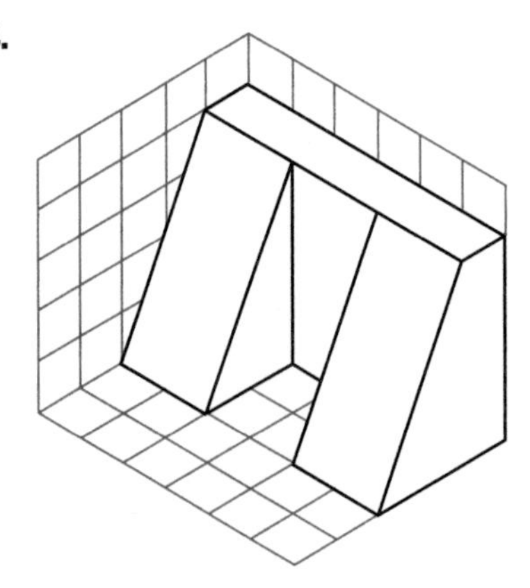

15.

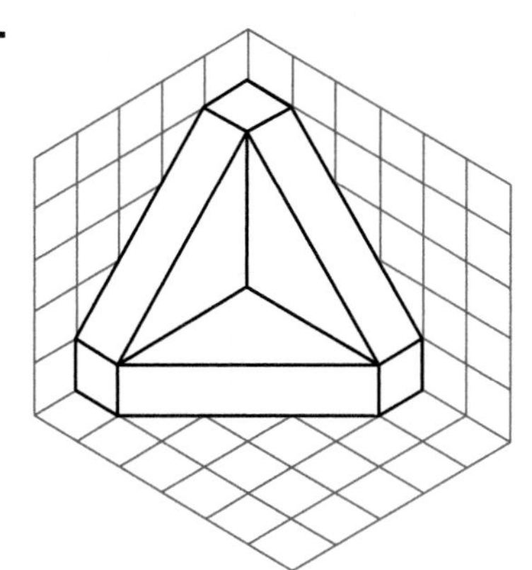

16.

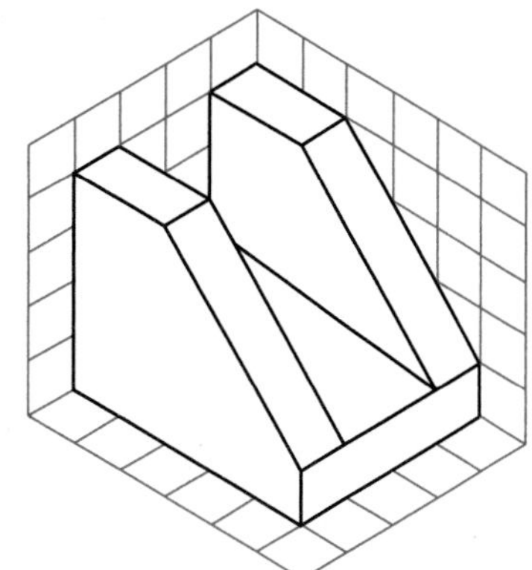

17.

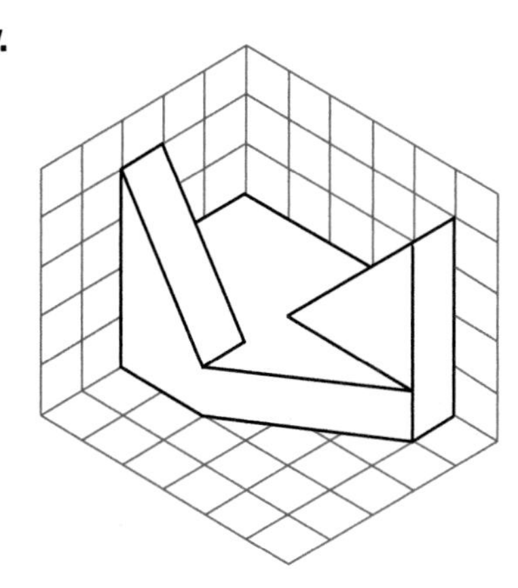

18.

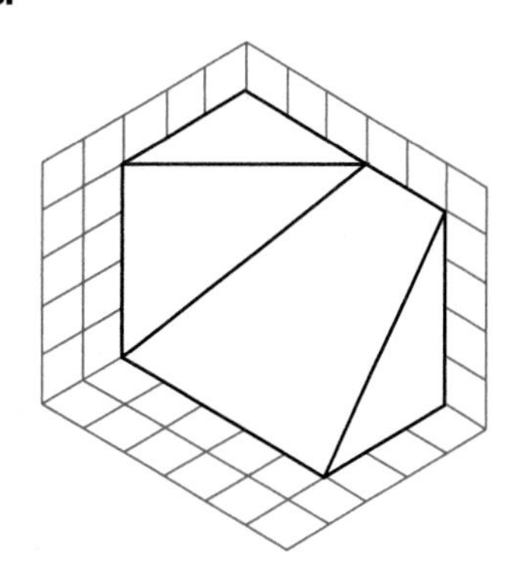

19.

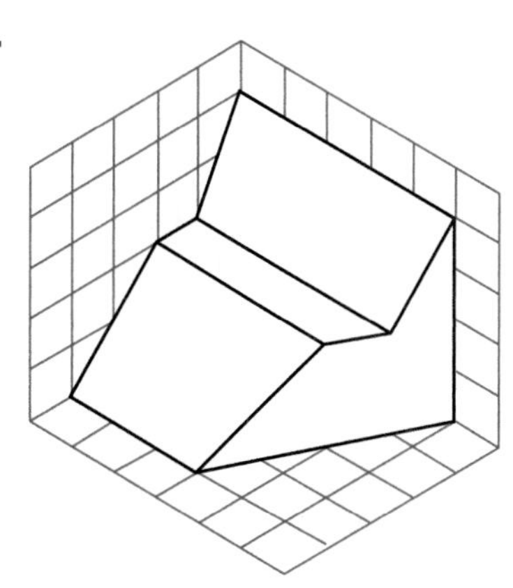

20.

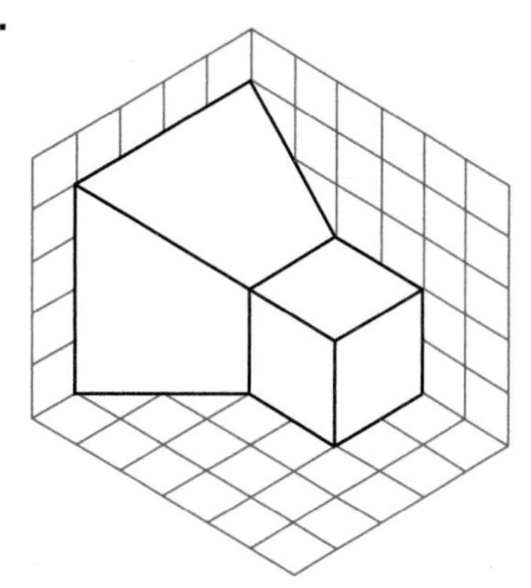

21.

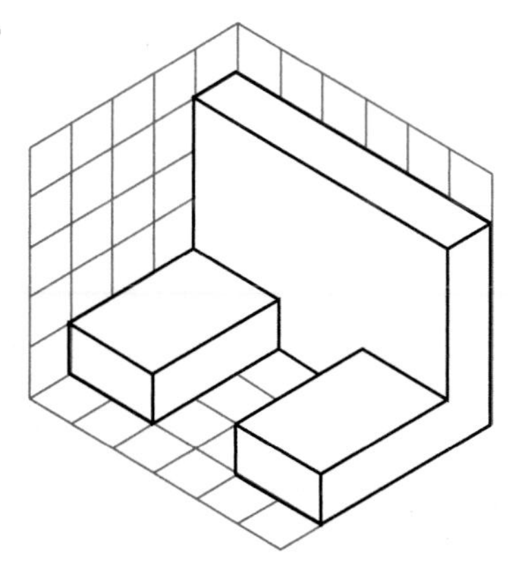

22.

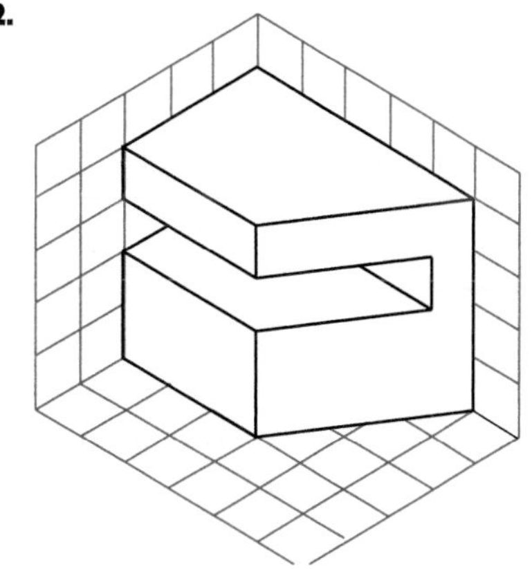

23.

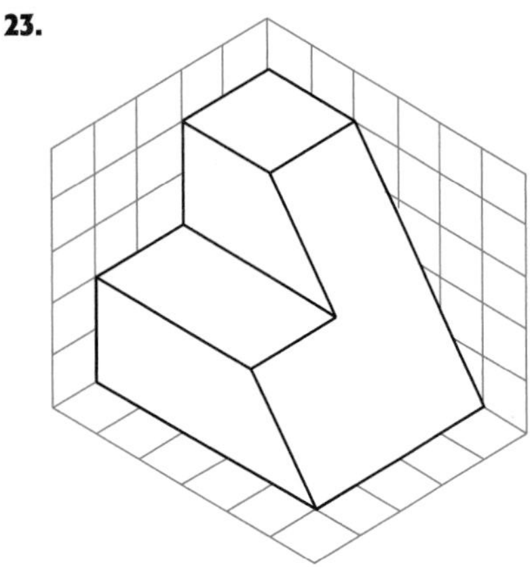

24.

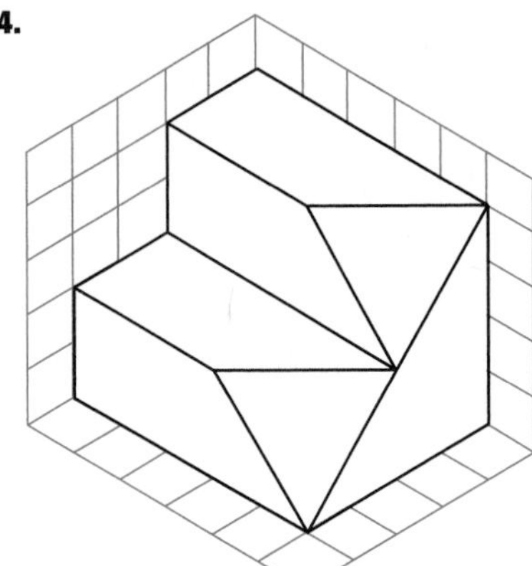

25.

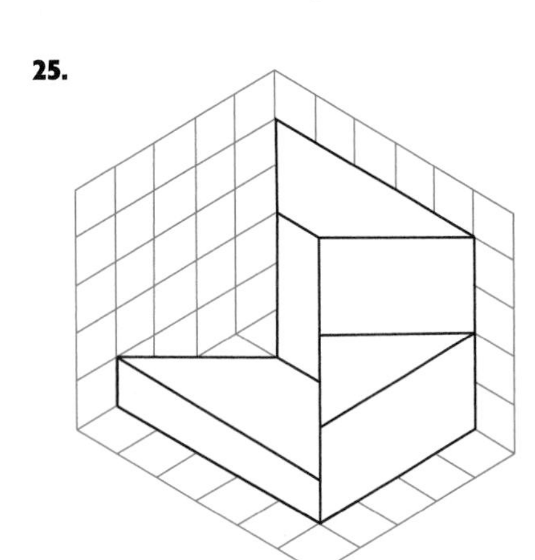

SKETCHING CURVED LINES

Since about 90 percent of the shapes found in engineering can be formed by a combination of rectangular prisms, cylinders, and cones, we should be able to sketch a circle or an arc. You already know something about circles: You know the equation describing a circle, and you have probably drawn many circles with a compass, a circle template, or a computer drawing program. The problem now is to sketch a circle that looks like a circle. There are techniques to use that will give reasonable results.

A sketch, by its very nature, will not show a perfect circle. It is, after all, drawn freehand. The radius, even if it is known, will not be exact, but no one is going to measure it anyway. The exact size is not important at this stage. A circle is a circle, no matter what size. If you showed your work to strangers and asked them to identify the shape, they would immediately say, "A circle."

We know what a circle looks like and that:

- all circles have a centre,
- all circles have a radius, and
- any tangent to the circle is 90° to the radius.

Nothing very mysterious here, but can we use this information to construct a circle?

The first step is to locate the centre in the desired location. The centre is specified by two lines intersecting at 90°. Since these lines identify the centre of a circle, they are drawn as centrelines. Size is defined by a square with sides equal to twice the radius, drawn around the centre point. Diagonals are added as a check on the size of the square (Figure 2.13).

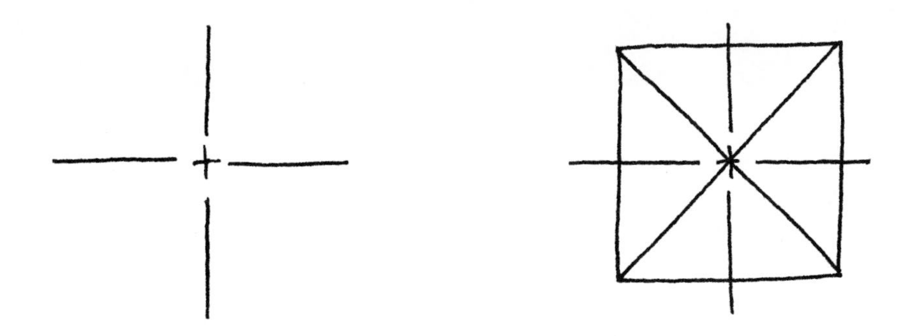

Figure 2.13 Initial steps for sketching a circle are locating the centre and defining the size

The circle will be tangent to the square at the four points identified in Figure 2.13, and the direction of the tangent at these points is known (90° to the centrelines).

All that remains now is to draw the circle.

It is difficult to draw the entire circumference all at once; so, draw one quadrant at a time. Position the paper so your drawing hand is comfortable, and draw the first quadrant as a light construction line between two tangent points. This line must be an equal distance from the centre at all points. You can add more points on the diagonals, if desired, by locating them at a distance r from the centre. The distance is estimated by eye. Move the pencil between points without touching the paper and, after a little practice, draw a light construction line between the points (Figure 2.14). The direction of the arc must be perpendicular to the centrelines. Repeat this step in the other three quadrants.

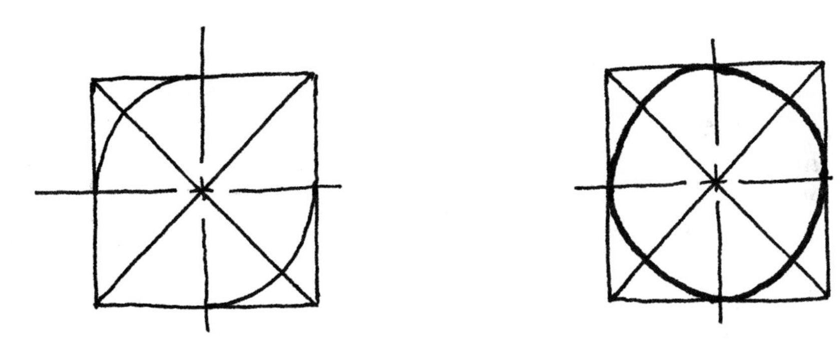

Figure 2.14 Quadrants are sketched separately and the final outline darkened

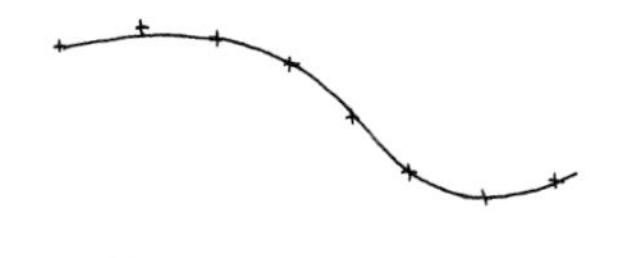

Figure 2.15 Irregular curve drawn through points

Darken the outline to distinguish it from construction lines. There is no need to erase construction lines.

You can follow the same process when sketching an arc.

An irregular curve can be drawn by drawing a smooth line through points on the curve. The more points used to define the curve, the easier it is to sketch. Figure 2.15 shows an irregular curve constructed through several points.

Example 2.2

The shaft support bracket, shown in **Figure 2.16**, supports a long shaft that passes through the large hole. The four holes in the base are used to secure the bracket. The bracket will be made of welded steel. A sketch of the bracket is required. No dimensions will be added.

1. Even though the bracket in Figure 2.16 is a fairly simple object, a breakdown of the basic shapes is useful in understanding the problem. **Figure 2.17** shows these basic shapes.
2. Choose the front view so that the right-side view can be used, and the size and the location of holes and other features blocked out. Show centres of circles and arcs by centrelines, and define sizes by squares or, in the case of the arc, by a rectangle. Draw hole centrelines in all views. **Figure 2.18** shows the location and size of all holes blocked out.

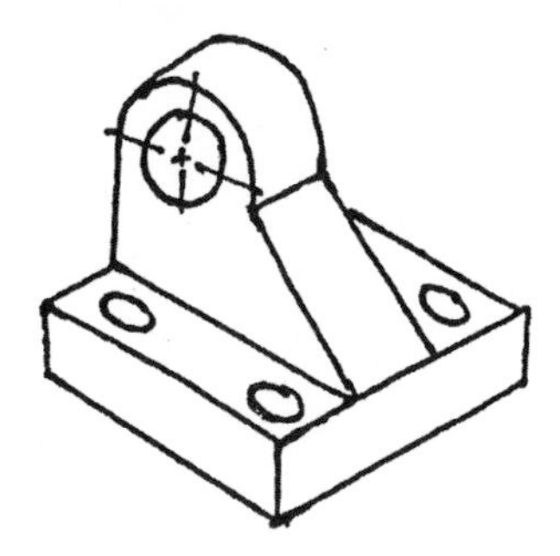

Figure 2.16 Pictorial of object with plane and curved surface

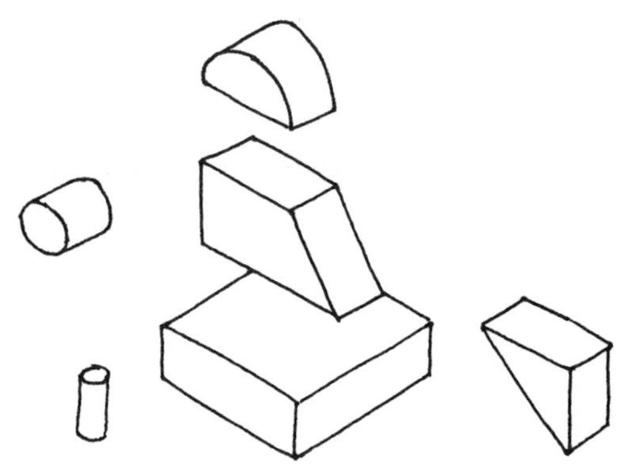

Figure 2.17 The object broken into the basic shapes that make it up

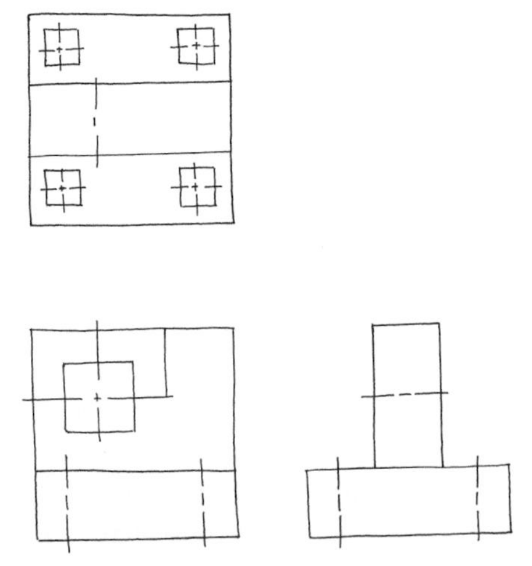

Figure 2.18 Location and size of holes blocked out

3. Add hidden lines as construction lines. If the shapes and sizes are correct, darken the outlines. **Figure 2.19** shows the final sketch.

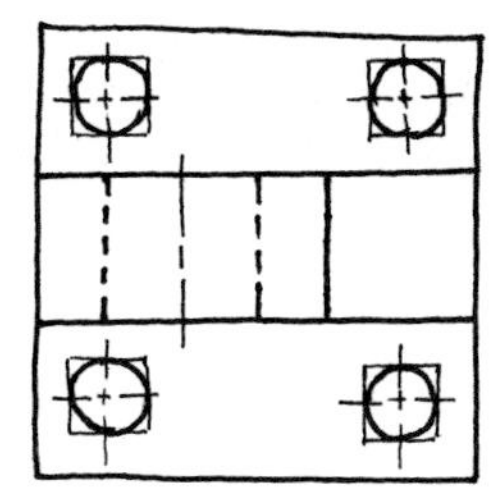

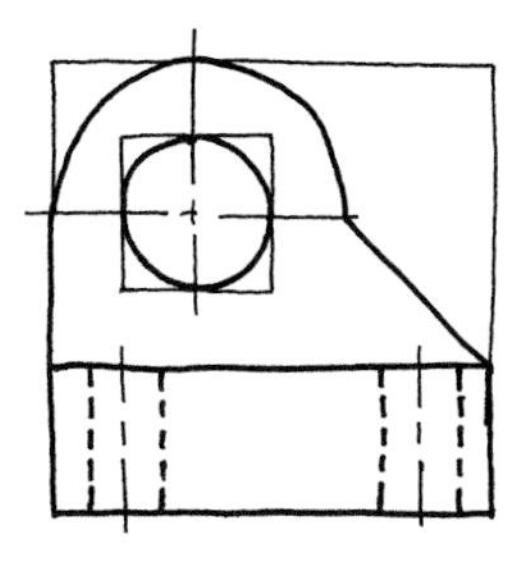

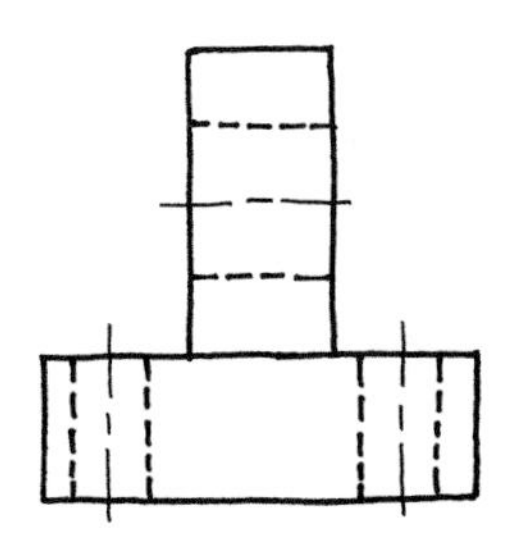

Figure 2.19 Finished three-view sketch with outlines darkened

Example 2.2

Problems

Sketch orthographic views of the objects below. Use two or three views to describe the objects. Choose the best view for each object.

26.

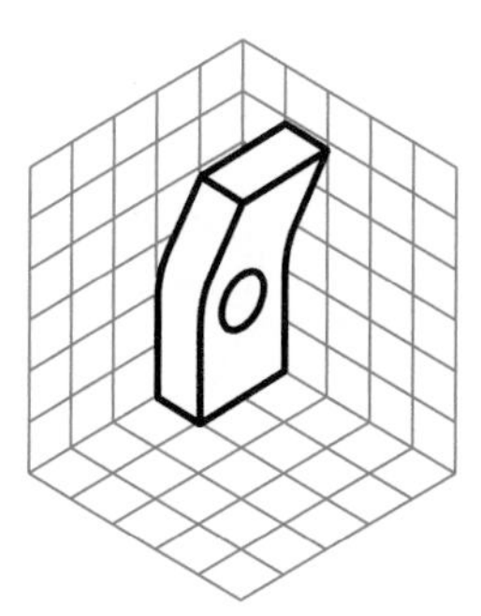

27.

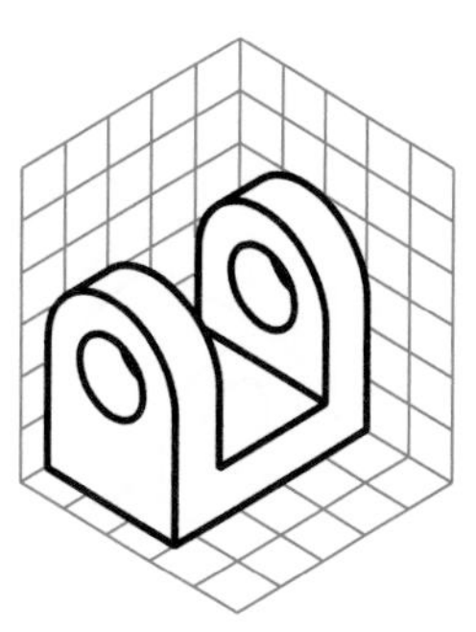

28.

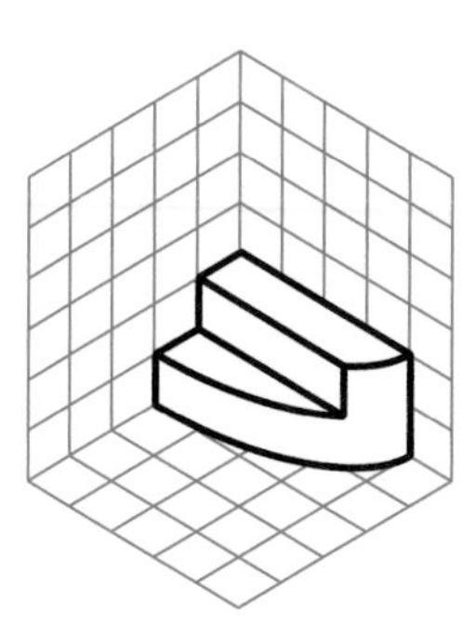

29.

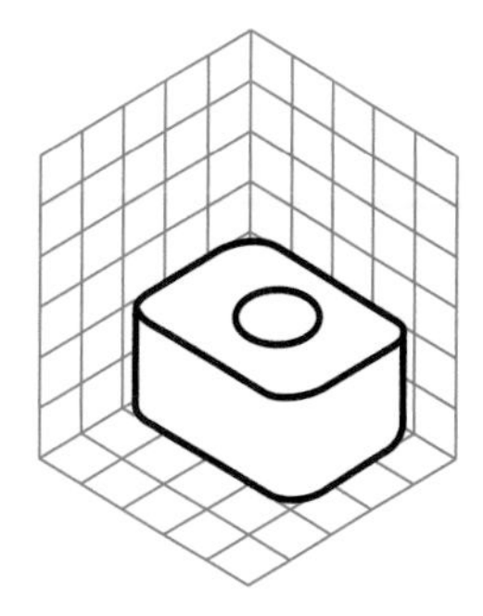

30.

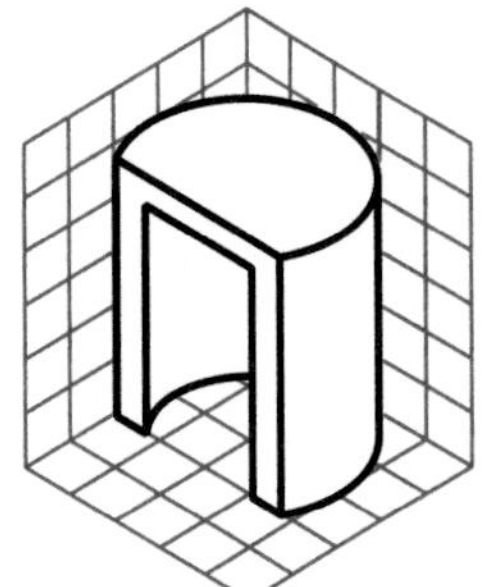

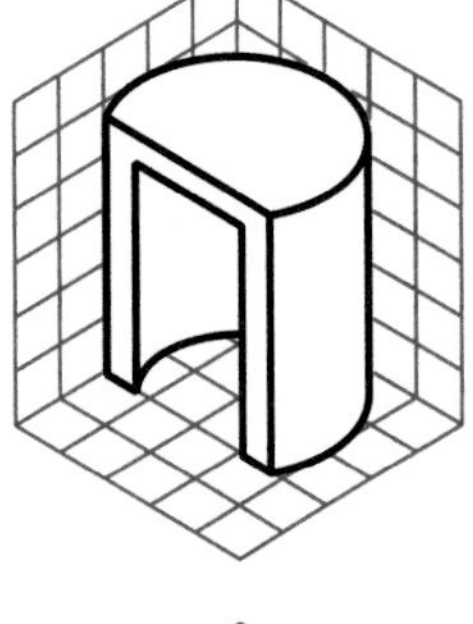

31.

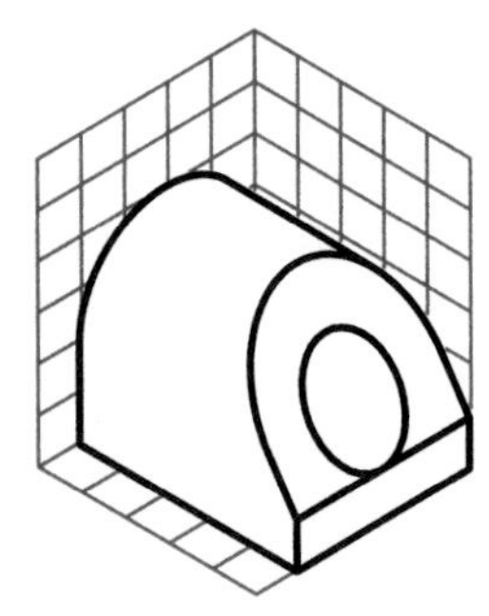

32.

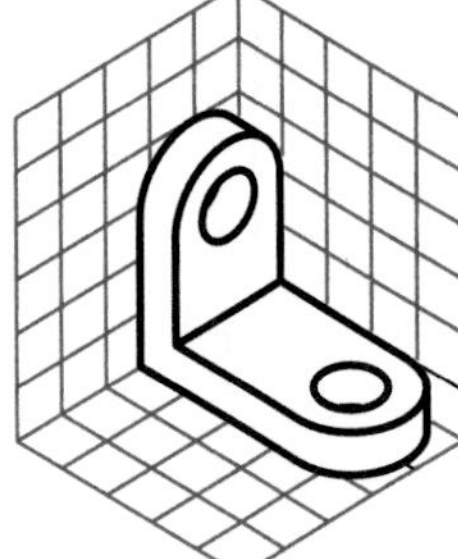

33.

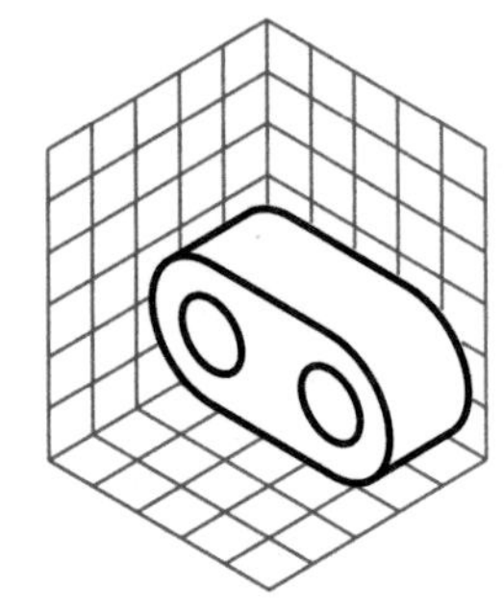

34.

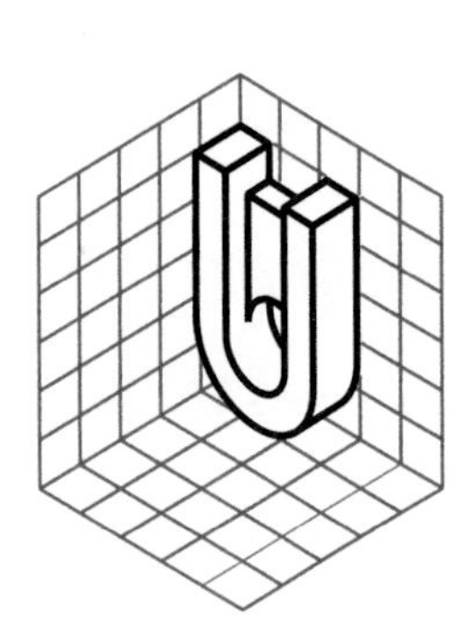

35.

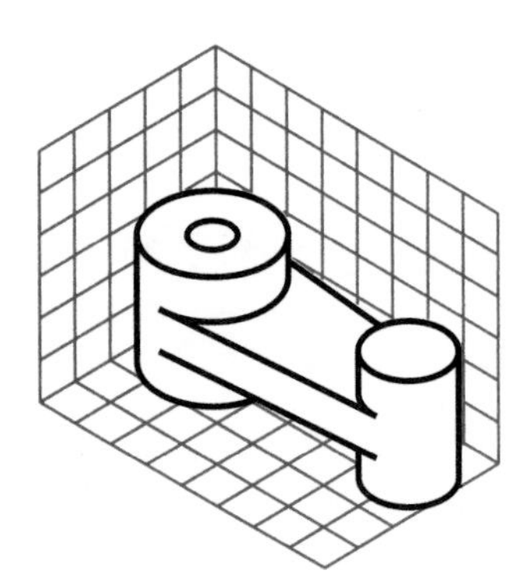

36.

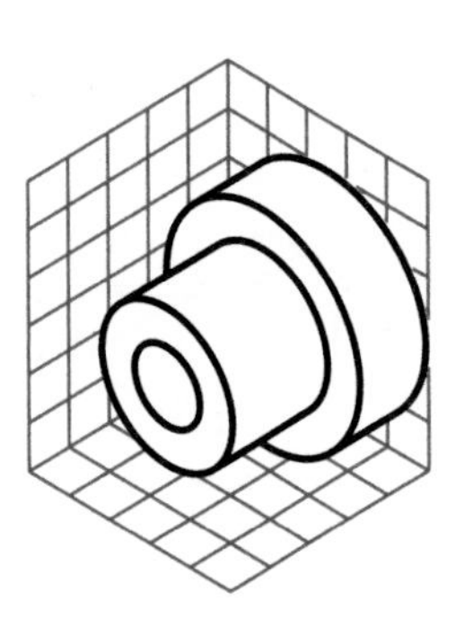

37.

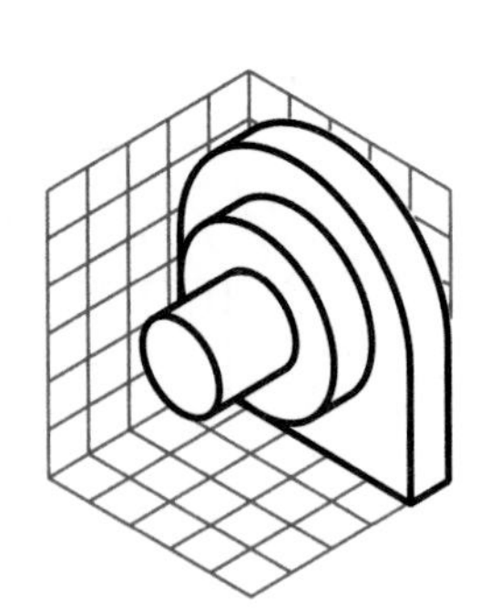

38.

39.

40.

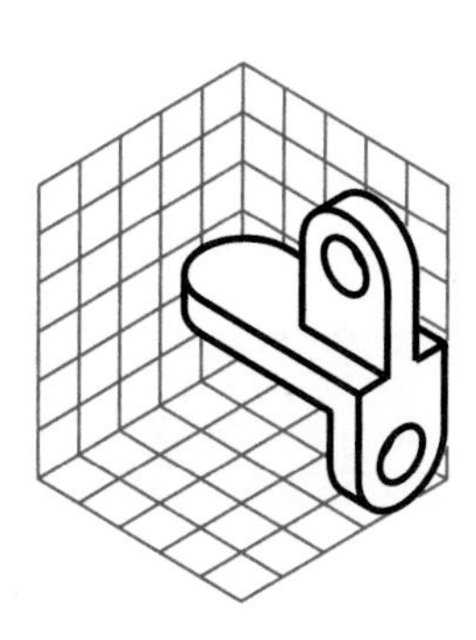

41.

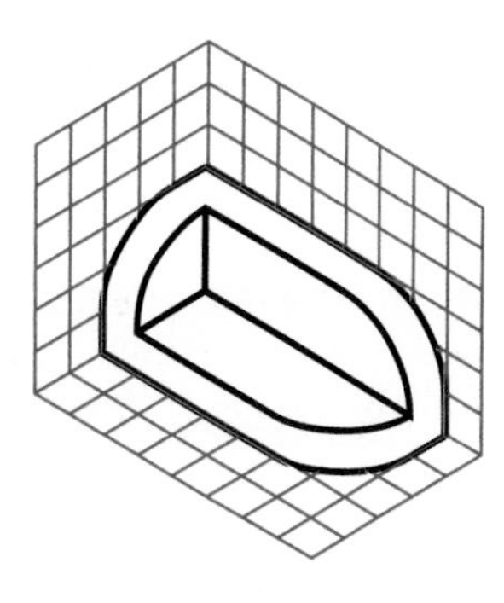

42.

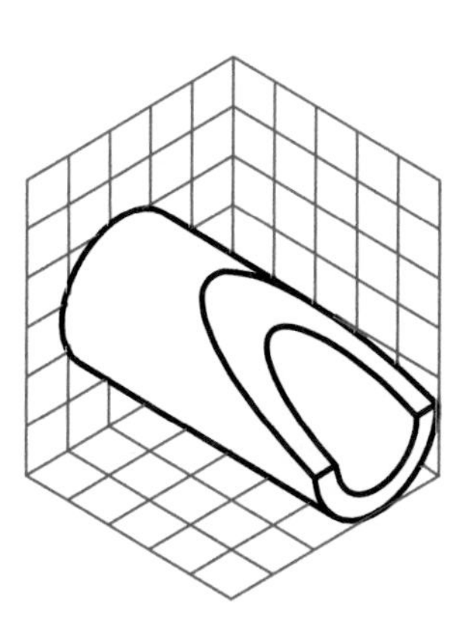

43.

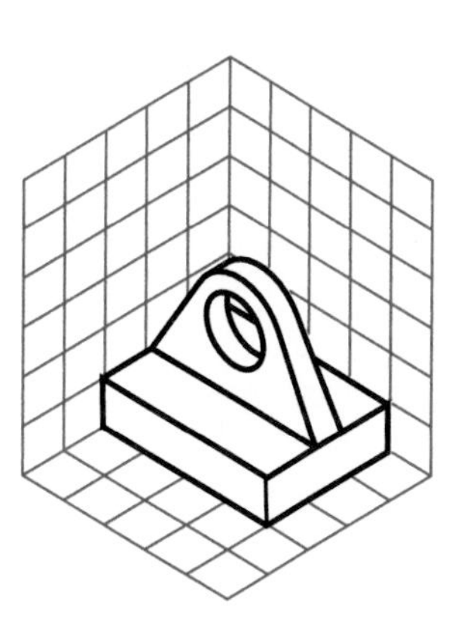

44.

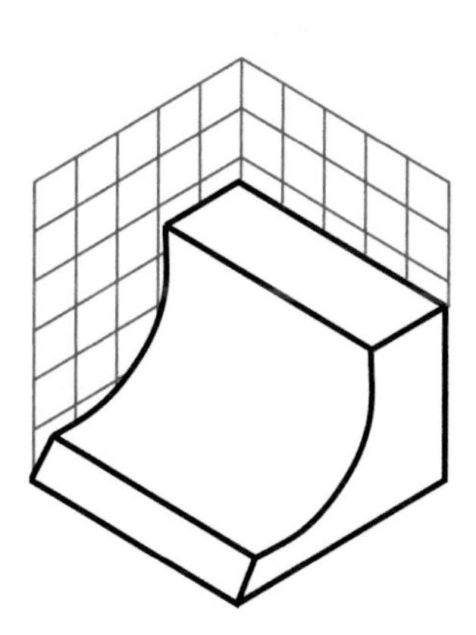

45.

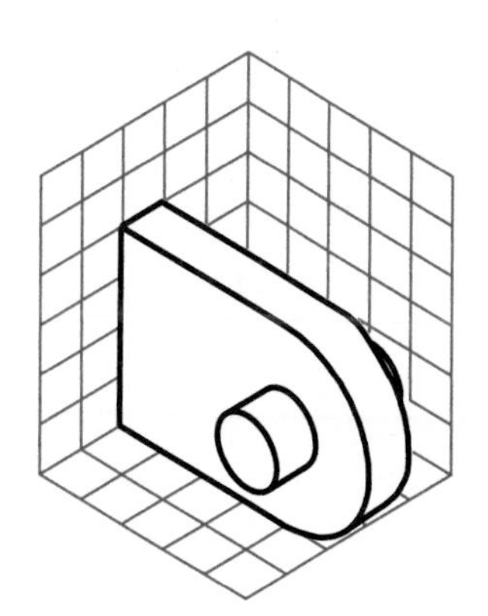

USING PICTORIAL SKETCHES

Pictorials are often used to aid visualization. Some computer programs can create a pictorial from orthographic views, and vice versa, so that the designer can work with whatever is more appropriate. Sometimes, it is easier and faster to develop your ideas by sketching a pictorial and then drawing detailed orthographic views. Computer solid modelling programs show pictorial views of objects created by combining different volumes, called primitives. The object can be rotated and viewed from any direction, or any orthographic view can be seen at any time. Some computer programs have methods for "sketching." The start and end points of a line do not have to be specified exactly, and corners do not have to meet. The program will clean up the drawing according to programmed instructions.

Pictorial sketches, like orthographic sketches, are a combination of straight lines, curved lines, and circular arcs. Straight lines can be drawn by locating end points and joining them. Circular arcs require the location of the centre and a radius. A circle appears as an ellipse when drawn as an isometric or other type of pictorial.

Example 2.3

The objective is to sketch an isometric pictorial of the object shown in **Figure 2.20** so that it can be easily visualized by someone who is not familiar with engineering conventions. Proportions must be correct, and the shape must be clearly shown.

1. Block out the "space" occupied by the shape, with an enclosing box to define the width, height, and depth. Determine proportions by eye. The height is about two-thirds the width, and height and depth are equal. Block out the "space" with light construction lines as shown in **Figure 2.21**.
2. Check the sketch to confirm that proportions are correct. Does the shape of the box that defines overall size correctly indicate the relative lengths of the three sides? Make the necessary corrections before proceeding.
3. The sides of the enclosing box represent the planes of the front (the *x-z* plane), top (the *x-y* plane), and side (the *y-z* plane) views. Transfer the orthographic views to the three sides of the box as shown in **Figure 2.22**. The box has been "exploded" to show this more clearly.

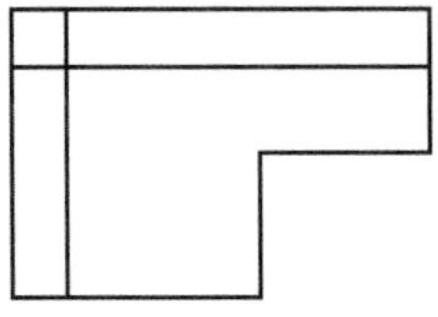

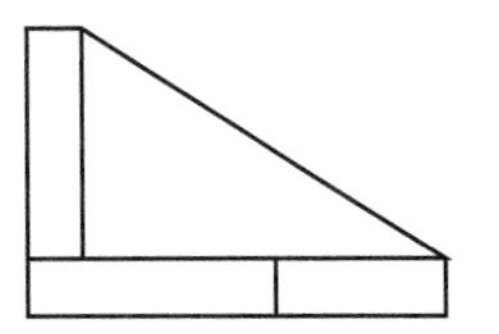

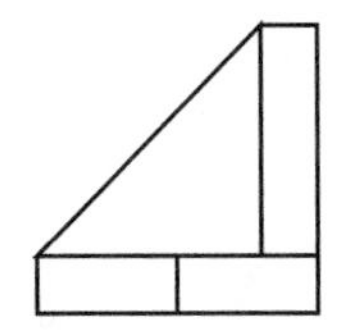

Figure 2.20 Three-view drawing

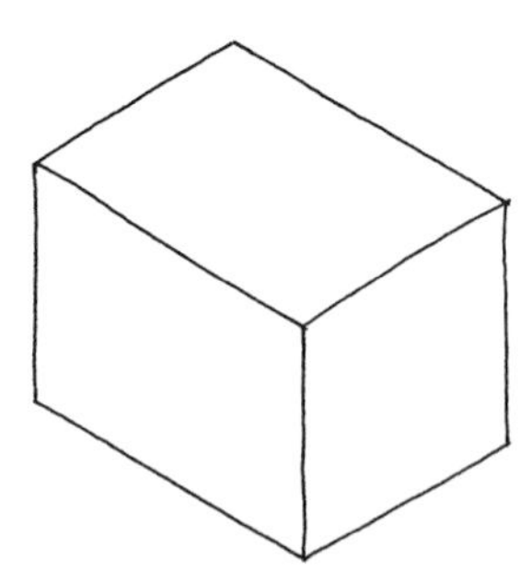

Figure 2.21 Overall size defined by an enclosing box

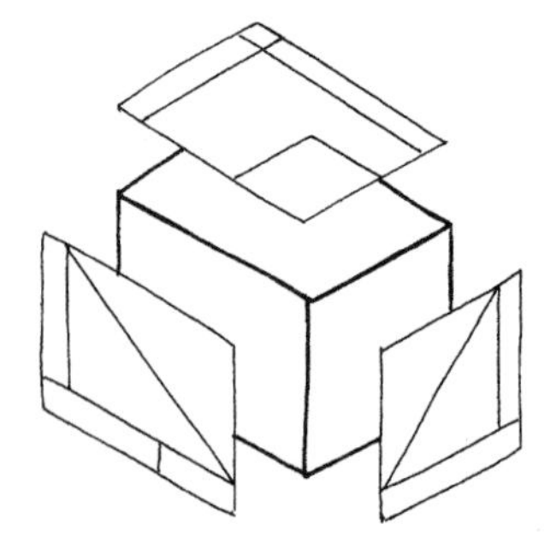

Figure 2.22 Orthographic views drawn on enclosing box

4. Combine the coordinates on the *x-y*, *x-z*, and *y-z* planes to give coordinates (*x,y,z*) in space. Project points from the projection planes to the object. This is illustrated in **Figure 2.23** with points on the base. (Only the base is shown in Figure 2.23.) End points of lines on the base are determined and joined by straight lines.
5. Draw non-isometric lines (lines not parallel to any axis) by locating the end points and joining them. The completed object is shown in **Figure 2.24**. Note that there are several groups of parallel lines in this object. Parallel lines always appear parallel; so, if they do not appear parallel in any view, it is an indication that an error has been made.

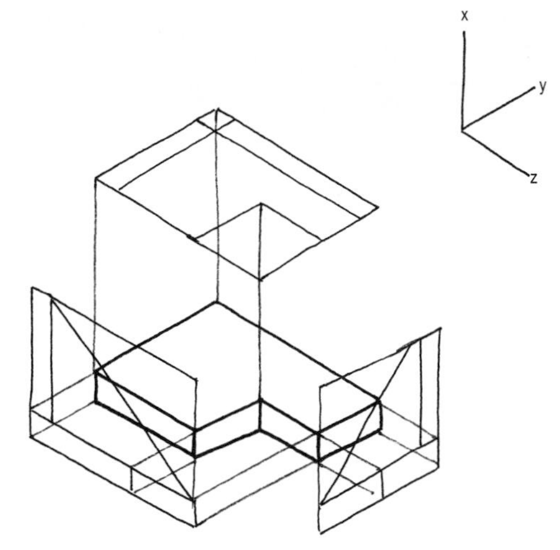

Figure 2.23 Projection from orthographic views to the isometric box

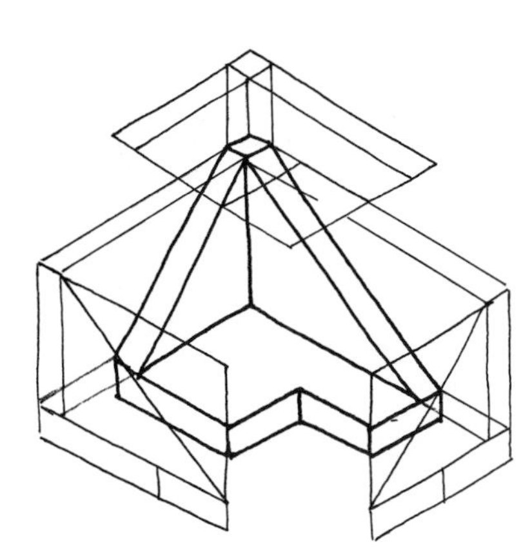

Figure 2.24 Completed isometric with outlines darkened

Example 2.4

Another way of sketching an isometric from orthographic views is to put basic shapes together. This is the method you would use with a solid modelling program. **Figure 2.25** shows the same three views of the object used in Example 2.3. The goal is to represent this object with an isometric sketch. This will be done by combining basic shapes.

1. Determine the shapes that make up the object. **Figure 2.26** shows a base, two triangular prisms, and two rectangular prisms, one of which is a negative volume.
2. Begin with the base. **Figure 2.27** shows an enclosing box, which defines the space and the base. Only visible lines are shown, and some of these will be hidden as other parts are added.
3. The first triangular prism is added at one end of the base (**Figure 2.28**).

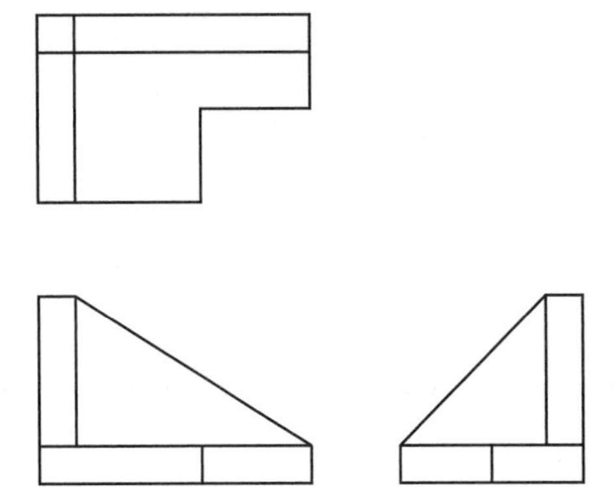

Figure 2.25 Three-view drawing

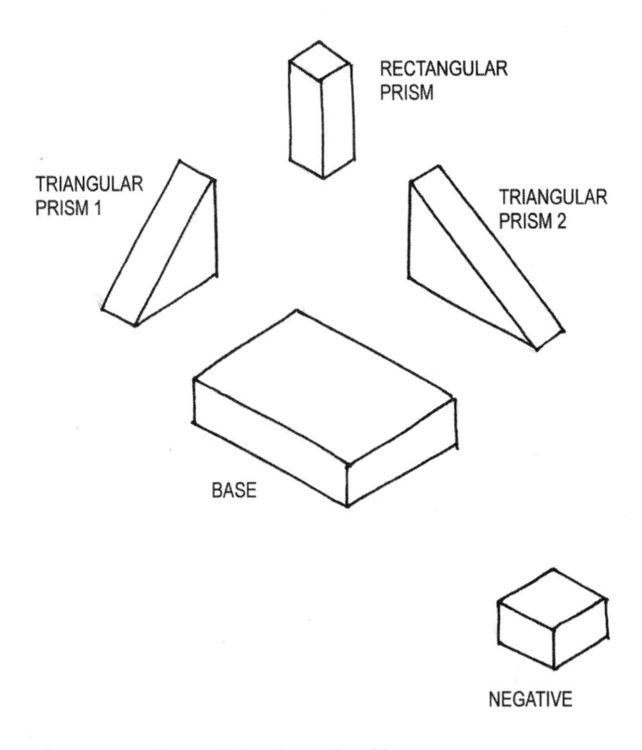

Figure 2.26 Shapes that make up the object

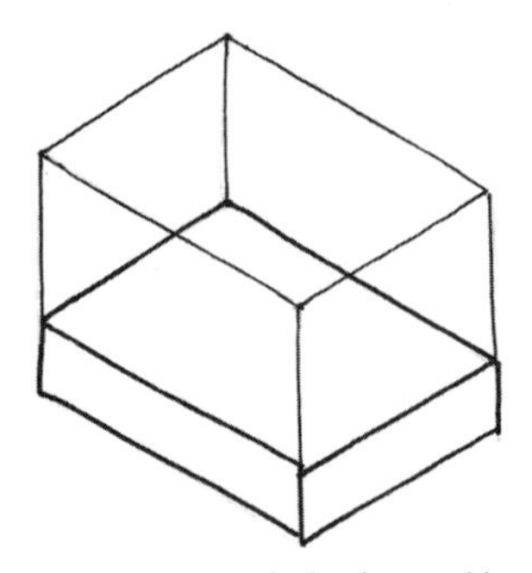

Figure 2.27 Enclosing box and base

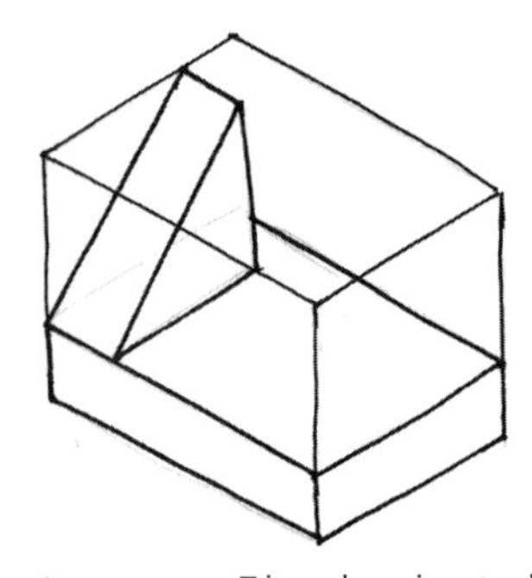

Figure 2.28 Triangular prism 1 added

4. The second triangular prism and the positive rectangular prism are added (**Figure 2.29**). Only the top surface of the rectangular prism can be seen.
5. The negative volume is removed from the base, and outlines are darkened (**Figure 2.30**). There is no need to erase other lines.

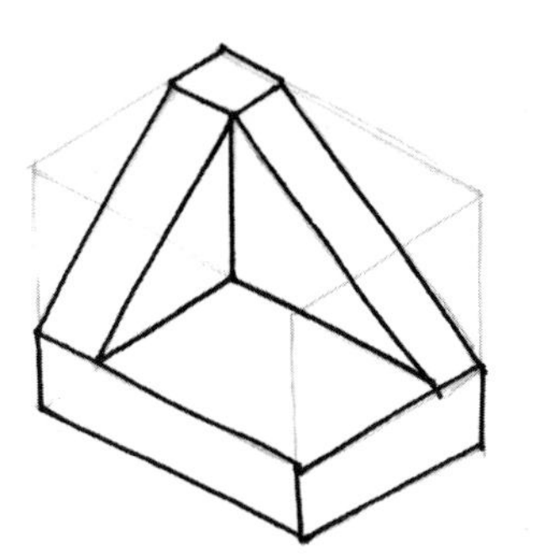

Figure 2.29 Second triangular prism and rectangular prism added

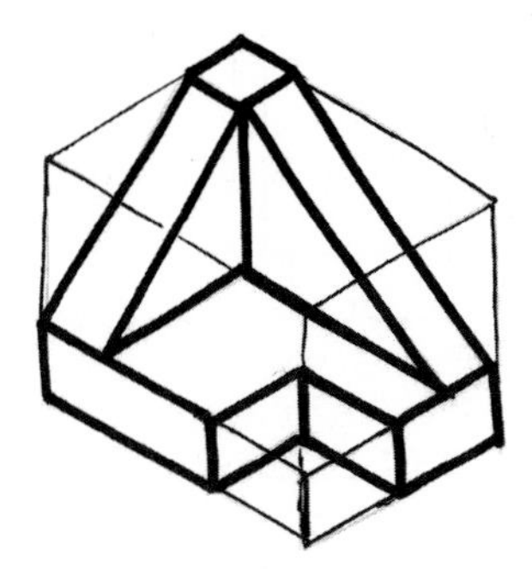

Figure 2.30 Negative volume removed and outlines darkened

Problems

Sketch isometrics, approximately full size, of the objects shown below. The grid may be taken as 5 mm.

46.

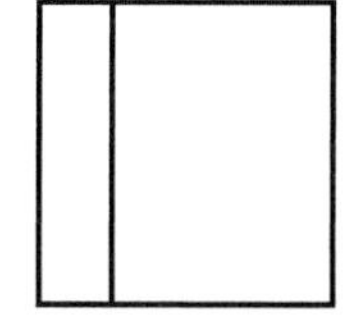

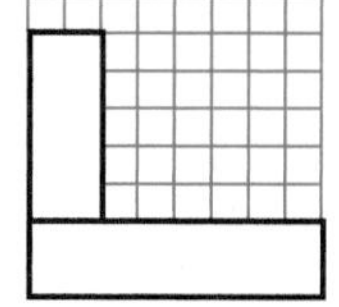

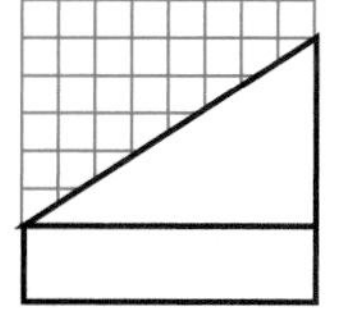

47.

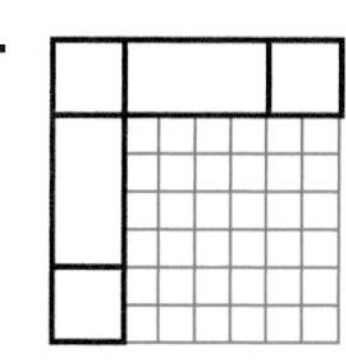

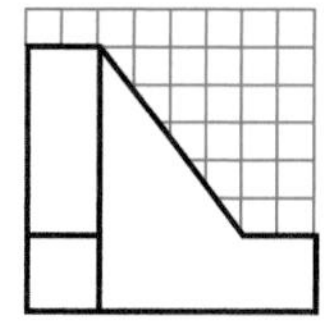

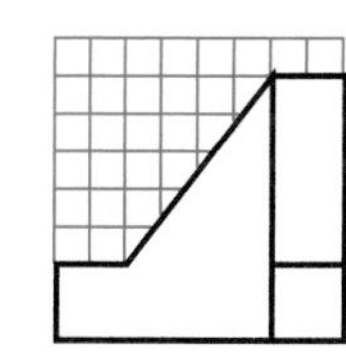

48.

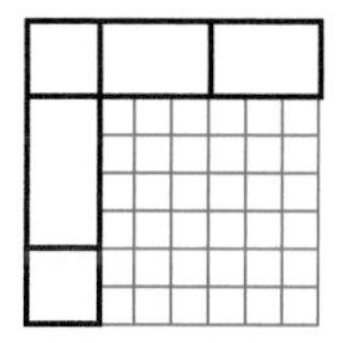

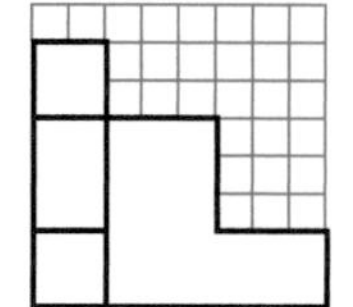

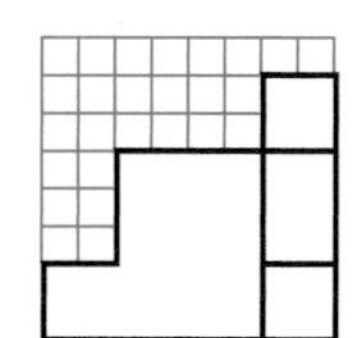

49.

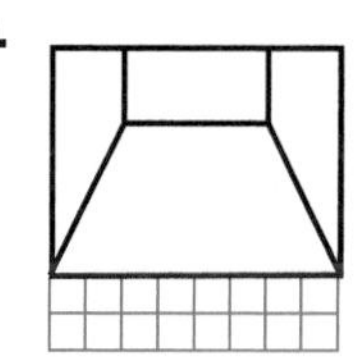

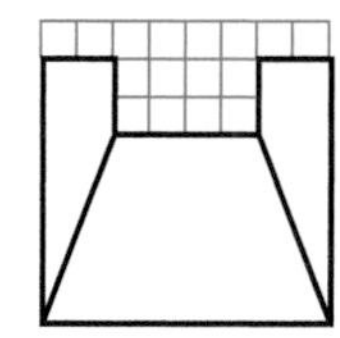

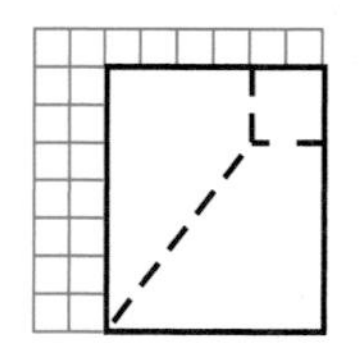

50.

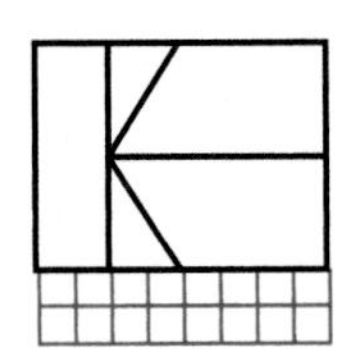

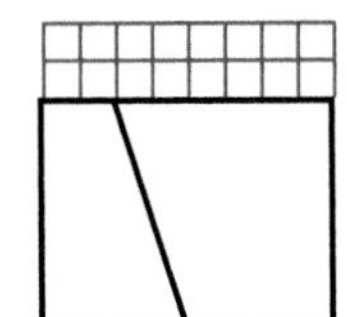

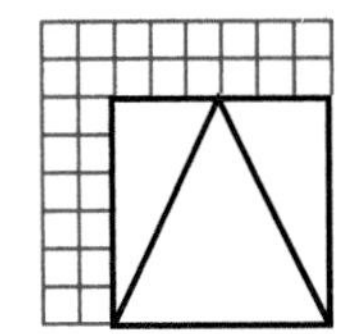

51.

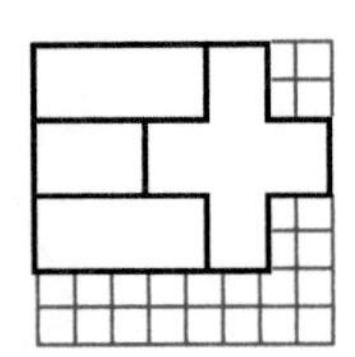

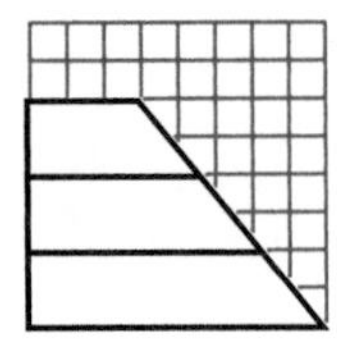

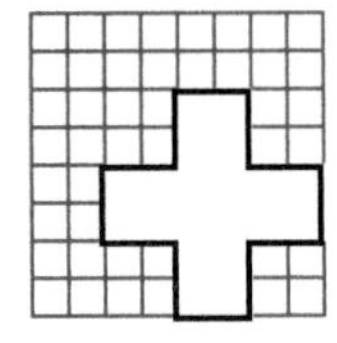

52.

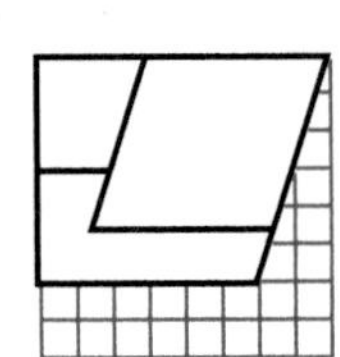

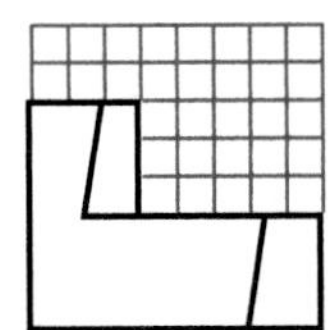

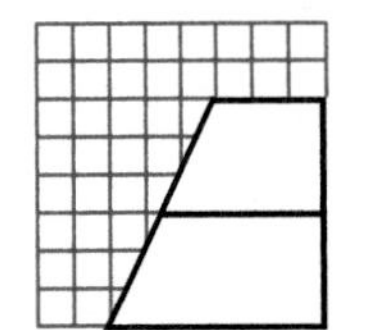

53.

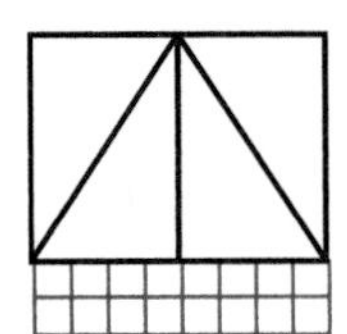

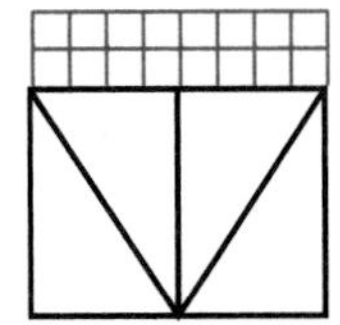

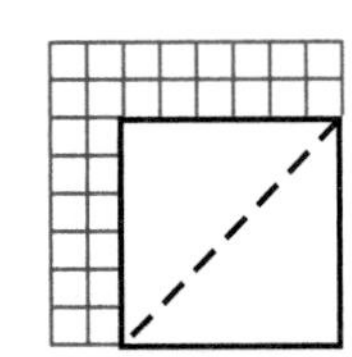

54.

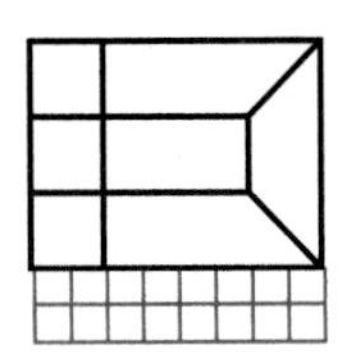

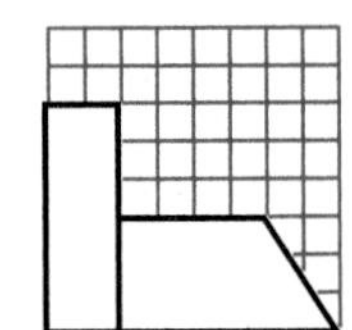

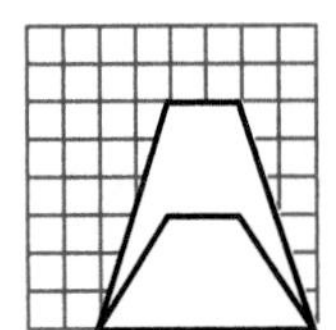

55.

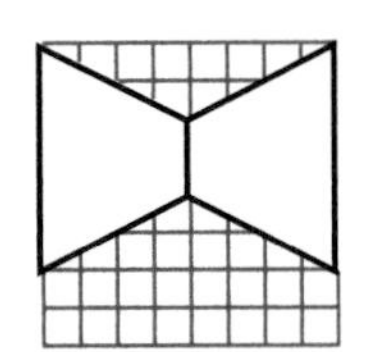

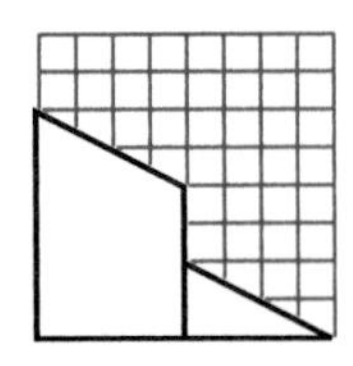

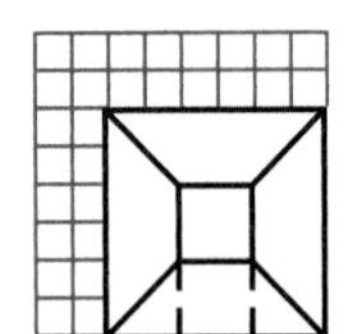

56.

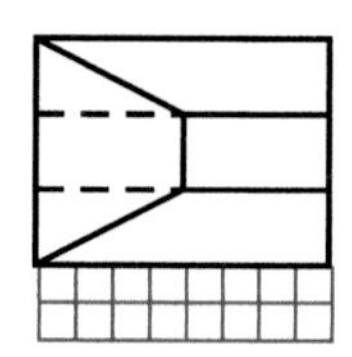

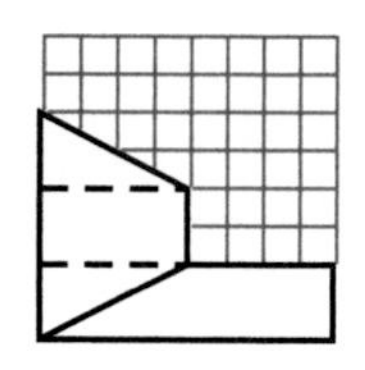

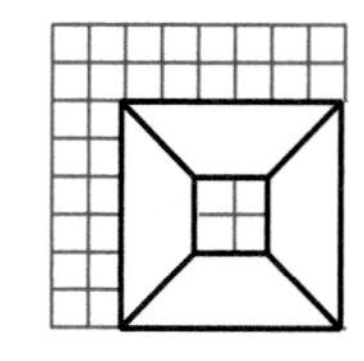

57.

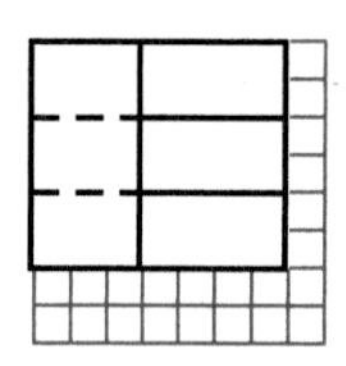

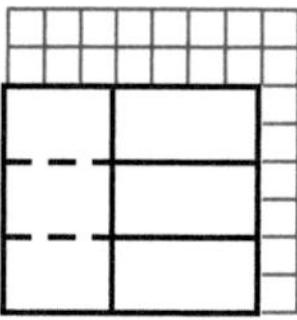

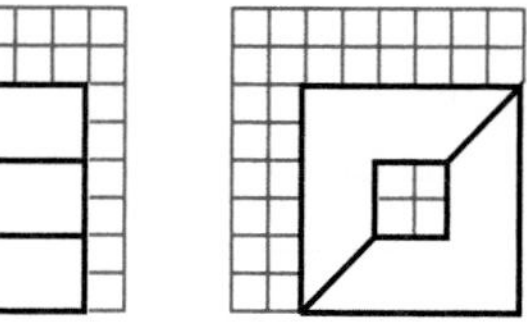

58.

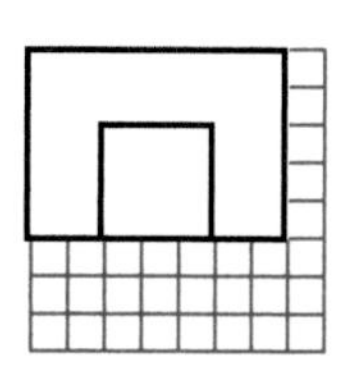

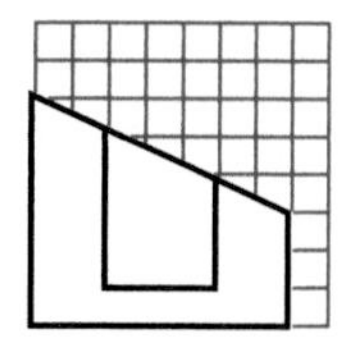

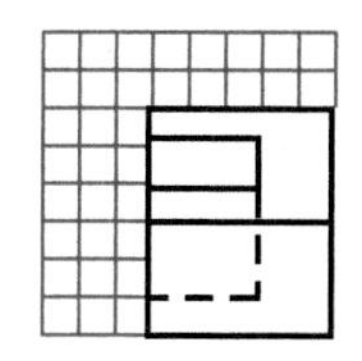

59.

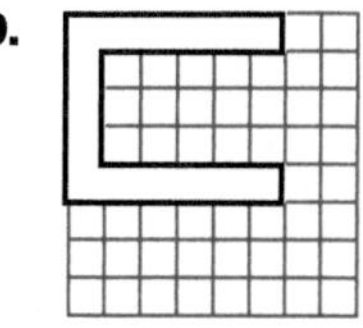

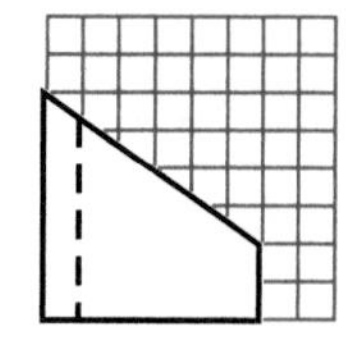

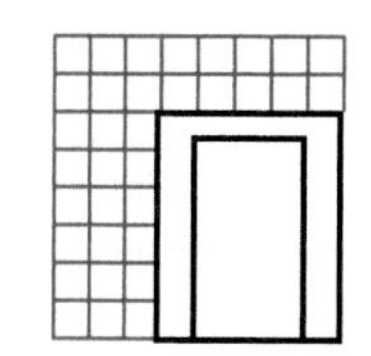

60.

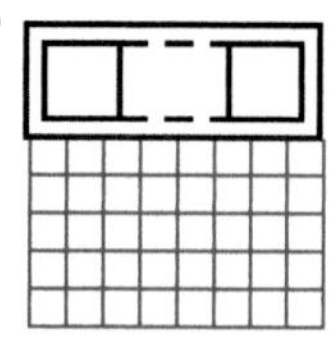

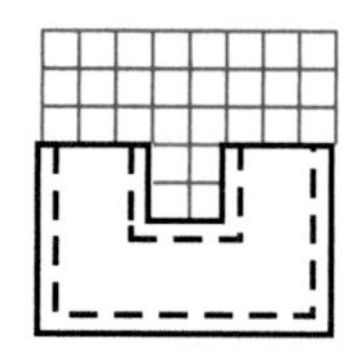

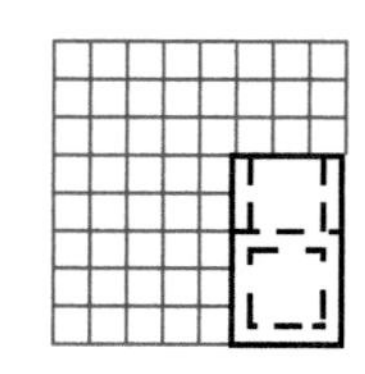

61.

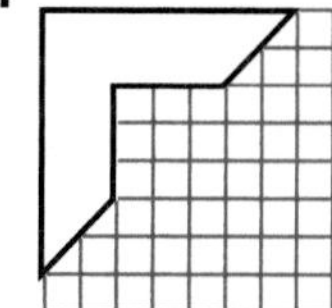

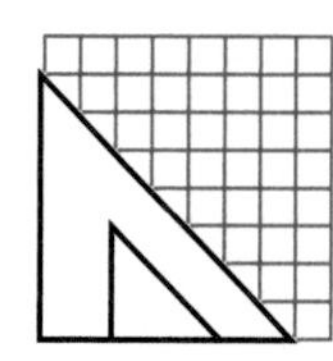

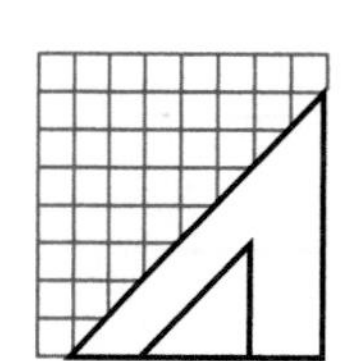

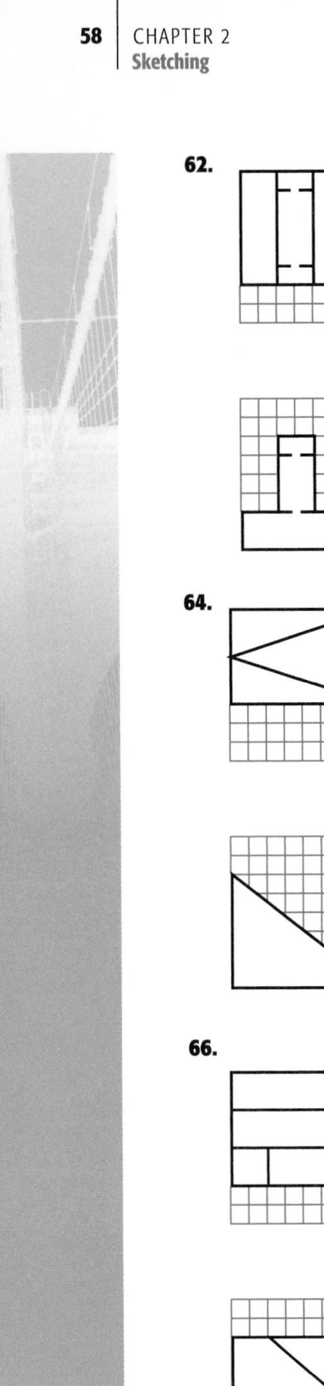

62.

63.

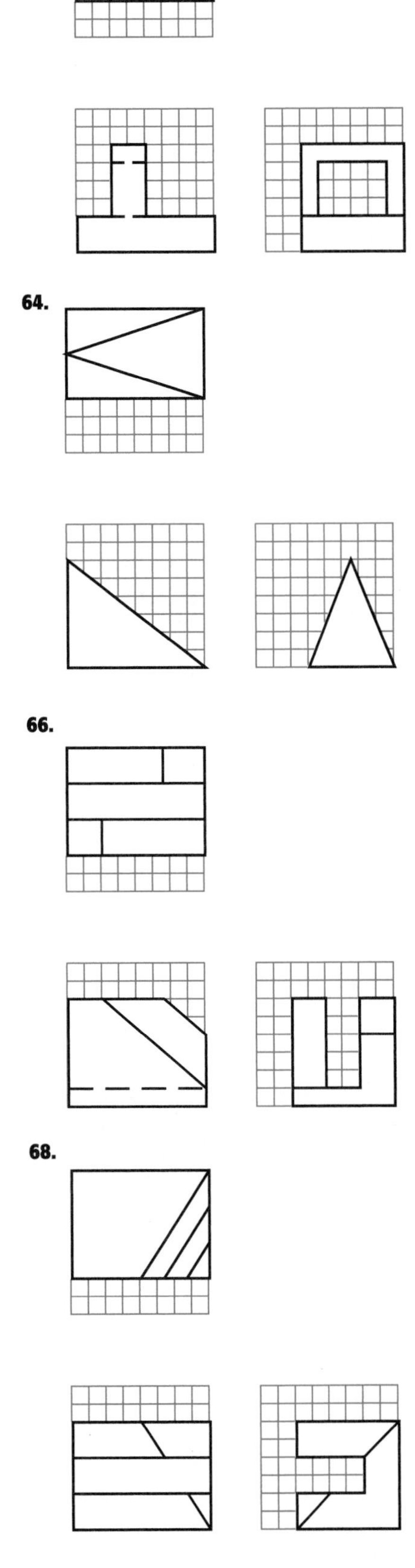

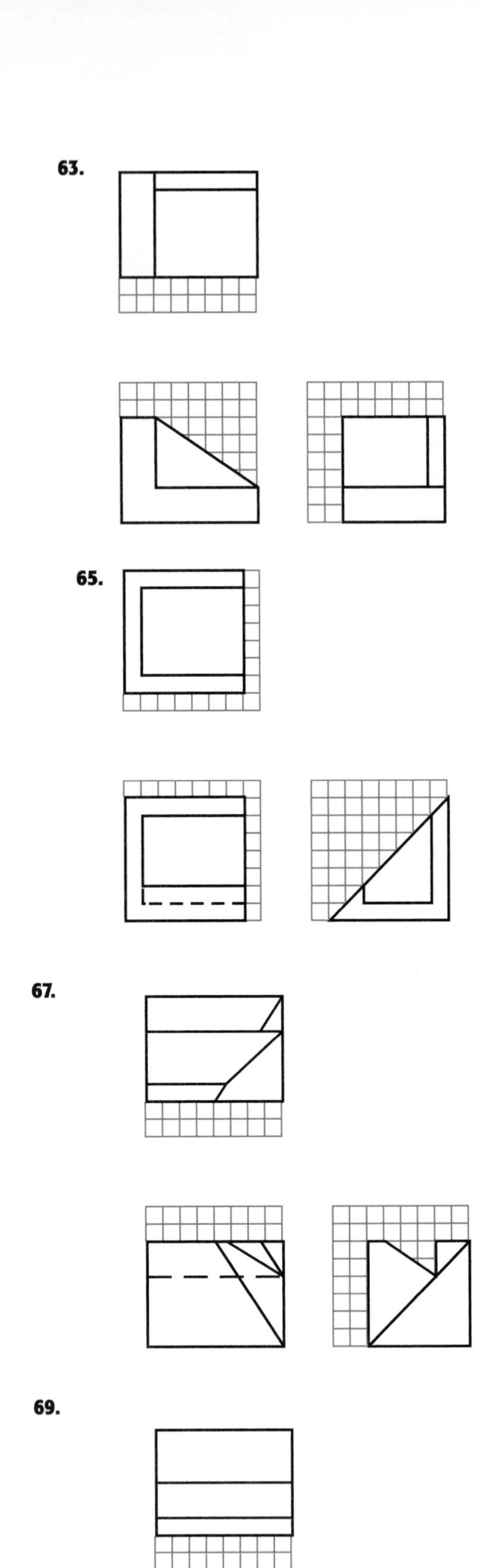

64.

65.

66.

67.

68.

69.

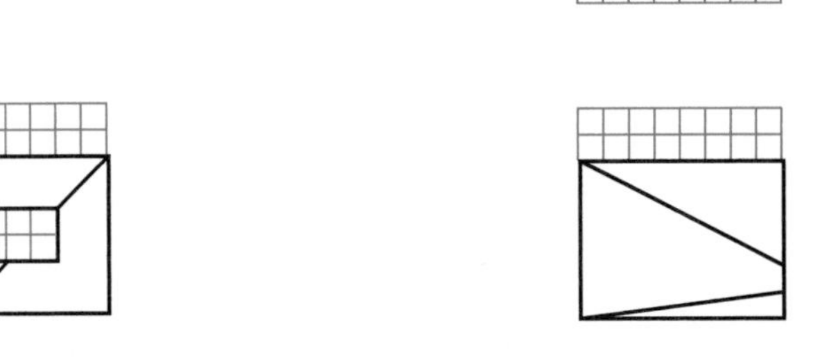

70.

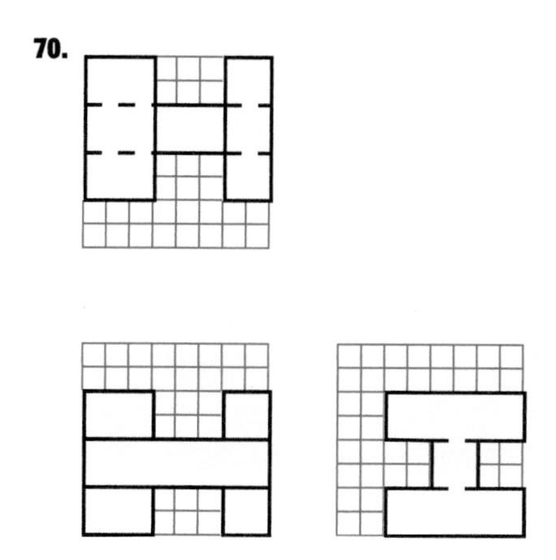

SKETCHING CIRCLES IN ISOMETRIC

A circle appears as an ellipse in isometric. To draw a circle in isometric, follow the same procedure used to sketch an orthographic view—that is, locate tangent points, and draw one quadrant at a time. The enclosing box is drawn in isometric rather than in orthographic. Figure 2.31 shows orthographic and isometric views of the enclosing box. The location of the centre can be found by joining the corners and drawing diagonals. The diagonals are not equal in an isometric. Tangent points are located at the midpoint of each side.

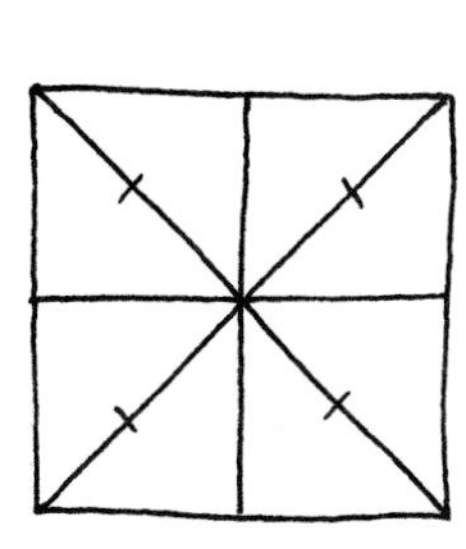

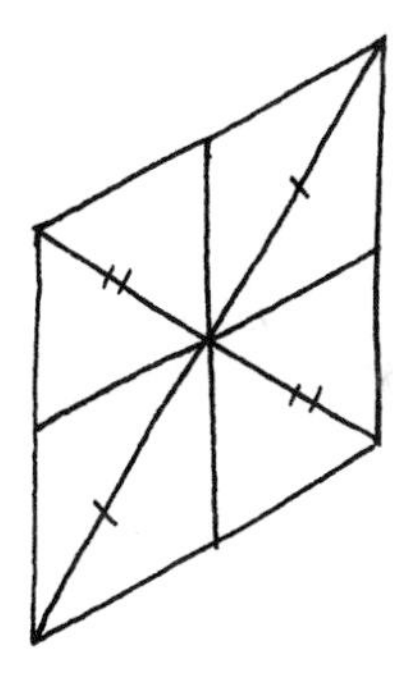

Figure 2.31 Orthographic and isometric sketches of a square

Each quadrant is drawn separately, but the outline is not a constant distance from the centre. Drawing the outline as a dashed line can make it easier to sketch the correct shape. The final line is darkened to distinguish it from construction lines. Before darkening the outline, the results should be checked to ensure that the isometric circle looks as it should. Figure 2.32 shows the process and the finished circle.

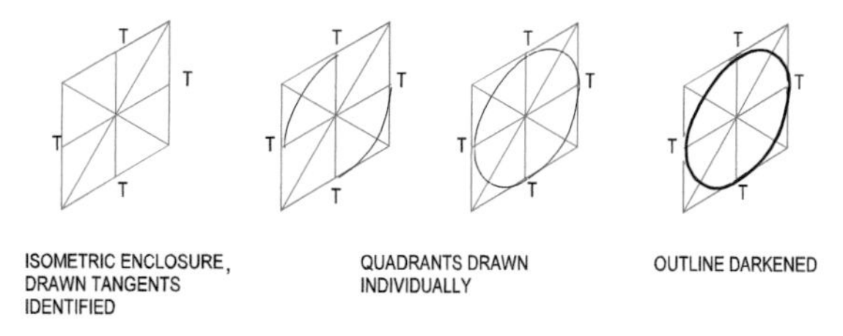

Figure 2.32 Steps in creating an isometric sketch of a circle

Now let's move on to look at sketching cylinders.

SKETCHING CYLINDERS

Circles are associated with cylindrical components, such as shafts and holes, that are drawn with a combination of straight and curved lines. Cylinders are specified by diameter, not radius, because a diameter is easily measured. (How would the radius of a hole be measured?)

Example 2.5

Sketch a 50 mm diameter shaft, 150 mm long, with one end stepped to 25 mm diameter for a length of 25 mm.

1. Position the object to show the stepped end clearly. Two orientations–a good choice and a poor choice–are shown in **Figure 2.33**. The illustration on the right is a poor choice because it does not clearly show the stepped end.
2. Block out the space with light construction lines. The construction, a combination of four circles representing the ends of the cylinders, is shown in **Figure 2.34**.
3. The finished sketch, with hidden lines removed and outlines darkened, is shown in **Figure 2.35**.

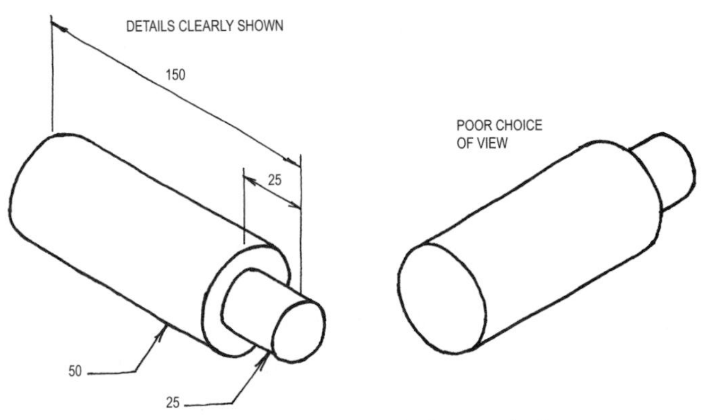

Figure 2.33 The sketch on the left shows the features clearly

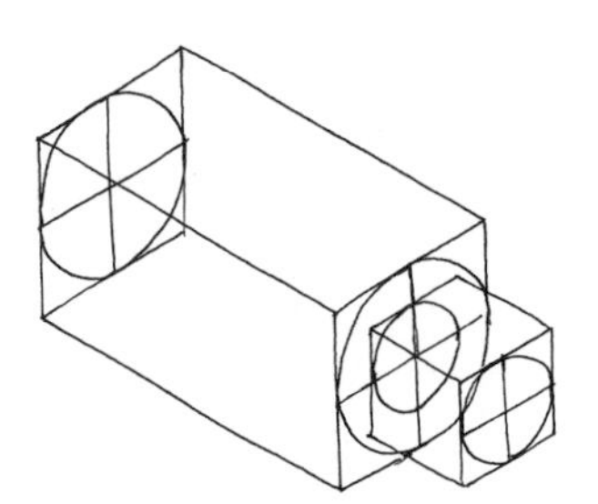

Figure 2.34 The size of each cylinder is defined by a rectangular prism

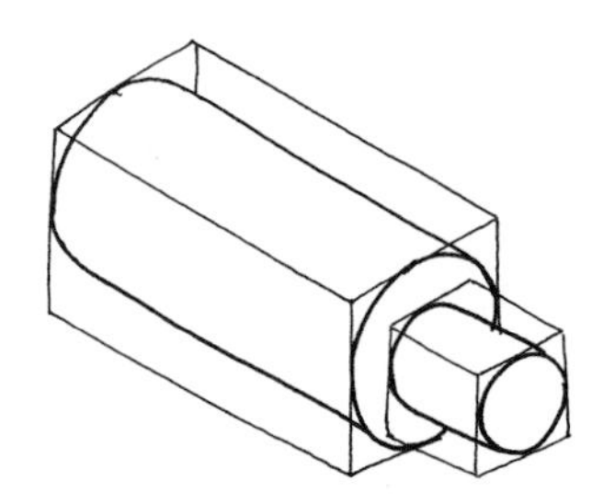

Figure 2.35 Finished sketch with outlines darkened

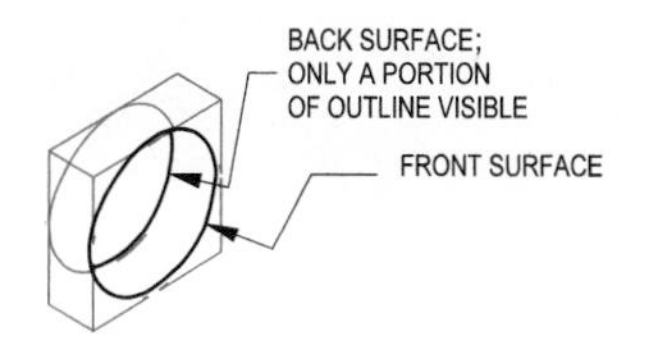

Figure 2.36 The visible portion of the rear of the hole is darkened to show both ends

A round hole is a negative cylinder and is constructed in the same way as a positive cylinder. The difference between a hole and a cylinder is that often the bottom end (or rear surface) of the hole is not visible. If the material is thin, a portion of the other end of the hole may be seen on the rear surface. In this case, only part of the outline is visible. An enclosing box is drawn for both front and rear surfaces, as shown in Figure 2.36. Only the visible portion of the circle on the rear surface is darkened in the final step of the construction.

Now let's move to look at sketching intersecting surfaces.

SKETCHING INTERSECTING SURFACES

When two plane surfaces meet at an angle, the line of intersection is a straight line. If one or both surfaces are curved, there is still a line of intersection, but it may not be shown, depending on how the surfaces intersect. If two curved surfaces share a common tangent, the line of intersection is shown in an orthographic view only if the tangent is vertical or horizontal.

If curved surfaces have a common tangent, there will be no abrupt change in the surface, and the intersection will not be shown unless the common tangent is vertical or horizontal. This is shown in Figure 2.37. The line of intersection is not drawn in the pictorial.

The common tangent would not be drawn in the front view; it is shown here for illustration only. The line of intersection is seen in the plan view, but not in the side view.

Figure 2.38 shows two curved surfaces with a common tangent that is neither vertical nor horizontal. The line of intersection is not drawn in the pictorial view or in any orthographic view. The common tangent is shown in the front view for illustration only; it would not be drawn in practice.

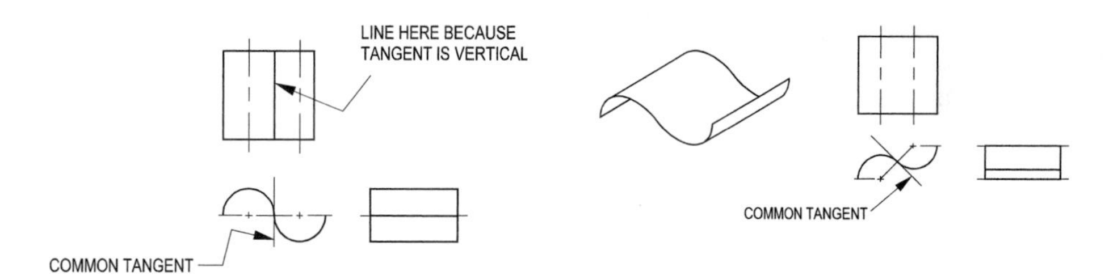

Figure 2.37 The method of drawing two curved surfaces with a common vertical tangent

Figure 2.38 The method of drawing the intersection of curved surfaces

When a plane surface is tangent to a curved surface, as shown in Figure 2.39, a line of intersection is not drawn in the isometric or orthographic views.

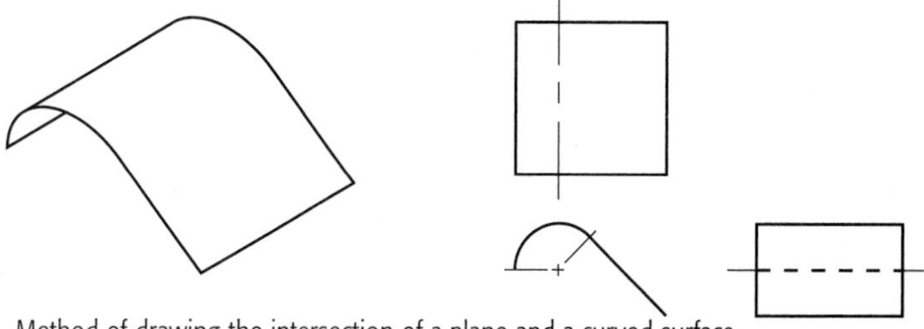

Figure 2.39 Method of drawing the intersection of a plane and a curved surface

Example 2.6

Sketch an isometric pictorial of the object shown in **Figure 2.40**. Different parts of the object have been labelled for reference.

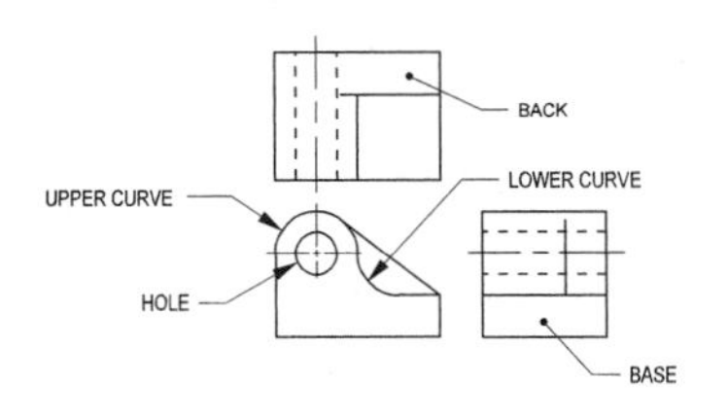

Figure 2.40 Three-view drawing

These shapes define overall size, proportions, and features and are used to locate centres and radii.

1. Define overall size by blocking out an isometric frame defining width, height, and depth.

   Take height as 75 percent of width and equal to depth.

   The width of the back is seen in the side view and is taken as one-third of the width.

   These proportions are determined "by eye" and are sufficiently accurate for a sketch. The space occupied by this solid is blocked out using these lengths (**Figure 2.41**).
2. Block features into this frame using the basic shapes previously identified.

   The upper and lower curves have a common vertical tangent, indicated by the line of intersection in the plan view.

   The hole and the upper curve have a common centre.
3. Block the locations of the centres for the circular features and their sizes as shown in **Figure 2.42**. Only the lower portion of the rear is blocked in because only the location of the bottom is known.
4. Sketch a semicircle, representing the upper curve, and a quarter circle, representing the lower curve, using the method previously described. Darken the outline on the front face to distinguish it from construction lines. The result is shown in **Figure 2.43**.
5. Project the curved surfaces to the rear to complete it. The finished sketch is shown in **Figure 2.44**, with outlines darkened to differentiate them from construction lines.
6. An oblique drawing of this object is shown in **Figure 2.45**. The front view as seen in the three-view drawing is used as the starting point, and additions are made to represent depth. There are curved lines that are not on the front surface, but these are duplicates of those on the front surface (as with the isometric). They are simply copied in the appropriate location. Because there are curved lines that are not on the front surface, the choice between an isometric and an oblique is not clear cut but depends on personal preference and the purpose of the sketch.

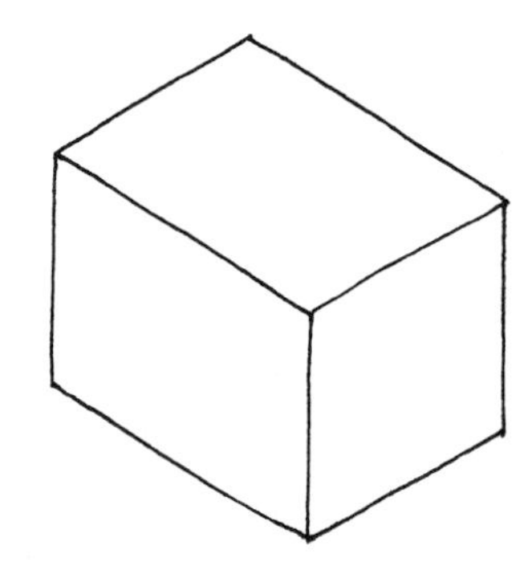

Figure 2.41 Overall size defined by a rectangular prism

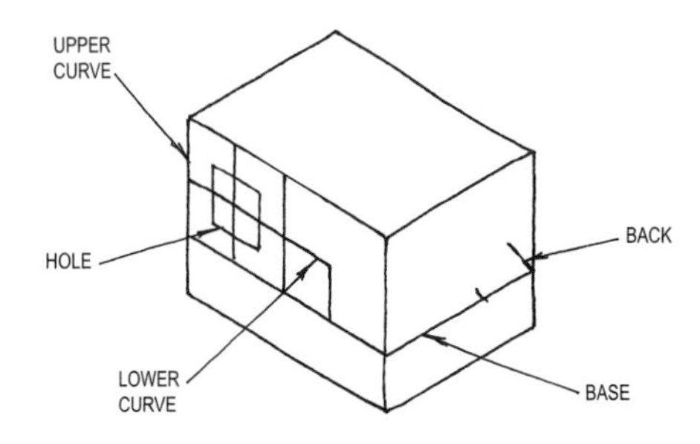

Figure 2.42 Centres and sizes of circular features blocked in using light construction lines

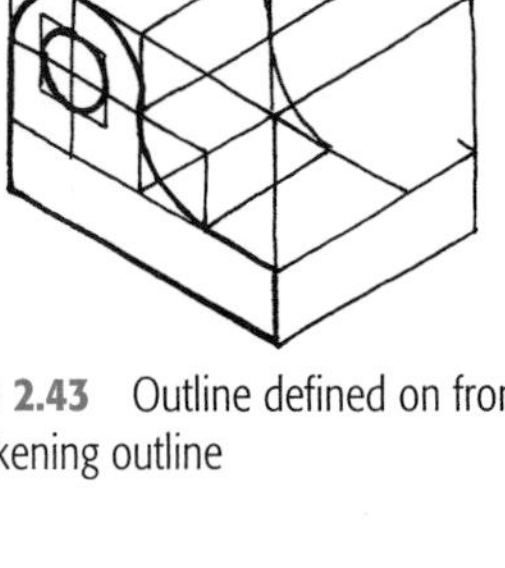

Figure 2.43 Outline defined on front face by darkening outline

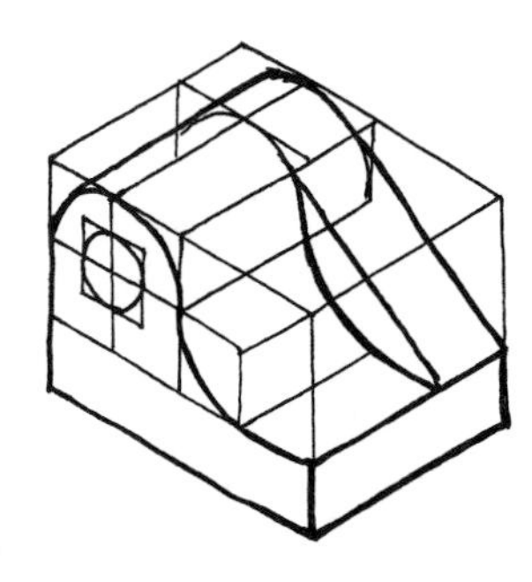

Figure 2.44 Outlines of object darkened to stand out from construction lines

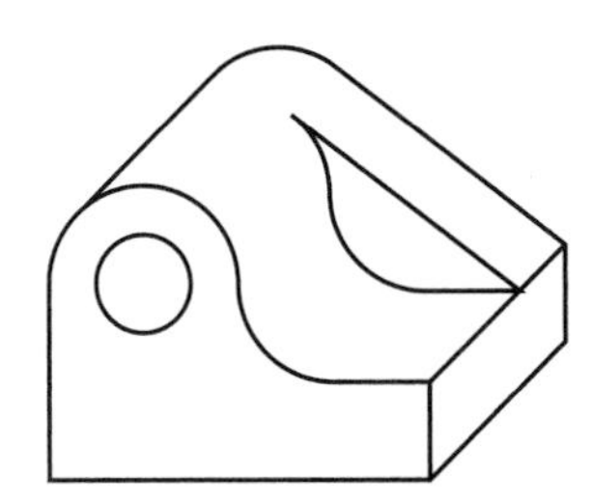

Figure 2.45 Oblique drawing shows the front face in true shape

Now that you know a little more about sketching curved lines, using pictorial sketches, sketching circles in isometric, and sketching cylinders and intersecting surfaces, try the following problems.

Example 2.7

An isometric sketch can be done by identifying combining basic shapes. A cylinder is the most common shape you will use. A negative cylinder is a hole, or a portion of a curved object as shown in **Figure 2.46**. A positive cylinder can be a portion of a curved object.

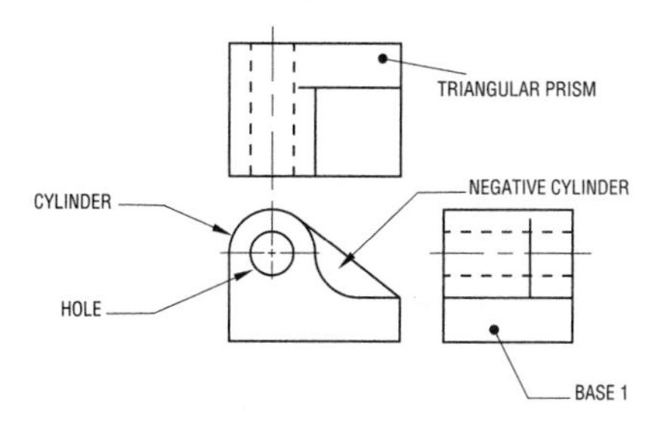

Figure 2.46 Three-view drawing

Three cylinders, two rectangular prisms, and one triangular prism are combined to make this object. Two cylinders are negative, and the hole and the curve are shown. Only one base is identified. The other is combined with the positive cylinder. **Figure 2.47** shows an exploded view of these shapes.

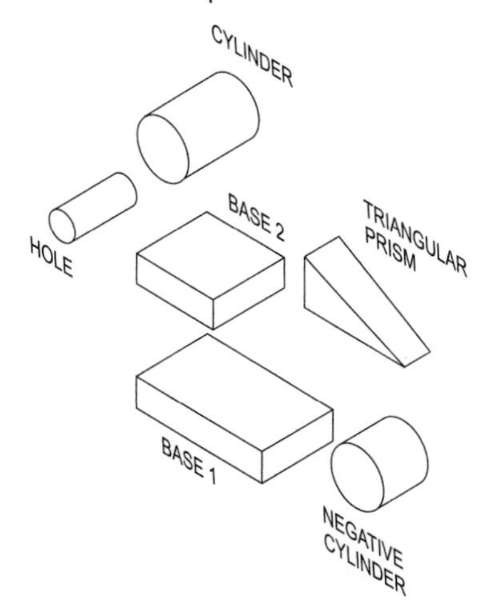

Figure 2.47 Shape breakdown

The positive cylinder is combined with base 2 on the left side and the negative cylinder on the right side to create the transition from the curved surface to the base. These cylinders have the same diameter, but the negative cylinder is shorter than the positive cylinder. The length of base 2 is three times the radius of the cylinders, and the height is equal to the cylinder radius. **Figure 2.48** shows the result of combining base 1 and 2. Hidden lines are not shown.

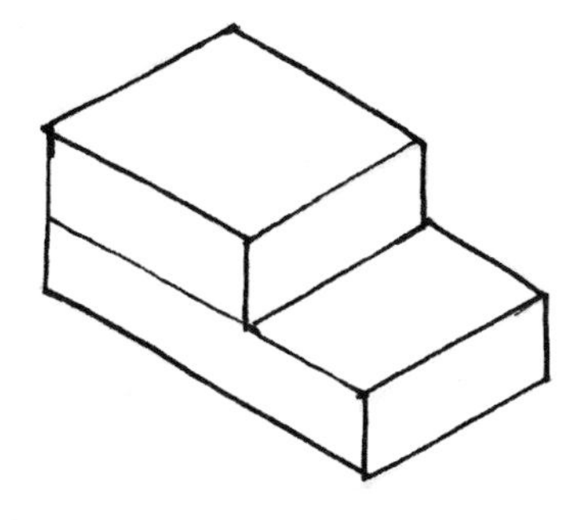

Figure 2.48 Base 1 combined with base 2

The positive cylinder is combined with base 2 in **Figure 2.49**. The line of intersection between the cylinder and the base is shown to illustrate the shape in this figure.

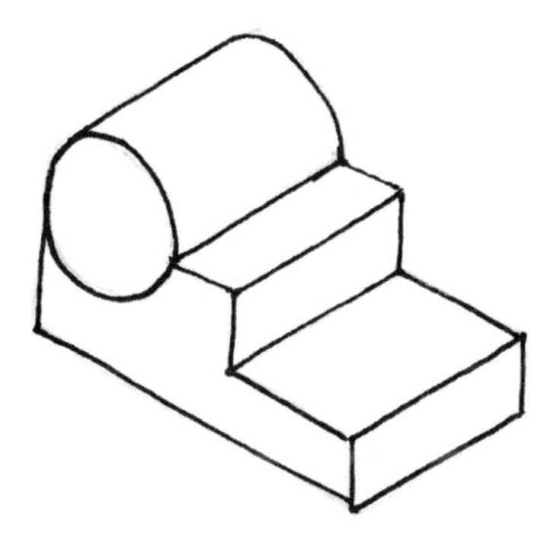

Figure 2.49 Cylinder combined with base 2

The triangular prism is then combined with all three shapes as shown in **Figure 2.50**. The length of the triangular prism is such that the sloping side is tangent to the cylinder.

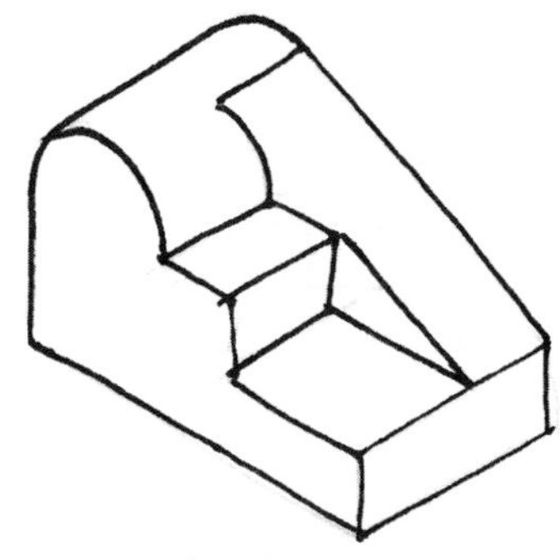

Figure 2.50 Triangular prism combined with base 1, base 2, and cylinder

The negative cylinder is then added to base 2 to create the transition between the cylinder and base 1. The result is shown in **Figure 2.51**. The lines of intersection between the negative cylinder, base 1, and the cylinder are shown for illustration, but these are smooth transitions, and in practice, there would be not be any actual lines.

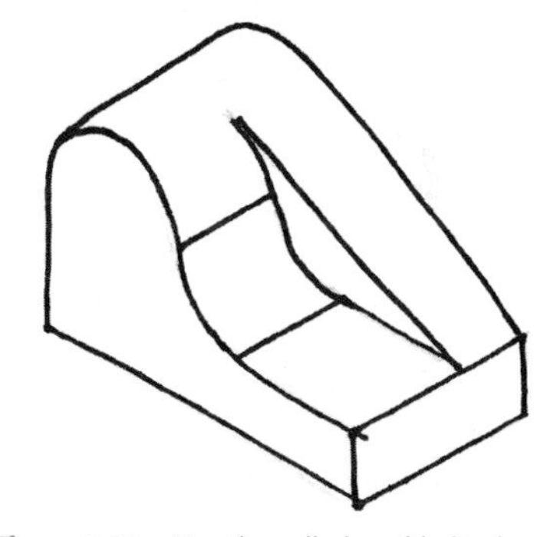

Figure 2.51 Negative cylinder added to base 2

The hole is now added to the cylinder to complete the object (**Figure 2.52**).

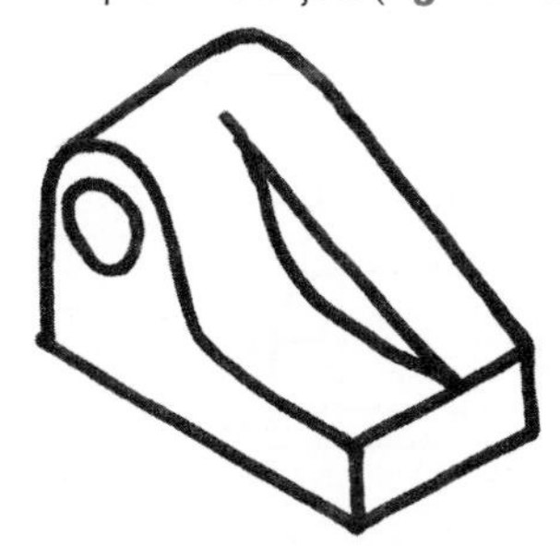

Figure 2.52 Hole added and outline darkened

Creating an isometric by combining basic shapes is similar to the process used with a solid modelling program. When you design objects with a solid modelling program you should plan your work with a few sketches first.

Problems

Sketch the best isometric view of the following objects:

71.

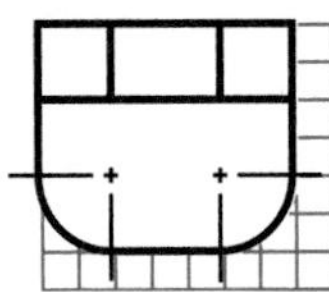

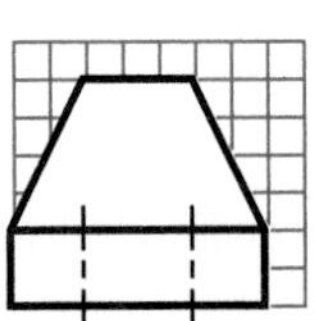

72.

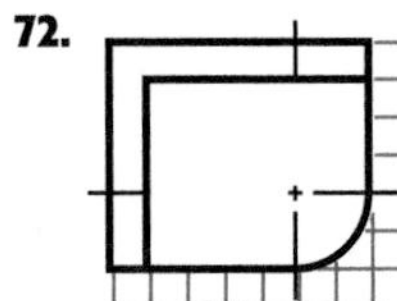

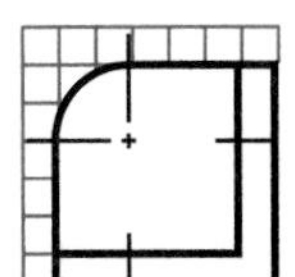

73.

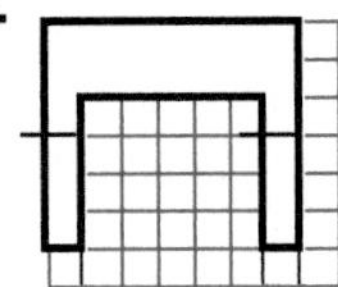

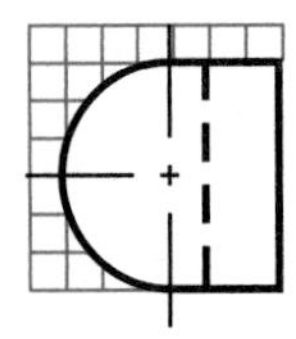

74.

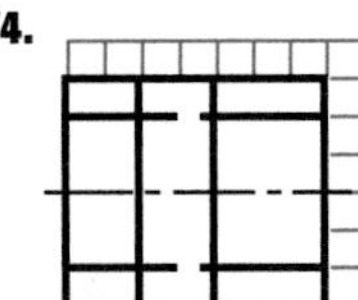

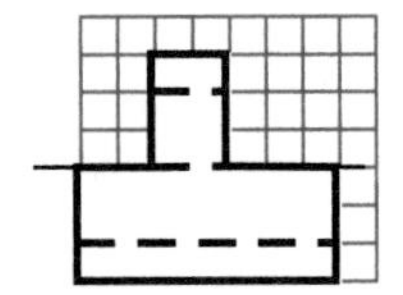

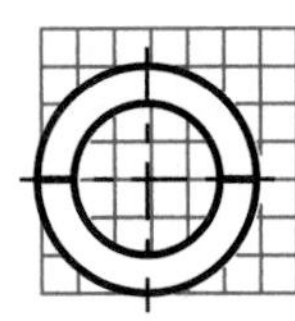

75.

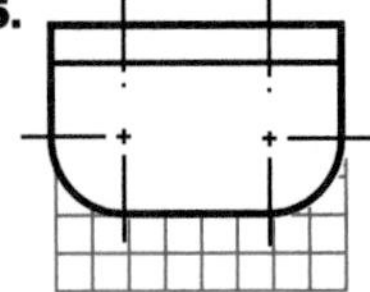

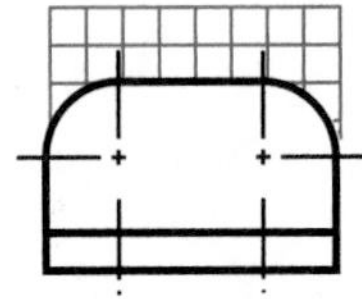

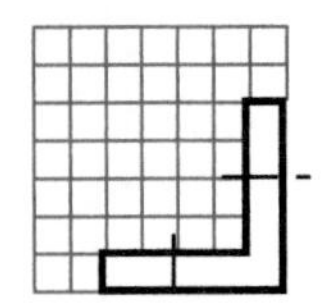

76.

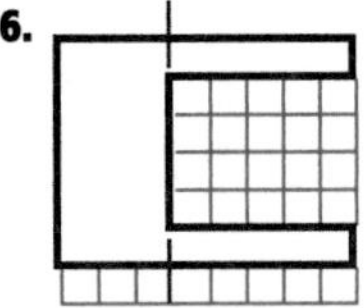

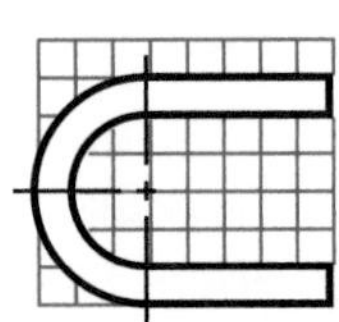

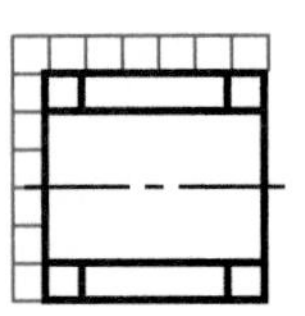

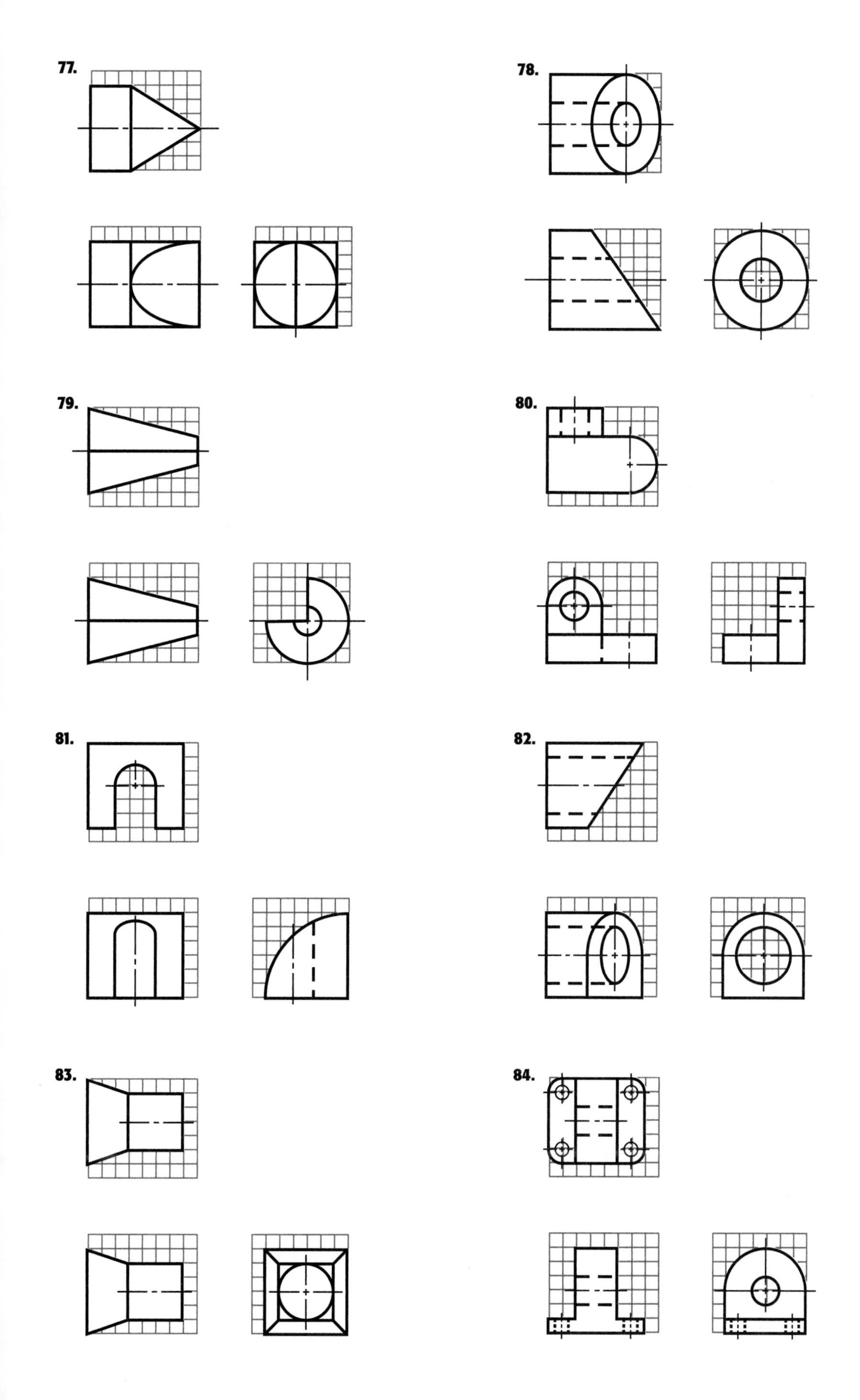
77.
78.
79.
80.
81.
82.
83.
84.

Problems

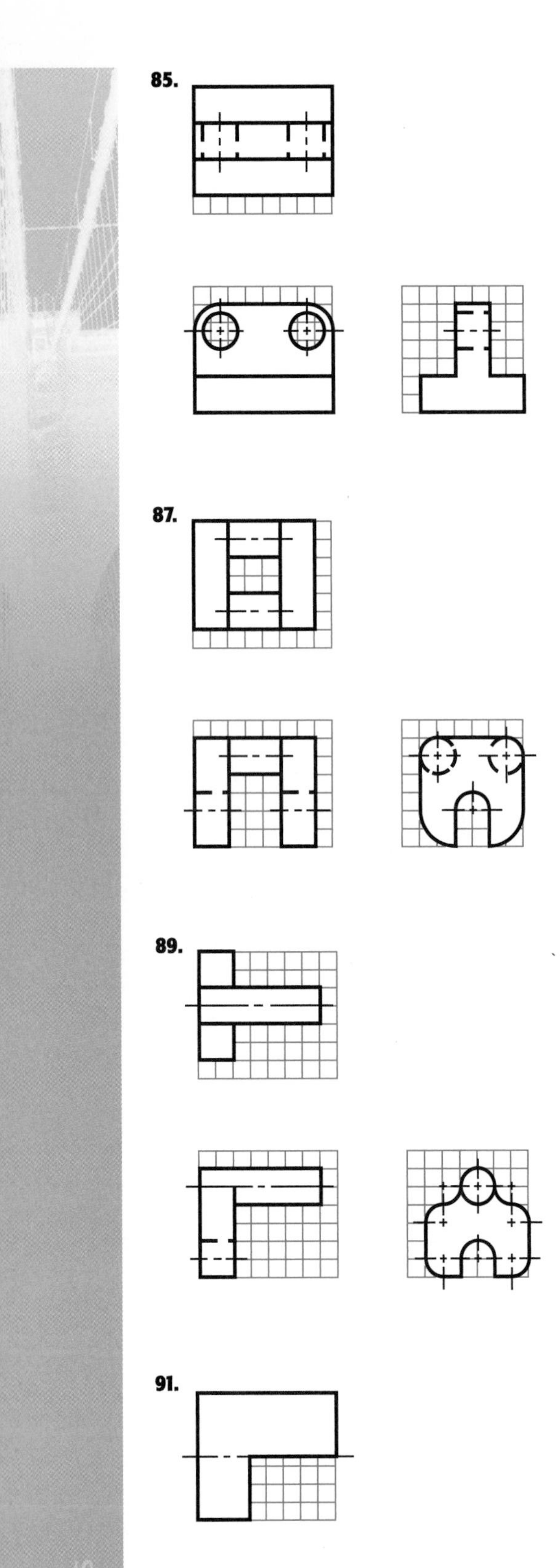
85.
87.
89.
91.

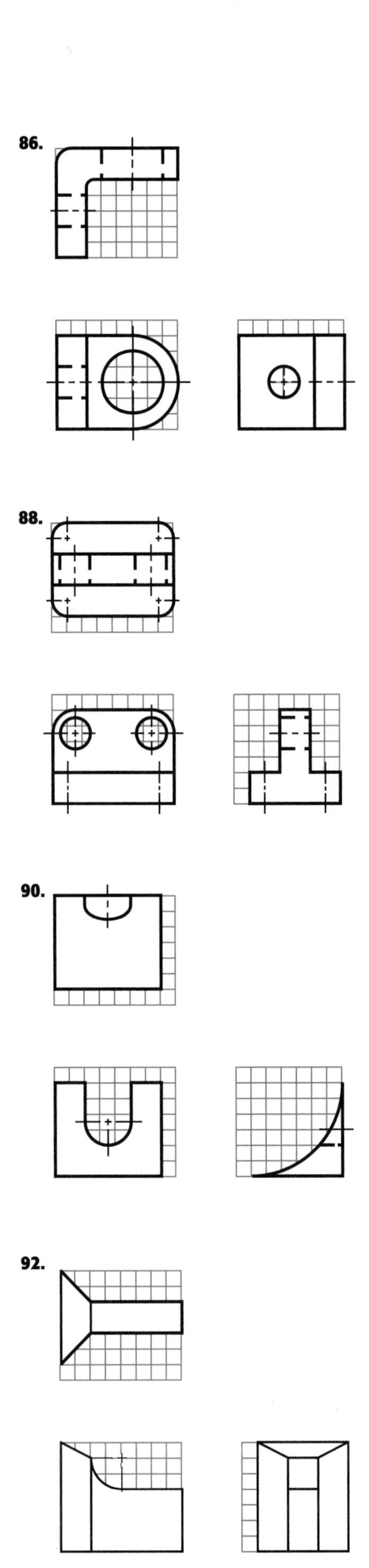
86.
88.
90.
92.

93.

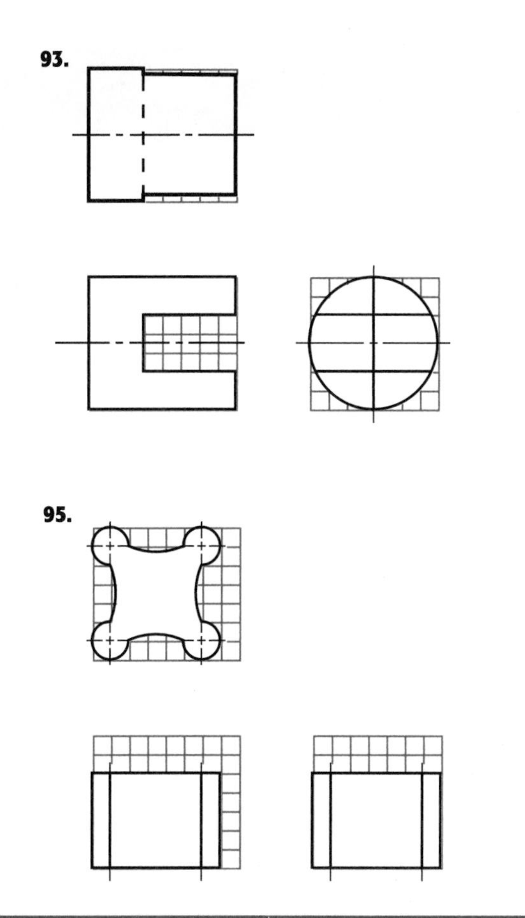

95.

94.

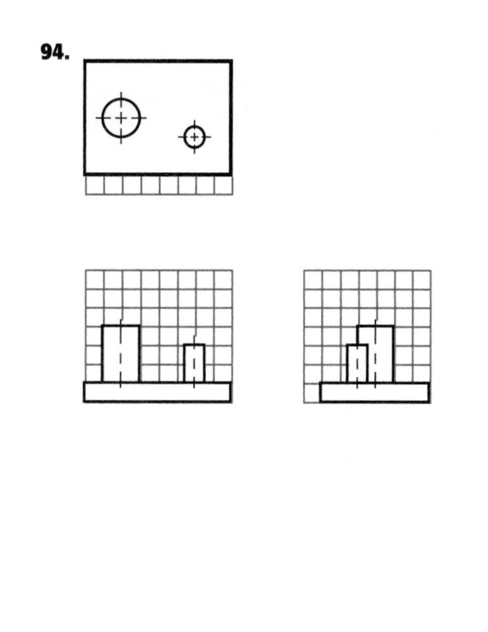

Now let's move on to look at how to determine what information is missing.

DETERMINING MISSING INFORMATION

Any missing information must be determined from what is given. With simple objects, you may be able to visualize what is missing, but complicated objects require a systematic approach. Pictorial and orthographic sketches, individually or together, can be used to determine what something looks like.

Example 2.8

Figure 2.53 shows two views of an object. The front view is incomplete, and the side view is complete. The front view must be completed and the plan view drawn. What is known?

From the orientation of the views, we know which are the right-side and front views. Because the right-side view is complete, nothing can be added.

Width, height, and depth are known.

1. Front View
   There is nothing in the front view to correspond with the bottom surface of this slot, marked A in **Figure 2.54**. Draw the horizontal line (a hidden line) in the front view to correspond with surface A. If this surface is not horizontal, more lines are needed, since both ends are visible. The front view is now complete.
2. Plan View
   Block out the plan view by projecting the width from the front view. Take the depth from the side view. Figure 2.54 shows work to this point.
3. Find the location of the slot relative to the front surface in the side view (length *s*) and transfer it to the plan view. Locate the slot length and the sloping surface by projecting points *b* and *c* from the front view. **Figure 2.55** shows these two steps.

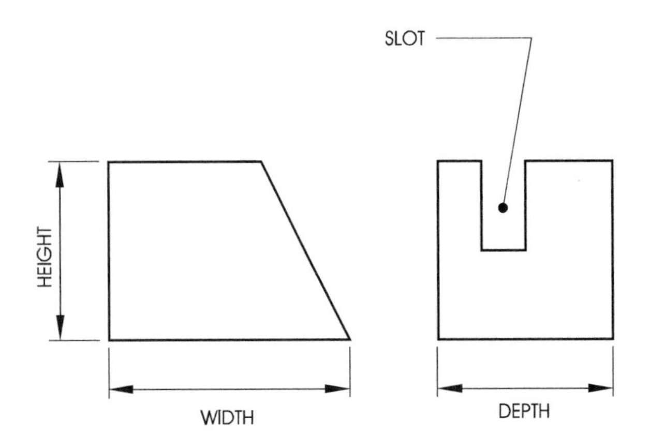

Figure 2.53 Missing information must be drawn in the front view

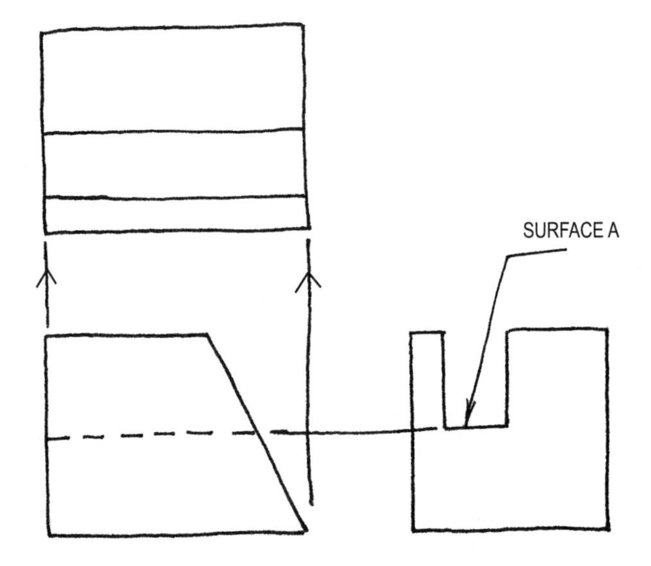

Figure 2.54 Front view completed and plan view blocked out

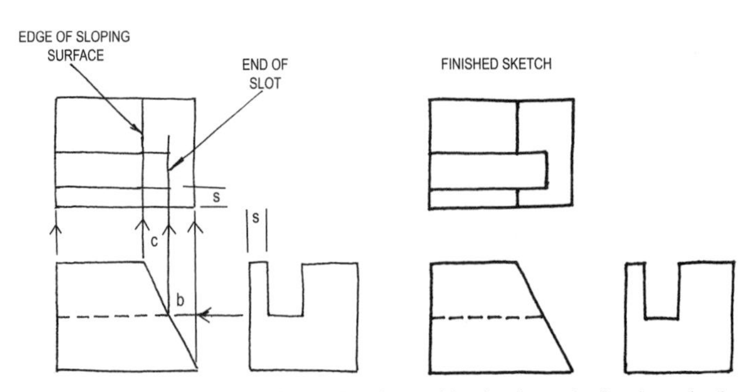

Figure 2.55 Slot and sloping surface located in plan by projecting from the front view

4. Use an isometric in conjunction with a three-view drawing by sketching orthographic views on the faces of an enclosing box. **Figure 2.56** shows the given information added to two faces of an isometric enclosure.
5. Lines identified as *a* and *b* represent the intersection of the vertical sides of the slot with the sloping face. Transfer them from the end to the sloping face. Draw lines representing the slot on the top surface. Transfer the plan view back to the multiview drawing. **Figure 2.57** shows the finished isometric sketch.

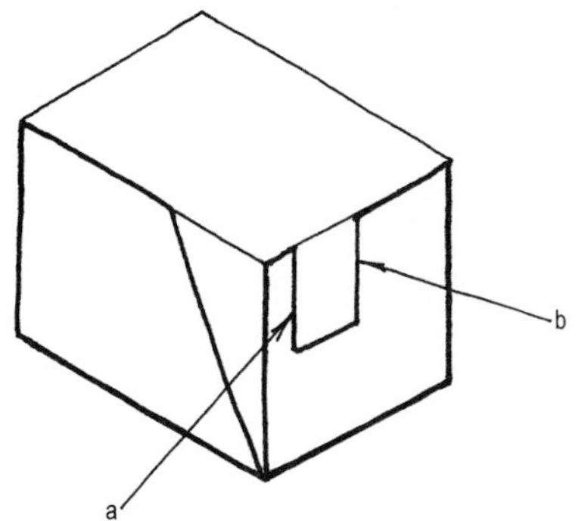

Figure 2.56 Orthographic views transferred to enclosing box

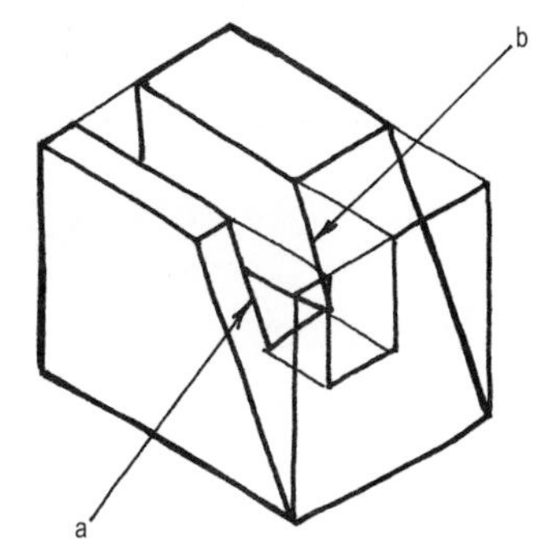

Figure 2.57 Plan view completed on enclosing box and outlines darkened

Another approach is to combine basic shapes. Create the object by combining:

- a rectangular prism (a positive volume),
- a smaller rectangular prism (the slot, a negative volume), and
- a triangular prism (a negative volume).

Figure 2.58 shows a rectangular prism removed to create the slot.

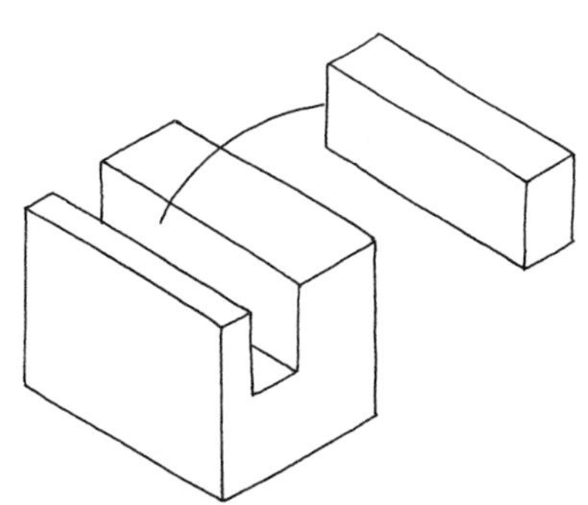

Figure 2.58 Rectangular prism removed to create a slot

Remove the triangular prism to form the shape shown in **Figure 2.59.** Add the line of intersection between this triangular prism and the slot. Show the triangular prism in phantom lines in its original position.

The order in which shapes are removed is not important. The same result is achieved if the triangular prism is removed first.

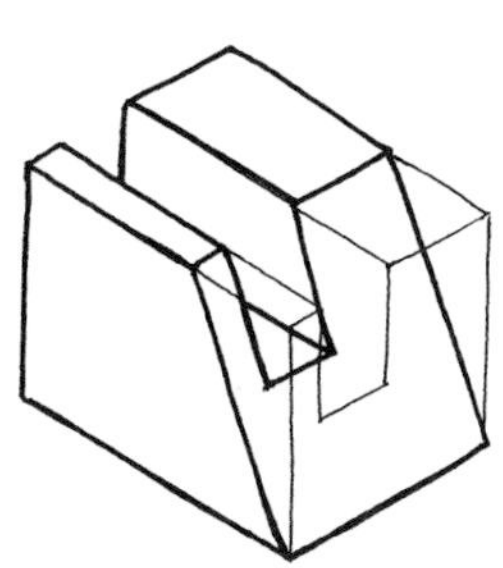

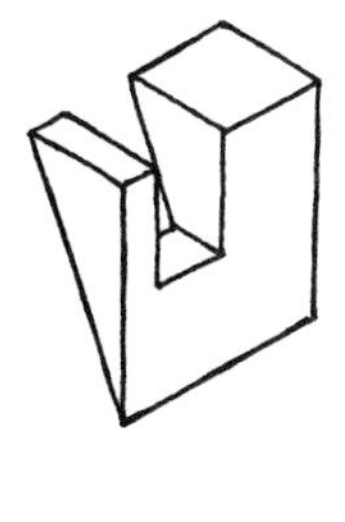

Figure 2.59 Triangular prism removed and line of intersection added to create the object

Now that you know how to determine missing information, try the following problems.

Problems

There are lines missing on some of the views shown below; however, there is sufficient information given to determine what the objects look like. Sketch the missing view or the missing lines and an isometric of each object. The objects are shown on a 5 mm grid.

96.

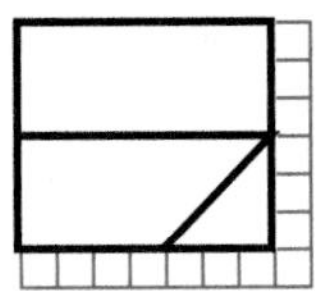

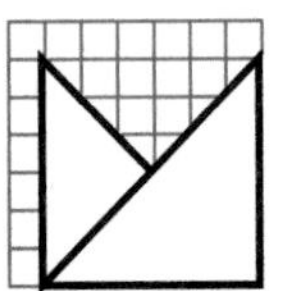

97.

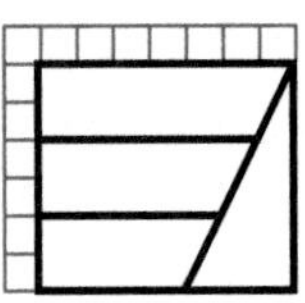

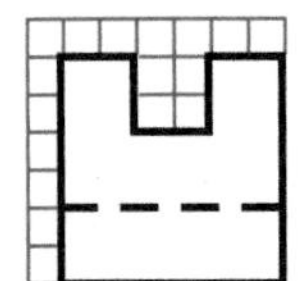

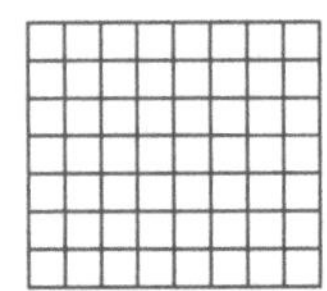

98.

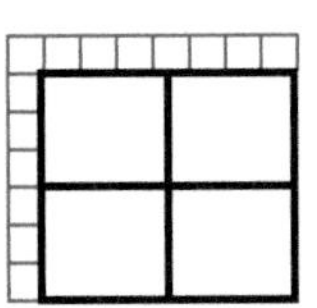

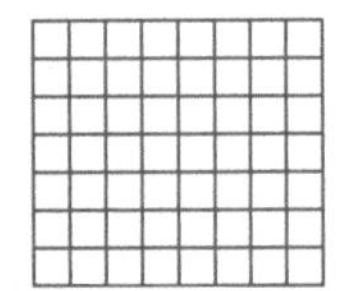

99.

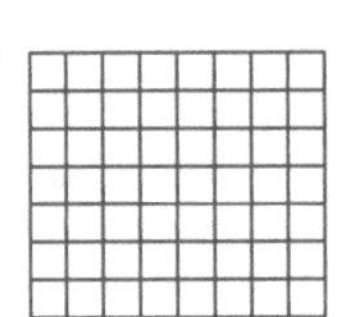

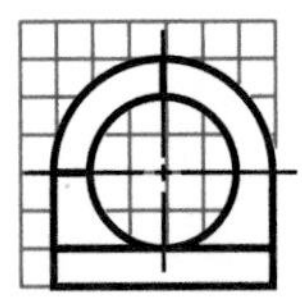

100.

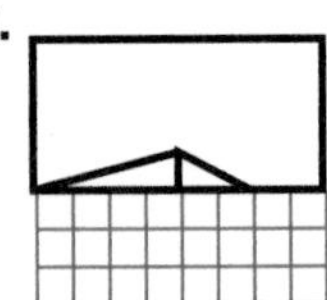

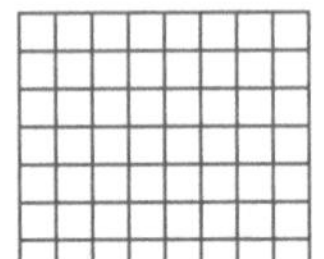

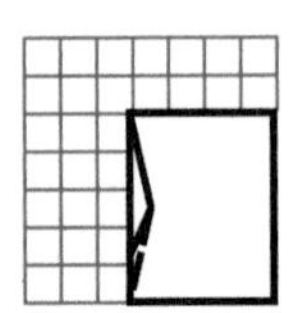

101.

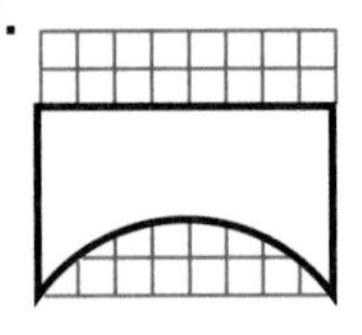

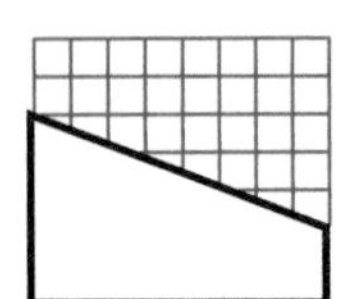

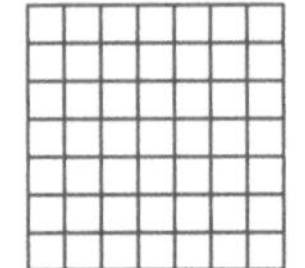

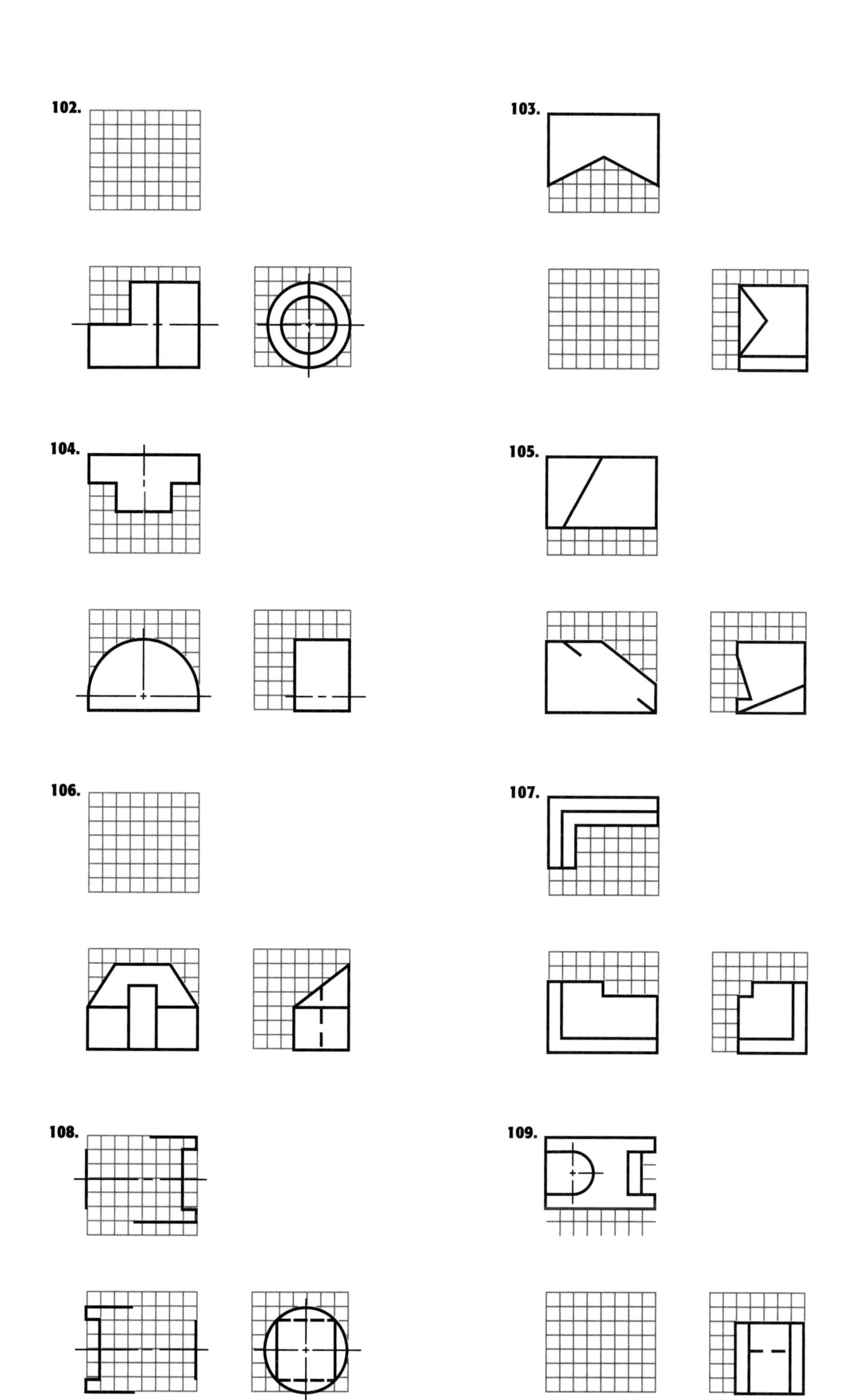
102.
103.
104.
105.
106.
107.
108.
109.

110.

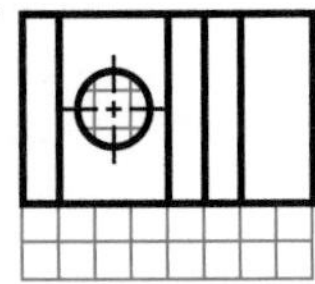

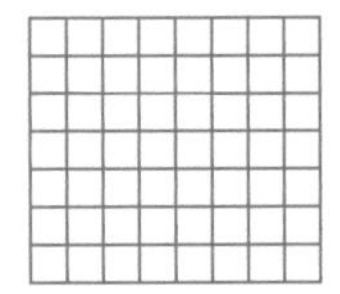

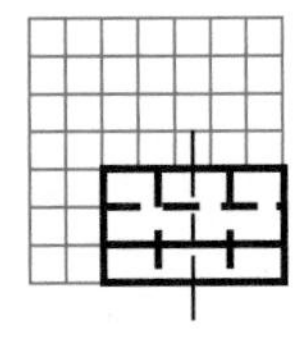

111.

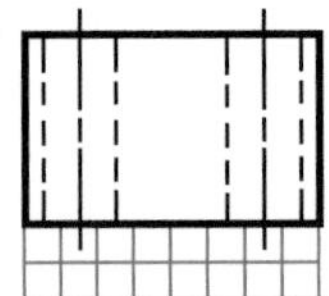

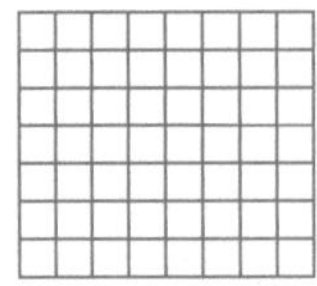

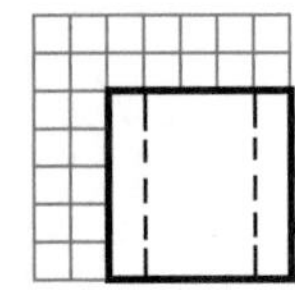

112.

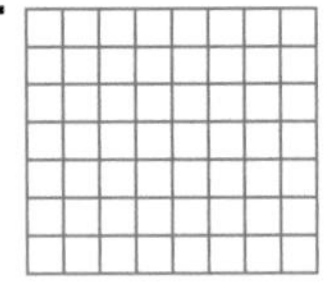

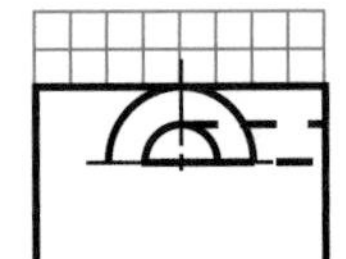

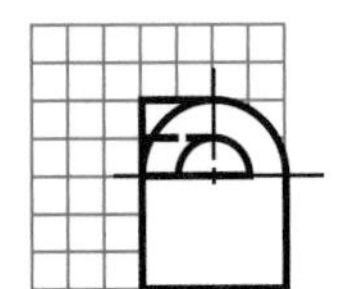

113.

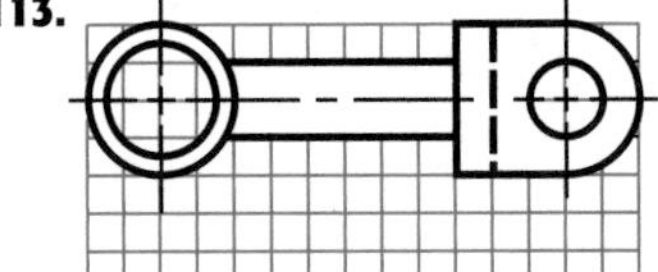

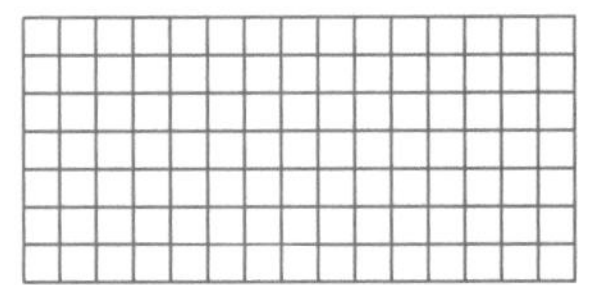

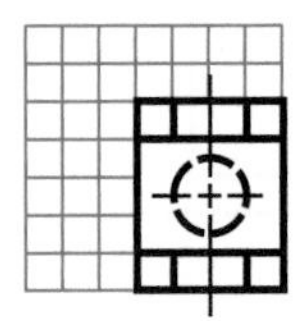

114.

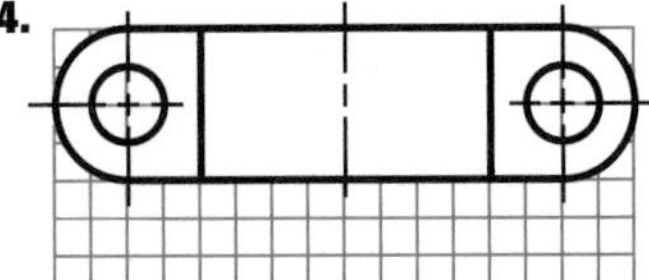

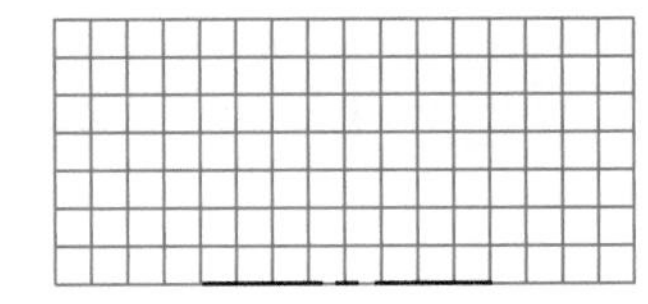

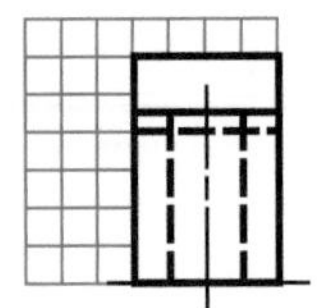

115.

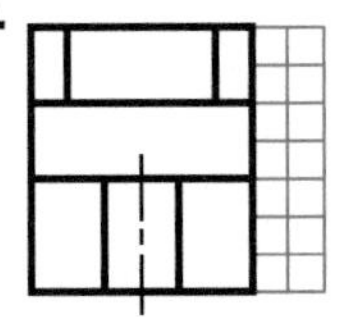

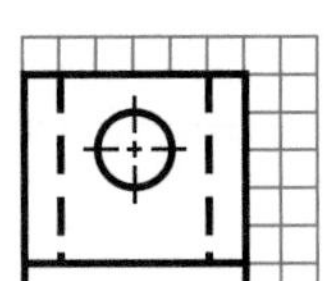

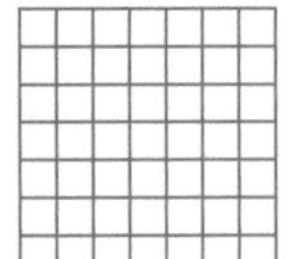

> *For those who find such drawings daunting, the process of understanding is not helped, initially, by the bewildering array of standard symbols designed to reduce drawing time and to avoid ambiguity. The thousands of standard symbols both increase the use of visual jargon and make it easier for initiates to decode drawings.*
>
> *Eugene S. Ferguson,* Engineering in the Mind's Eye, *The MIT Press, 1992*

CHAPTER

3 Sectioning

We have seen how the outside of an object can be shown, but what about internal features? Not all objects are solid, so internal details must somehow be specified. A simple example is shown in Figure 3.1. There is a hole in the block, but there is no way of knowing anything about the hole. It may only go to the middle of the block, for example. Hidden lines can be used to specify internal features, but they can be confusing, particularly if there are many of them.

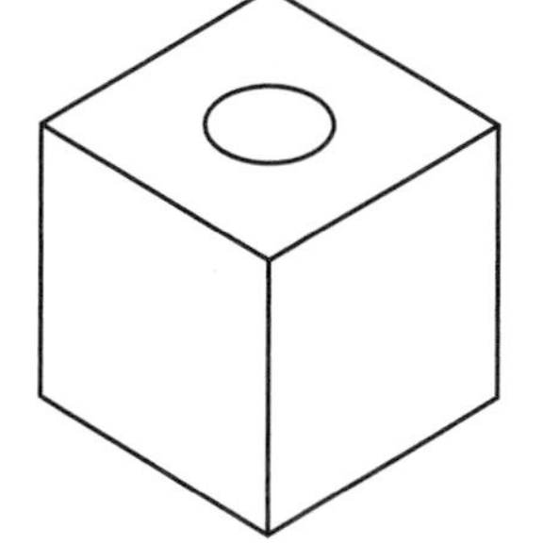

Figure 3.1 Object with internal features that cannot be seen

SECTIONING PROCESS

Internal details are shown with a process called *sectioning*. A section of the object is "removed" to show the internal details. Imagine that the block in Figure 3.1 is cut in half through the middle of the hole. Figure 3.2a shows the line on which the block is cut. The front half of the block is removed, and the back half is drawn as it would appear after the front half has been removed. Internal features, drawn as solid lines, are now clearly visible (see Figure 3.2b).

To indicate that the front half has been removed and that it is not really "half a block," **section lines** are added to the "surface" formed by the "cut" (Figure 3.3).

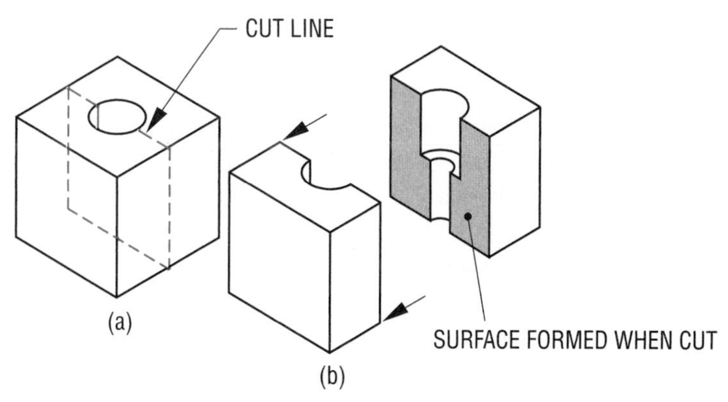

Figure 3.2 A portion of the object "removed" to show internal features

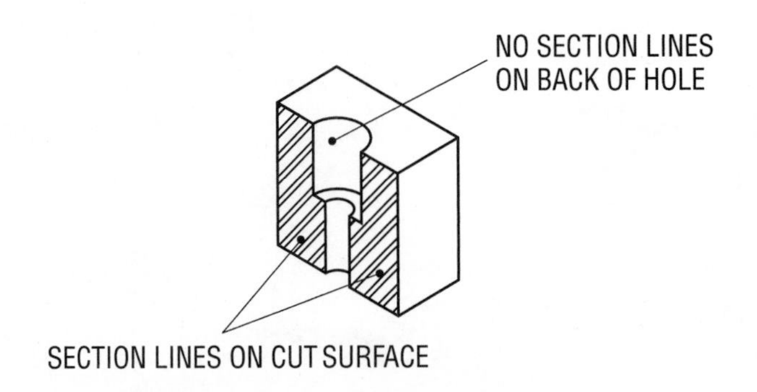

Figure 3.3 Section lines added to imaginary cut surface

Section lines are thin lines, drawn with a straight edge or a computer program (if you are sketching they can be drawn freehand). The pattern of the section lines identifies the material. The pattern in Figure 3.3 indicates steel.

The location of the imaginary cut is shown in the figure by a **cutting plane line** that shows the edge of the plane. A cutting plane line is a chain of thick and thin lines, with arrows at each end to show the viewing direction. Figure 3.4 shows the symbols used for two applications.

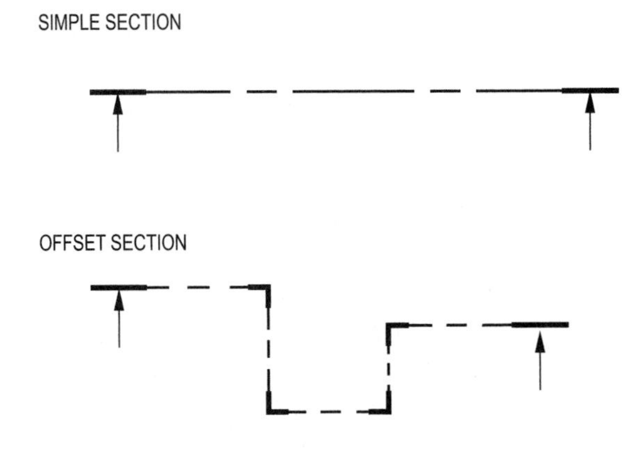

Figure 3.4 Symbols for a cutting plane line

Full Section

Figure 3.5 shows the top view of an object with the cutting plane through the centre. The front view is drawn as a **full section** because the section cuts through the whole block. Other types of section will be discussed later. The section is identified in the plan view by letters at each end of the cutting plane line, in this case A-A. The section is identified as "Section A-A" and is seen in the front view. The cutting plane line can often be omitted when it corresponds to the centreline of the part. We will put it in until we become more familiar with sectioning. For a simple section like this, the cutting plane line can go right across the object. For more complex section, it should terminate close to the outline (see Figure 3.10 on page 77). The cut surface is shown with a section line symbol, in this case, the symbol for steel. Section lines are shown only on the "surface" formed by the cutting plane line and not in the hole.

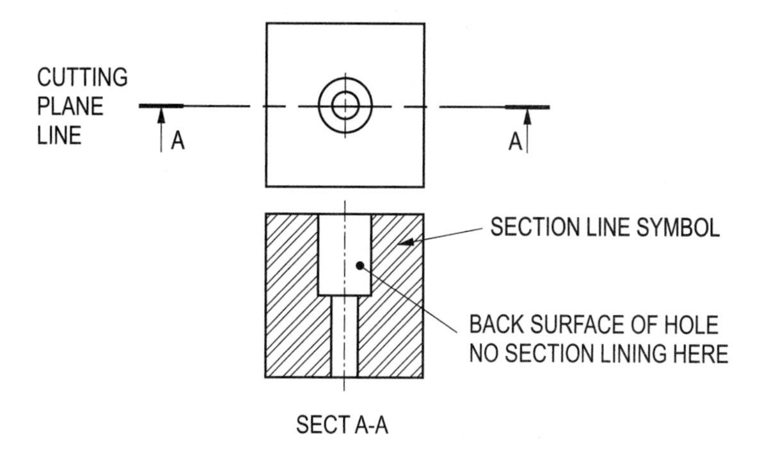

Figure 3.5 Front view as a full section to show internal details

The methods used for sectioning have been standardized. Understanding the conventions used in sectioning makes it easier to do and understand. The appropriate standards and conventions must be used so that an engineering drawing can be understood by others.

Let's, first of all, look at sectioning symbols.

SECTIONING SYMBOLS

A unique section line symbol (also called a **hatch pattern** when done with a computer program) can be used to indicate material; however, this is not always done. There are many sectioning symbols—too many for anyone to remember. A simple section line symbol is often used to show that a section has been taken. When section lines are added with a computer program, it is easy to use the correct symbol, but a simple pattern is often used, and the material is specified by a note. Figure 3.6 shows a simple section line symbol that is often used. This object is made from aluminum but the symbol used is not the one for aluminum. The material is specified by a note on the drawing.

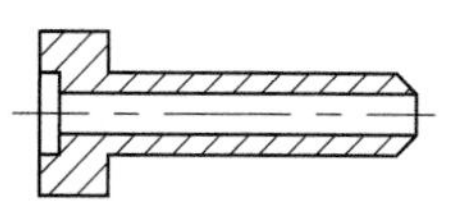

Figure 3.6 A simple section line symbol used when there is only one material

This symbol can be used if only one part is being sectioned. Sectioning symbols, as specified by the American National Standards Institute (ANSI), are given numbers, such as ANSI 32 (steel). The simple symbol is actually the symbol for cast iron (ANSI 31). Some common sectioning symbols and their designations are given in Appendix C. Computer graphics programs have a library of sectioning symbols, so you do not have to remember all of the symbols and their designations.

Section lines are not added randomly. They are placed so that there is an angle between the section lines and the outlines of the object. Section lines are drawn at 45° unless there is a good reason to use some other angle. The angle should be changed from 45° if section lines are parallel, or almost parallel, to any portion of the outline. Figure 3.7 shows two examples of section lines and the effect of different section line angles. This can sometimes be difficult to do depending on the object being sectioned, and the end result must be a compromise. The objective is to make the drawing as clear as possible.

A common error that occurs when sectioning is the omission of lines that are behind the cutting plane line. Lines on the object must be drawn if they are visible. Figure 3.8 shows a front view drawn as a full section to show details of a hole.

Lines at the top, bottom, and middle of the hole are on the back surface of the hole. They do not vanish when the block is sectioned, and so they must be shown. Figure 3.8 shows what the section would look like if these lines were omitted. There

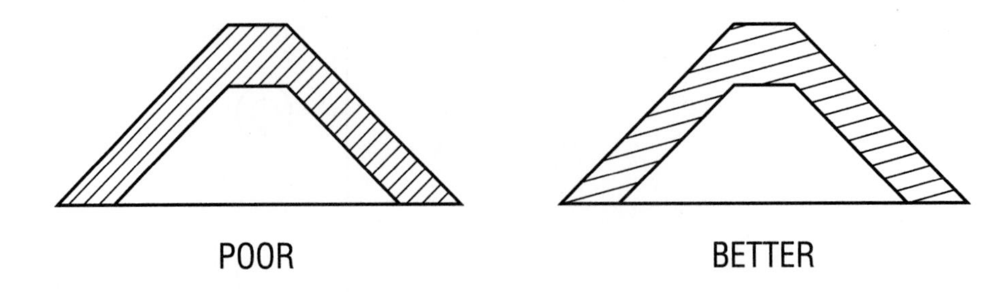

Figure 3.7 Angle of section lines changed to suit an unusual object

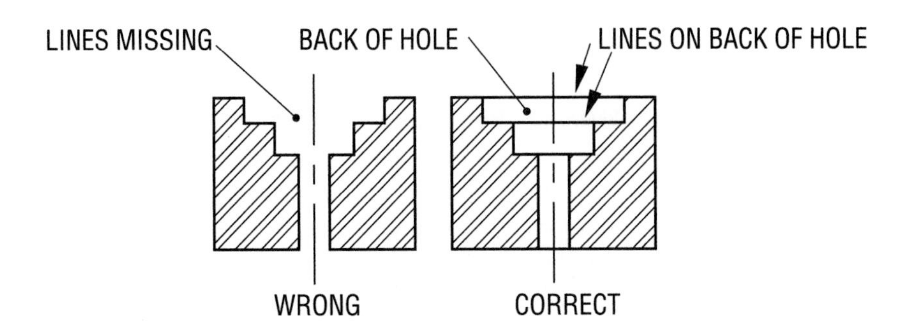

Figure 3.8 A common error is leaving out lines at the back of the hole

is nothing to connect the two sides of the block. Section lines are not put on the rear surface of the hole, since this surface is behind the cutting plane line.

Now let's move on to look at locating the section to show the required internal details.

TYPES OF SECTION VIEWS

The type of section should be chosen to show the required internal details. The previous example used a section through the middle of the hole, which happened to coincide with the centre of the block.

Offset Section

Sections do not have to be taken through the middle of the object. They can be taken anywhere. If, for example, internal details of a hole are required, the obvious place for the section is through the centre of the hole (see Figure 3.9).

CUTTING PLANES PASSING
THROUGH EACH HOLE

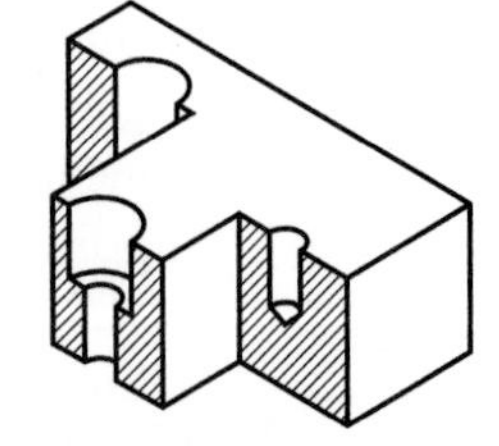

Figure 3.9 Example of offset section

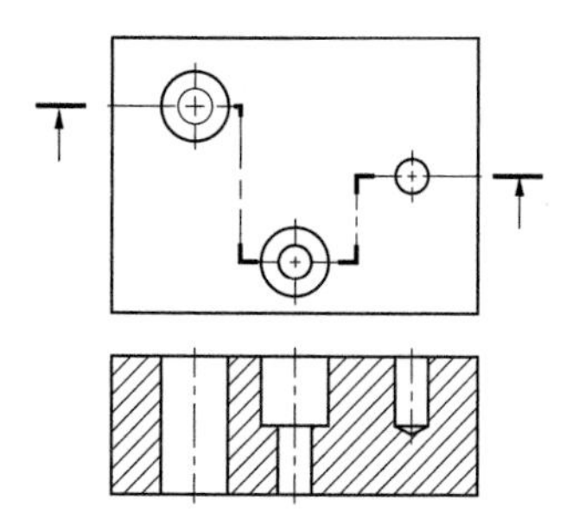

Figure 3.10 Offset section showing all holes as if they were in one plane

This plate has three holes that are not in line. Where should the section be taken to show the shape of each hole? The best location is through the centre of each hole, which requires three different cutting planes, as shown in Figure 3.9. Sections should be taken where they will provide the most information, so these are the correct locations; however, three drawings are not required. All sections can be shown on the same drawing as if the holes were in line. Figure 3.10 shows the cutting plane line (in the plan view) where each section is taken. This type of section is an **offset section.** When this is shown in the front view, there is no indication on the section that it is offset. Orthographic views are shown in Figure 3.10.

Because it is conventional practice to show offset sections in this way, the cutting plane line is sometimes omitted. When in doubt as to whether to show the cutting plane line, put it in.

Half Section

So far, all of the examples shown have been full sections—the section has been taken through the whole object. If there is an axis of symmetry, only one side need be drawn in section. An external view is drawn for the other half, and hidden lines are usually omitted on this portion. For obvious reasons, this is called a **half section**. A half section of a simple object is shown in Figure 3.11.

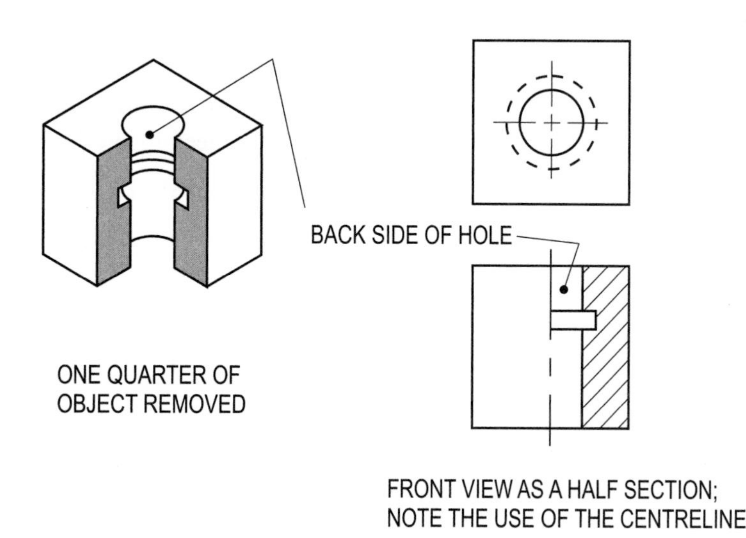

Figure 3.11 Half section used on a symmetrical object

Figure 3.12 shows a cylindrical part with internal features shown as a half section. A 90° segment of the cylinder is "removed," and the front view shows the interior details.

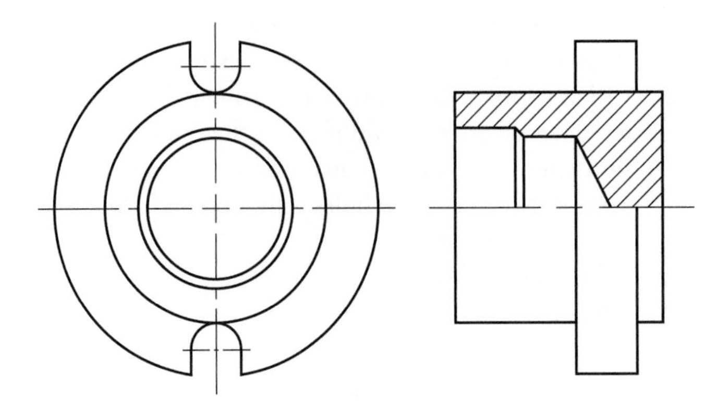

Figure 3.12 Half section

Broken-Out Section

Figure 3.13 shows a part with an axial hole and a small radial hole on the top. This could either be shown as a full section or as a half section. Since there are no other interior details, however, there is no reason to section the whole or even half the part. A section of the region around the holes is all that is required. A piece is "broken out" to show the holes. This is a **broken-out section**. Figure 3.13 shows plan and front views of the part. There is no need to show a cutting plane line, since its location is obvious.

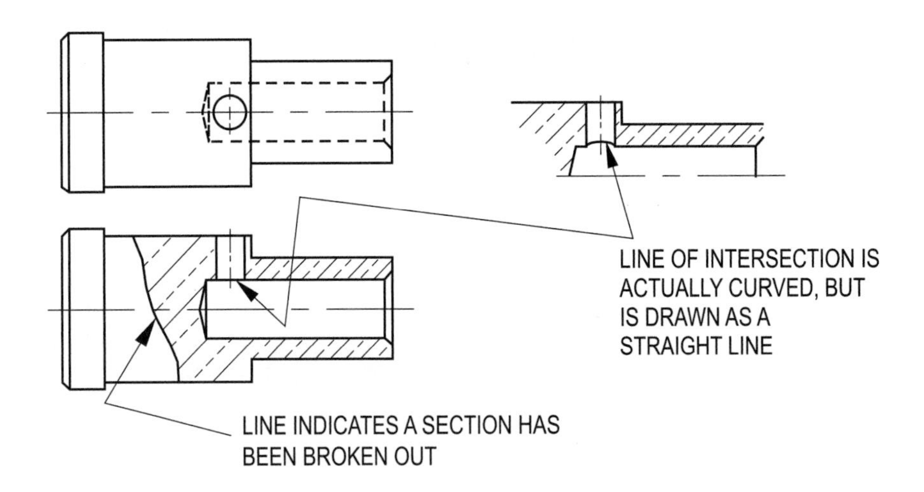

Figure 3.13 Broken-out section shows only the portion of the object with internal details

Revolved Section

A **revolved section** shows the cross-section revolved at 90° to the plane of the drawing. A revolved section can be drawn directly on the view, as shown in Figure 3.14. Only the shape where the section is taken is shown. A conventional side view could also be drawn, if required.

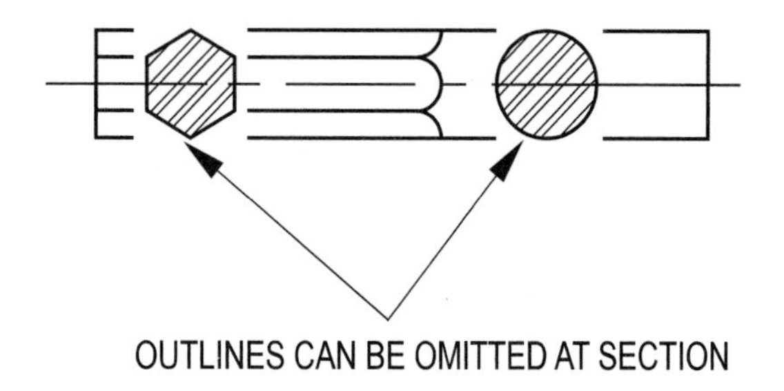

Figure 3.14 Revolved section showing the cross-section at two locations

Removed Section

A section view need not be located on the object itself, as with a revolved section. It can be located anywhere on the drawing or even on another sheet of paper. A **removed section** is simply a section located somewhere other than in a "normal" position. Figure 3.15 shows examples of removed sections. The cutting plane lines show where the section is taken and identify the section (Section A-A, Section B-B).

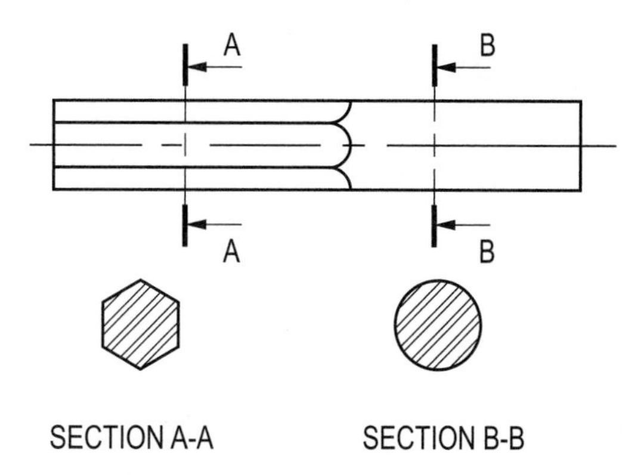

Figure 3.15 Removed sections

Sections of large items, such as buildings, are often drawn on another sheet of paper. The cutting plane line indicates where the section is taken. There is usually a note on the drawing with the cutting plane line indicating the drawing on which the section can be found. Figures 3.16 and 3.17 show examples of this.

Now let's move on to look at some standard drawing conventions that can be used in engineering graphics.

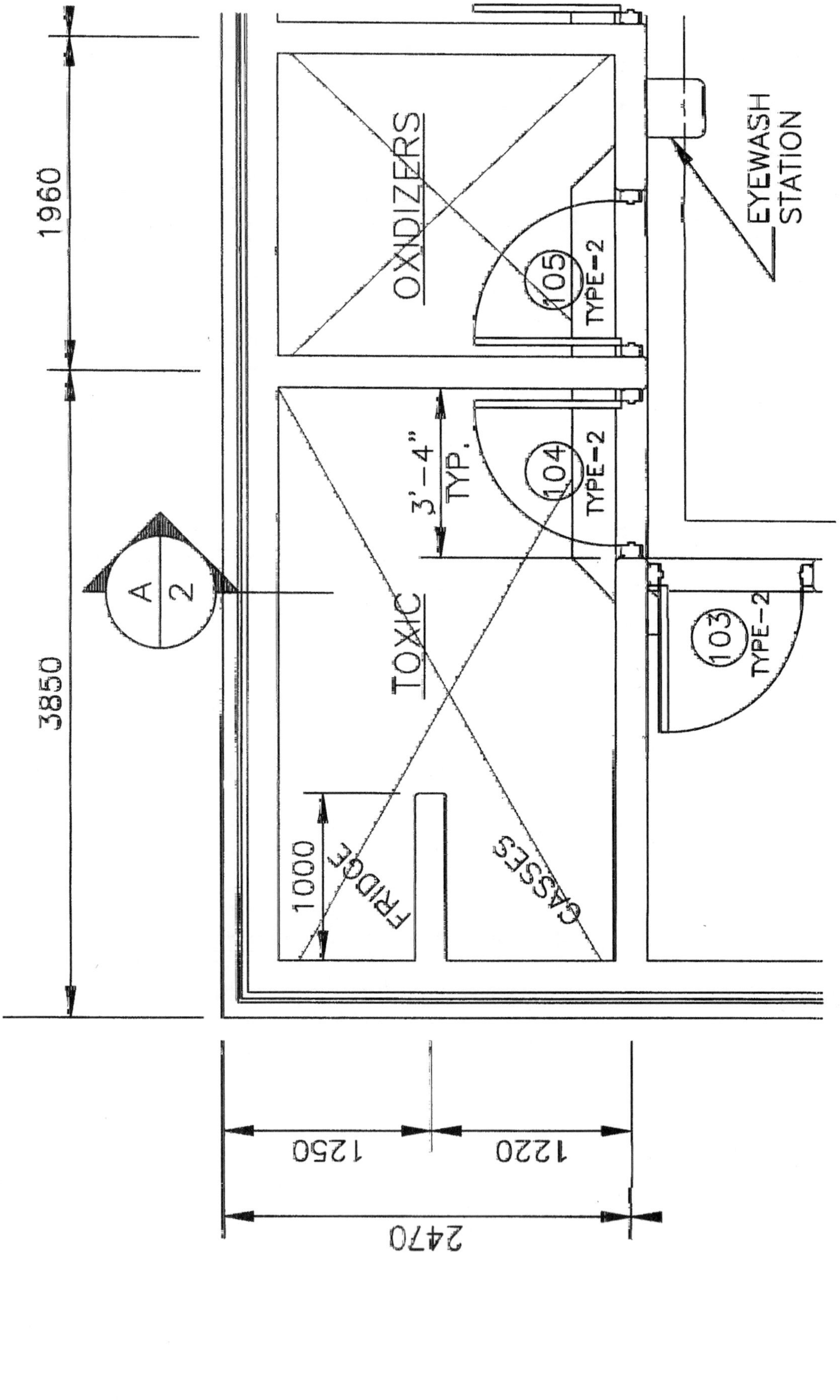

1960
3850
A
2
OXIDIZERS
105
TYPE-2
EYEWASH
STATION
3'-4"
TYP.
104
TYPE-2
TOXIC
103
TYPE-2
1000
FRIDGE
GASSES
2470
1250
1220

Figure 3.16 Plan view of a building showing where sections are taken

Figure 3.17 Section views of a building on another drawing

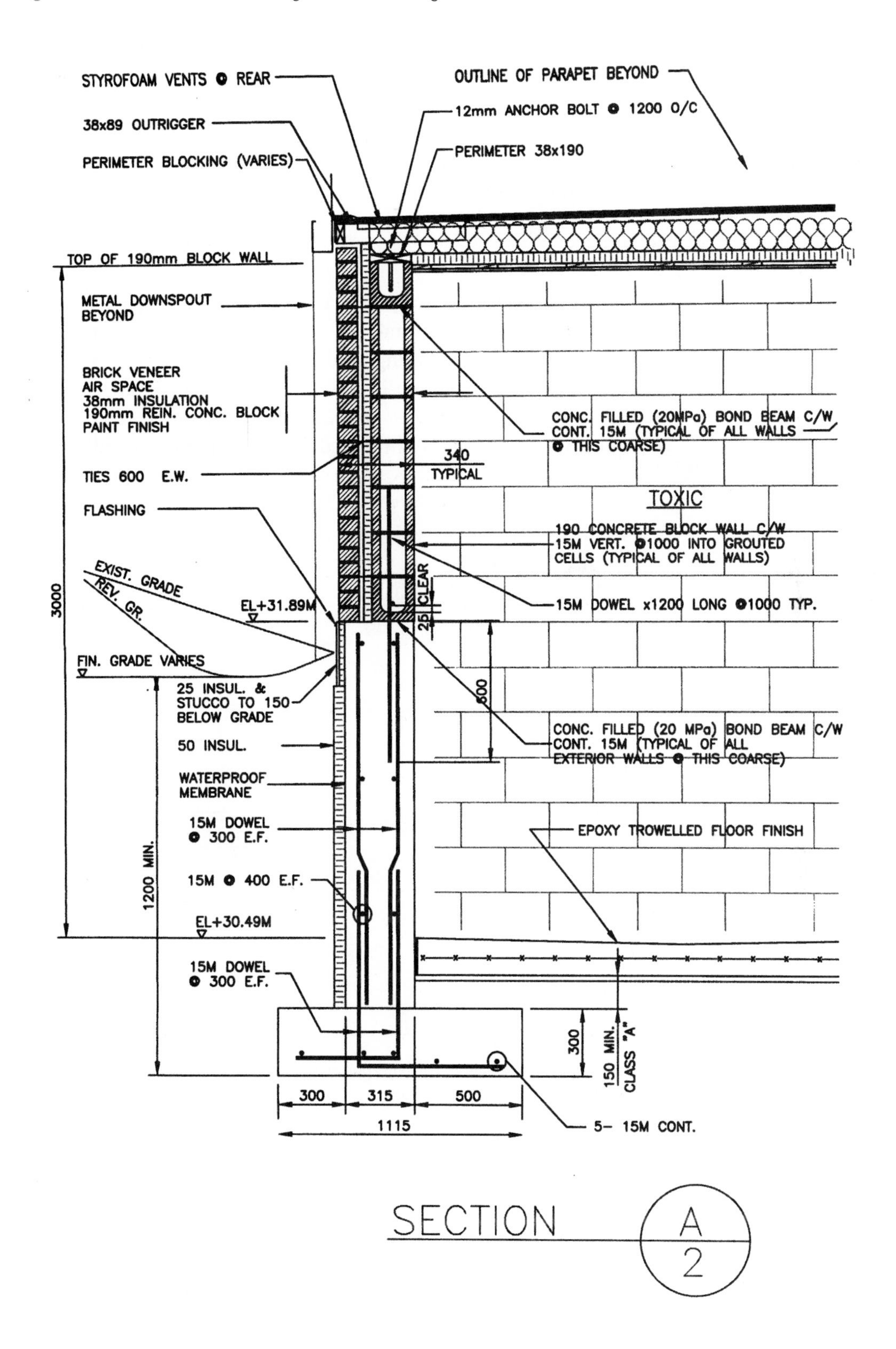
STYROFOAM VENTS @ REAR
OUTLINE OF PARAPET BEYOND
12mm ANCHOR BOLT @ 1200 O/C
38x89 OUTRIGGER
PERIMETER 38x190
PERIMETER BLOCKING (VARIES)
TOP OF 190mm BLOCK WALL
METAL DOWNSPOUT BEYOND
BRICK VENEER
AIR SPACE
38mm INSULATION
190mm REIN. CONC. BLOCK
PAINT FINISH
CONC. FILLED (20MPa) BOND BEAM C/W CONT. 15M (TYPICAL OF ALL WALLS @ THIS COARSE)
TIES 600 E.W.
340 TYPICAL
TOXIC
FLASHING
190 CONCRETE BLOCK WALL C/W 15M VERT. @1000 INTO GROUTED CELLS (TYPICAL OF ALL WALLS)
EXIST. GRADE
REV. GR.
EL+31.89M
25 CLEAR
15M DOWEL x1200 LONG @1000 TYP.
3000
FIN. GRADE VARIES
25 INSUL. & STUCCO TO 150 BELOW GRADE
600
CONC. FILLED (20 MPa) BOND BEAM C/W CONT. 15M (TYPICAL OF ALL EXTERIOR WALLS @ THIS COARSE)
50 INSUL.
WATERPROOF MEMBRANE
15M DOWEL @ 300 E.F.
EPOXY TROWELLED FLOOR FINISH
1200 MIN.
15M @ 400 E.F.
EL+30.49M
15M DOWEL @ 300 E.F.
300
150 MIN. CLASS "A"
300
315
500
1115
5- 15M CONT.
SECTION
A
2

STANDARD CONVENTIONS

In addition to providing standardization, conventions can make it easier to show certain features. Some features are not shown as they would actually appear but are simplified to make them easier to draw. To make sense of a drawing, both the person creating the drawing and the person reading it must know these conventions.

We will be looking at the standard conventions for conventional breaks, rotations, and thread symbols.

Breaks

If a part is long (a 1 m round shaft, for example), only the ends need to be drawn; however, there must be some way of showing that it is a long shaft. One could draw the whole shaft, but plotting a full-size drawing could present problems with regard to paper size. The solution is to draw the ends and a part of the centre with a **conventional break**. Examples of breaks used for different applications are shown in Figure 3.18. The length of the shaft is specified, but the full length is not drawn.

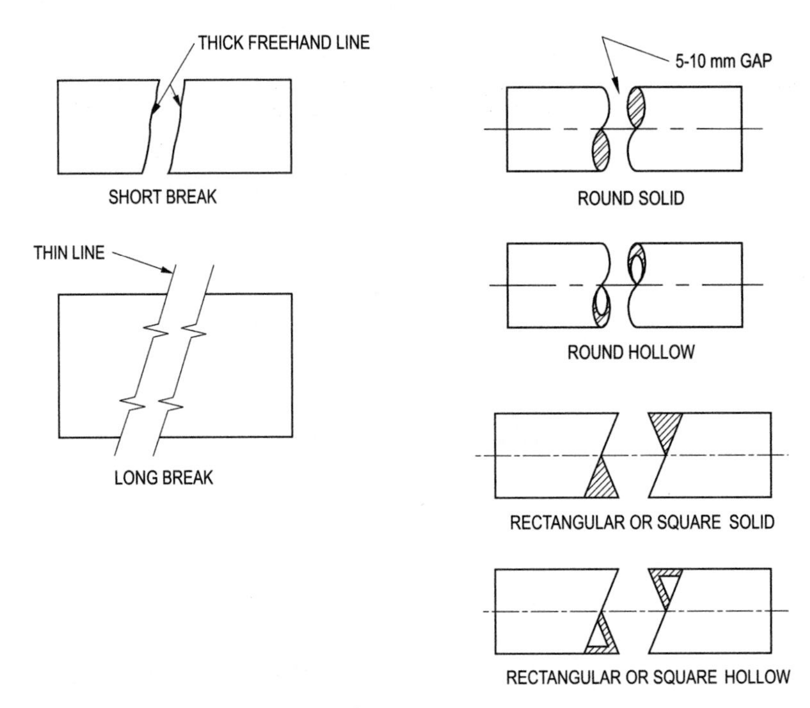

Figure 3.18 Break lines for different applications

The break line for a short break is a thick "wavy" line to distinguish it from outlines. If a computer is not used, break lines can be drawn freehand. It is not important that the shape be exact, as long as it is distinct from an outline. A long break line is a thin line.

Rotations

Figure 3.19 shows a circular plate with holes spaced at 120° (a common hole spacing). If the side view is drawn following the principles of projection, it would look as shown in Figure 3.19a. Since it is confusing to look at, the section is drawn as if the holes were rotated to a location where the true length of the hole diameter can be seen, as in Figure 3.19b. The true hole diameter is seen in both plan and section

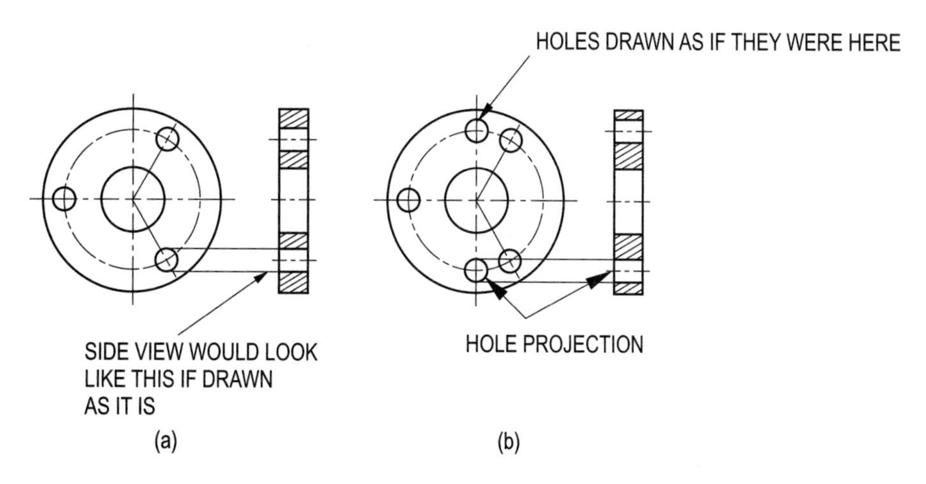

Figure 3.19 Rotation of holes, done to simplify drawing

views. The diameter at which the holes are located is seen in front and side views when drawn this way. Hidden lines are not included.

The same convention is used on other features arranged in a circular array. Figure 3.20 shows a common situation of supports (called **webs**) for a hub. The side view is confusing and time consuming to draw if corresponding points are projected from the front view. The side view is therefore simplified by rotating the webs so that they appear full size in the side view.

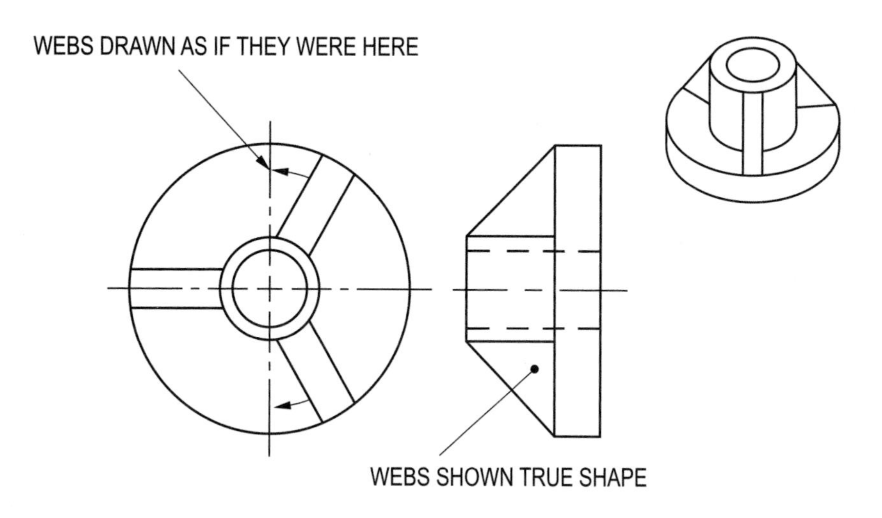

Figure 3.20 Rotation of webs to simplify drawing and show true shape of web

If a hub with webs is to be sectioned, the same convention applies. The right-side view is shown as a full section in Figure 3.21. The webs are rotated so that their true shape can be seen. Webs are not sectioned, but there are no internal details to show anyway. The cutting plane line is not shown in the front view because it passes through the centre of the object.

If there are webs as well as holes, both are rotated, as shown in Figure 3.22.

The spokes of a wheel present a similar case to webs. Spokes are not sectioned but are rotated so that they are seen in the section. A revolved section is used to show the cross-section of the spokes. Figure 3.23 shows a drawing of a handwheel with three spokes at 120°.

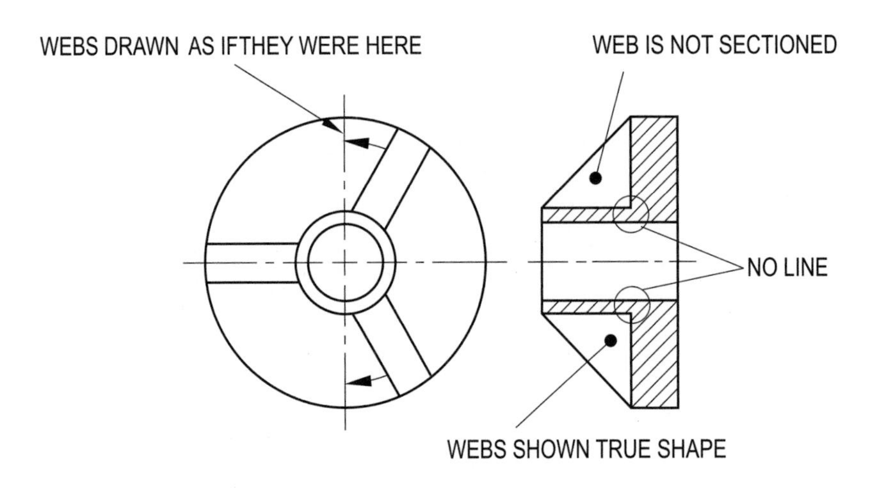

Figure 3.21 Sectioning of parts with webs. Webs are not sectioned

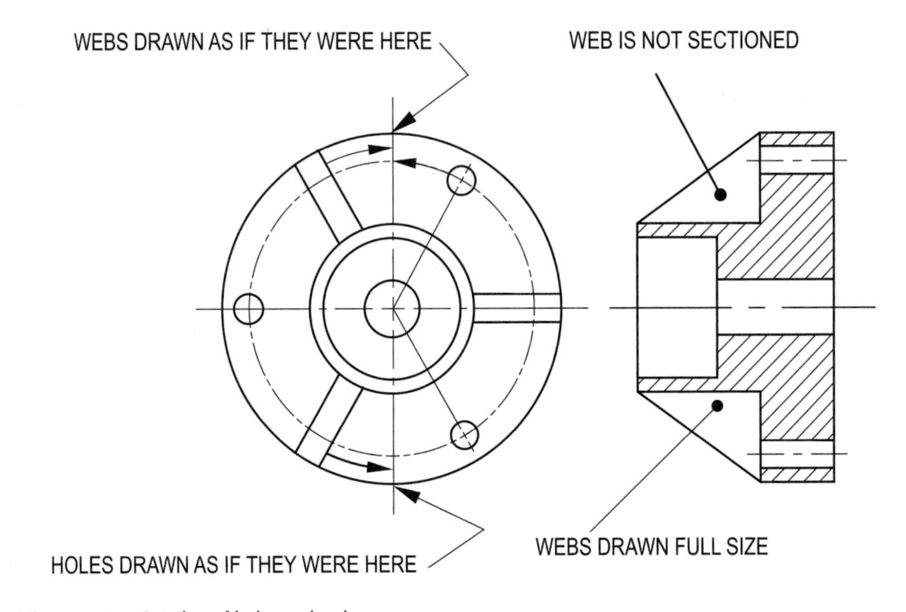

Figure 3.22 Rotation of holes and webs

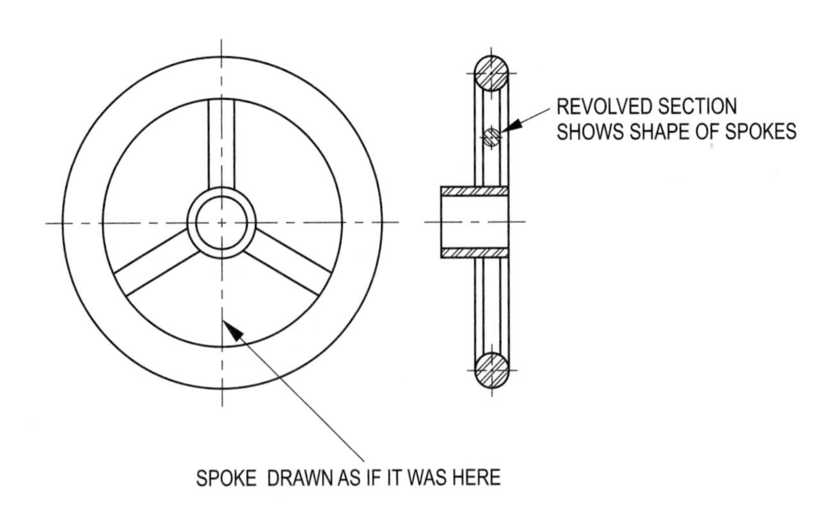

Figure 3.23 Rotation of spokes

The full section shows the cross-section of the rim and the hole in the hub as sections. The spokes are shown but not sectioned. Note that two spokes are shown because they are rotated. The cutting plane line has been omitted from the front view, since it is obvious where the section is taken. The section view would be the same no matter where the section was taken in this application.

Thread Symbols

For the sake of simplicity and speed, threads are represented by symbols so that each individual thread need not be drawn. The conventional way of representing threads is shown in Figure 3.24. Individual threads are not drawn. For an external thread, the outline is the outside diameter (called the **major diameter**), and a thin solid line represents the bottom of the thread (called the **minor diameter**). Threads are dealt with in more detail in Chapter 4; at this stage, all we want to do is show how they are represented in sections. The line representing the minor diameter is located to ensure that it "looks right," with respect to the outside diameter (Figure 3.24a). The distance between these two lines represents the height of the thread. This height can be found, but you would have to look it up. Thread specifications are given by a note.

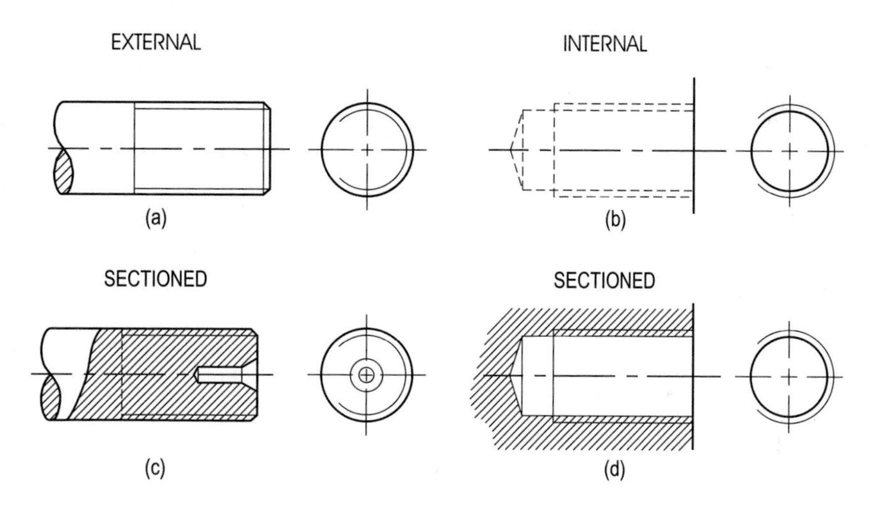

Figure 3.24 Conventional representation of threads

An internal thread (Figure 3.24b) is represented by broken lines for both major and minor diameters, since they are hidden. When an object with an external thread is sectioned to show interior details (Figure 3.24c), the minor diameter is a solid line. A section of an internal thread (Figure 3.24d) shows major and minor diameters as solid lines.

The thread is represented on the end view by an arc of about 270°, as in Figure 3.24. The arc, which is a thin line, represents the minor diameter of an external thread and the major diameter of an internal thread. The arc representing the threads should not start or end on a centreline.

The only information required to show a thread is the size and the length. Size is specified by a note giving the major diameter, the number of threads, and the length. Two other methods of representing threads are shown in Figure 3.25.

Schematic and simplified thread symbols are similar to conventional symbols, but dashed lines are used to represent major or minor diameters. The schematic symbol has more lines, which take longer to draw and plot, and the line type must be changed to draw hidden lines.

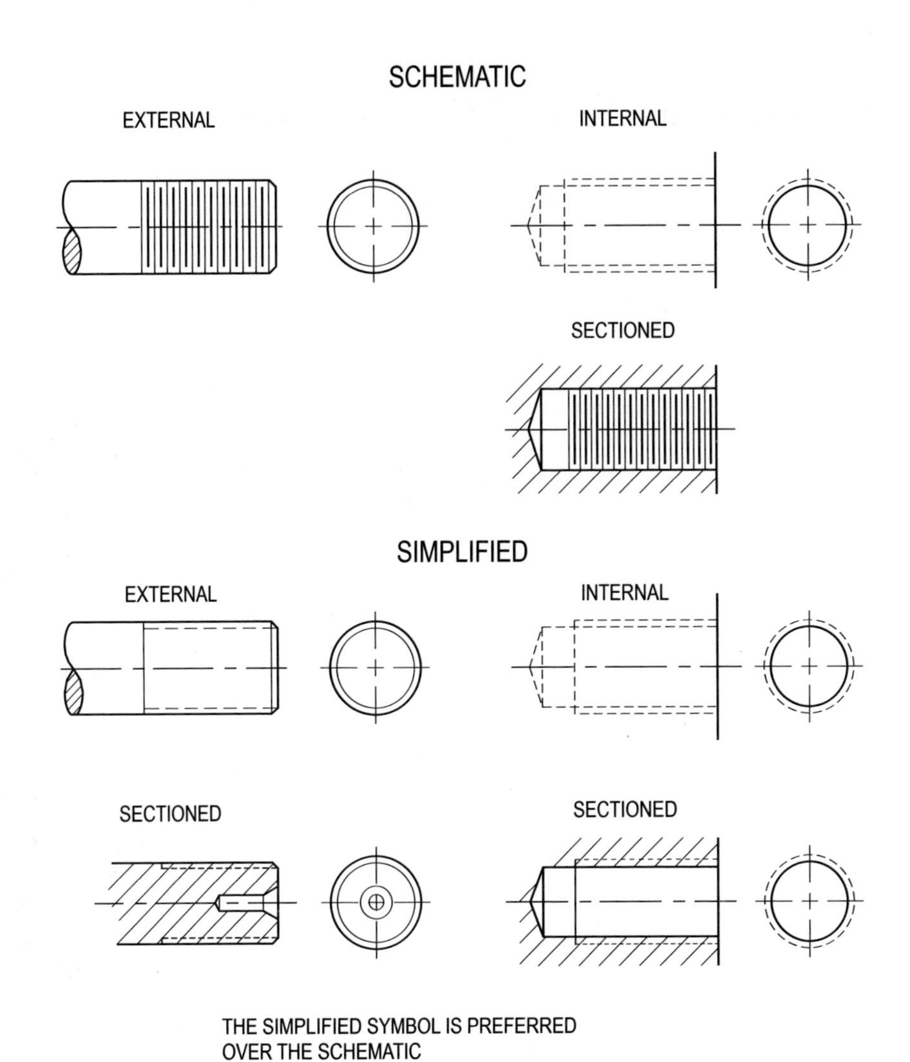

Figure 3.25 Schematic and simplified thread symbols

ASSEMBLIES

An assembly drawing shows how the parts are put together, and a section view shows internal details. Since there is more than one part involved, section lines must be varied to distinguish each part. The section lines on adjacent parts are drawn in different directions, as in Figure 3.26. Here, only one section line symbol is used, and material must be specified with a written note. Section line direction would be different on adjacent parts, even if different sectioning symbols were used. The section lines go in the same direction on the same part.

Not all parts of the assembly are sectioned. Sectioning standard parts or common parts, such as nuts and bolts, serves no purpose. Even if there are internal features, as in the case of a bearing (a standard part), they are of no concern to the person assembling the parts. Different section line symbols are used to show different materials, and the section line angle is different on adjacent parts.

Figure 3.27 shows the assembly of a shaft passing through the end of an aluminum pipe. The end assembly (steel) is bolted to a flange on the end of the pipe. The steel shaft slides in a bronze bushing held in the end assembly.

The full section shows all of the parts, but not all of them are sectioned. One part can have section lines in only one direction. The heavy black section between the alu-

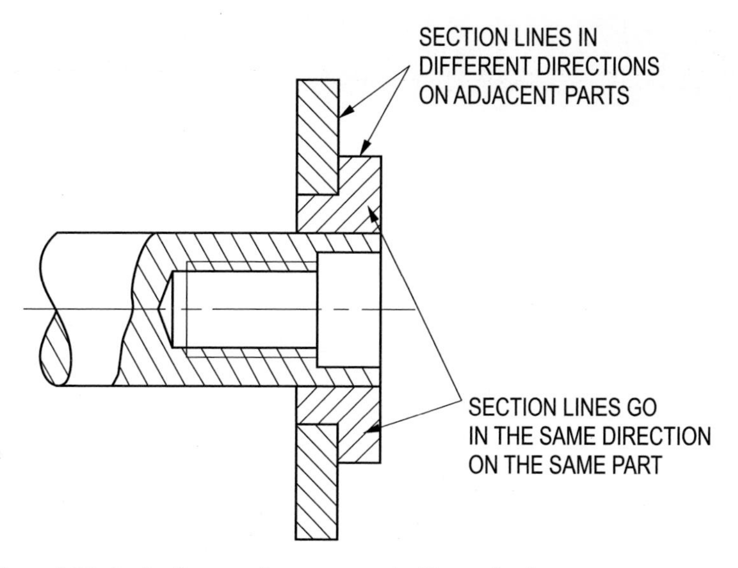

Figure 3.26 Section lines on adjacent parts go in different directions

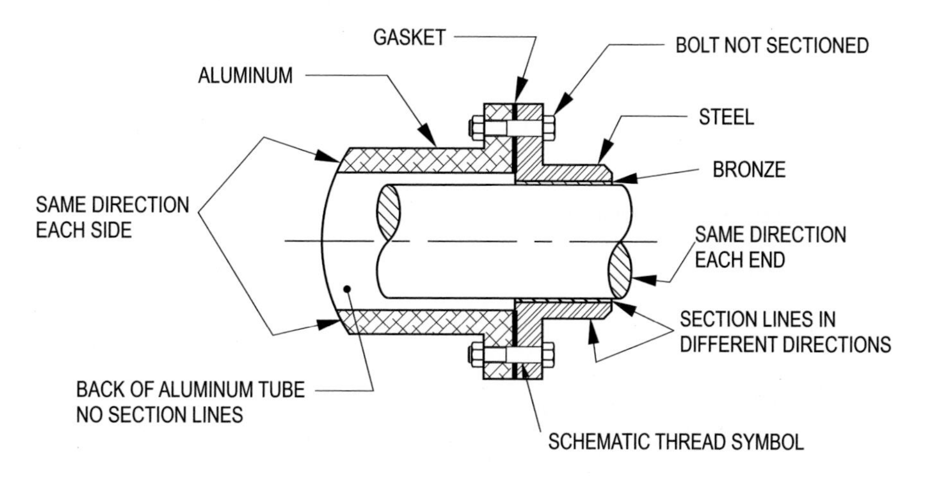

Figure 3.27 Some parts of an assembly are not sectioned

minum and the steel end cap represents a gasket to seal the end. It is too thin for any sectioning symbol, so it is simply filled in. This is the usual way of handling what is called a **thin section**.

Bolts holding the end cap to the aluminum flange are not sectioned. These are common standard parts, and there is no need to section them. The shaft is not sectioned, except where the conventional break on the left end indicates that the entire shaft is not shown.

Now that you know the basics of sectioning, try the following problems.

Problems

1. Draw the front view as a full section. Show the cutting plane line in the plan view. Material is steel.

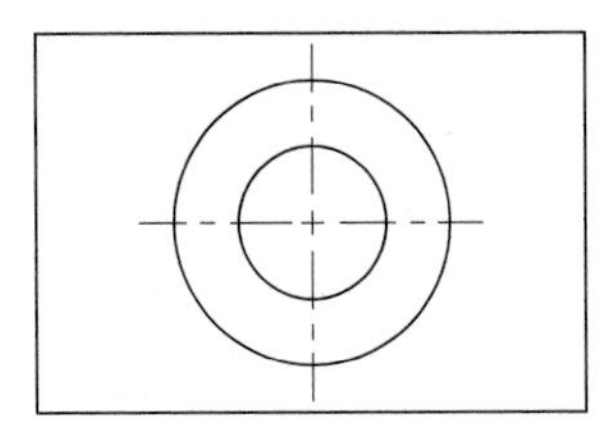

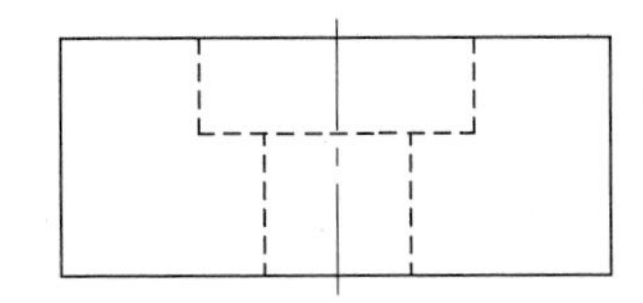

2. Draw a conventional break for the solid shaft.

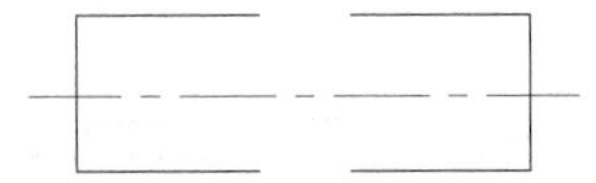

3. Draw a conventional break for the hollow shaft.

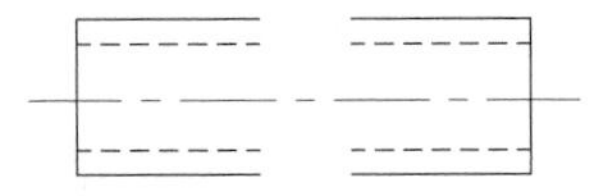

4. Draw the front view as a full section. Show the cutting plane line in the plan view. Material is brass.

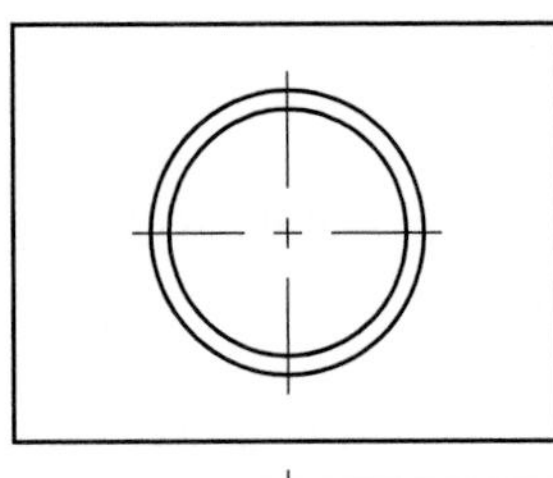

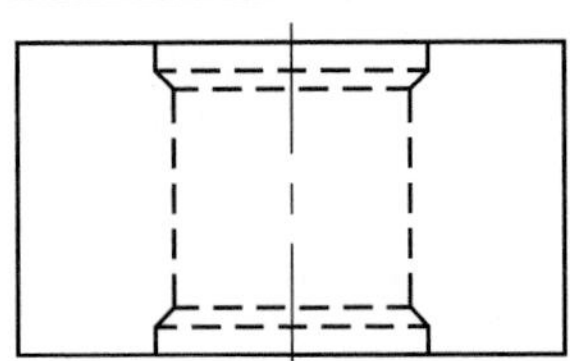

5. Complete the left-side view as a half section. Material is brass.

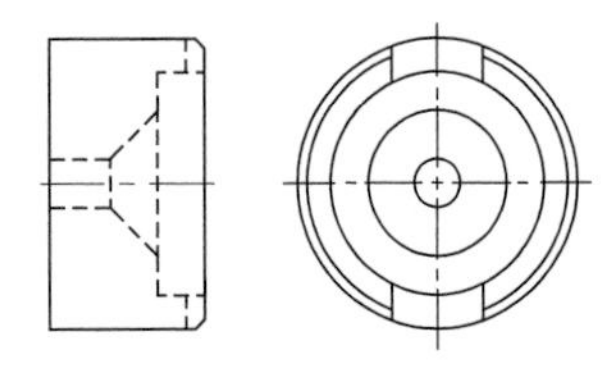

6. The plate has three holes at 120°. Complete the side view as a full section.

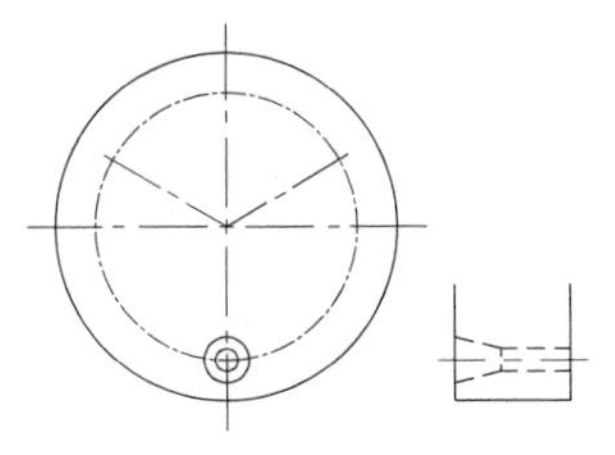

7. Add a broken-out section to show the oil hole in the top of the shaft. Material is steel.

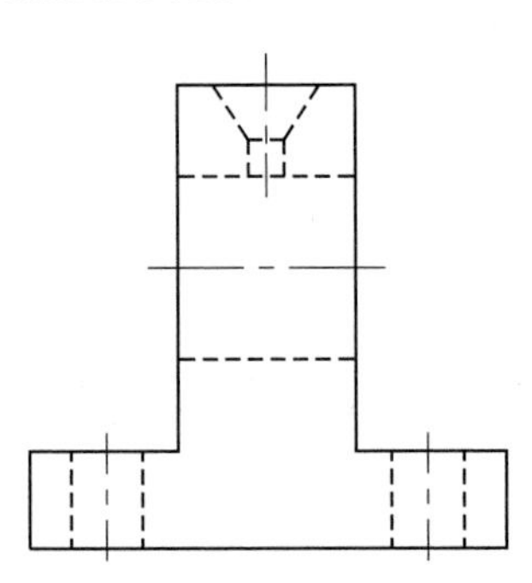

8. Draw the plan view with a broken-out section to show details of the axial and radial holes. The radial hole goes half way through the part.

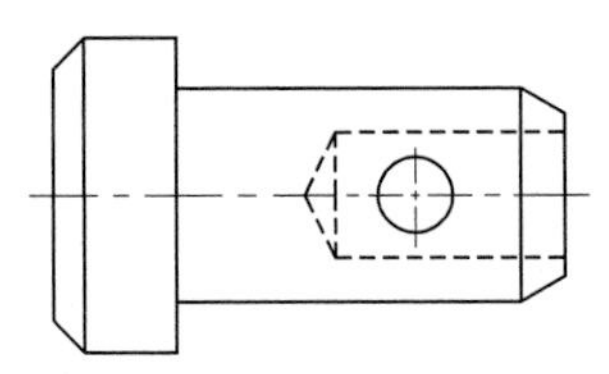

9. Complete the right-side view as a half section. Material is brass.

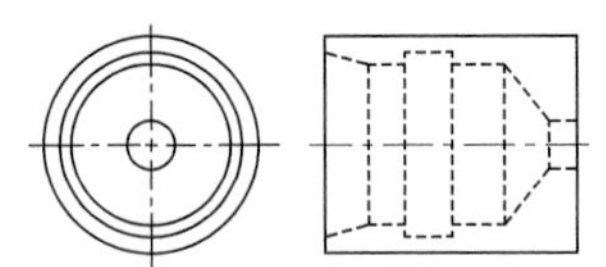

10. Draw the front view as a full section. Show the cutting plane line in the plan view. Material is aluminum.

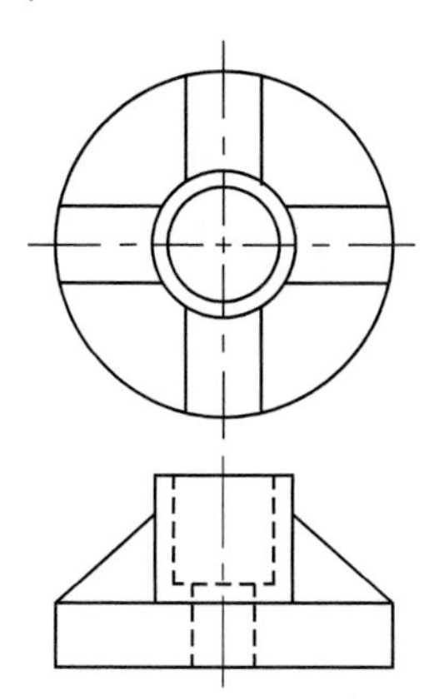

11. Draw the right-side view as a full section. Show the cutting plane line in the front view. Material is steel.

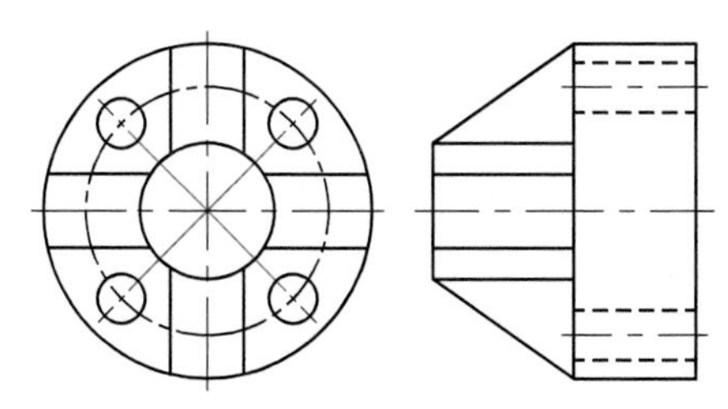

12. Complete the left-side view as a full section. Material is steel.

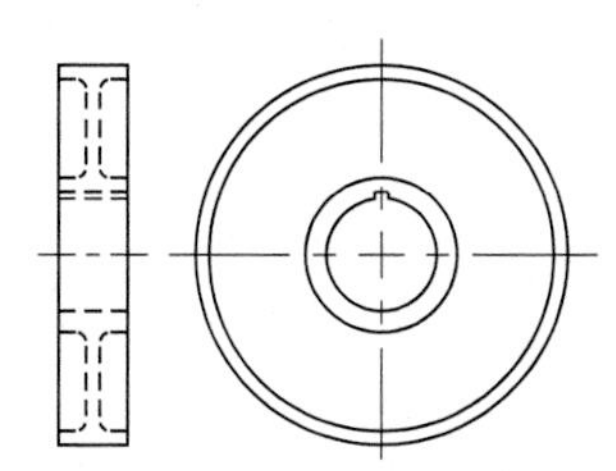

13. Complete the right-side view as a full section. Show the circular cross-section of the spokes with a revolved section.

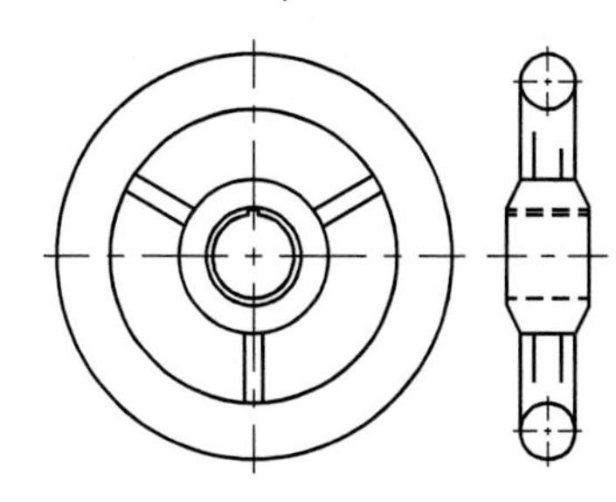

14. Complete the left-side view as a full section.

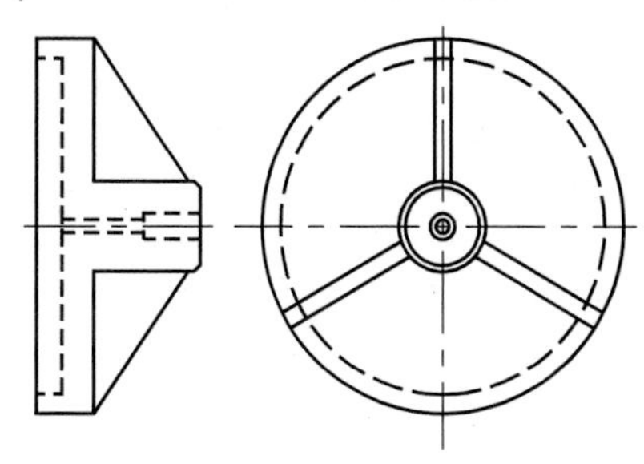

15. Complete the front view as an offset section to show the three holes. Show the cutting plane line in the plan view. Material is aluminum.

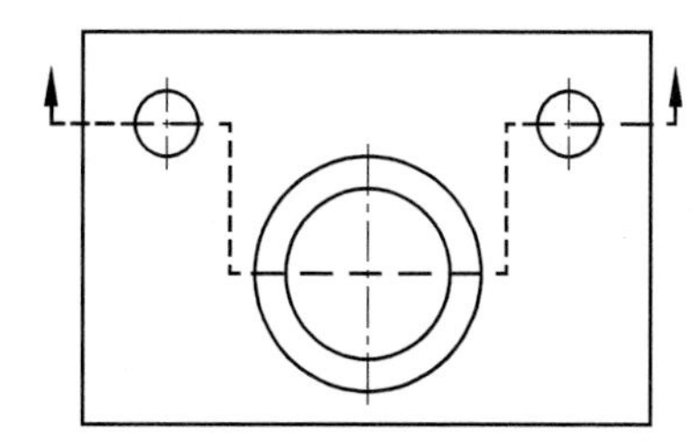

16. Draw the right-side view as a full section showing the internal and external threads.

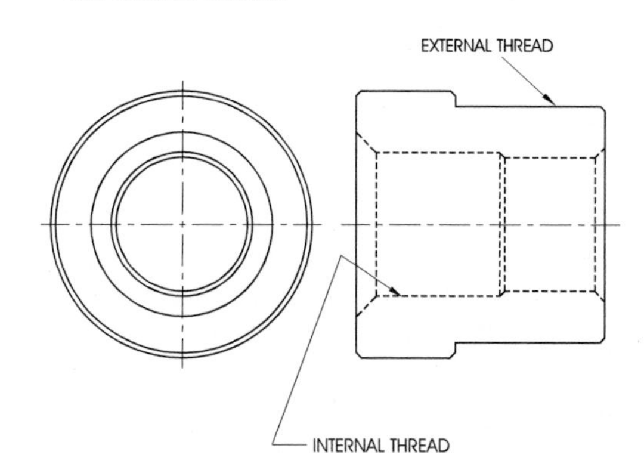

17. Draw the right-side view as a full section to show the holes.

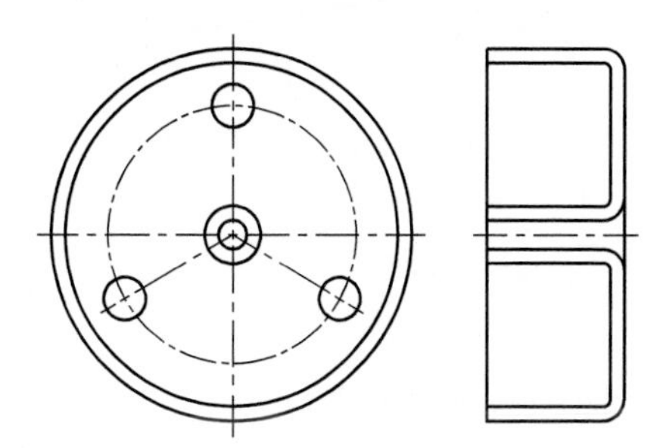

18. Complete the front view as a full section to show the horizontal holes (three).

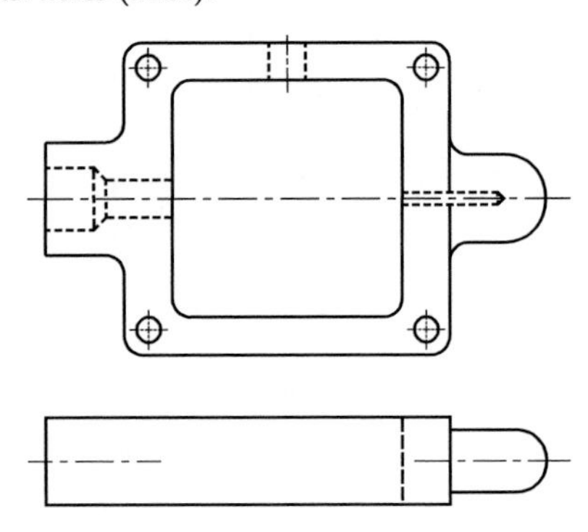

19. Complete the front view as a full section to show the shapes of the three holes.

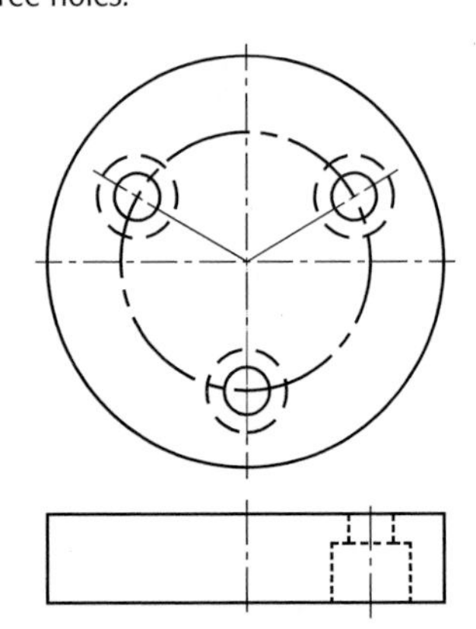

20. Complete the front view with the appropriate sections to show the oil holes in each end and the rectangular cross-section of the centre section. Material is steel.

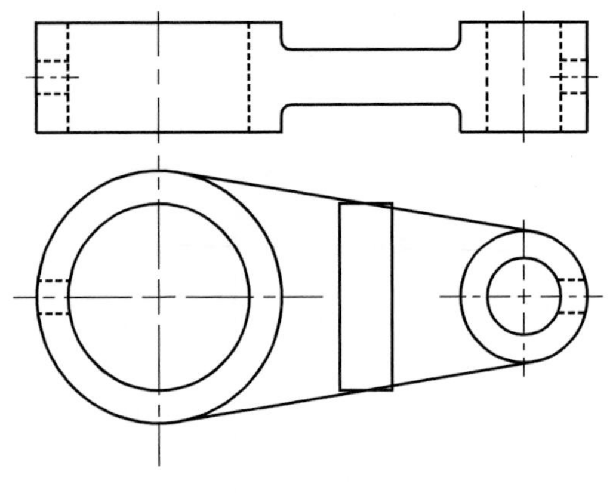

21. Complete the side view as a full section.

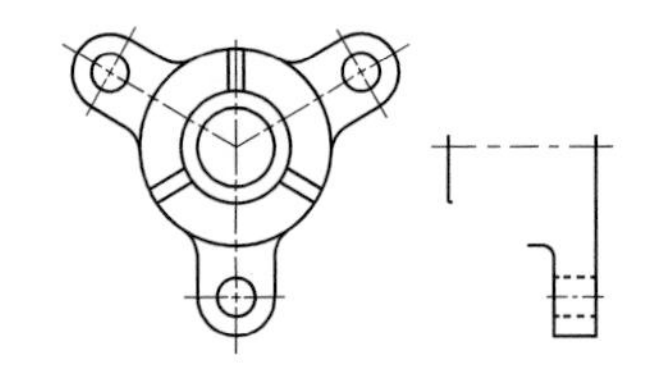

22. Complete the front view as a section view to show the holes. Show the offset cutting plane in the plan view.

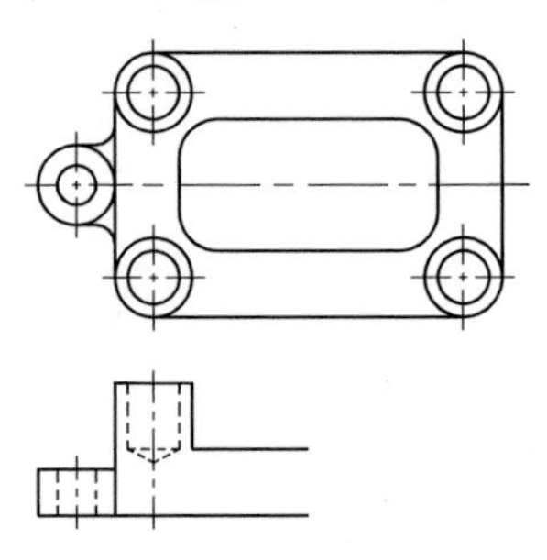

23. Complete the front view as a full section.

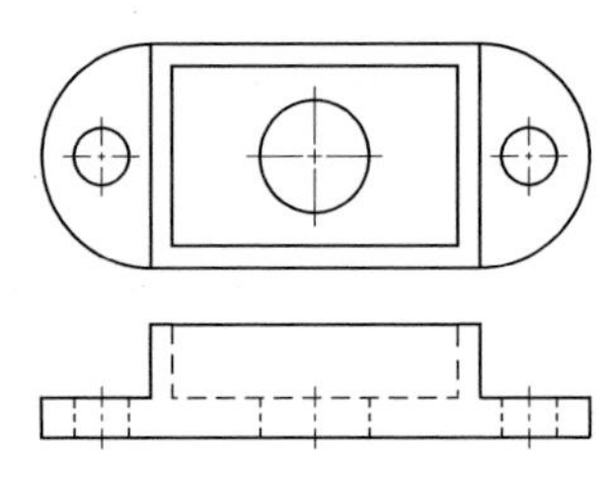

24. Draw section lines on the assembly. The body is steel, and the packing nut is brass.

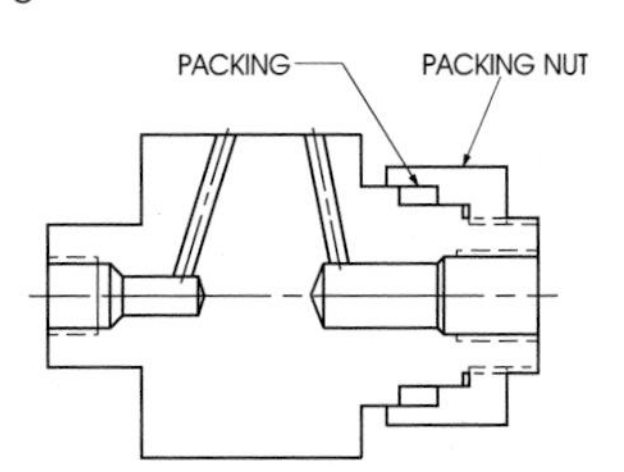

25. Section the body of the safety valve and the spring. The body is brass, and the spring is steel.

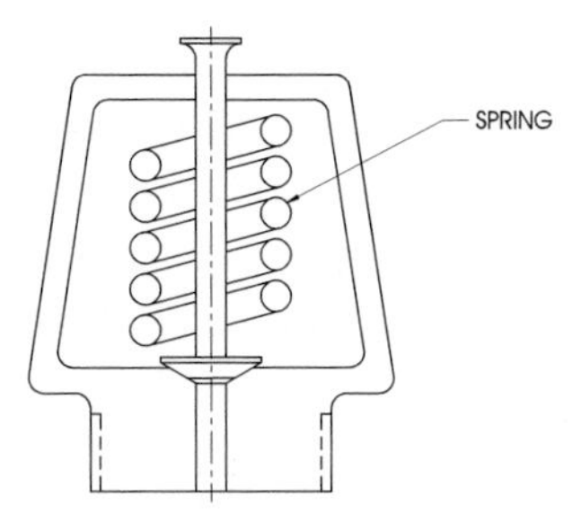

26. Complete the front view as a full section. Note that a portion is rotated. Draw revolved sections to show the elliptical shape of the arms.

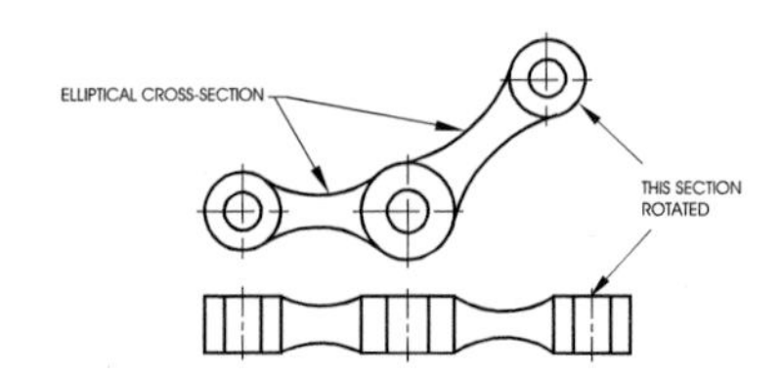

27. Use broken-out sections in the front view to show the details of the holes. Use a revolved section to show the rectangular centre section.

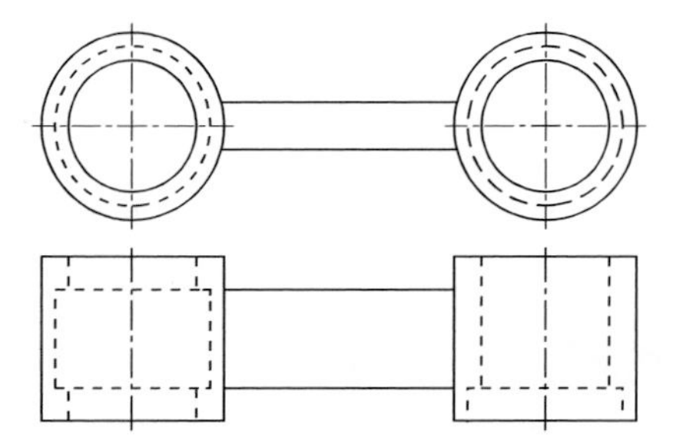

28. Complete the front view as a full section. Show the cutting plane line in the plan view. Material is brass.

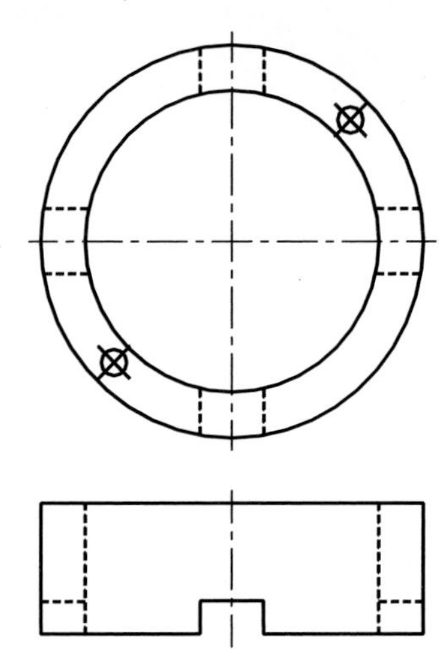

29. Complete the side view as a full section to show details of the hub and rim.

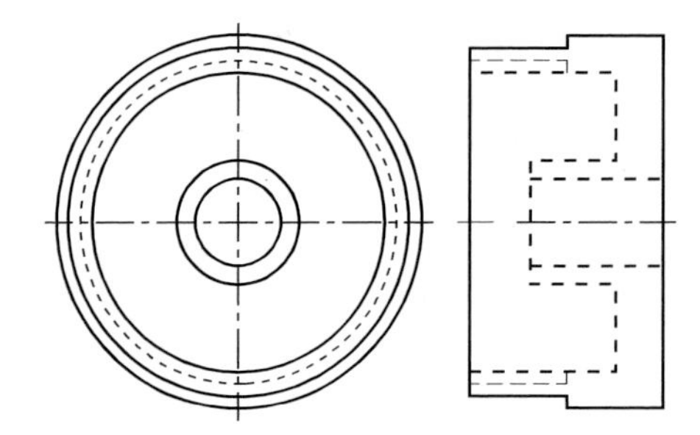

If the following problems are done by hand, copy the pages so that you do not have to redraw. There is enough information given to draw the required views. Dimensions may be scaled from these drawings. Some details are not given; use your judgment to determine these, based on what is given. If the material is stated, use the correct section line symbol.

The type of section used will depend on what is to be shown. You must decide the appropriate section technique to use. Some can be done in different ways. Use your judgment to decide on the best way, based on ease of drawing and communication of the information.

30. Use broken-out sections in the front view to show the nylon bushing in each end. The centre section is hollow and the walls are 2 mm thick. Use a revolved section to show the cross-section of the centre section. Material is steel.

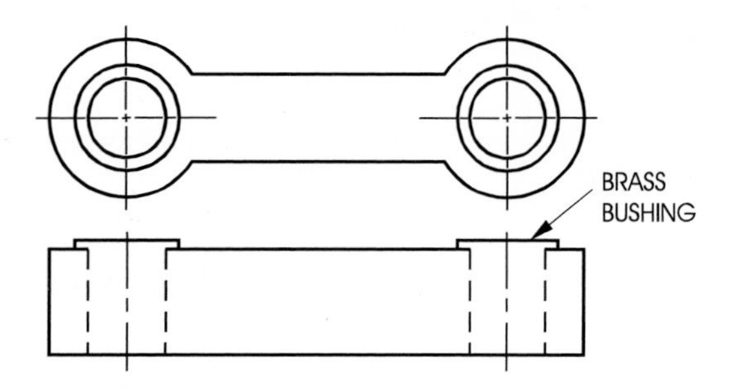

31. Complete the front view to show details of the hole and the slot.

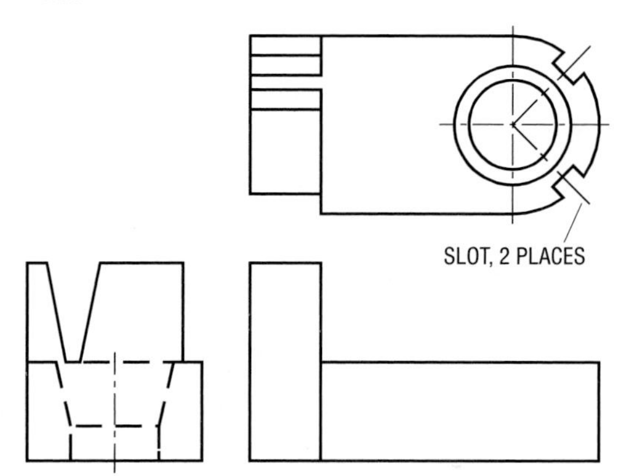

32. Complete the front view as a full section. Material is steel.

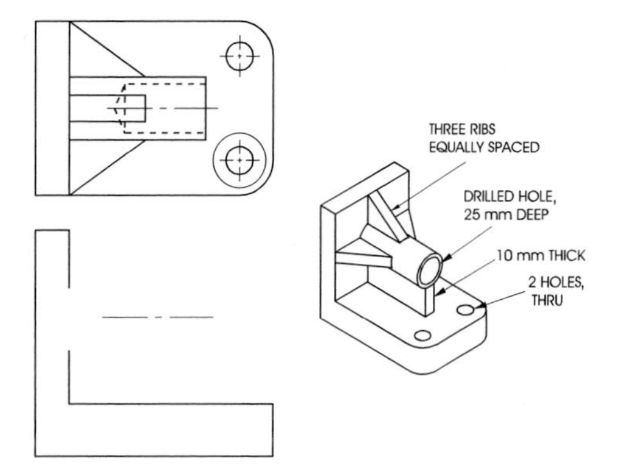

.....systematic dimensioning and tolerancing on drawings is necessary in any situation where design and manufacture are completely separated, so parts can be made from the information given and assembled without the need of fitting...

P. J. Booker, A History of Engineering Drawing, *Chatto & Windus, 1963*

CHAPTER

4 Dimensioning

Placing dimensions on engineering drawings and related documents is not a random process. It requires standards so that the information given can be understood by anyone reading the drawing.

STANDARDS

Various organizations publish standards that establish uniform methods of dimensioning and tolerancing for engineering and related documents. A **tolerance** is the maximum amount by which a specific dimension can vary and still be acceptable. The American Society of Mechanical Engineers Standard Dimensioning and Tolerancing, ASME Y14.5M, and the Canadian Standards Association Dimensioning and Tolerancing, CSA B78.2, are two examples of standards that apply to engineering drawings. The dimensioning sections of ASME Y14.5M and CSA-B78.2 are essentially the same, but there are minor differences. Other national standards that apply to various aspects of engineering documents are set by the ASME, the CSA, and other organizations. A list of some of these standards is given in Appendix D.

Computer programs have the capability to place dimensions; however, the user must specify what is to be dimensioned, where the dimension is to be placed, and how the dimensions are to be formatted in accordance with the applicable standard. Dimensions that are placed automatically by a computer program must often be edited to meet the accepted standard.

In addition to standards, it is important that you understand measurement units used on engineering drawings.

UNITS

SI Units

The common linear unit is the millimetre (spelled millimeter in the United States). The symbol for millimetre—mm—is used only in combination with a number (13 mm) and is always lowercase. The number and symbol are separated by a space.

Rules for using SI units and symbols are given in Appendix I.

U.S. Customary Units

The U.S. customary linear unit used on engineering drawings is the decimal inch.

If all dimensions on a drawing are in either millimetres or inches, you do not need to put the symbol after each dimension. The drawing should contain a note stating ALL DIMENSIONS IN MILLIMETRES (or INCHES).

DIMENSIONING TERMS

Many terms are used in specifying dimensions and tolerances; not all are dealt with here.

Figure 4.1 illustrates some of the terms that you will need.

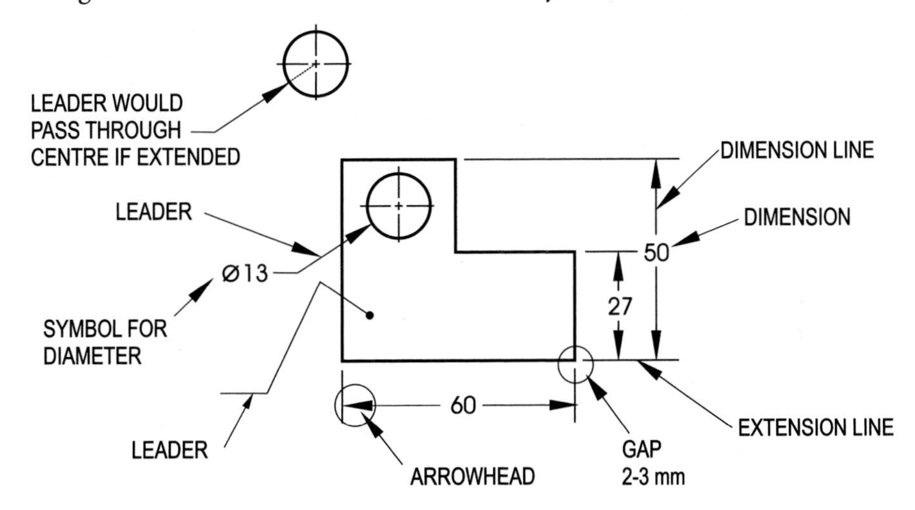

Figure 4.1 Dimensioning terms

Extension lines, also called **witness lines**, indicate the length to which the dimension given applies. They are short, light lines extending from the ends of the feature being dimensioned. They do not touch the object. There is a small, visible gap (about 2–3 mm) between the object and the extension line. Extension lines extend about 2–3 mm beyond the dimension line for the last related dimension.

Extension lines should not cross other extension lines or dimension lines, but this is not always possible. If an extension line must cross an outline, there is no break in the extension line. If an extension line must cross another extension line or a dimension line, the extension line can be broken.

Extension lines are usually perpendicular to the length being specified but can be at an oblique angle if there is some reason why they cannot be perpendicular (for example, crowding).

Dimension lines show the extent of the dimension. They are thin lines running parallel to the length being dimensioned. They can be straight (linear dimension) or curved (angular dimension). Dimension lines should not be closer than 10 mm from the object and should be at least 6 mm apart. They should not cross other dimension lines or the outline and should be placed outside of the object (although this is not always possible).

An **arrowhead** is usually placed at each end of a dimension line, although other terminators are sometimes used. The length of an arrowhead should equal the height of the numerals used for dimensions, and the width should be about one-third the length. Arrowheads may be filled or open, but the same style should be used throughout the drawing.

Dimensions specify the numerical value of the entity being dimensioned. Numbers should not be smaller than 3.5 mm high. The dimension should preferably be placed in a break in the dimension line, but may be placed above, and parallel to, the dimension line.

Notes give information about other things that must be specified, such as the material and quantity required. Notes are always printed in uppercase letters. To avoid misunderstanding, there are standards for handwritten letters and figures. Notes are usually placed on the right side of the drawing.

Leaders point to a specific feature on the object. If a leader identifies a circular feature, the part of the leader touching the feature must be radial. (It would pass through the centre if extended.) The horizontal portion of the leader points to the midpoint of the text or symbols that describe the feature. Leaders pointing to a feature terminate with an arrowhead, and those identifying a surface terminate with a dot. Figure 4.1 illustrates both of these terminations.

Now let's look at linear dimensions and tolerances.

LINEAR DIMENSIONS AND TOLERANCES

Linear Dimensions

Linear dimensions apply to straight lines or distances and specify the distance between two points. Dimensions can be in any direction. The direction—horizontal or vertical—refers to the paper: a horizontal line is parallel to the bottom of the paper; a vertical line is parallel to the side.

Figure 4.2 shows two methods of linear dimensioning: **chain (or continuous)** and **common point (or baseline) dimensioning**. The reason for choosing one method over the other will become apparent when tolerances are considered.

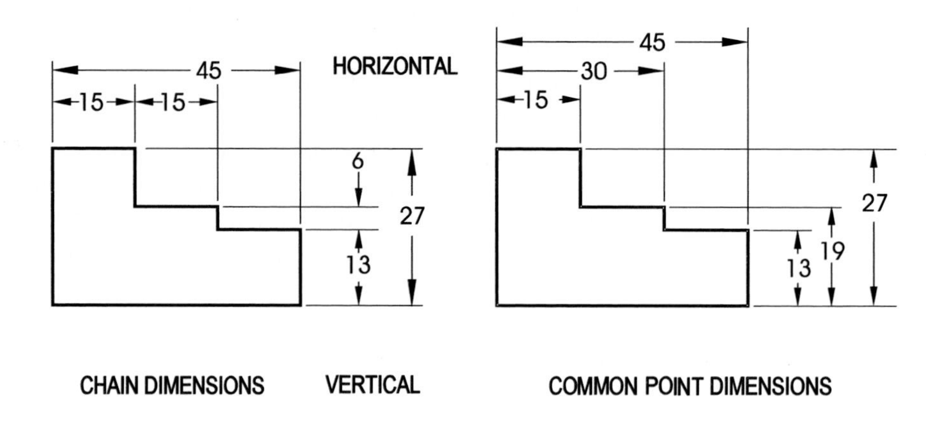

Figure 4.2 Chain and common point dimensioning

Chain dimensions are point-to-point dimensions. The starting point for one dimension is the end point of the previous dimension. Common point dimensions are referenced from one point, in this case the left edge. Note that one dimension is not given on this object. It is not necessary, since the dimensions that are given completely define the shape.

Tolerances

A **tolerance** is the maximum amount by which a measurement can vary and still be acceptable. The acceptable tolerance depends on the application. If you were cutting logs, a variation of 100 mm may be acceptable. In other applications, even a variation of 0.001 mm would be too large.

All measurements have some tolerance, however small. Even the most accurate atomic clocks have a variation of less than one part in 10 billion. It is advisable to allow the maximum tolerance possible, since costs increase as tolerances decrease.

Tolerances are important because parts must fit together. A simple example of how tolerances affect an assembly can be illustrated with a shaft that must fit in a hole. Figure 4.3 shows a round shaft, with diameter specified as 30 +/– 0.5 mm. Shaft diameter can be 0.5 mm greater (30.5 mm) or 0.5 mm smaller (29.5 mm) and still be acceptable. This method of tolerancing is called **plus and minus tolerancing**, and this is an example of **bilateral tolerancing**, since the diameter can increase or decrease. The tolerance is the maximum variation, in this case $2 \times 0.5 = 1$ mm. (For illustrative purposes, this is a large tolerance; dimensions are usually given to more than one decimal place.) It is assumed for the moment that the shaft has the correct form. That is, it is a perfect cylinder and that the sides are parallel. We will see how tolerance can be applied to the form later in this chapter.

This shaft must fit into a hole, which also has a tolerance of 1 mm.

The hole diameter is specified as 30 +/– 0.5 mm. The maximum hole diameter is 30.5 mm, and the minimum is 29.5 mm.

If the shaft is the maximum size, 30.5 mm (the **maximum material condition, MMC**) and the hole is the minimum size (29.5 mm) (the MMC for the hole), the shaft will not fit into the hole, as illustrated in Figure 4.3. This negative clearance is called **interference**. The shaft must be forced into the hole. Once this is done, it is difficult to take them apart. If only one assembly is required, the hole can be enlarged or the shaft diameter reduced so that they fit, but if there are thousands of assemblies, this is not practical. The smallest shaft, 29.5 mm (the **least material condition, LMC**), will fit into the largest hole, 30.5 mm (the LMC for the hole), but the clearance, 1 mm on the diameter, may be too large. With these tolerances, some shafts will not fit. The tolerance on the hole and shaft must be changed if every shaft must fit in any hole. Figure 4.4 shows how tolerance on the hole could be changed to allow every shaft to fit.

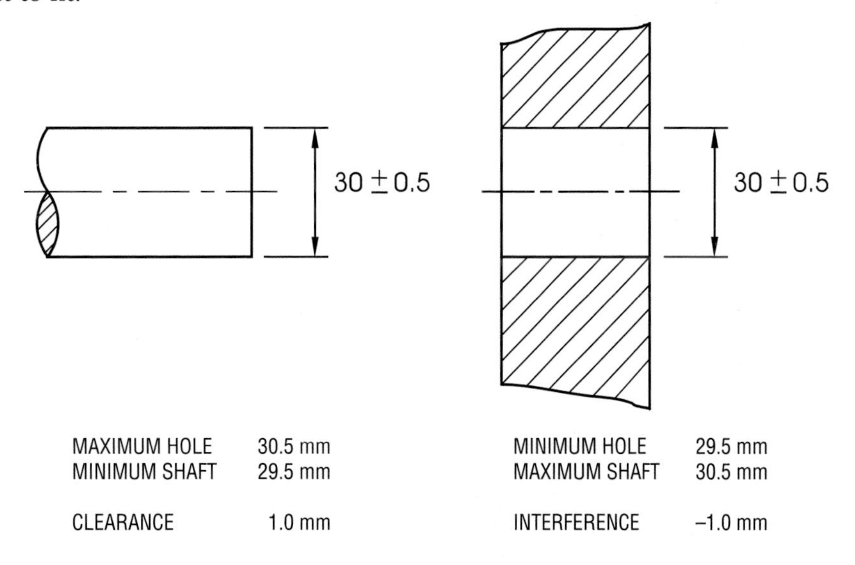

Figure 4.3 Tolerance on shaft and hole are such that some shafts will not fit

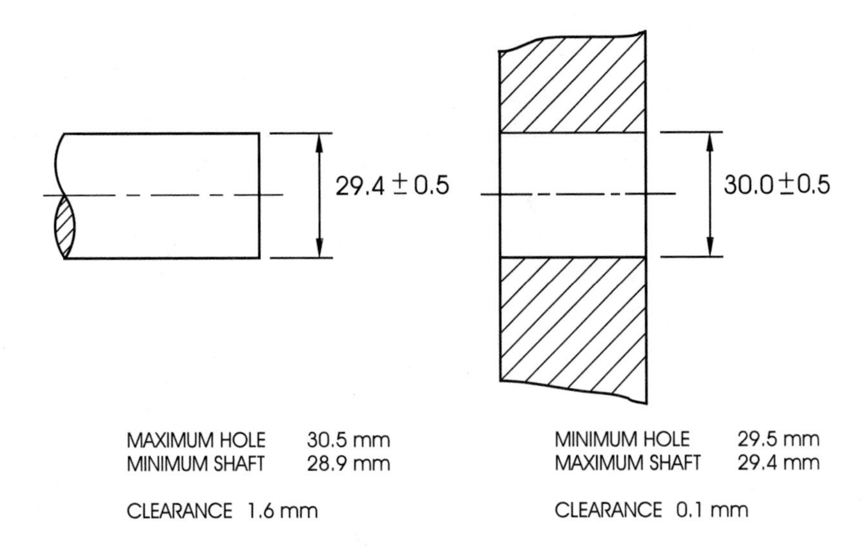

Figure 4.4 Tolerance on shaft adjusted so all shafts will fit

The shaft diameter is specified as 29.4 + 0.0 – 0.5. This is called a **unilateral tolerance**, since it can vary in only one "direction." The smallest shaft (28.9 mm) will have a clearance of 1.6 mm in the largest hole (30.5 mm diameter). This is the maximum clearance. The maximum diameter shaft (29.4 mm) will fit the smallest hole (29.5 mm) with a clearance of 0.1 mm on the diameter. This is the minimum clearance.

Not only are there tolerances on size dimensions, but there are also tolerances on location dimensions. If tolerances "add up," and sometimes they will, some parts will not fit together, while others may be too loose. This is called **tolerance accumulation** and can be minimized by choosing the type of dimensioning system used. Figure 4.5 shows how tolerances can accumulate to affect length and position using chain dimensions. The dimension of each length is given, starting at the left side and proceeding to the right.

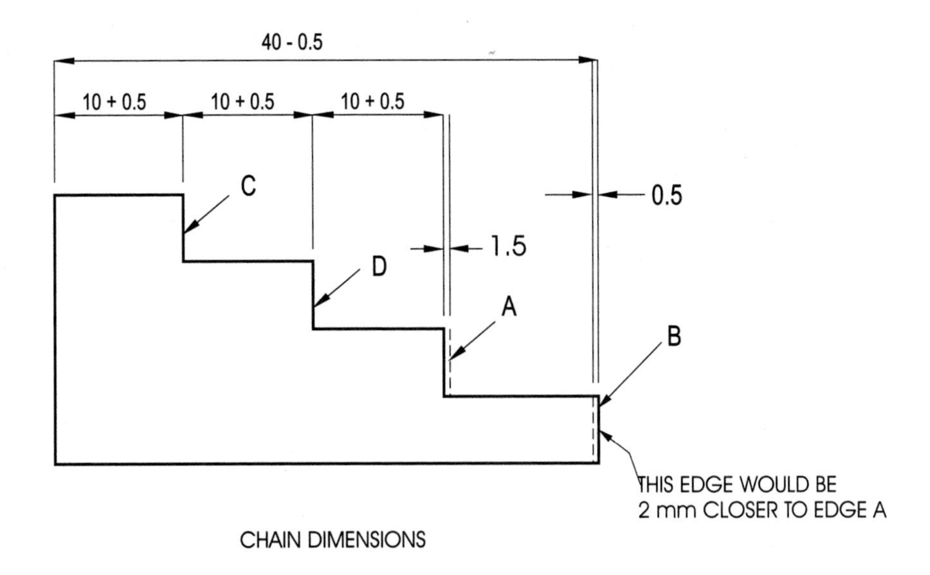

Figure 4.5 Chain dimensions give the greatest accumulation of tolerances

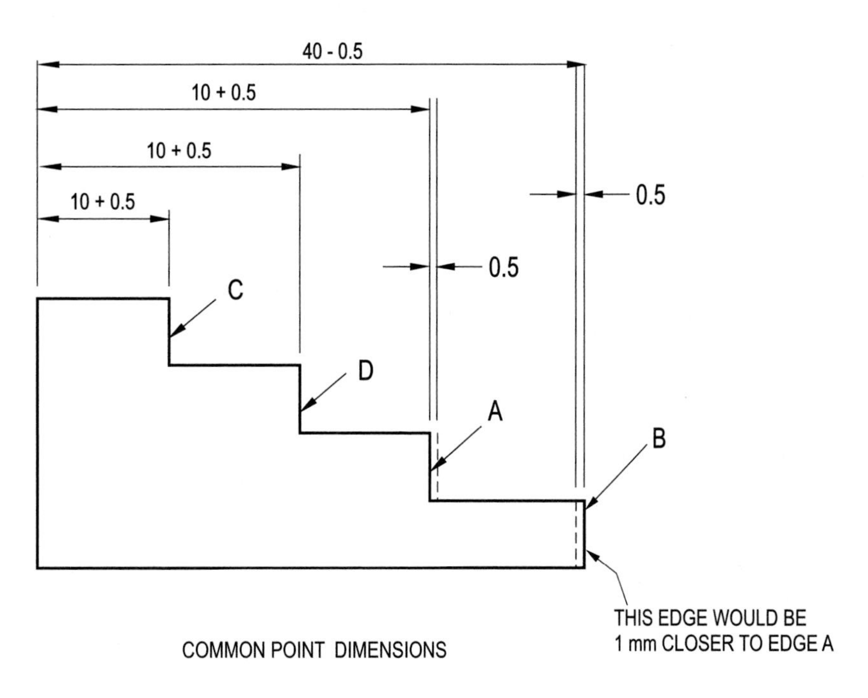

Figure 4.6 Common point dimensioning reduces tolerance accumulation

If each length has a tolerance of 1 mm (+/– 0.5 mm), edge B could be 2 mm (1.5 mm on the left and 0.5 mm on the right) closer to edge A if chain dimensioning is used, as in Figure 4.5. Tolerance accumulation is reduced if common point dimensioning is used, as in Figure 4.6, where edge B could be 1 mm closer to edge A using common point dimensioning.

Tolerances on dimensions are called **direct tolerancing**, and tolerances on shapes are called **geometric tolerancing**, for example, how flat is flat, how round is round, or how straight is straight. Geometric tolerancing will be dealt with in more detail later. We will look at how direct tolerances are called up on drawings.

Direct Tolerancing

Limit Dimensions If a dimension can vary between some maximum and some minimum, both dimensions are given in the dimension line. The most common way to do this is to place the maximum limit above the minimum, as shown in Figure 4.7a, but they can be written in a single line, as shown in 4.7b. In this case, the lower limit and upper limit are separated by a dash. Both upper and lower limits must have the same number of decimal places.

In some cases, there is only one limit if only one length is important. An example is the depth of a hole. The actual depth is not critical as long as it is deep enough to serve the intended purpose, that is, some minimum depth. The dimension is given and identified as either MAX or MIN as shown in Figure 4.7c.

Plus and Minus Tolerances We saw this in the previous example in Figure 4.4. The length is specified, and the deviations, either greater or less than this value, are given. The length given is called the **basic size**. Figure 4.8 shows two ways to do this. The first method (Figure 4.8a) shows a bilateral tolerance. The basic size is 25.50 mm, and this can increase by 0.10 mm or decrease by 0.30 mm and still be acceptable. If the positive and negative deviations are the same, the base size and tolerance are as shown in Figure 4.8b. When a bilateral tolerance is used, both the plus and minus

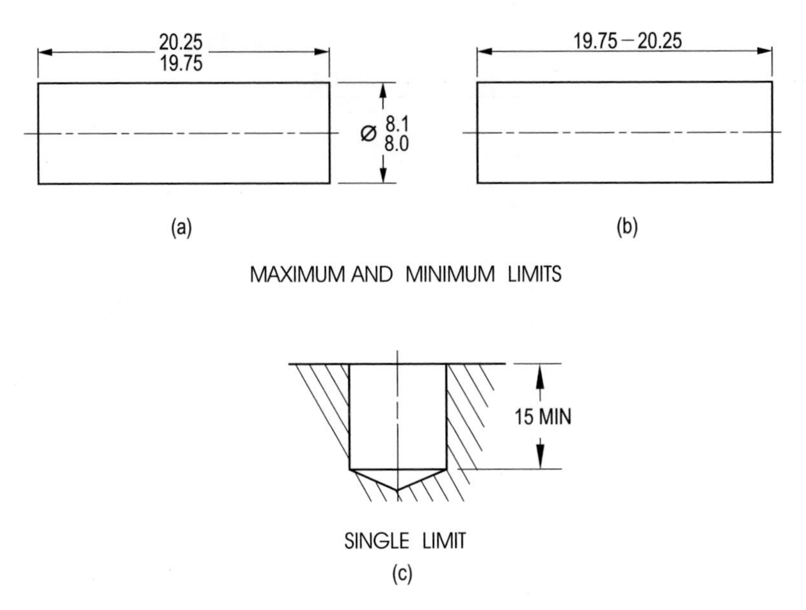

Figure 4.7 Limit tolerancing gives maximum and minimum limits

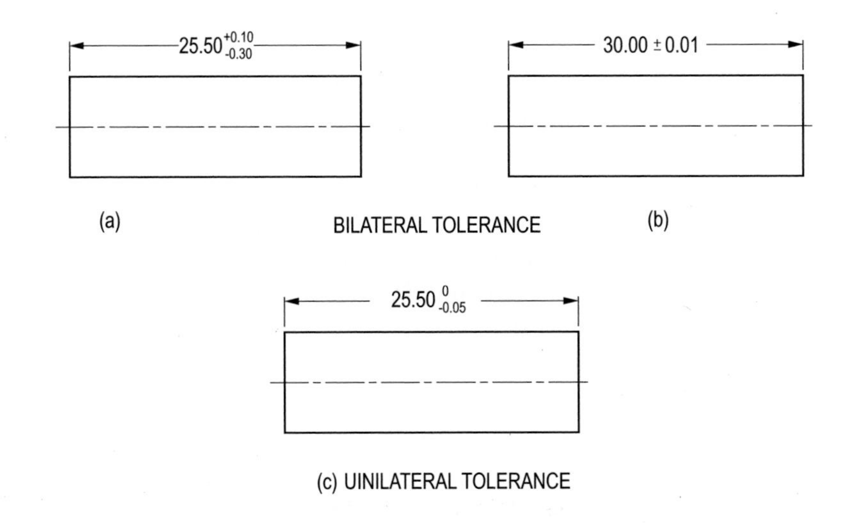

Figure 4.8 Specifying bilateral and unilateral tolerances

values have the same number of decimal places, using zeros where necessary. When a unilateral tolerance is used, a single zero is shown without a plus or minus sign, as in Figure 4.8c.

While all dimensions have a tolerance, it is not necessary to include a tolerance on every dimension on a drawing. Tolerances can be specified by a general note on a drawing. An example is:

> THE FOLLOWING TOLERANCES SHALL APPLY FOR ALL DIMENSIONS UNLESS OTHERWISE SPECIFIED.
>
> LINEAR DIMENSIONS +/– 0.2 mm
>
> ANGLES +/– 2°

This would cover all dimensions except those for which a specific tolerance was required.

There is much more to tolerancing than what is covered here. Our intention here is to give you an introduction and show how direct tolerances appear on engineering drawing.

RULES FOR DIMENSIONING

Because dimensioning must be done in a uniform way, rules and conventions have been established to ensure that everyone uses the same language. The rules given here are based on ASME Y14.5M and CSA-B78.2 standards.

1. Dimensioning must be complete. There must be no information missing. The user must not be required to make any assumptions or measure anything directly on the drawing. (This is called **scaling the drawing.**) There must be only one possible interpretation.

   Because information is missing from the left-hand drawing in Figure 4.9, it is possible to interpret the dimensions in two different ways.

   In Figure 4.9a, the width of the cutout is given (30 mm), but the depth is not. The dimensions across the top give an overall width of 60 mm, but the width stated at the bottom is 55 mm. This is ambiguous. Note how much clearer Figure 4.9b is.

2. Do not add extra dimensions. You cannot dimension everything and "hope for the best." The overall length is always given. Some dimensions are not given explicitly. There was example of this is in Figure 4.2 (page 94), where one dimension was not specified, because other dimensions completely defined the shape. In Figure 4.10a, the dimension on the sloping side of the

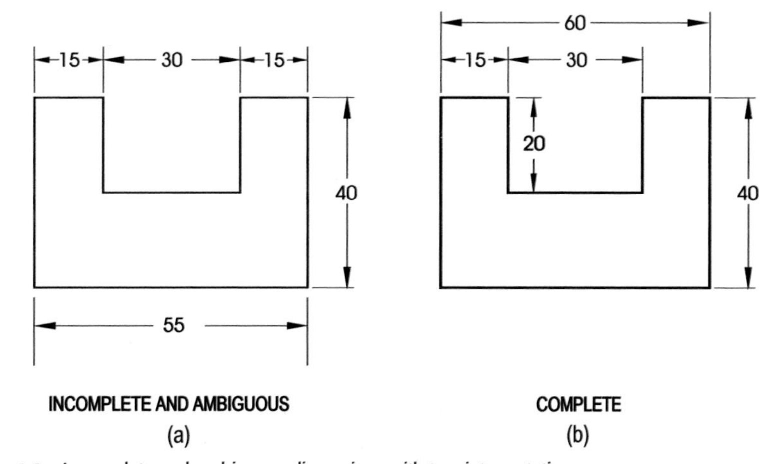

Figure 4.9 Incomplete and ambiguous dimensions with two interpretations

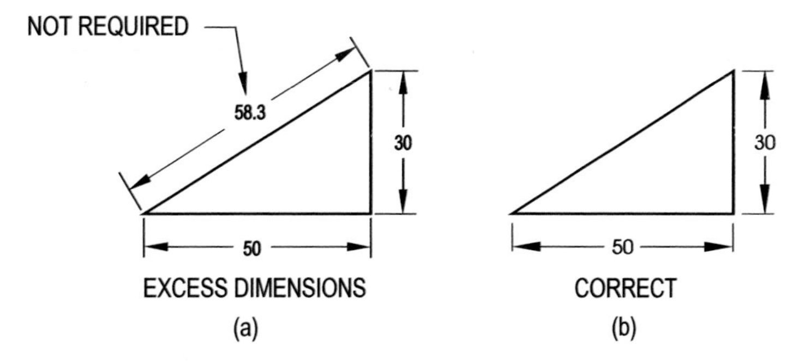

Figure 4.10 Too many dimensions given

drawing (58.3 mm) is not necessary, since the information required is given by the vertical (30 mm) and horizontal (50 mm) dimensions. (In some situations, extra dimensions are helpful. This will be dealt with later.)

3. Show dimensions on the true profile view, and refer to visible outlines. Do not dimension to hidden lines. There are ways to avoid dimensioning to hidden lines. One is to locate the dimension on another view; another is to use a section view.

   The only length that can be seen clearly in the side view (Figure 4.11) is the thickness (12 mm). The size of the hole (20 mm) should not be given on the side view because it is dimensioned to a hidden line. The distance from the bottom to the centreline of the hole (20 mm)—or is it the distance to the corner?—should be given where the shape is seen in profile, in the front view (Figure 4.12). The location of the hole centreline should be given where the hole is seen as a circle. The height (40 mm) can be seen in the side view, but there is no way of telling what the shape is in this view. The shape of the part is not seen in profile in this view. Height should be shown on the front view.

4. Dimensions should be arranged for maximum readability. Figure 4.13 shows the same item dimensioned in two ways. All information is given for both drawings, but the information is disorganized on Figure 4.13a.

   Note that the disorganized dimensions are placed randomly. Some horizontal dimensions are on the top of the part and others on the bottom. When dimensions are organized, as in Figure 4.13b, they are easier to read and understand. Horizontal dimensions can be grouped together at either the top or the bottom of the part. The vertical dimensions should be grouped on the right side. That is where the features (the "steps") are. If there were a right-side view, the dimensions would be between the front and side views. Vertical dimensions are usually located between the front and side views so that they can easily be applied to both views.

5. Parts should be described without specifying the method of manufacture. The diameter of a hole would be given without specifying how it is to be made. (There are a number of ways to make a cylindrical hole.) In a situation where a hole must be made in a specific way, that should be specified.

6. There should be no redundant dimensions, but it is often helpful to have extra information which can be added if identified as extra dimensions. For example, it is often useful to know the overall size. Extra dimensions should be identified as **reference dimensions** by enclosing them in brackets. Reference dimensions are secondary to other dimensions; if there is disagreement, the reference dimension is ignored. Figure 4.14 (page 102) shows how reference dimensions are identified. Another way of identifying a reference dimension is to add the letters REF after the number (17 REF). The use of reference dimensions should be kept to a minimum.

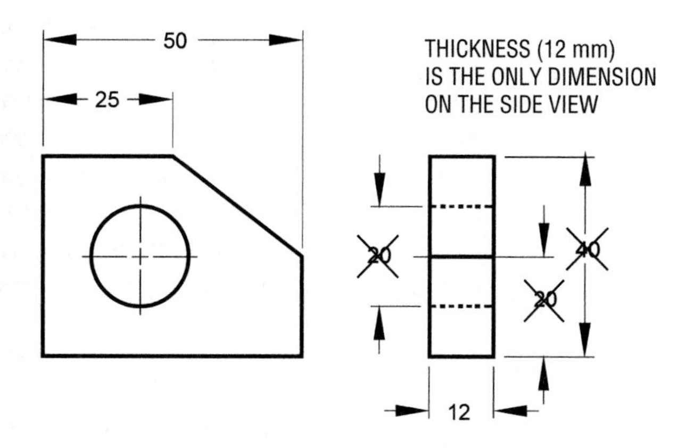

Figure 4.11 Dimensions put on a view (side) that does not show the shape

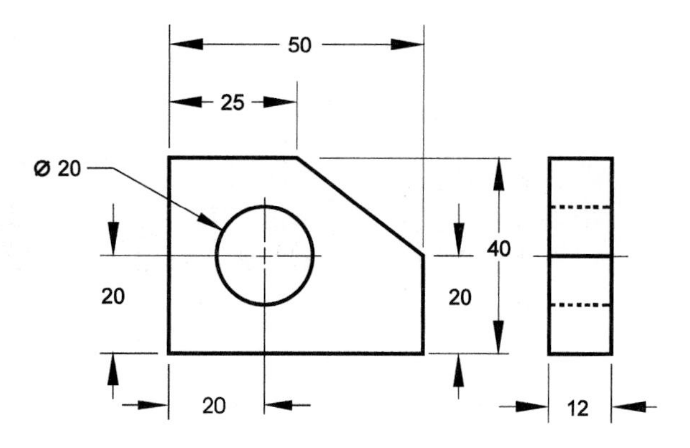

Figure 4.12 Dimensions should be on the view that shows the shape

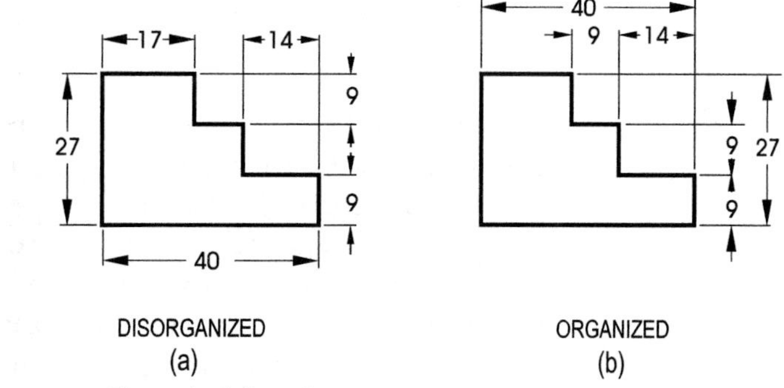

Figure 4.13 Disorganized dimensions

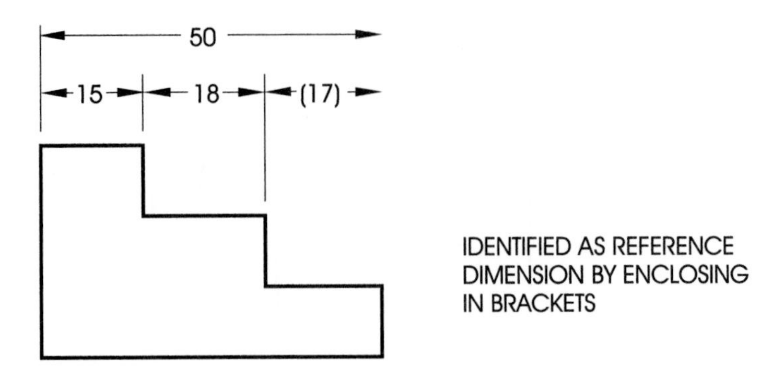

Figure 4.14 An extra dimension identified as a reference dimension

There are other rules for using dimensions, but those outlined above give a good guide to good practice.

Now that you know the basics of dimensioning standards and rules, try the following problems.

Problems

These problems cover size dimensions. They may be done as sketches or as finished drawings on a computer or by any other means.

Dimension the following objects in accordance with the guidelines given in this chapter. The objects are drawn on a square grid. The size can be specified by the instructor or the student.

The original sketch or drawing can be passed to another member of your work group (a "checker") for checking. The checker marks all errors and omissions and returns it to the originator for correction.

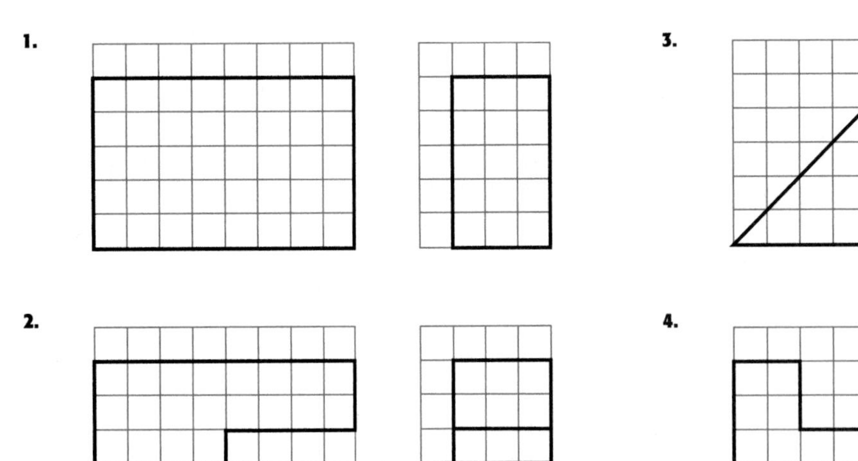

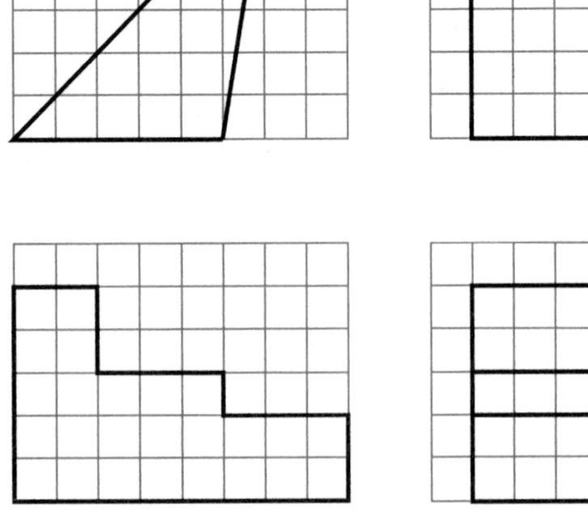

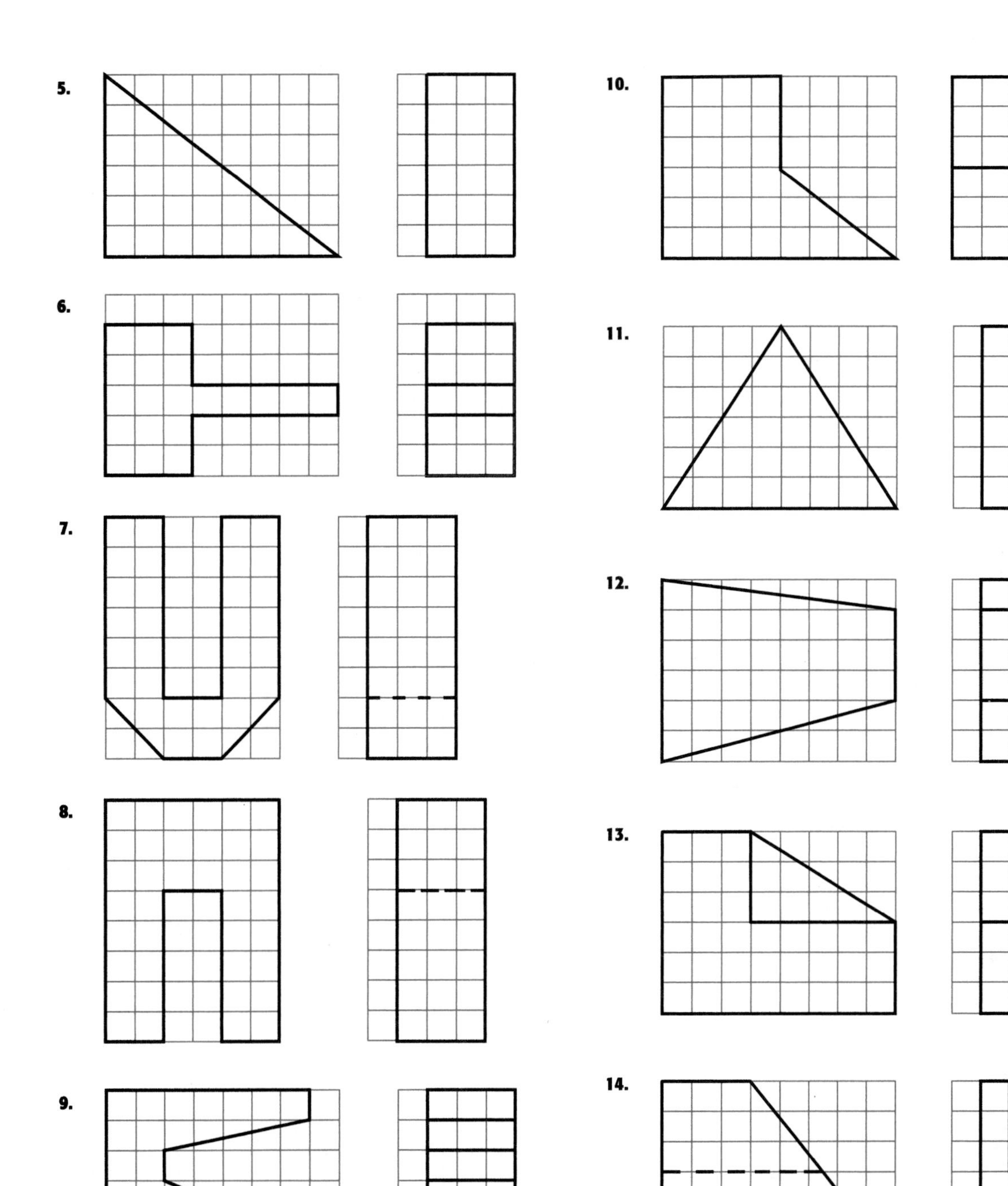
5.
6.
7.
8.
9.
10.
11.
12.
13.
14.

15.

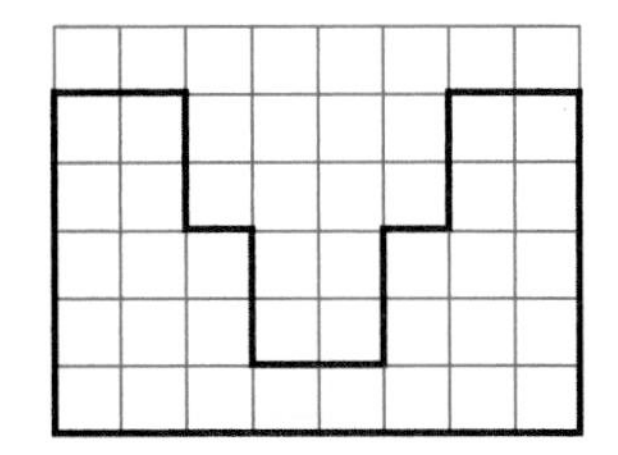

16.

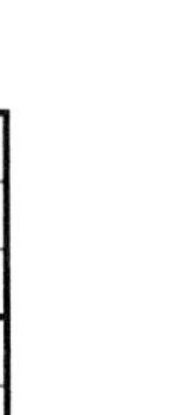

17.

18.

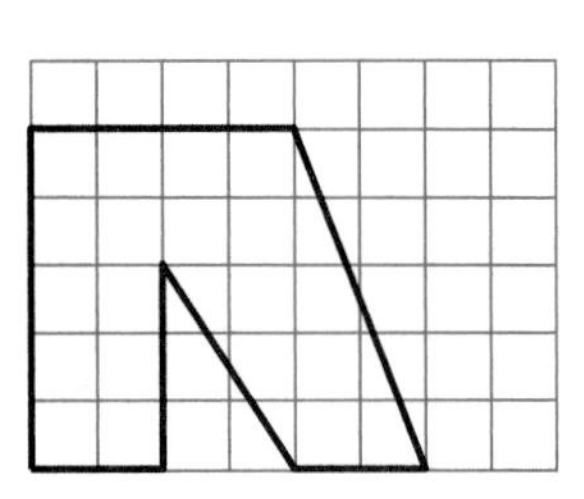

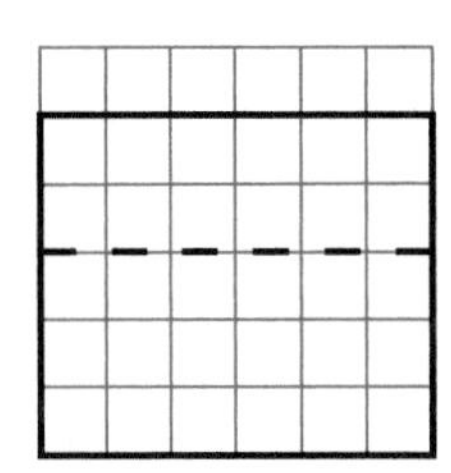

19.

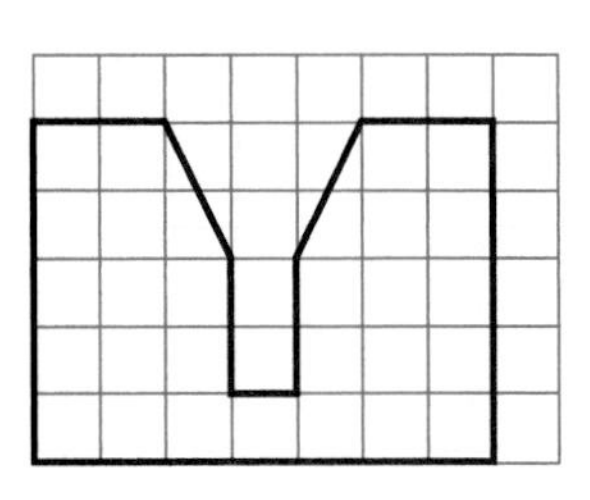

20.

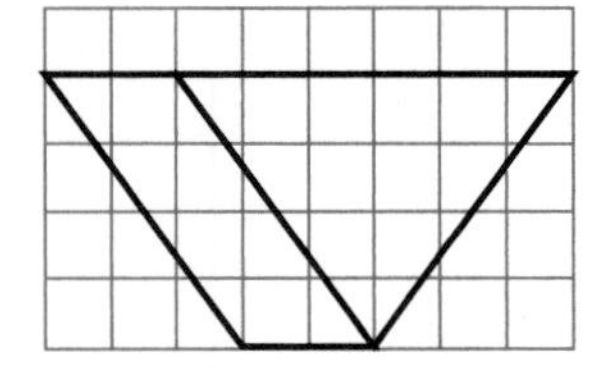

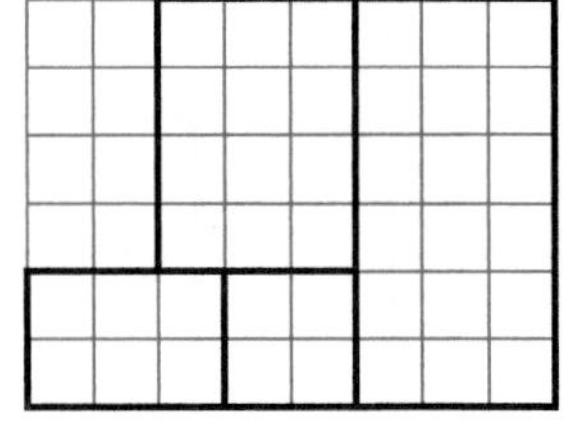

21.

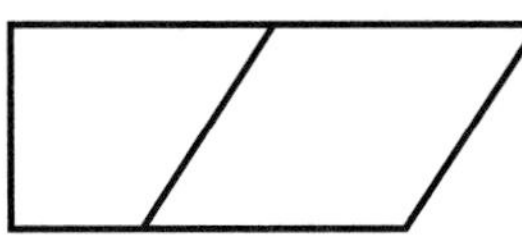

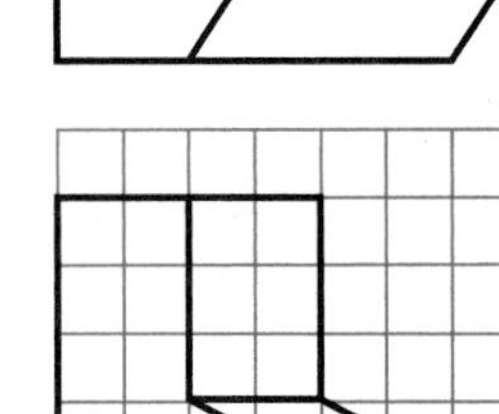

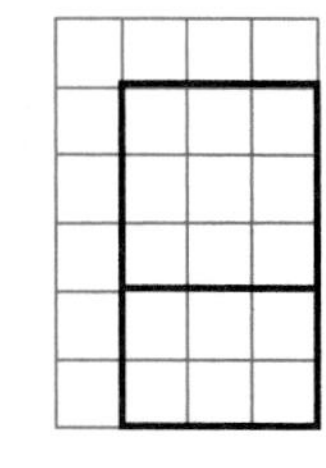

22.

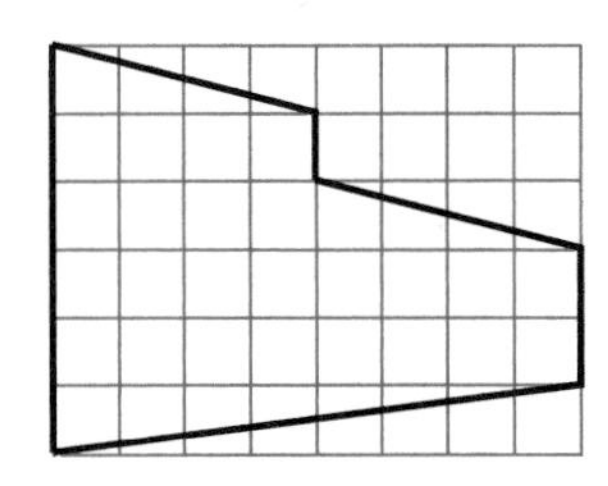

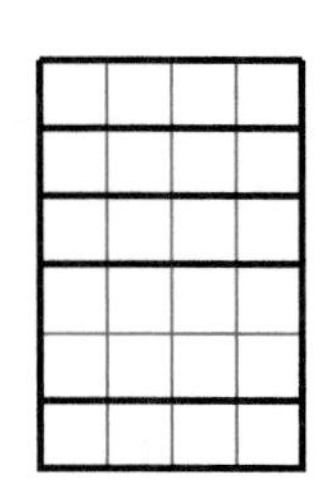

23.

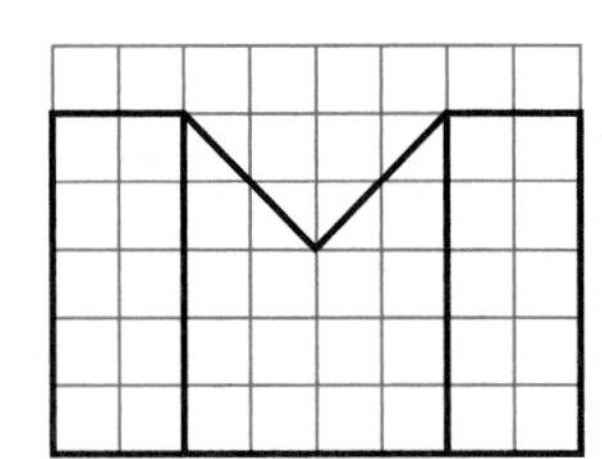

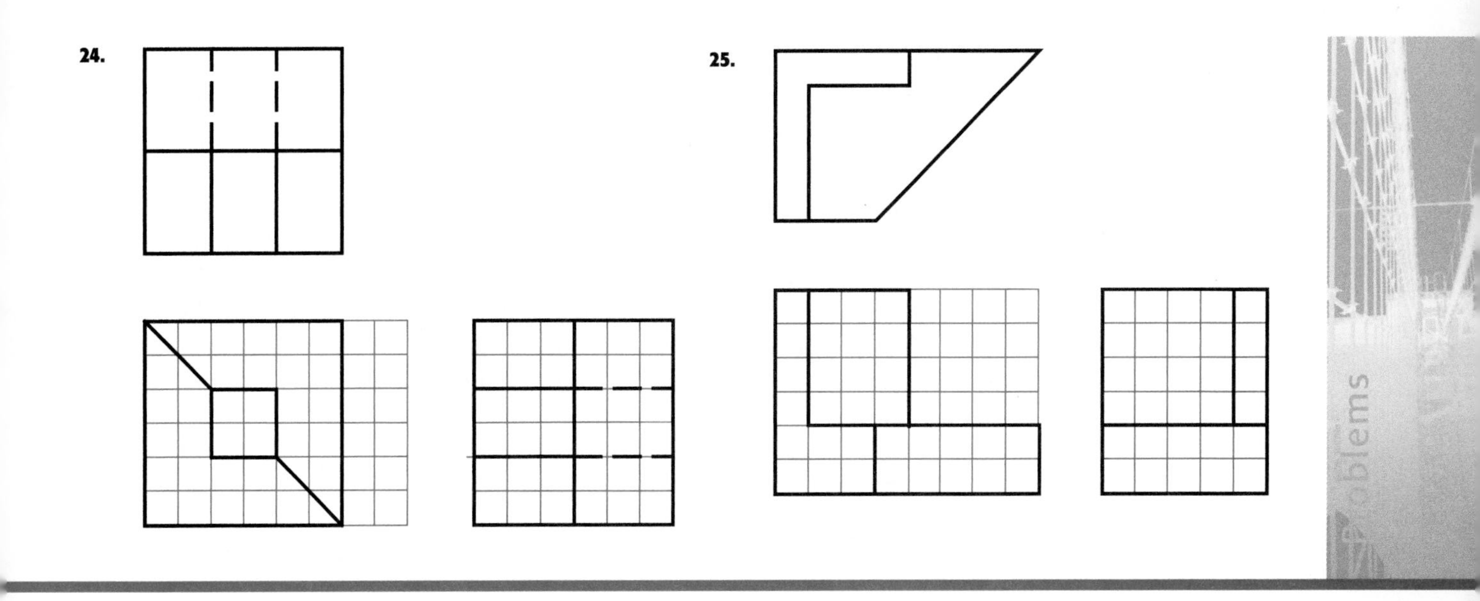

DIMENSIONING FEATURES

There are accepted methods of specifying dimensions for different features. Using these methods enables you to communicate with other engineers and technical personnel.

In this section, we will look at angular and circular dimensions.

Angular Dimensions

Angular dimensions specify the angle between two lines or directions. The dimension line is an arc with the centre at the apex of the angle and terminating with arrowheads at extension lines (extensions of the two sides of the angle). Angles can be specified in degrees-minutes-seconds or in decimal degrees, as shown in Figure 4.15.

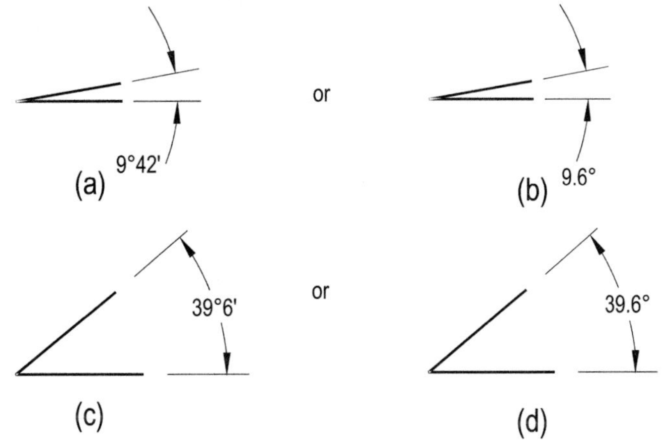

Figure 4.15 Different ways of dimensioning angles

Circular Dimensions

Circular features are defined by specifying the location of the centre and either a radius or a diameter. A centre and a radius are used for an arc, and a centre and a diameter are used for a circle. There are several ways to specify this information depending on the size of the circular feature and whether it is a solid cylinder or a hole.

Diameter

A solid cylinder is dimensioned where both dimensions—diameter and length—are in the same view as visible outlines. A hole (a negative cylinder) is dimensioned where the circular shape is seen.

Figure 4.16 shows a pulley with three diameters and a hole for mounting on a drive shaft. The diameter and length of each section are shown in the front view. The diameter of the hole is given on the side view where the circular shape is seen. The hole is assumed to go all the way through if there is nothing to indicate the contrary. A note "THRU" is added after the diameter to show that the hole goes all the way through the pulley. The hidden line in the front view also shows that the hole goes through the pulley.

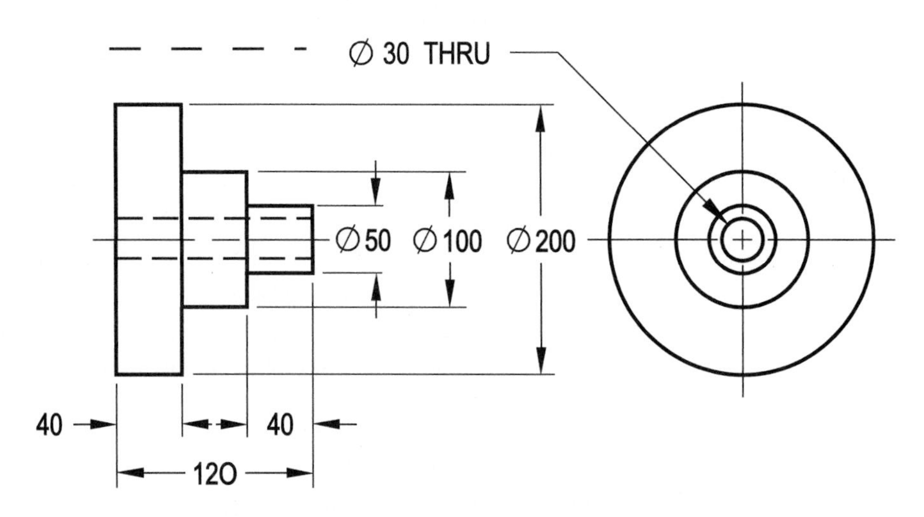

Figure 4.16 Length and diameter of cylinders put on the same view

If a hole does not go all the way through the part (called a **blind hole**), the depth must be specified, as shown in Figure 4.17. The depth does not include the conical portion at the bottom. This conical section results from the shape of the drill used to make the hole.

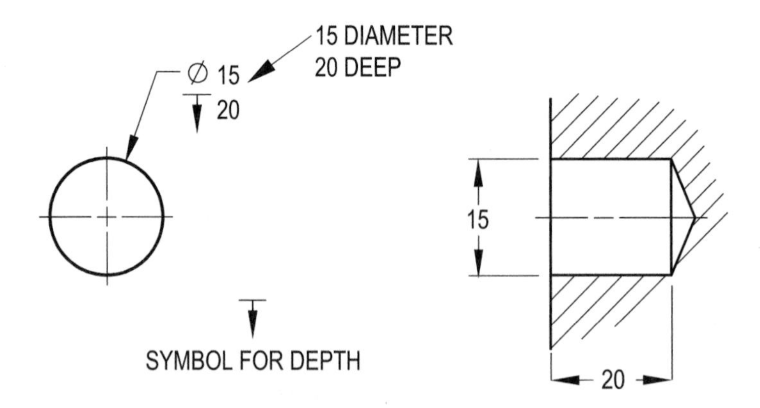

Figure 4.17 Symbols for specifying a blind hole and their meaning

Small holes are dimensioned by a note specifying the diameter and depth, with a leader pointing to the hole, as in Figure 4.17. The leader is radial and, if extended, would pass through the centre of the circle.

Large diameter holes are dimensioned as shown in Figure 4.18. A circular hole is always specified by the diameter.

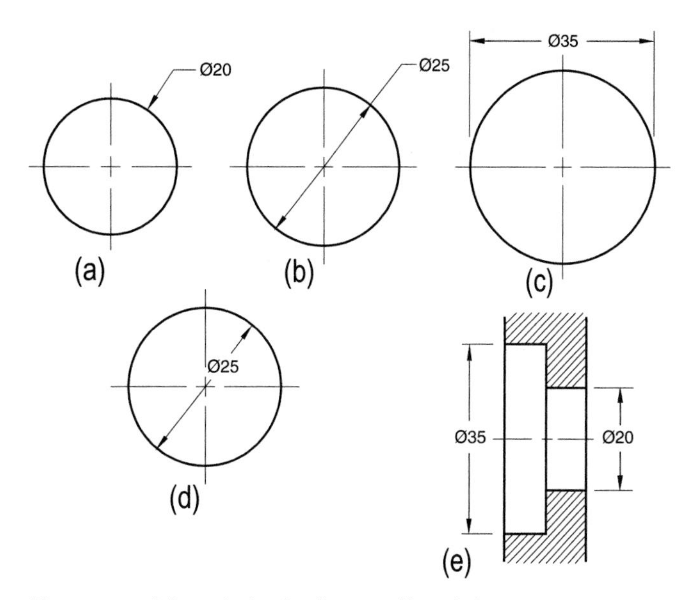

Figure 4.18 Different ways of dimensioning the diameter of large holes

Radius

If a circular feature is not a complete circle, it is specified by the location of the centre, the starting point, the end point, and the radius. Other features often define the end points.

To draw an arc, you must know its centre, but the location of the centre may not be specified on the drawing. Other information, such as tangent points, must be given to locate the centre. This information is usually not explicit but is obvious from the drawing. If, for some reason, it is not obvious, the location must be given so that it can be located correctly. Figure 4.19 shows different ways to specify a radius without giving a centre.

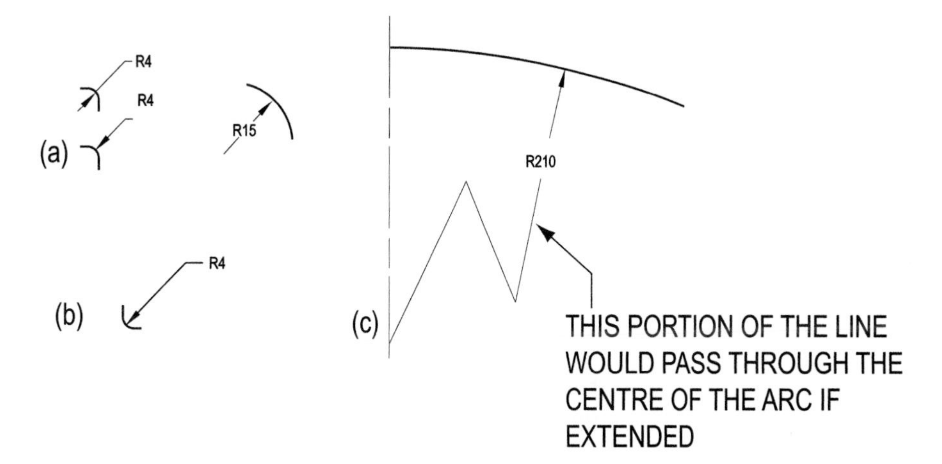

Figure 4.19 Dimensioning a radius without explicitly locating the centre

The radius is given as a dimension with a line from the centre, even though the centre is not marked. If the arc is not large enough to include the dimension, a leader is used either "outside" or "inside" the arc, as shown in Figure 4.19a and 4.19b.

Depending on the scale of the drawing, the centre of a large arc could be "off the paper" when it is printed. The centre is specified, as shown in Figure 4.19c. The portion of the dimension line touching the arc is a radial line that would pass through the actual centre if extended.

It is not good design practice to have sharp corners. A crack will often start at a sharp corner. Sharp corners are eliminated by forming a small radius at each corner, as in Figure 4.19b. This feature is called a **fillet** at an inside corner and a **round** at an outside corner. (It can also be called a fillet at an outside corner.) Dimensioning filleted corners can be done with a note, such as ALL FILLETS AND ROUNDS 2 mm.

When an outline consists of several arcs, not all centre locations may be specified. Locations of the centres are determined by the tangent points between arcs. Some centre locations must be specified so that others can be determined. Geometric construction is required to locate intermediate centre locations and tangent points, or the SNAP commands on your CAD program can be used. Figure 4.20 shows an example of an outline comprising several arcs. Arc location is controlled by other features, in this case tangent locations.

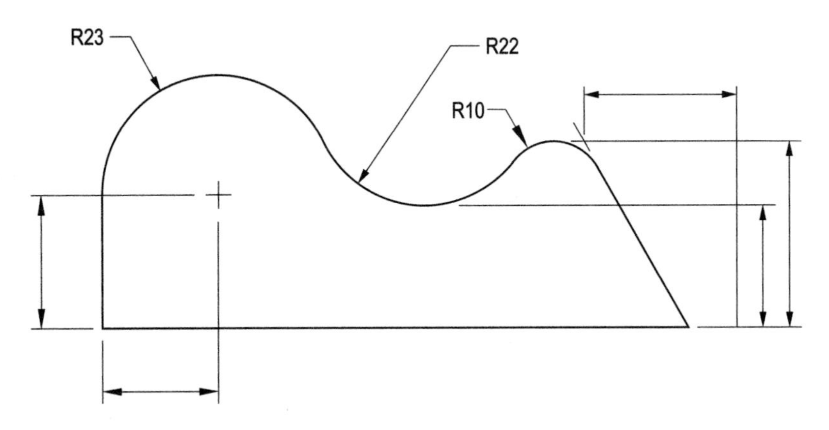

Figure 4.20 Dimensioning an outline composed of several arcs

Now let's move on to look at specification and dimensioning of threaded fasteners.

THREADED FASTENERS

Threaded fasteners, such as nuts, bolts, and screws, are used to hold parts together. A threaded connection has two components: one with an external thread and one with an internal thread. The internal thread may be in a threaded hole or a nut. The design of threaded connections is beyond the scope of this book: we will deal only with specification and dimensioning of threaded fasteners.

Thread Terms

Figure 4.21 shows thread terms on the cross-section of an external and internal thread.

The **major diameter** is the outside diameter of an external thread or the maximum diameter of an internal thread.

The **minor diameter** is the diameter at the root of the thread of an external thread and the minimum diameter of an internal thread.

The **crest** is the top of a thread and the **root** is the bottom.

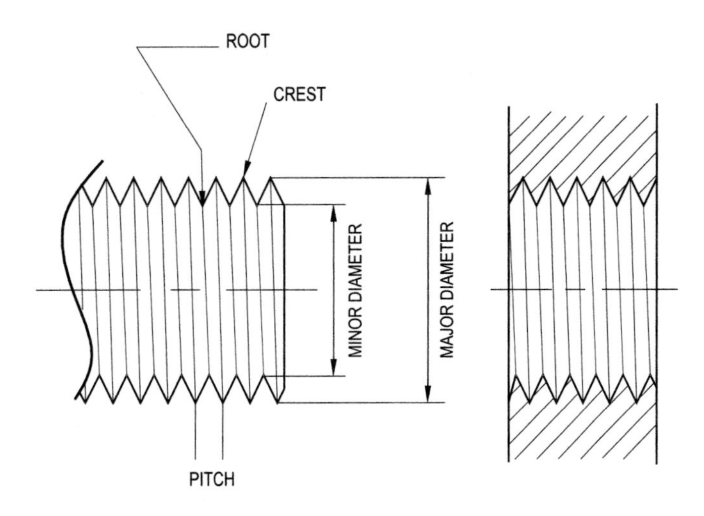

Figure 4.21 Common thread terms

Pitch is the distance from a point on the crest of one thread to the corresponding point on the crest of the adjacent thread.

Other thread terms will be described as required.

Thread Specifications

Some common thread profiles are shown in Figure 4.22.

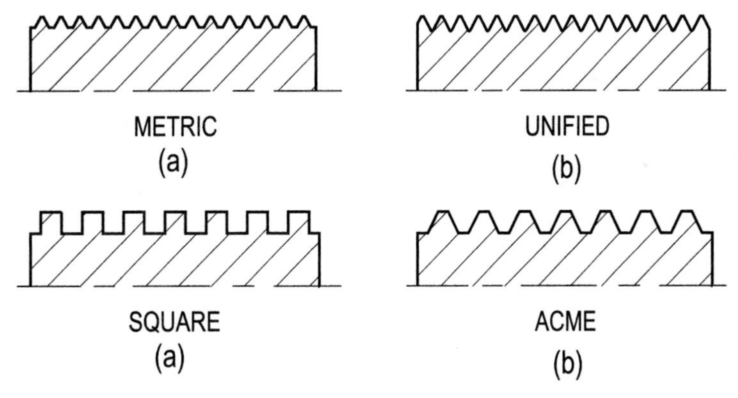

Figure 4.22 Common thread profiles

Metric threads are the international standard thread. **Unified threads** are the standard threads in the United States and Canada. Square and Acme threads are used for power transmission.

Since thread dimensions are set by international and national standards, there is no need to give the dimensions. You only need to give the thread series, diameter, length, and the **class of fit** (a tight or loose fit) to specify a thread. There is also no need to draw the thread profile, as shown in Figure 4.22. Threads are represented by a symbol that is easy to draw (Figure 4.24).

Metric Threads

General purpose metric threads are specified by the major diameter and the pitch (Figure 4.23). The designation M10 × 1.5, for example, means a 10 mm diameter metric thread with a 1.5 mm pitch. The M signifies metric.

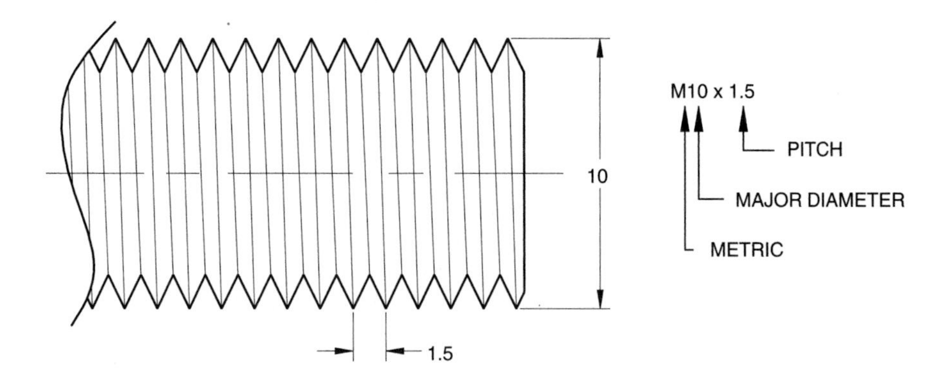

Figure 4.23 Metric threads are specified by diameter and pitch

Pitch and diameter are related so that once the diameter is selected, there are only a few possible pitches. For example, a 10 mm thread can be "coarse" (1.5 mm pitch) or "fine" (1.25 mm pitch). There are other possible pitches for a 10 mm thread, but 1.5 mm and 1.25 mm are the common ones. If pitch is not specified, it is assumed to be a coarse pitch. For example, M10 means M10 × 1.5.

A complete specification would include the thread tolerance class comprising tolerance grade and tolerance position. For example, M10 × 1.5 – 6g, where 6 is the tolerance grade and g is the tolerance position symbol (Figure 4.24). The lowercase letter for tolerance grade means that the tolerance applies to an external thread. An uppercase letter indicates an internal thread. A complete discussion of tolerance class is beyond the scope of this text. Further information on tolerance grades and position symbols can be found in ISO Standard 965/1 and ANSI/ASME B1.13M Metric Screw Threads for Commercial Mechanical Fasteners—Boundary Profile Defined.

Thread length—the distance from the end of the part to the last full thread—can be specified as a note or dimensioned on the drawing.

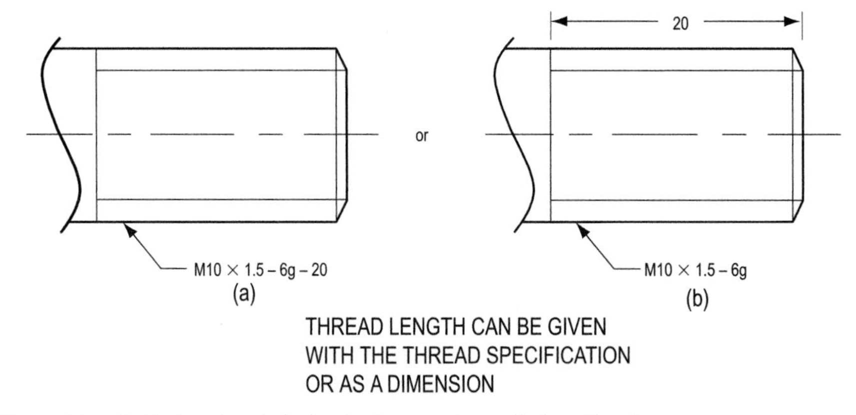

Figure 4.24 Metric thread symbols showing two ways to specify thread length

The thread is represented by a symbol. The outside diameter is equal to the major diameter, and the inside line represents the minor diameter. The line representing the minor diameter is located "by eye" and there is no need to determine the minor diameter (which you must look up).

Unified Threads

A Unified thread is specified by the major diameter, number of threads per inch, thread series, whether the thread is coarse or fine, and the class of fit (a tight or loose fit). The notation 1-8UNC-2A specifies a 1-inch diameter, 8 threads per inch (tpi), Unified National Coarse thread, with a class 2A fit.

There are several possible combinations of diameter and threads per inch depending on the thread form; for example, 1-12UNF-2A specifies a 1-inch diameter, 12 threads per inch, Unified National Fine thread, with a class 2A fit.

There are six classes of fit in the Unified system. Classes 1A (loose), 2A (standard), and 3A (tight) apply to external threads, and classes 1B, 2B, and 3B apply to internal threads. Classes 2A (external threads) and 2B (internal threads) apply to commercial bolts, nuts, screws, and other fasteners.

In addition to the size and type of thread, length must also be specified, either by a dimension on the drawing or by a note. Since fasteners are standard parts, they are often specified with a note. Figure 4.25 shows how length can be specified.

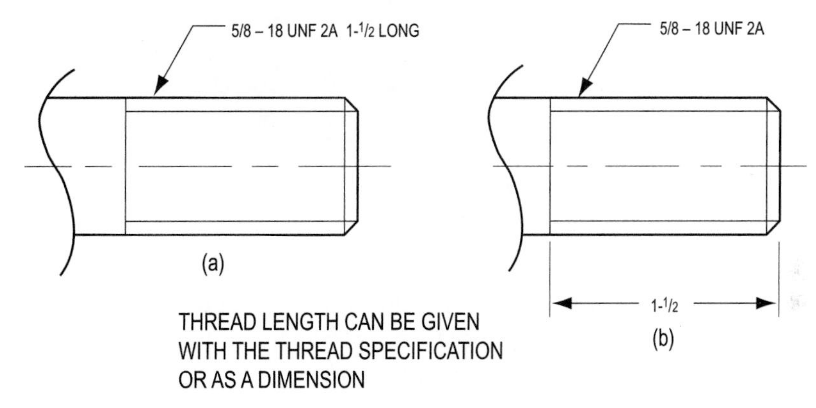

Figure 4.25 Unified thread symbols with two ways of specifying thread length

Two thread symbols are used with the Unified system. The thread symbol used in Figure 4.26, called a **simplified symbol**, is similar to that used with metric dimensions. The dotted line represents the minor diameter, which is located by eye. Figure 4.26 shows both thread symbols used with the Unified system.

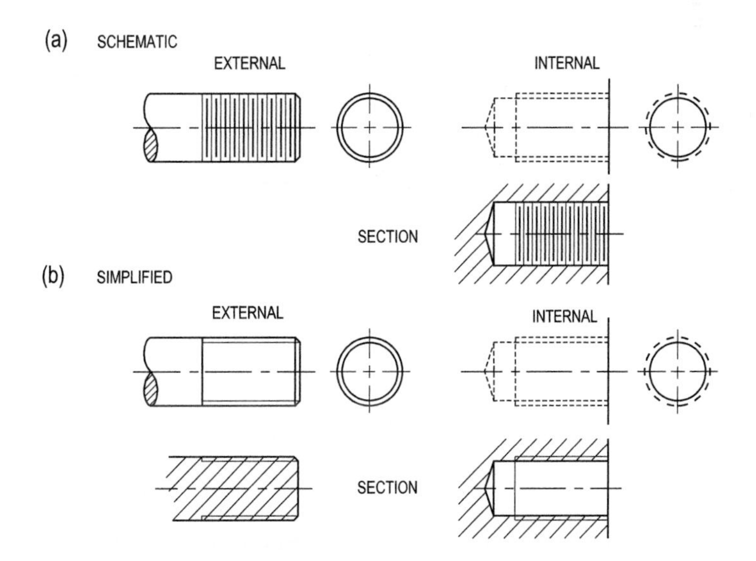

Figure 4.26 Thread symbols used in different applications with the Unified system

The distance between the lines in the schematic symbol represents the pitch, which, since the number of threads per inch is known, can be found.

Nuts and Bolts

Nuts and bolts are illustrated in Figure 4.27. The heads of nuts and bolts are generally hexagonal, but square heads and square nuts are available.

Figure 4.27 An assortment of nuts and bolts showing different heads and sizes

Carriage bolts have round heads. Figure 4.28 illustrates square and hexagonal bolt and nut heads. Dimensions of nuts and bolts are generally not given (except bolt length), since they are standard parts, but dimensions must be known to check clearances and to draw them.

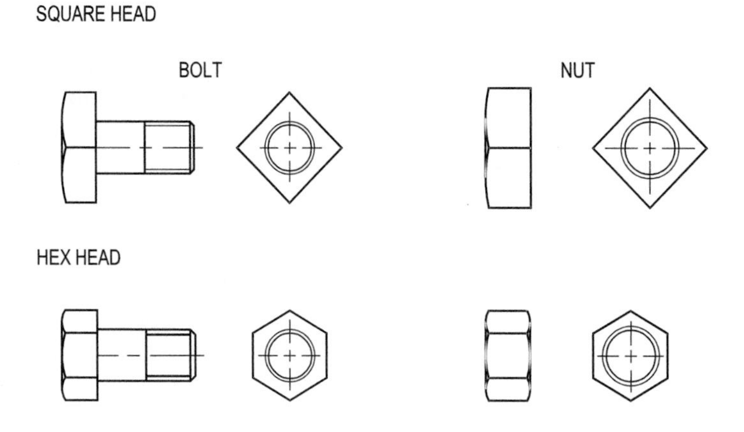

Figure 4.28 Square and hexagonal nuts and bolts

Nut and bolt dimensions can be found in Appendix E.

Cap Screws and Machine Screws

Cap screws and machine screws are similar to bolts but have different types of heads. Cap screws and machine screws are generally screwed into threaded holes. Figure 4.29 shows the different head designs available.

Machine screws are smaller in diameter than cap screws and come in sizes from #0000 (0.0210 inch diameter) to 3/4 inch diameter. Sizes less than 1/4 inch are designated with a number. Cap screws are greater than 1/4 inch diameter.

Standard dimensions can be found in the tables in Appendix E.

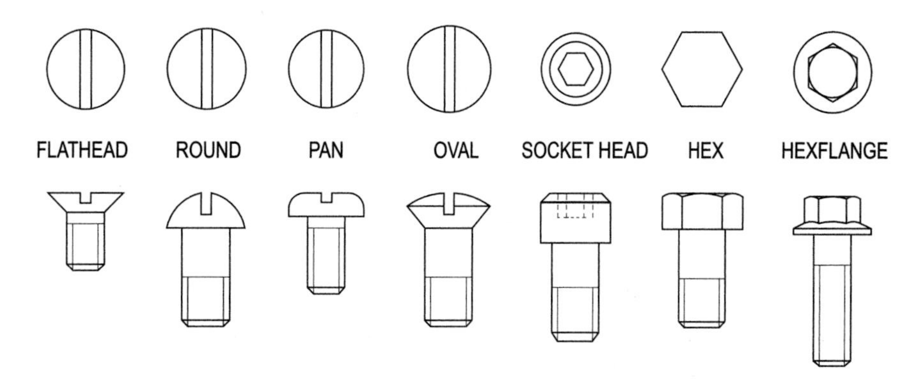

Figure 4.29 Many different heads are available for cap and machine screws

Countersunk, Counterbored, and Counterdrilled Holes

These names refer to the different shapes of the top of a threaded hole. The top of the threaded hole must be modified so that the fastener is either flush with the surface or below it. All three holes are shown in Figure 4.30, along with the threaded fastener that would go in the hole.

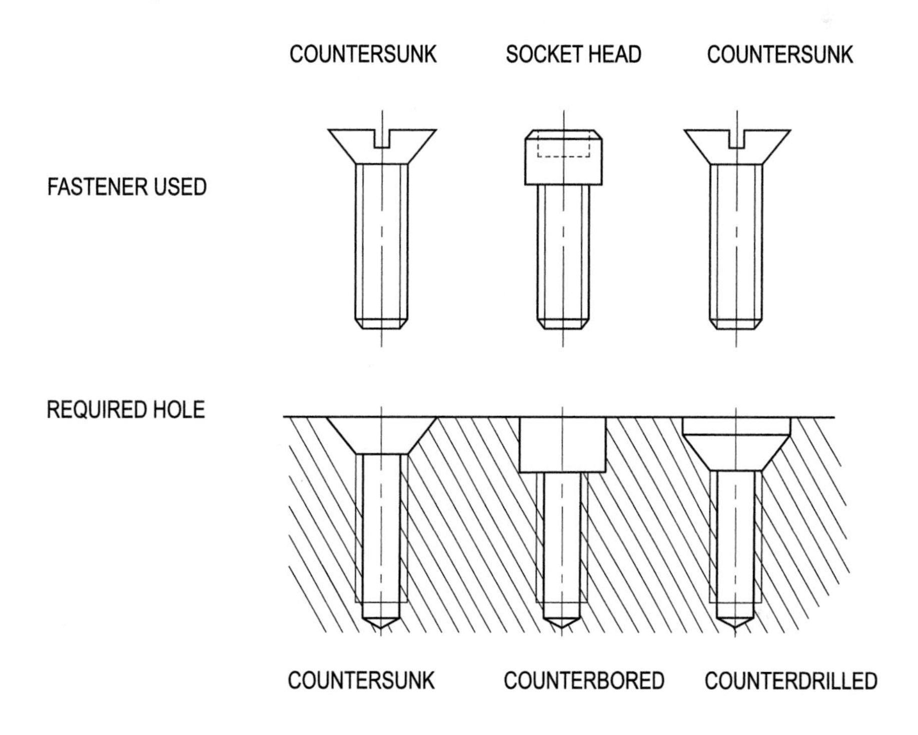

Figure 4.30 Different holes are required if the fastener must be below the surface

Dimensions are given with symbols as in Figure 4.31a or by a combination of symbols and notes as shown in Figure 4.31b.

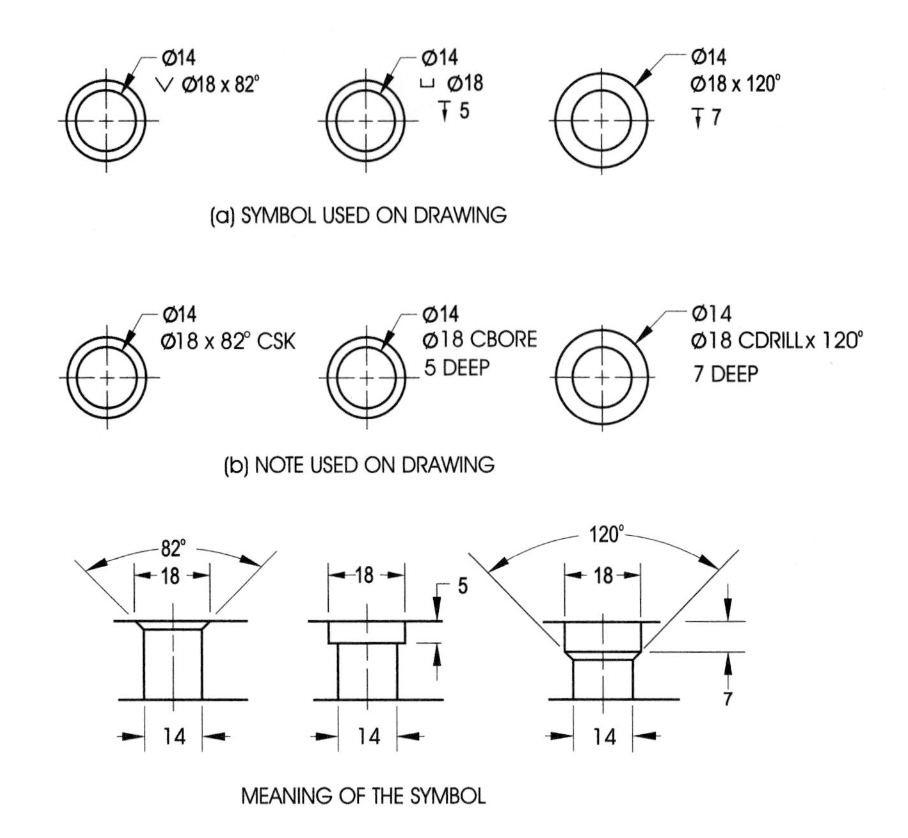

Figure 4.31 Methods of specifying dimensions for countersunk, counterbored, and counterdrilled holes

Spotfaced Holes

In many applications, a bolt head could be above the surface, but the bolt head must sit on a flat surface. If the surface is uneven or rough, the bolt will not seat evenly and high loads can result.

The area around the top of the hole is machined to make it flat and smooth. This is called **spotfacing**, and the result is a **spotfaced hole**. Figure 4.32 shows a bolt in contact with a sloping surface. Spotfacing the hole allows the washer to rest on a flat surface. It is similar to a counterbored hole except only a small amount of material is removed and no depth is specified.

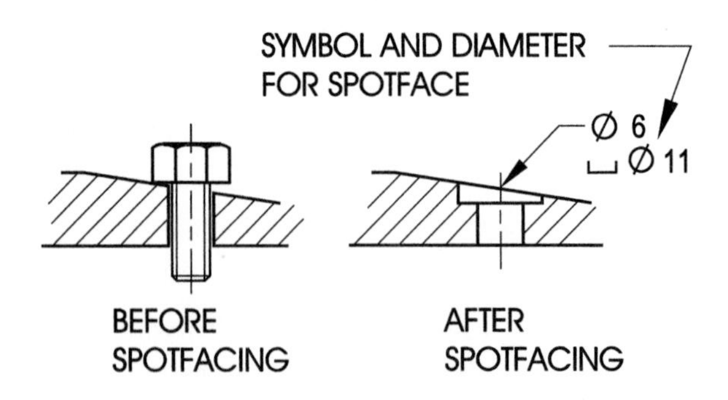

Figure 4.32 A spotfaced hole is required if the bolt is to seat properly

Slotted Holes

If the position of a piece of equipment must be adjusted and it is secured with fasteners, the fasteners must be removed for repositioning. If the fasteners pass through slots, any repositioning requires only that they be loosened. Slotted holes usually have rounded ends and straight sides. The length and width of the slot must be specified. Three ways of doing this are illustrated in Figure 4.33. The radius of the arc is indicated at each end (2X R), but not dimensioned.

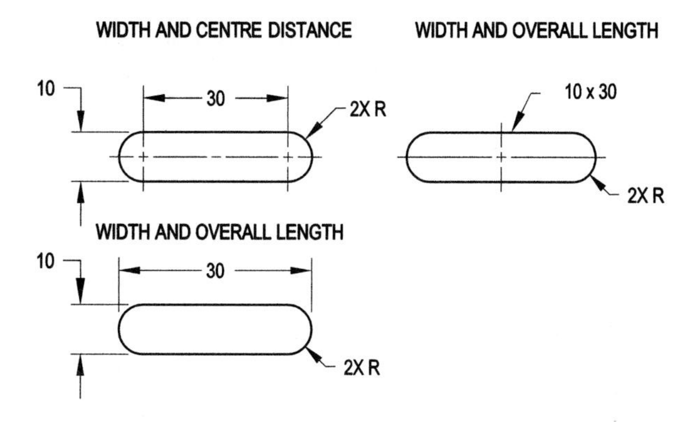

Figure 4.33 Different ways of dimensioning a slotted hole

Chamfers

When two surfaces meet at a sharp angle, it is common to bevel the edge. This is called a **chamfer**. Chamfers are used on round shafts, holes, and plane surfaces. Chamfers are dimensioned by an angle and a linear dimension, or by two linear dimensions. The linear dimension is the distance from the end of the object to the start of the chamfer. If the chamfer is at 45°, a note can be used, since the linear dimension is the same in both directions. Figure 4.34 shows how a chamfer could be specified.

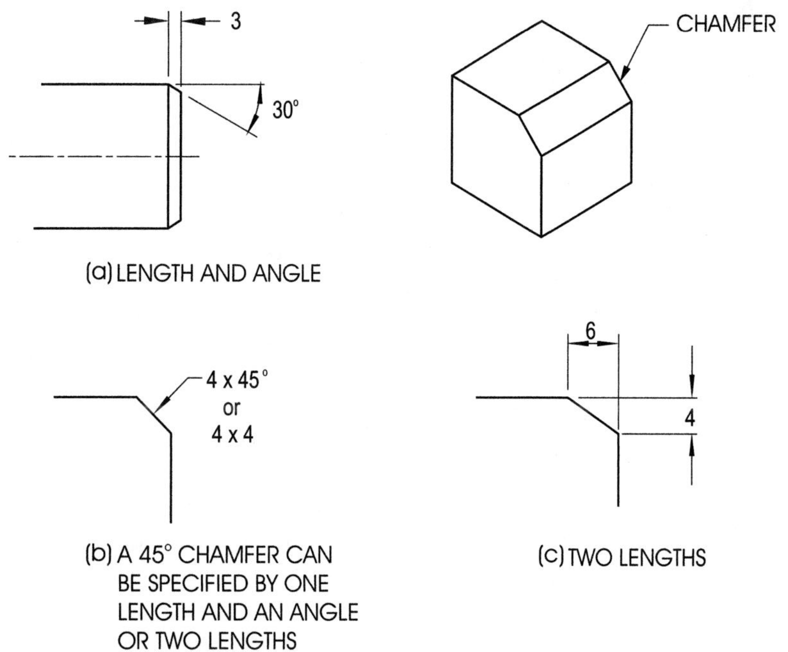

Figure 4.34 Three ways to dimension a chamfer

Keyseats

When a hub is mounted on a shaft, some method of transferring torque without slipping is needed. An axial slot, called a **keyseat**, is cut in the shaft and hub, and a piece of metal, called a **key**, is put in this slot. The key provides a positive means of transmitting torque between the shaft and the hub. Figure 4.35 shows how this is done. The key shown here is square, but there are other shapes. The size and shape of the key depends on the diameter of the shaft. Square keys are used for shafts up to about 150 mm. Figure 4.35 shows how keyseats in the shaft and hole are dimensioned. Width, depth, location, and length are specified. Dimensions and tolerances on keys and keyseats depend on the diameter of the shaft and tightness of the fit. There are three classes of fit; ranging from loose to tight. A loose fit allows the hub to slide over the shaft. Keys and keyseats are specified in CSA B232-75 (R2002) and ANSI standard ANSI B17.1-1967, R1989.

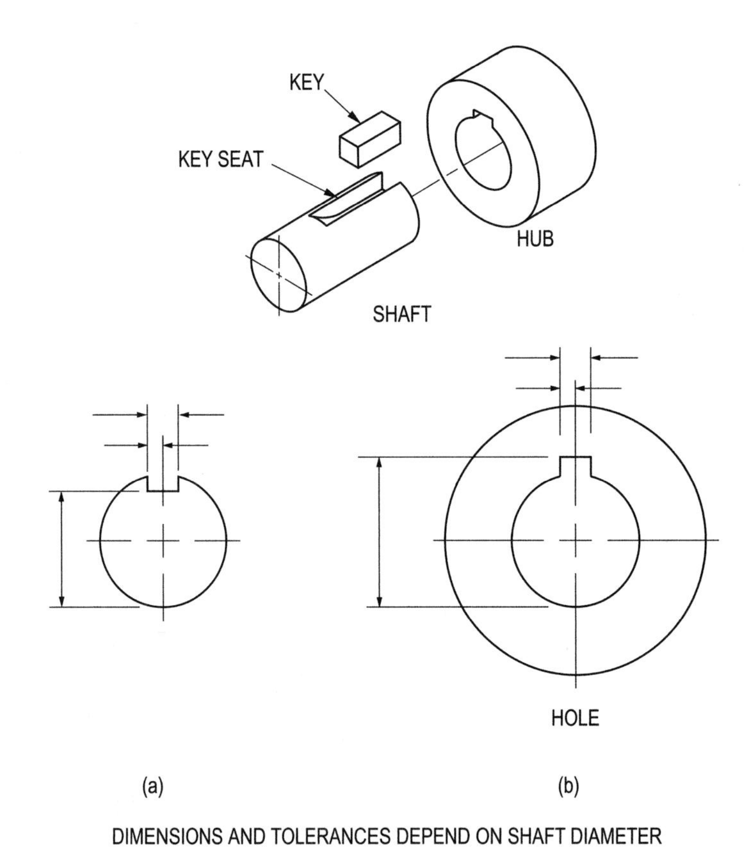

Figure 4.35 Dimensioning a keyseat in a shaft and hole

Tapers

A **taper** is the ratio of the difference in diameters of two cross-sections, perpendicular to the axis of a cone to the distance between these two sections. It is the change in diameter per unit length. A taper should not be confused with a slope. Figure 4.36 shows various methods of specifying a taper. This figure shows dimensions of a standard machine taper, but any taper would be dimensioned in this way. Since this is a standard taper, dimensions are given in inches. There are several standard tapers. Two other common ones are **Morse tapers** and **Jacobs tapers**. Dimensions of these tapers can be found in references such as *A Mechanical Engineers Handbook*, or *Machinery's Handbook*. Tapers are used to mount tool holders in machine tools.

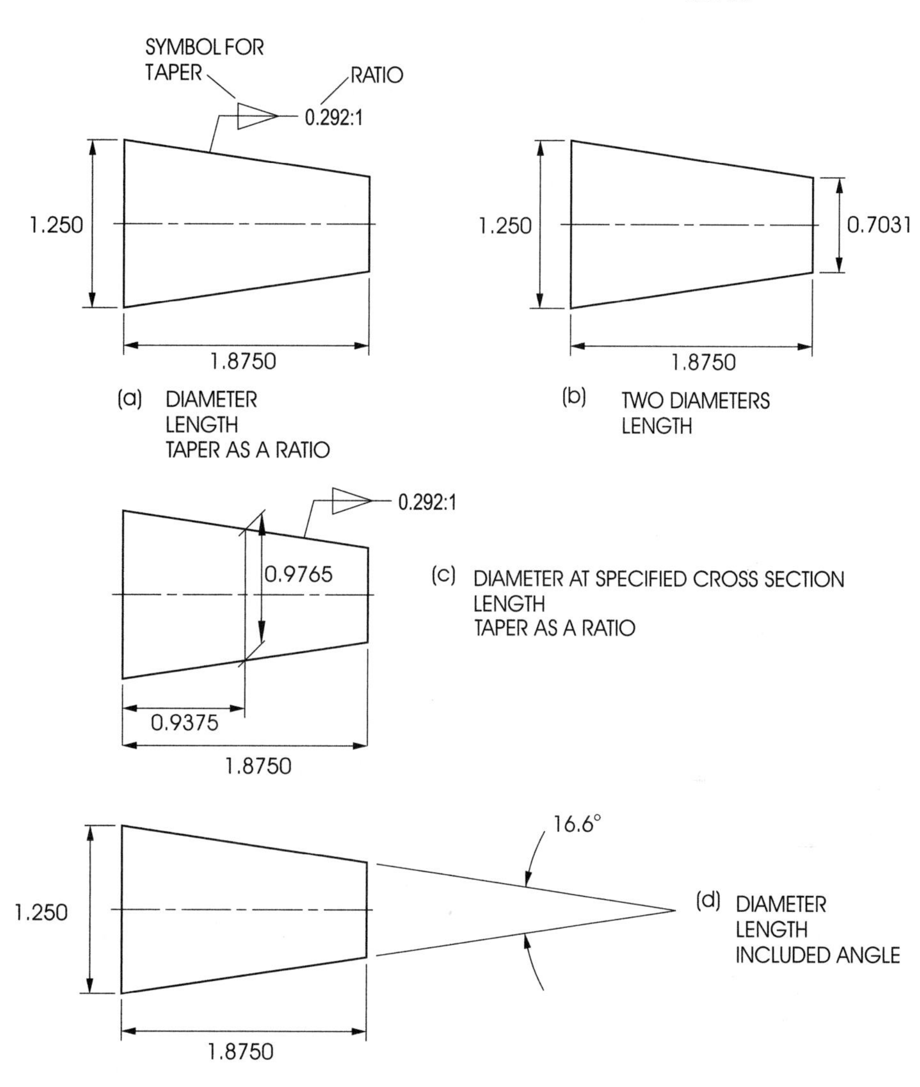

Figure 4.36 Four ways to dimension a taper

Slope

A **slope** is the inclination of a line representing an inclined surface. It is expressed as the ratio of the difference in heights, at right angles to the base line, at a specified distance apart, that is, rise/run. A slope can be specified as an angle, the ratio of the height difference to the distance between the heights, or dimensions showing the heights and distance between them. Figure 4.37 shows different ways of dimensioning slope.

A slope can be also be stated as a **grade** (Figure 4.37d). The grade is the tangent of the slope angle expressed as a percent. This method of specifying a slope is used by surveyors and by civil engineers. The slope of a road is specified by a grade.

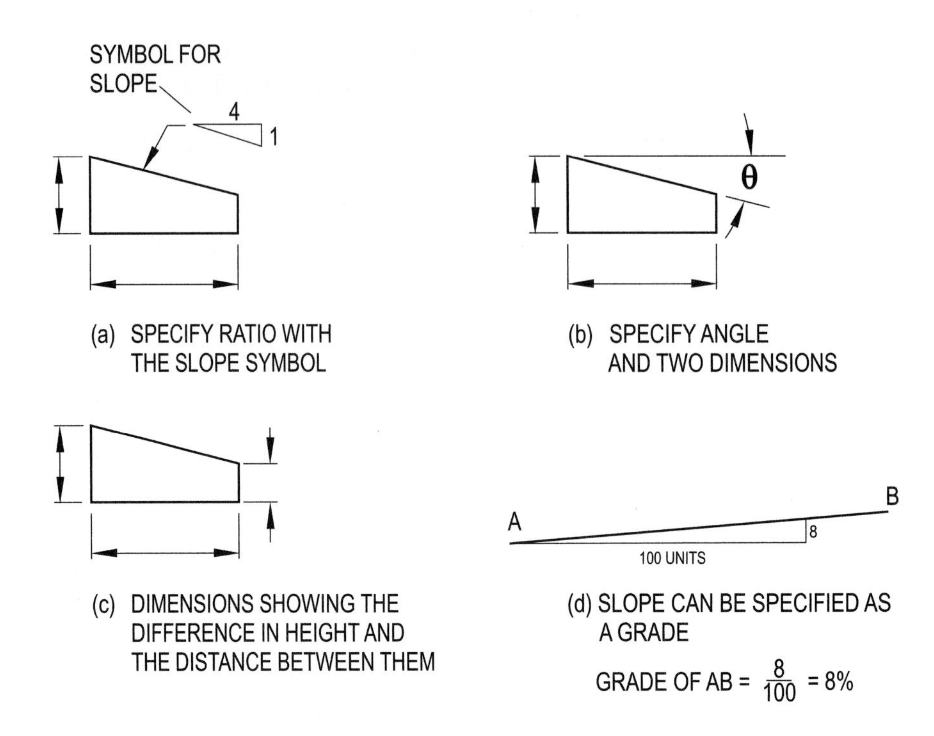

Figure 4.37 Four ways to dimension slope

Undercut

When one section of a shaft is a different diameter from that of another, it is common to reduce the small diameter at the shoulder (where the two diameters meet). This is called an **undercut**. If, for example, the smaller diameter is threaded, it is common to undercut to the root of the thread at the shoulder. An undercut is dimensioned by specifying the width and diameter. The radius at the bottom of the undercut should be half the width unless specified. Figure 4.38 shows how plain and radius undercuts are dimensioned.

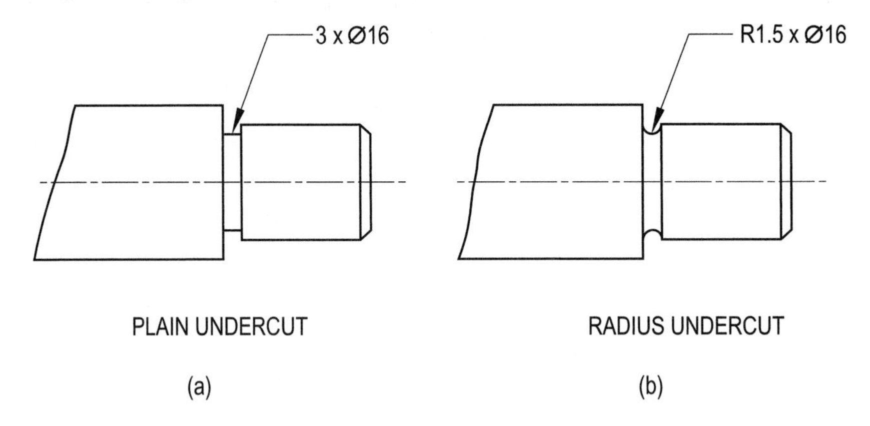

Figure 4.38 Dimensioning plain and radius undercuts

Now let's move on to look at location dimensions.

LOCATION DIMENSIONS

Location and size must be specified to completely dimension a feature. The location of a corner is specified by two linear dimensions. The location of features that are not on the outline are dimensioned in a similar way, except that extension lines are different from those used on the outline. Multiple copies are usually dimensioned only once.

Arcs

Arcs are dimensioned on the view in which they are seen as arcs. When the location centre is specified, it is identified by a small cross. The position of this cross is specified by extension lines and dimensions. Figure 4.39 shows how the centre is defined and specified.

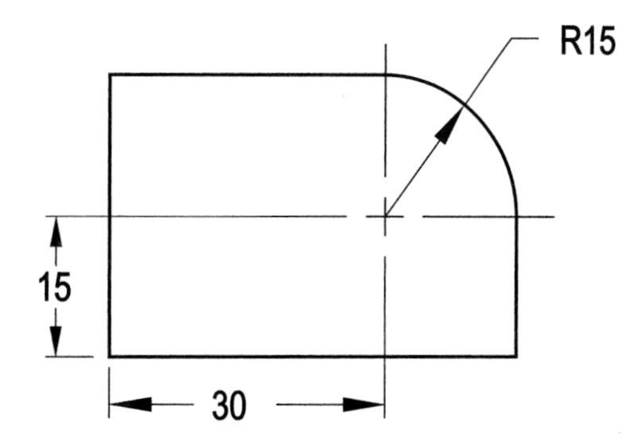

Figure 4.39 Locating the centre of an arc

Holes

The location of a hole is generally given on the view in which it appears as a circle. Depth cannot be seen in this view, and so it must be specified in a different way or dimensioned on another view (but not as a hidden line). Figure 4.40 shows how to dimension the location of a hole. The location of the centre is specified by extension lines that are extensions of centrelines. There is no gap between the centreline and extension lines. Extension lines at the other end of the dimension line have a gap between the line and the outline.

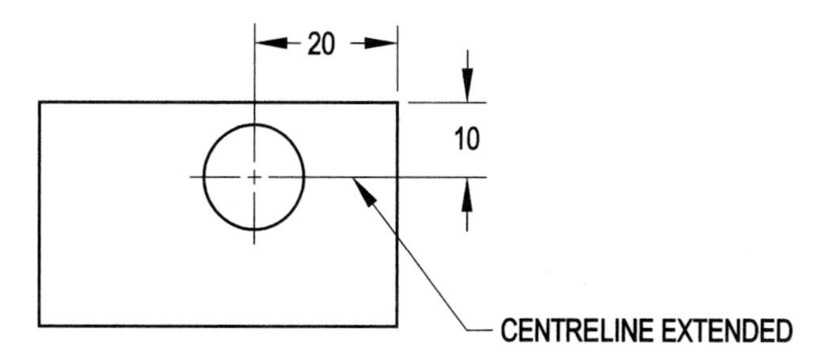

Figure 4.40 Dimension the location of a hole where it shows as a circle

REPEATED FEATURES

It is common to use one hole size or other feature more than once in a part. If four holes are needed in one part, they would usually be made all the same size unless there was a good reason to use different sizes. The same tool can then be used to make all of them and setup time and costs are reduced.

When a feature is repeated, there is no need to put the same dimension on each one. Size dimensions and sometimes location dimensions are specified only once, and the number of repetitions is given. Figure 4.41 shows an example with repeated features—holes and radii—with a common centre.

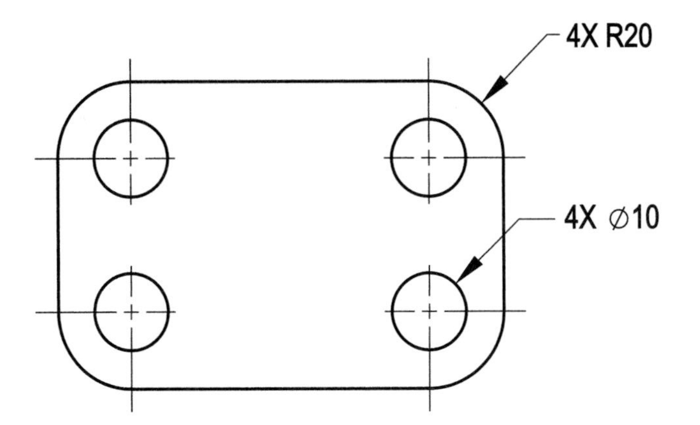

Figure 4.41 Dimension repeated features only once

The symbol 4X Ø10 means that there are four holes with a diameter of 10 mm, and 4X R20 means that there are four corners with a radius of 20 mm. It should be obvious by looking at the drawing where these features are.

If some feature is repeated at a regular spacing, there is no need to dimension every length or draw the feature many times, although this is easy to do with a computer program. The same circle drawn a hundred times takes more time to draw and increases plotting time. Figure 4.42 shows how an array (rows and columns) is dimensioned.

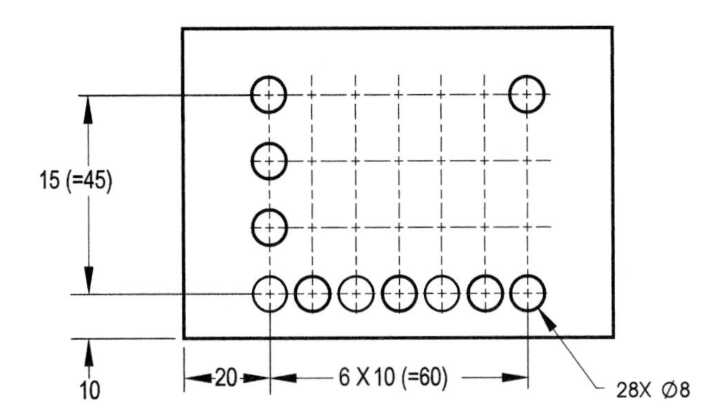

Figure 4.42 Dimensioning arrays

The drawing indicates that there are four rows and seven columns of 8 mm diameter holes. The holes are spaced 15 mm apart vertically and 10 mm apart horizontally.

The same feature can be repeated in a circular direction. In this case, angular dimensions may be more convenient than linear ones, although linear dimensions could be used. Figure 4.43 shows an array of holes arranged in a circle.

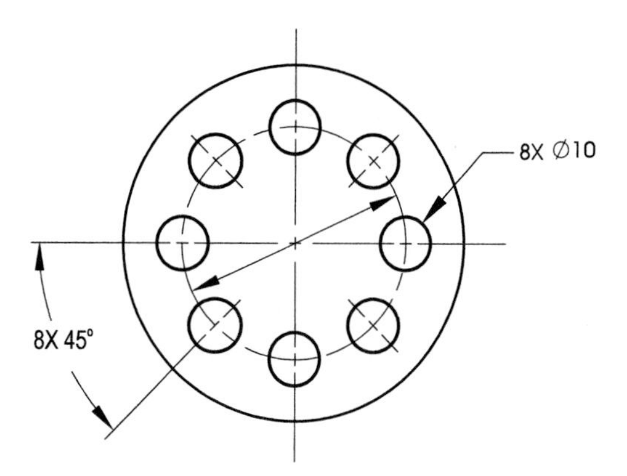

Figure 4.43 Angular dimensioning of a repeated feature

The same symbols are used to specify the number of repetitions and the spacing:

- 8X Ø10 means eight holes with a diameter of 10 mm; and
- 8X 45° means eight spaces of 45° (the holes are spaced at 45°).

It is common to locate the centre of each hole at the same radial distance. This distance is indicated by a centreline, and radial lines are drawn at some, or all, of the hole positions. The circle on which holes are located is sometimes called the **bolt circle diameter** (BCD), assuming bolts are used in the holes. At least one hole must be drawn to indicate the starting point. Circular arrays can be drawn easily with a computer program.

Linear dimensions can be used instead of radial dimensions to locate holes or other circular features, as shown in Figure 4.44.

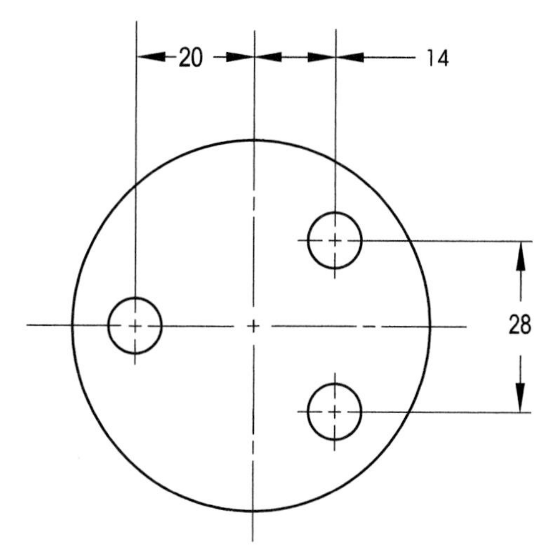

Figure 4. 44 Dimensioning radial features with linear dimensions

SCALES

Drawings are drawn "full size" with a computer, but they cannot be plotted full size if the object is large (an airplane, for example). A reduction scale, specified by the user, is applied when the drawing is plotted. One unit drawn on paper represents a larger number of units on the real object. For example, 1 mm on the paper represents 100 mm on the real object (1:100 scale). Dimensions shown on the drawing are actual dimensions.

Scales used on engineering drawings are not arbitrary: only certain ones are used. The scale must always be stated on the drawing.

Metric Scales

Metric scales used on engineering drawings are specified by national standards. Metric scales are expressed as:

1:1 1 mm on paper = 1 mm (this is full size)
1:2 1 mm on paper = 2 mm
1:5 1 mm on paper = 5 mm

Larger ratios follow the same sequence (1, 2, 5):

1:10	1:100	1:1000	1:10 000
1:20	1:200	1:2000	1:20 000
1:50	1:500	1:5000	1:50 000

A metric scale will typically have 1:5, 1:10, 1:20, 1:50, 1:100, 1:200, 1:500, and 1:1000 scales. Other scales can be used by simply adding or subtracting a zero.

When an engineering drawing is not done using a computer, it is drawn at the desired scale. A special ruler, called a **scale**, is used to lay out lines. Length can be read directly from the scale, and no calculations are required. Figure 4.45 shows a typical metric scale.

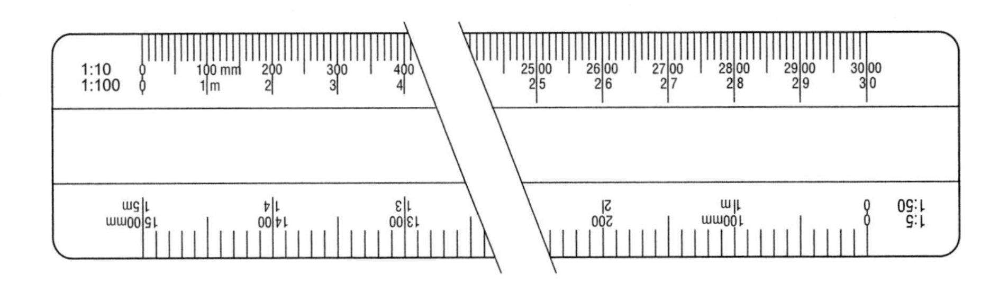

Figure 4.45 Typical metric scale

A length is marked off by reading the actual length on the scale and marking the end point. Figure 4.46 shows how lengths of 16 400 mm at a scale of 1:200 and 48.5 m at 1:500 are laid out. The desired length is read on the appropriate scale and marked on the line.

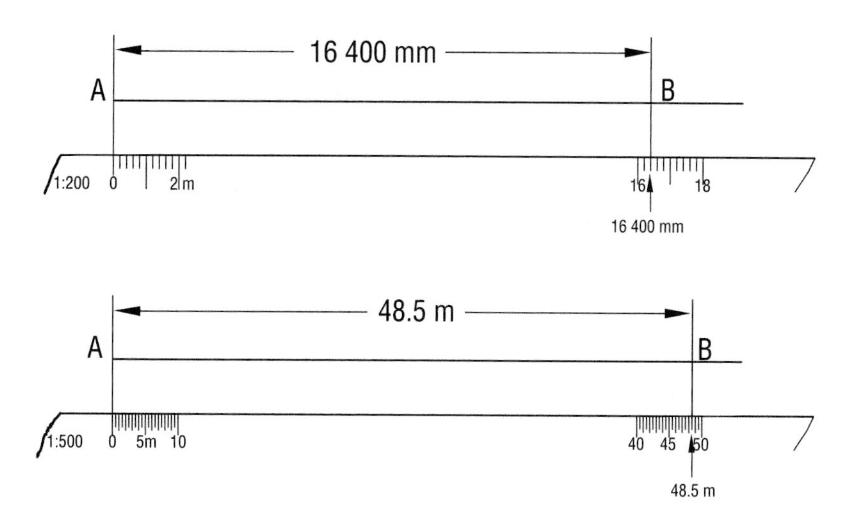

Figure 4.46 Laying out a length with a metric scale

Inch Scales

Scales, other than metric, are used on many engineering drawings, and you should be familiar with them. The common types of scales are an architect's scale and an engineer's scale (sometimes call a civil engineer's scale, although its use is not confined to civil engineering).

Architect's Scale

An architect's scale is divided into divisions representing feet and inches. You are familiar with a full-size scale, where a foot is divided into 12 inches and inches are divided into sixteenths. This full-size scale is suitable for drawing small objects but cannot be used to draw a building.

Standard architect's drawing scales are given below. The actual scale is given here for comparison only and is not stated on the drawing.

		Actual scale
3/32"	= 1'- 0"	(1:128) (one inch equals 128 inches)
1/8"	= 1'- 0"	(1:96)
3/16"	= 1'- 0"	(1:64)
3/8"	= 1'- 0"	(1:32)
1/2"	= 1'- 0"	(1:24)
3/4"	= 1'- 0"	(1:16)
1"	= 1'- 0"	(1:12)
1 1/2"	= 1'- 0"	(1:8)
3"	= 1'- 0"	(1:4)
Full Size	(1:1)	divided into sixteenths

The scale is stated on the drawing as 3/4" = 1'- 0". The actual scale (1:16, in this case) is not stated on the drawing.

A 1" = 1' - 0" scale is shown in Figure 4.47. The first inch on the left end, is divided into 12 major divisions, each representing one inch. The other divisions (1, 2, 3...), reading from the left, each represents one foot. The right end has a scale of 1/2"= 1' - 0", and the first half inch is divided into 12 major divisions, each representing one inch. The other 1/2 inch divisions (1, 2, 3), reading from the right, each represents one foot at a scale of 1/2" = 1' - 0".

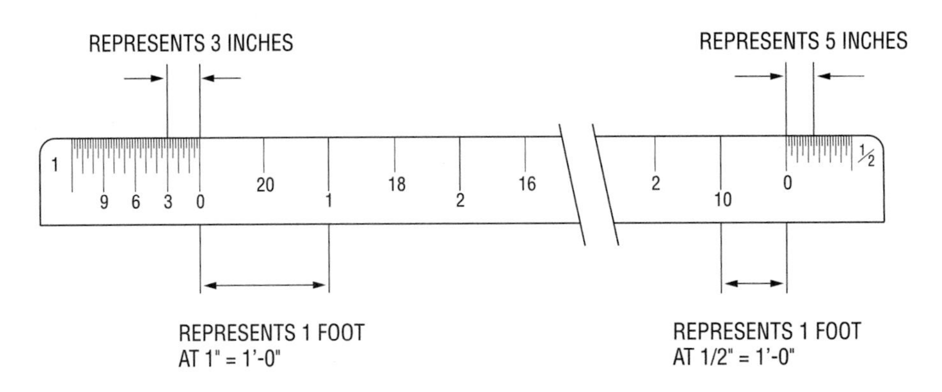

Figure 4.47 Architect's scale 1" = 1" - 0" (left end) and 1/2" = 1' - 0" (right end)

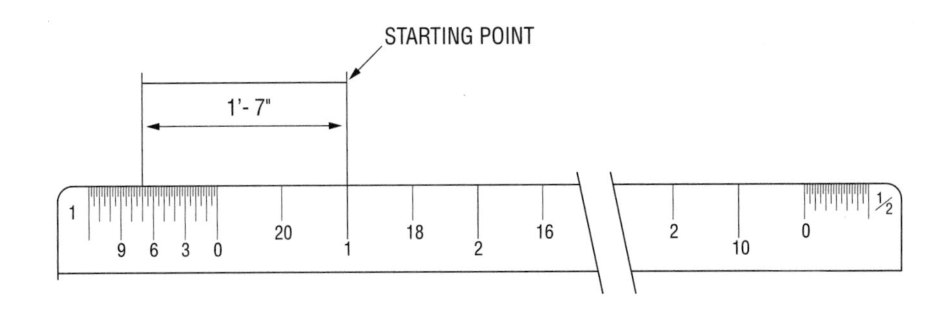

Figure 4.48 Laying out a length of 1' - 7" with an architect's scale, 1" = 1' - 0"

A line 1' - 7" long would be laid out starting from the right end and measuring to the left, as shown in Figure 4.48.

A larger reduction would be used to lay out longer lengths. Figure 4.49 shows a length of 4' - 8" laid out at a scale of 1/2" = 1' - 0". One half inch on the drawing represents one foot on the object (see also Figure 4.50).

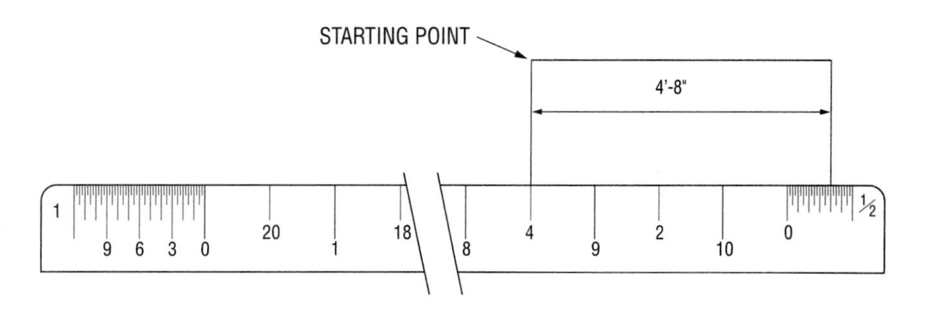

Figure 4.49 Laying out a length of 4′ - 8″ with an architect's, scale, 1/2″ = 1′ - 0″

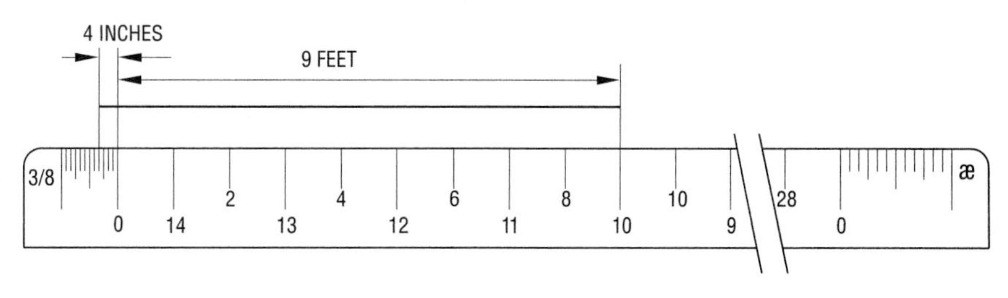

Figure 4.50 A length of 9′ - 4″ laid out at a scale of 3/8″ = 1′ - 0″

Engineer's Scale

The major divisions on an engineer's scale are one inch, but each scale has a different number of divisions per inch.

The scales are:

10	10 divisions per inch
20	20 divisions per inch
30	30 divisions per inch
40	40 divisions per inch
50	50 divisions per inch
60	60 divisions per inch

Figure 4.51 shows a 10 scale. The scale could be 1" = 10 ft, in which case the length represents 26 ft. The scale chosen could be 1" = 10 miles, in which case the length represents 26 miles.

Figure 4.52 shows a length laid out using the 50 scale. This length represents 190 kN at 1" = 50 kN.

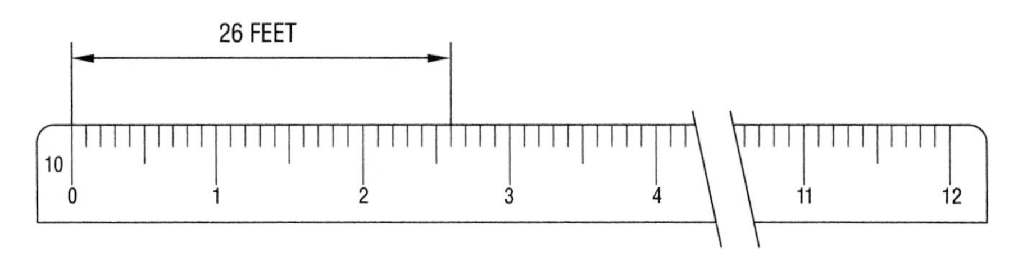

Figure 4.51 An engineer's scale used to lay out a length representing 26 ft at 1" = 10 ft

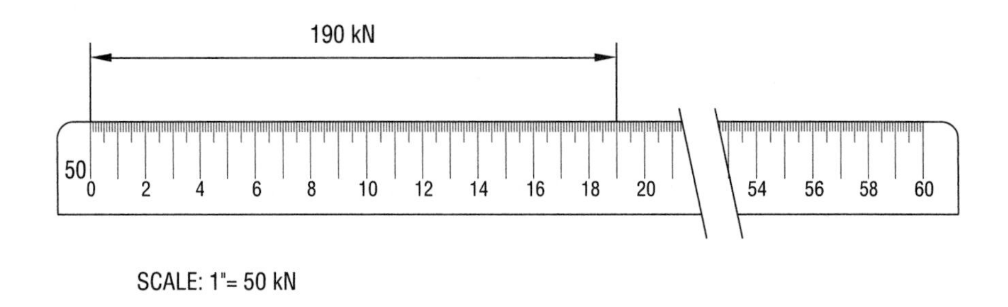

Figure 4.52 An engineer's scale used to lay out a length representing 190 kN at 1" = 50 kN

GEOMETRIC TOLERANCING

As we discussed in this chapter, tolerances are important because nothing can be made exactly, and parts must fit together. We have seen how tolerances on holes and shafts can prevent some shafts from fitting into some holes. We have also seen how tolerances can accumulate.

So far, we have made the assumption that when the diameter and tolerance of a shape, such as a cylinder, are specified, the cylinder is perfect and the sides are parallel. While the diameter could be within the tolerance, the axis could be curved. In other words, the form of the cylinder could be imperfect. Complete tolerancing requires tolerances on size, form, orientation, profile, and runout. These terms will be described later in this chapter.

Geometric tolerancing addresses the relationships between the shapes and positions of features to account for the fact that nothing can be made to an exact shape or located exactly. The following discussion gives a broad overview of some of the key features of geometric tolerancing. For more information on geometric tolerancing, please refer to standards such as CAN-CSA-B78-2, ASME Y14.5, and others listed in Appendix D.

Tolerance on Location

The following example will show how tolerances, when applied to the location of a hole, can result in problems with assembly, even though the hole is located within the tolerances specified. **Direct tolerancing** applies tolerances to the dimensions for locating the centre of a hole (Figure 4.53). With this form of tolerancing, the centre could be up to 0.7 mm from the desired location if it is at the corner of the enclosing square. This could cause problems with assembly.

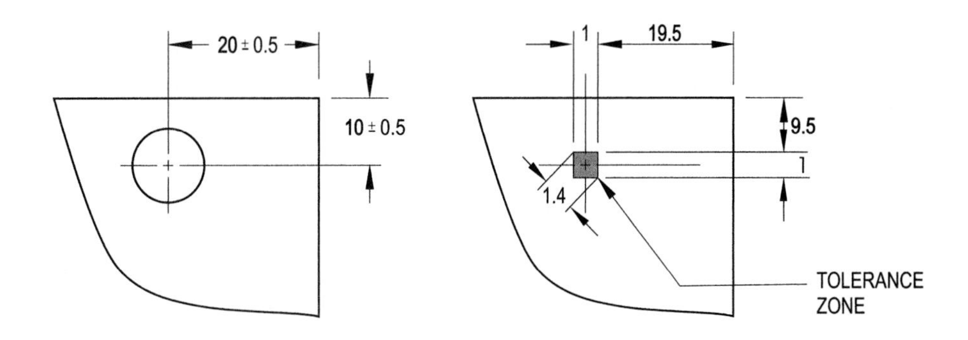

Figure 4.53 An error in location can result using direct tolerancing

A tolerance on the location of the centreline is used instead of a tolerance on the location dimensions. This is called **positional tolerancing**, and is shown in Figure 4.54. The centre of the hole must be inside the specified tolerance zone. A **basic dimension** locates the theoretical centre of the hole. The basic dimension is a numerical value used to describe the exact theoretical size. There are no tolerances on basic dimensions. They are the basis from which permissible variations are established by tolerances on other dimensions.

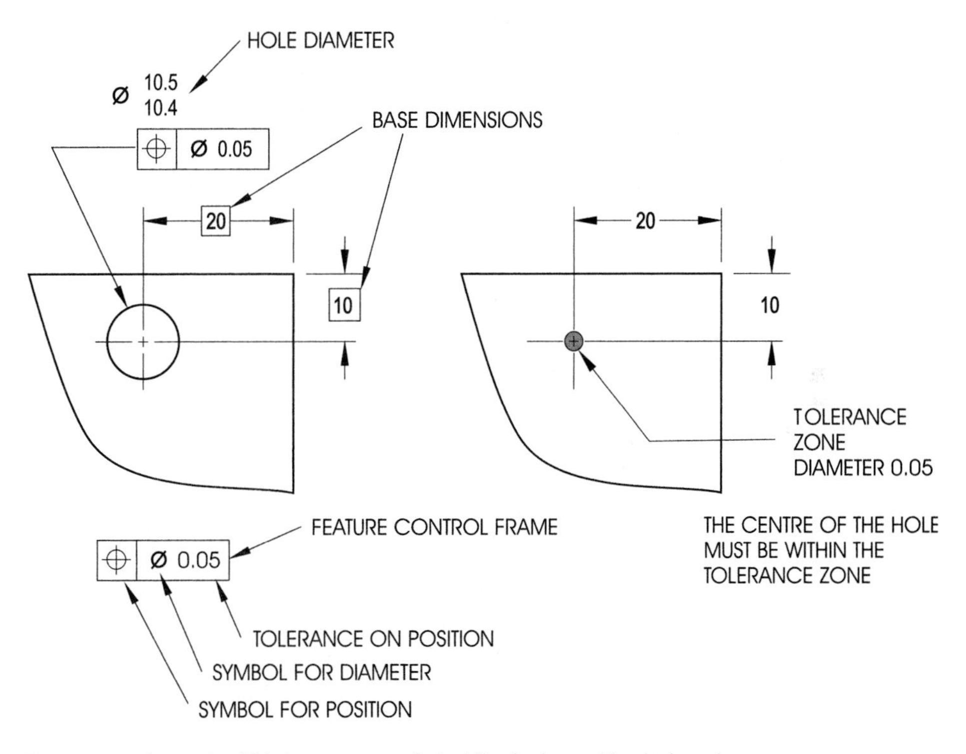

Figure 4.54 A zone in which the centre must lie is defined using positional tolerancing

Figure 4.54 shows how the tolerance is specified on the drawing (called the **drawing callout**), and an explanation of what the symbols mean. Basic dimensions locate the theoretical position of the centre, and a tolerance of 0.05 means that the centre must be within a circle of 0.05 mm diameter. Geometric tolerancing symbols are show in Figure 4.55.

	TYPE OF TOLERANCE	CHARACTERISTIC	SYMBOL
FOR SINGLE FEATURES	FORM	STRAIGHTNESS	—
		FLATNESS	⏥
		CIRCULARITY (ROUNDNESS)	○
		CYLINDRICITY	⌭
FOR SINGLE CORRELATED FEATURES	PROFILE	PROFILE OF A LINE	⌒
		PROFILE OF A SURFACE	⌓
FOR RELATED FEATURES	ORIENTATION	ANGULARITY	∠
		PERPENDICULARITY	⊥
		PARALLELISM	//
	LOCATION	POSITION	⌖
		CONCENTRICITY** (COAXIALITY)	◎
		SYMMETRY**	⌯
	RUNOUT	CIRCULAR RUNOUT*	↗
		TOTAL RUNOUT*	⌰

*ARROWHEAD(S) MAY BEFILLED IN.

**"CONCENTRICITY AND SYMMETRY" REQUIREMENTS MAY ALSO BE INDICATED BY A "POSITIONAL" TOLERANCE.

Figure 4.55 Symbols used for geometric tolerancing

(Source: With the permission of Canadian Standards Association, material is reproduced from CSA Standard, CAN/CSA-B78.2-M91, Dimensioning and Tolerancing of Technical Drawings, which is copyrighted by Canadian Standards Association, 178 Rexdale Blvd., Toronto, Ontario, M9W 1R3, www.shopcsa.ca. While use of this material has been authorized, CSA shall not be responsible for the manner in which the information is presented, nor for any interpretations thereof.)

Tolerance on Form

Form includes straightness, flatness, circularity, and cylindricity. A round shaft may not be perfectly straight, and this would prevent it from passing through a hole. Similarly, if the shaft is not perfectly round, it may not fit into the hole depending on the tolerances on the hole. It is not unusual for structural steel members to be bowed (although they are within allowable tolerances), and this must be taken into account if a long steel section is to be welded to other sections.

Straightness Tolerance

A **straightness tolerance** specifies a zone within which all points on the axis, or surface, must lie. Straightness tolerance can be applied to a centreline, or a **generator line** of a curved surface, or a flat surface. A generator line is a line on the surface of a solid (cylinder or cone, for example), which generates the surface when rotated about the axis. Generator lines for a cylinder are parallel to the axis and are rotated around the axis to generate the surface of the cylinder. Figure 4.56 shows how this is specified on a drawing and what it means. The leader on the control feature box points to the feature to which the tolerance applies, in this case the surface.

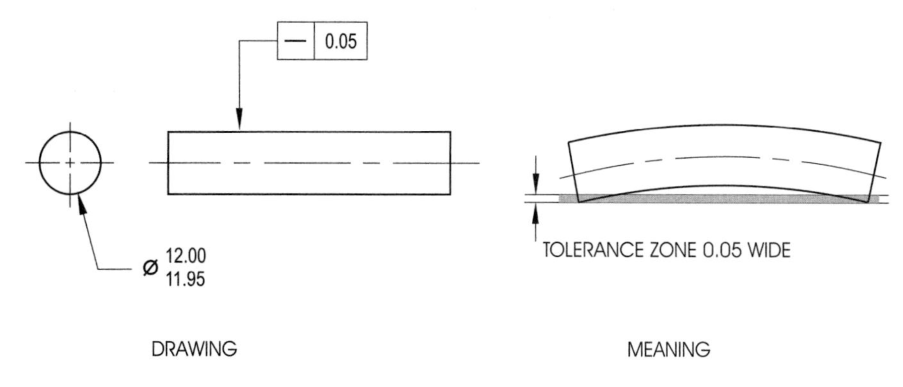

Figure 4.56 All lines in the surface must lie in the tolerance zone

Flatness Tolerance

A **flatness tolerance** specifies a zone between two parallel planes in which all lines in the surface must lie. Figure 4.57 shows how flatness is specified on a drawing and the meaning of the symbols.

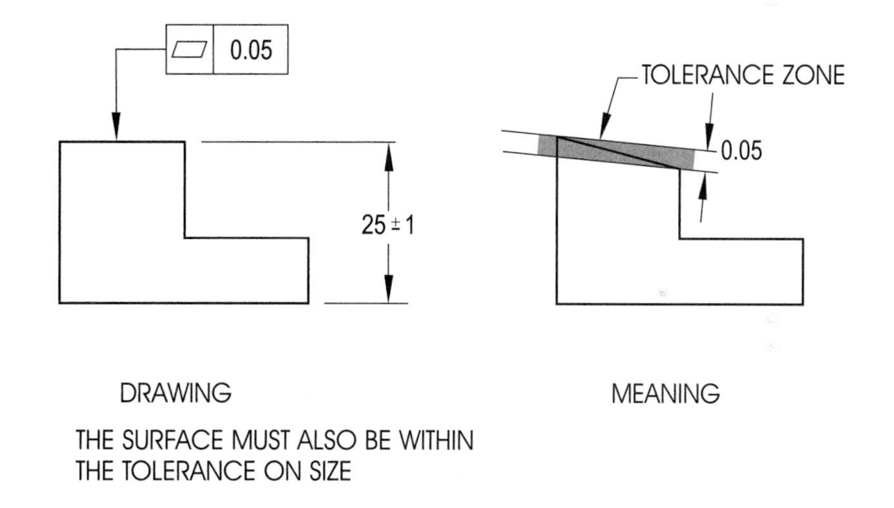

Figure 4.57 All lines in the surface must lie in the tolerance zone defined by two parallel planes

Circularity Tolerance

Circularity tolerance is a tolerance on roundness. All points on the surface of the circular feature must lie within the tolerance zone defined by two concentric circles, which define an annulus equal to the specified tolerance. Figure 4.58 shows the symbols and their meaning. Circularity tolerance for a cylinder applies over the length of the cylinder, and the tolerance symbol can be shown either on the end view or on the view in which the length of the cylinder is seen. The tolerance on diameter must also be met.

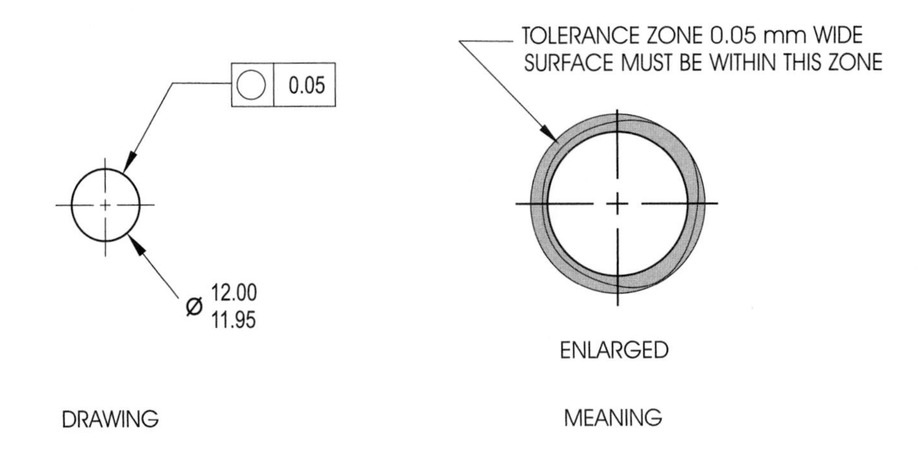

Figure 4.58 All points on the perimeter of the circular feature must lie in the shaded zone

Cylindricity Tolerance

Cylindricity tolerance applies to cylinders and defines a tolerance zone in which all points on the surface must lie. The tolerance zone applies over the length of the cylinder, so the tolerance zone can be shown on the view showing the end of the cylinder, or the view showing cylinder length. Figure 4.59 shows how the tolerance zone is specified and the meaning of the symbols. The tolerance on the diameter must also be satisfied.

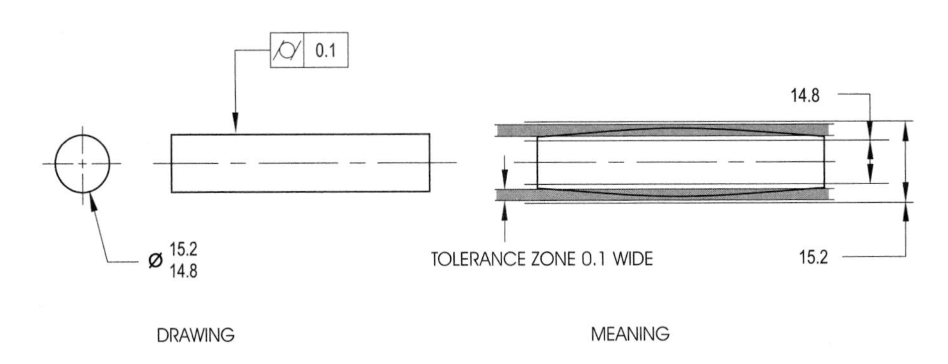

Figure 4.59 All points on the surface must lie in an annulus defined by two concentric cylinders

Orientation Tolerance

Orientation tolerance deals with the orientation of one feature relative to another. There are tolerances on angularity, how much a angle can vary, parallellism, how far two surfaces can vary from being parallel, and perpedicularity. Since orientation tolerances apply to one feature, relative to another, a **datum** must be specified (e.g., an angle must be measured between two planes). A datum is a theoretical exact reference to which toleranced features are related.

Angularity Tolerance

Angularity tolerance defines a zone between two parallel planes in which all lines in the surface must lie. The parallel planes are at the base angle. Figure 4.60 shows how this is specified on a drawing and the meaning of the symbols.

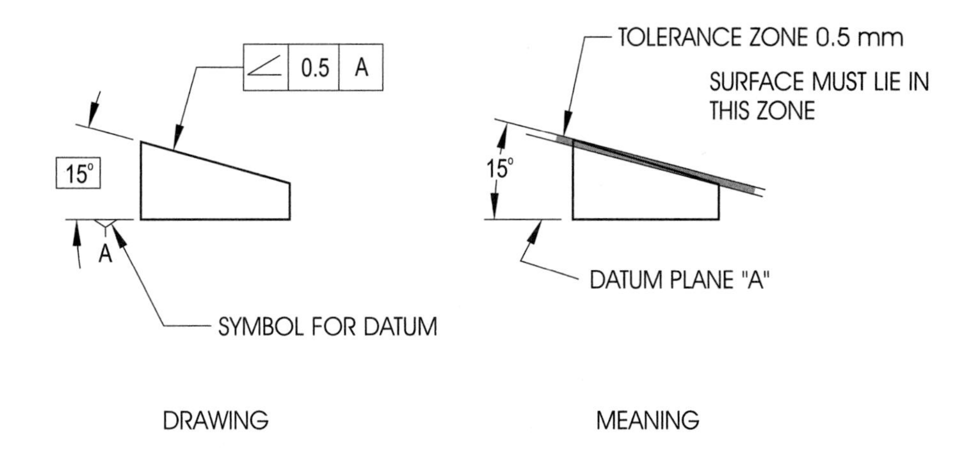

Figure 4.60 Tolerance zone is defined by two parallel planes at the base angle

Parallelism Tolerance

Parallelism tolerance specifies a tolerance zone defined by two planes, or lines, parallel to a datum surface, or an axis. All elements of the surface, or centreline, must lie in this zone. Figure 4.61 shows how this tolerance is specified for a surface and an axis. The specified surface must be in a zone 0.5 mm wide, parallel to a specified datum, "A" in this example.

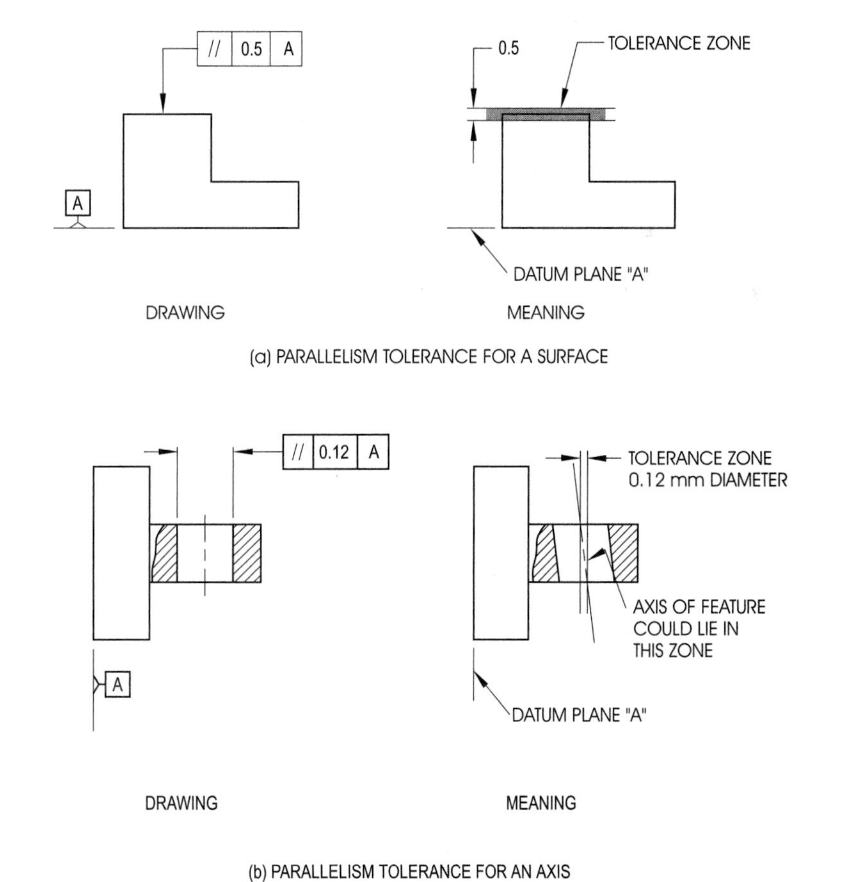

Figure 4.61 Specifying parallelism tolerance for a surface and an axis

A parallelism tolerance applied to a centreline of a feature, say, a hole, defines a cylindrical tolerance zone whose sides are parallel to a specified datum.

Perpendicularity Tolerance

Perpendicularity refers to the condition of one feature being at a right angle to another feature or datum, plane, or axis. Figure 4.62 shows how perpendicularity is specified for a surface, a median plane, and an axis.

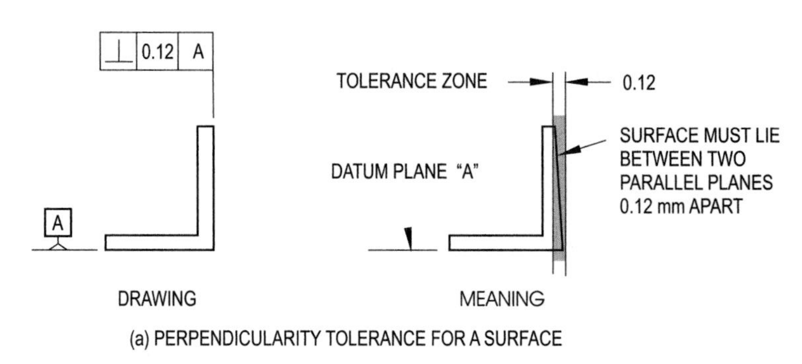

(a) PERPENDICULARITY TOLERANCE FOR A SURFACE

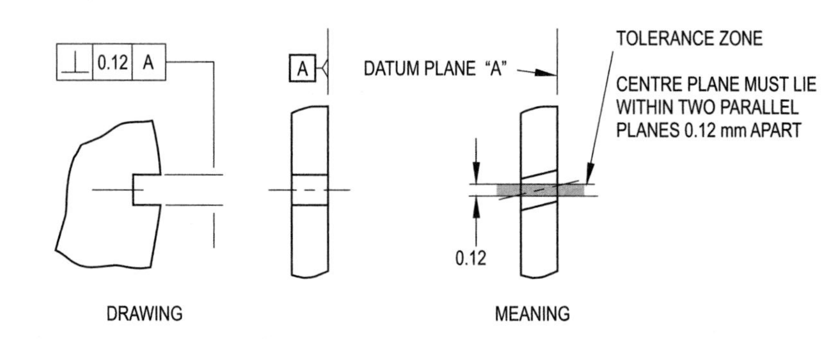

(b) PERPENDICULARITY TOLERANCE FOR MEDIAN PLANE OF A FEATURE

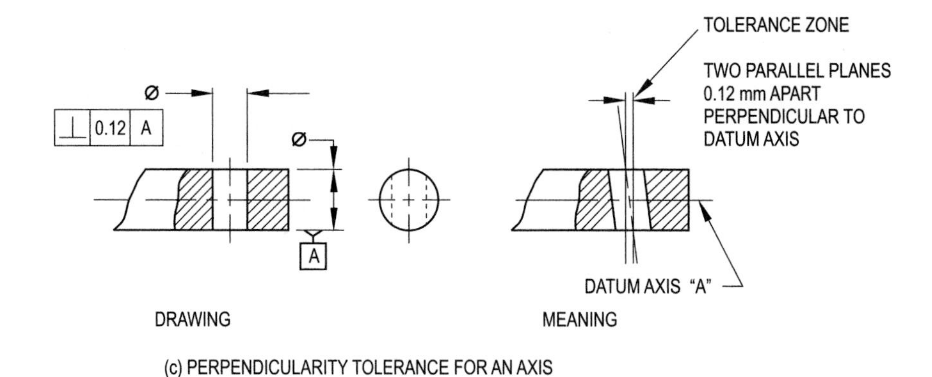

(c) PERPENDICULARITY TOLERANCE FOR AN AXIS

Figure 4.62 Perpendicularity tolerance for a surface, a median plane, and an axis

Profile Tolerance

The **profile tolerance** specifies a uniform boundary along the true profile. All elements of the surface must lie within this boundary. The profile tolerance controls only surface elements. Figure 4.63 shows how profile tolerance is specified and the meaning of the symbols.

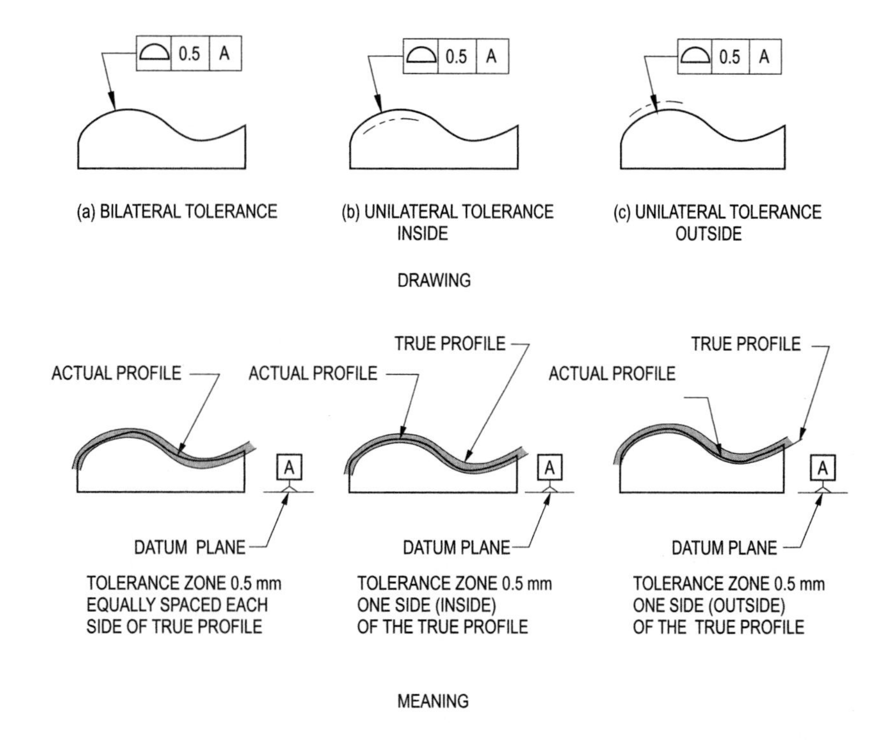

Figure 4.63 Specifying profile tolerance

The true surface is in the middle of the tolerance zone for a bilateral tolarance (Figure 4.63a). The true surface is at the top of the tolerance zone if the unilateral tolerance is inside (Figure 4.63b) and at the bottom if it is outside (Figure 4.63c).

Runout

Runout is a composite tolerance used to control the functional relationship of one or more features of a part to a datum axis. If a rotating shaft is not perfect, one end, or the centre, will wobble when the shaft rotates. The magnitude of this wobble is called *runout.* The **runout tolerance** is a zone within which the outer surface of the shaft must be as it rotates. An **indicator gauge** is used to measure the lateral movement as the shaft rotates. Figure 4.64 shows how runout is called up and the meaning of the symbols.

0.04 A

A

ØXXX

DRAWING

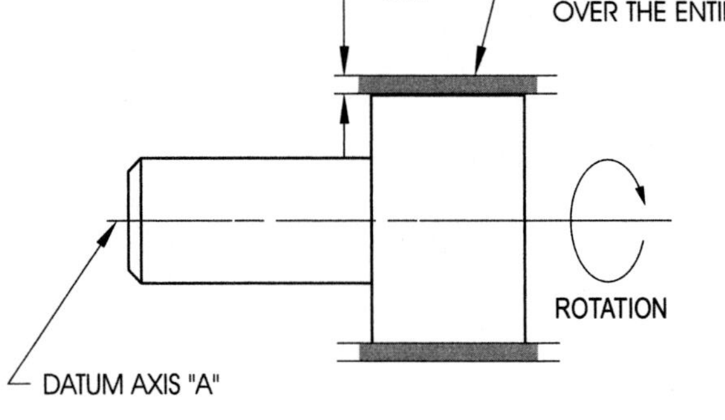

MEANING

THE ENTIRE SURFACE MUST BE WITHIN THE TOLERANCE ZONE
WHEN THE PART IS ROTATED 360° ABOUT THE DATUM AXIS

Figure 4.64 Total runout specified to a datum diameter

Problems

Dimension the following objects in accordance with the guidelines given in this chapter. Dimensions are found by scaling the drawings and rounding dimensions to the nearest millimetre, or as specified by the instructor. The scale must be specified on the drawing.

The finished drawing (or sketch) can be passed to another member of your work group for checking. The "checker" marks all errors and omissions and returns it to the originator for correction.

26.

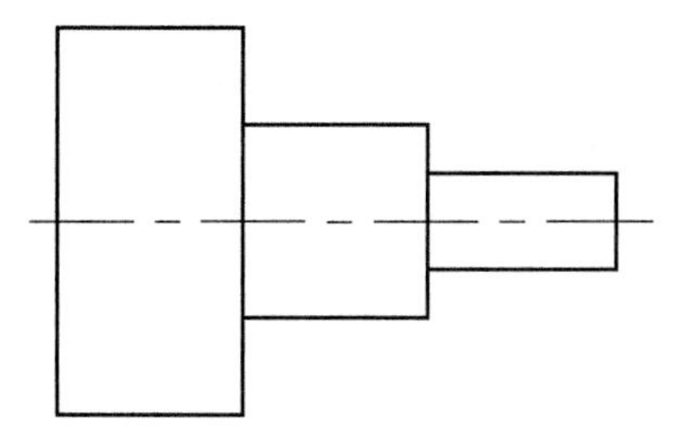

27.

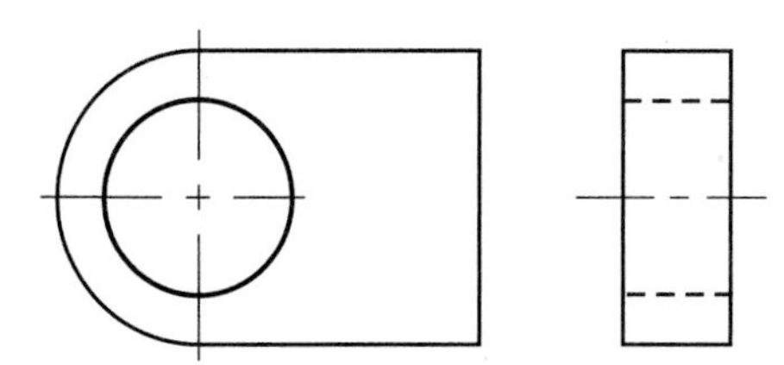

28.

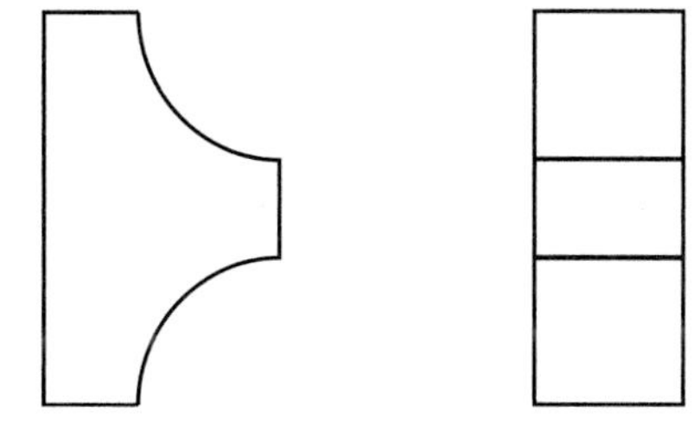

29.

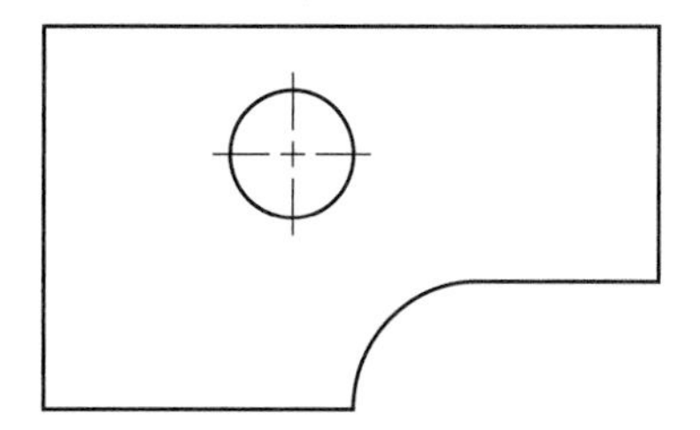

30.

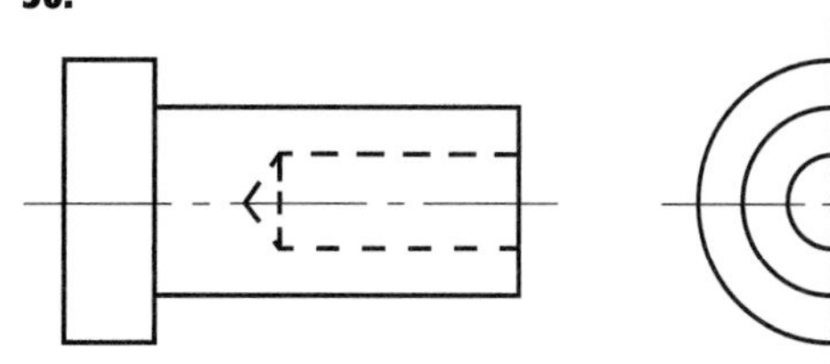

31.

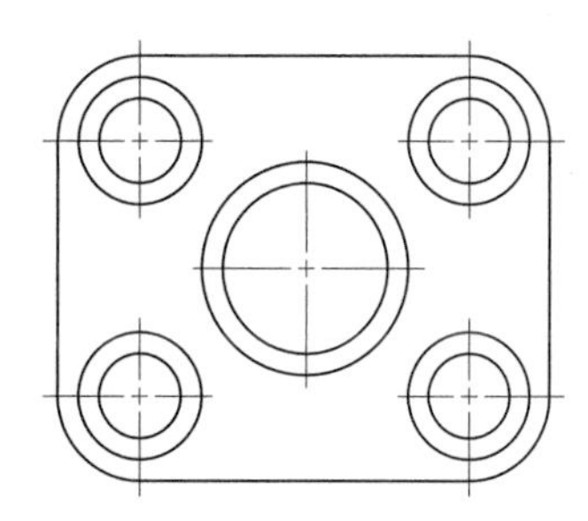

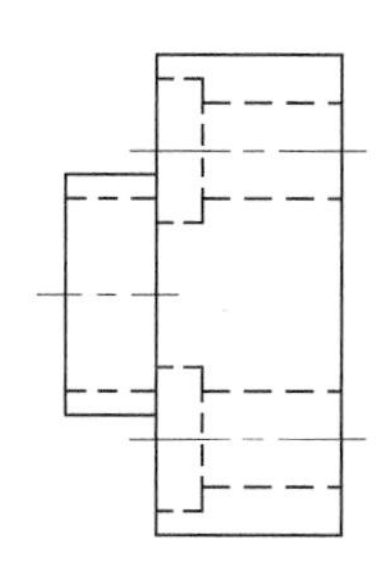

32.

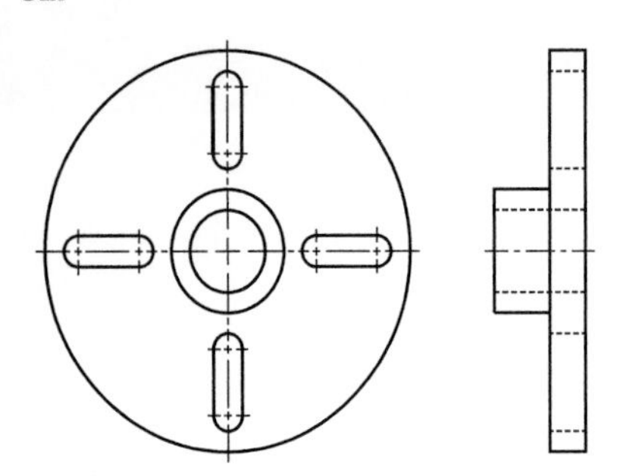

33.

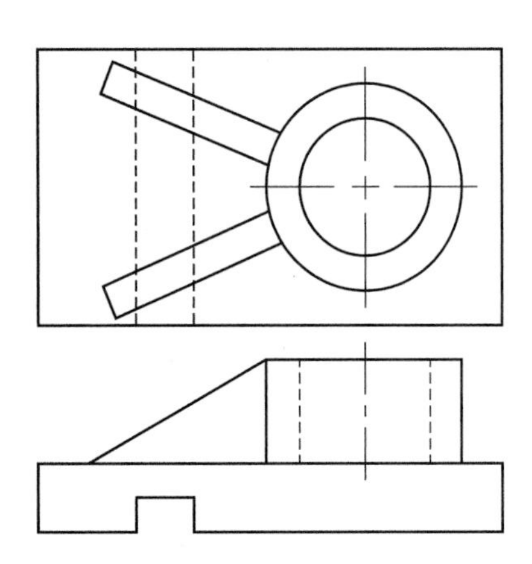

34.

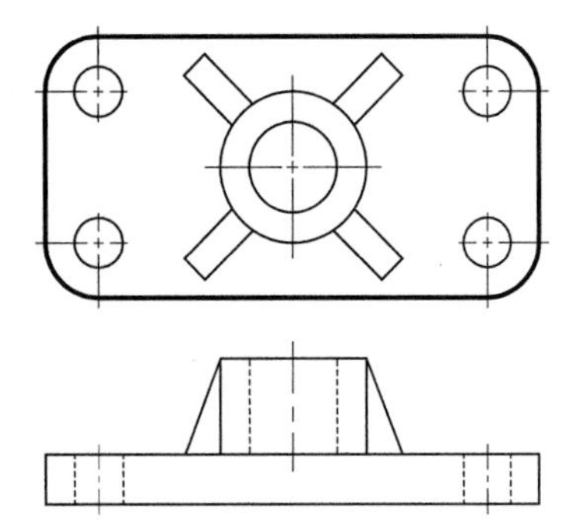

35.

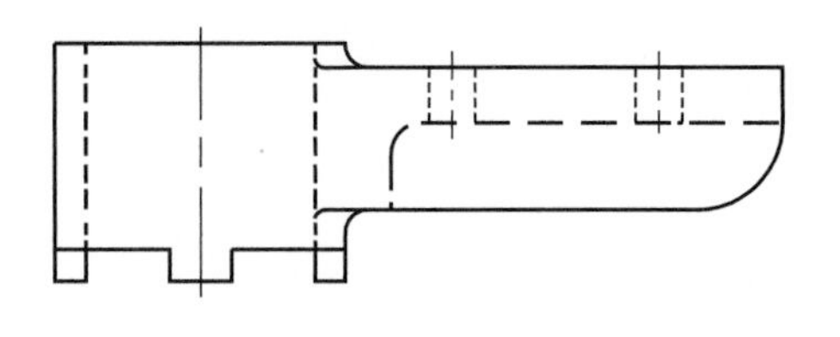

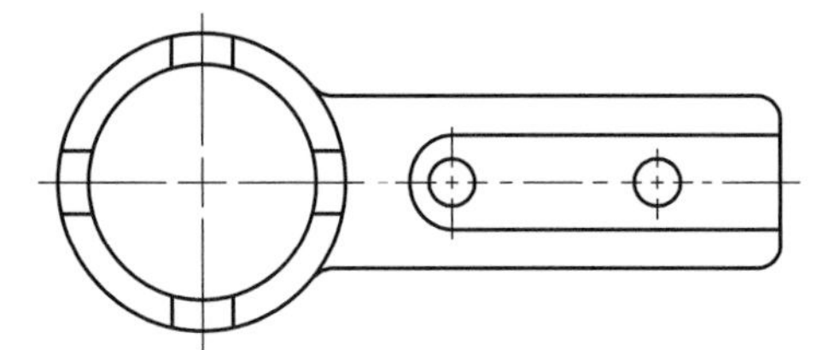

36.

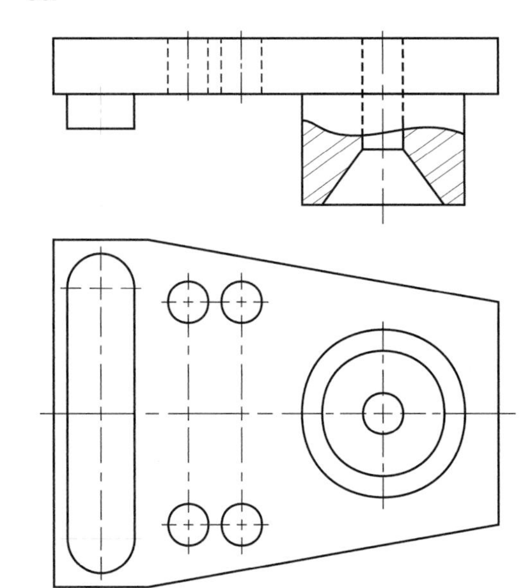

37.

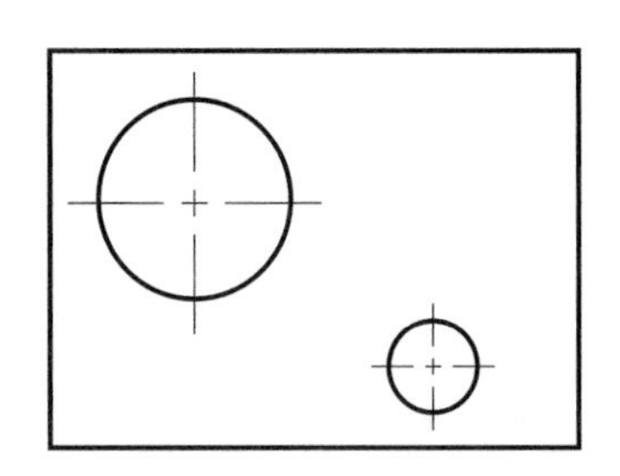

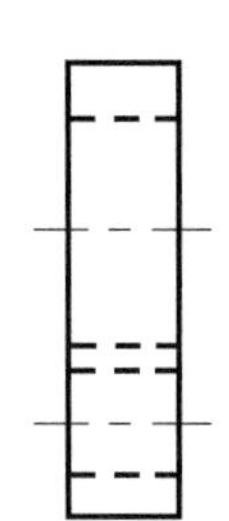

38.

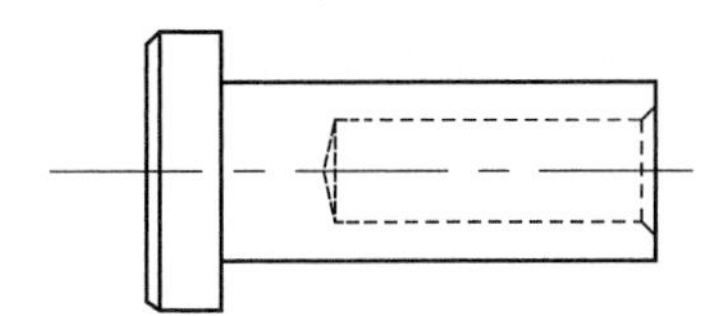

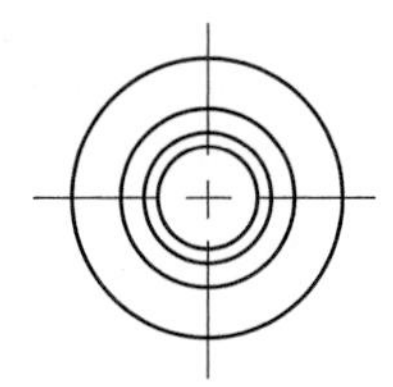

39.

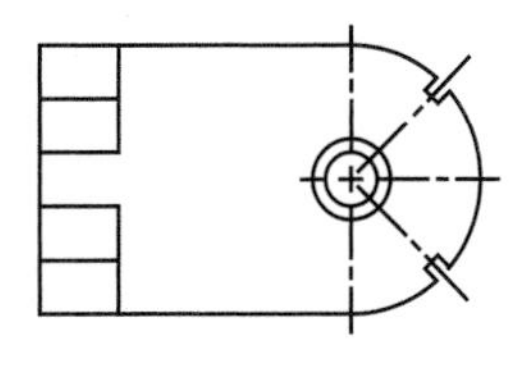

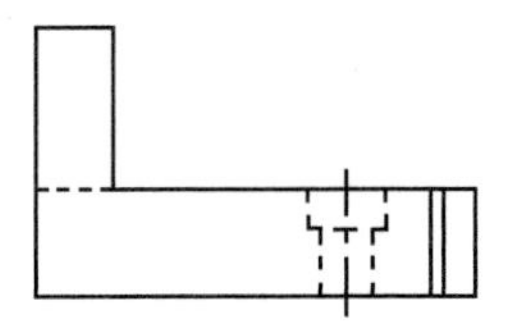

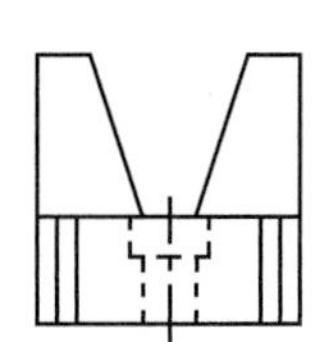

40.

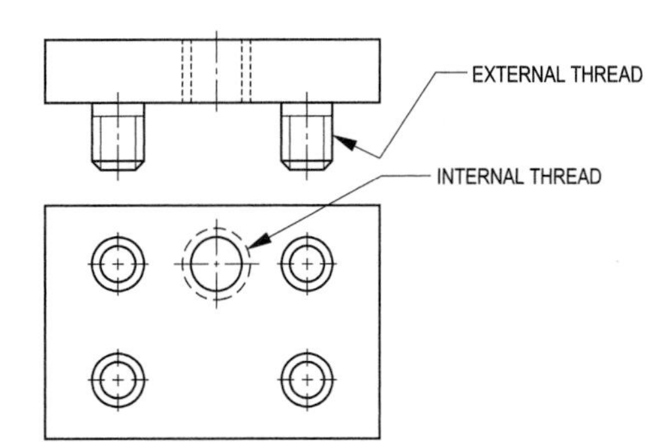

41.

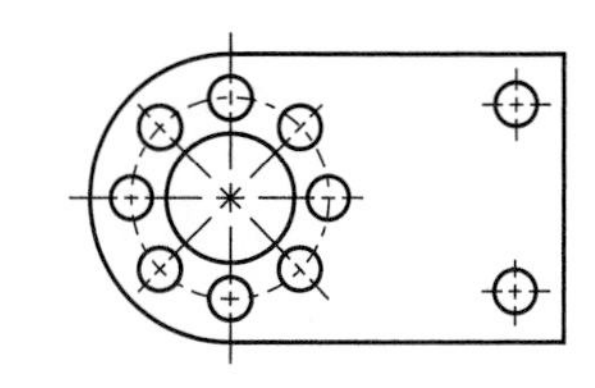

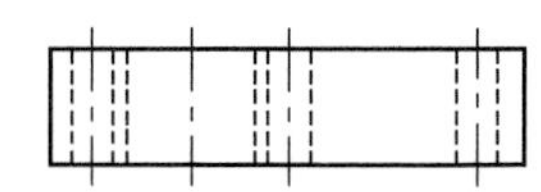

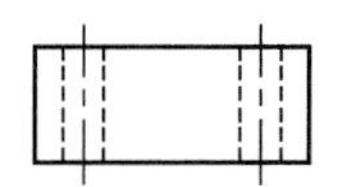

42. Create a dimensioned drawing of a 20 mm diameter by 250 mm long shaft threaded at both ends with a coarse thread. Thread length is 50 mm.

43. Create a dimensioned drawing of a shaft 20 mm in diameter by 200 mm long and threaded at one end with a fine thread. Thread length is 50 mm.

44. Detail the end connections for a W 610 × 140 beam. The end plate beam connectors will have 4 20 mm bolts arranged vertically. You will have to look up the dimensions of the beam and connector.

Draw dimensioned working drawings for all parts of the objects shown in the following figures. Dimensions are given in millimetres unless noted otherwise. Because these are isometric drawings, the locations of the dimensions are not as they would be shown on a working drawing. Dimensioning on your orthographic drawings should follow applicable standards. All dimensions are not given explicitly; you will have to determine some dimensions from the information given.

45.

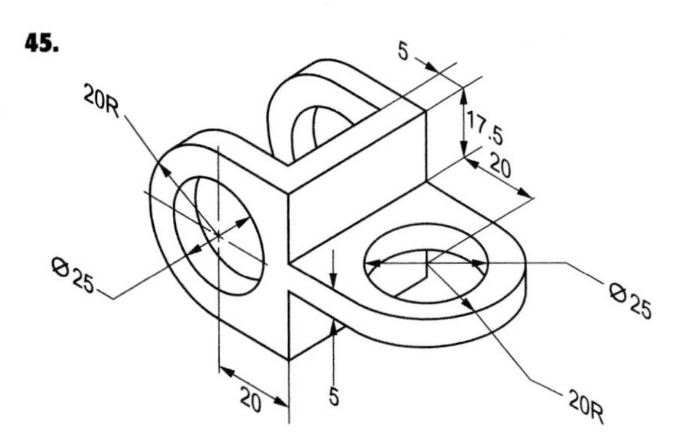

46.

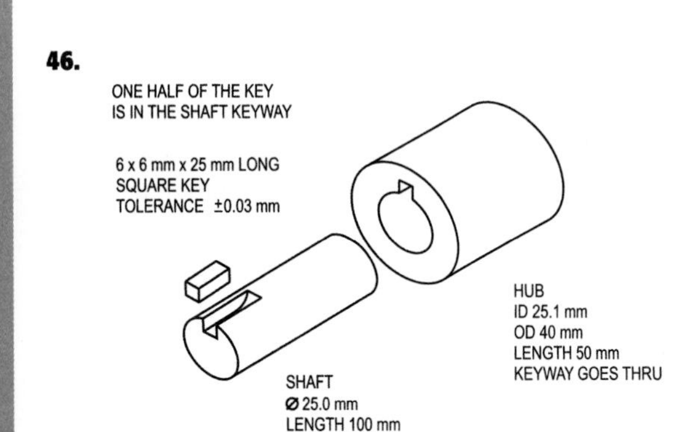

47.

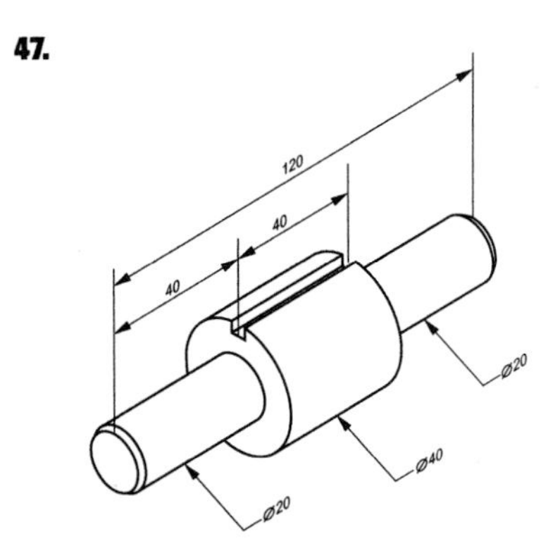

CHAMFER EACH END
2 x 45°

UNDERCUT SHOULDER
RADIUS UNDERCUT
1 x 18 2X

KEYWAY FOR 12 x 8 KEY
(12 mm HIGH x 8 mm WIDE)
DEPTH OF KEYWAY EQUALS ½ KEY

48.

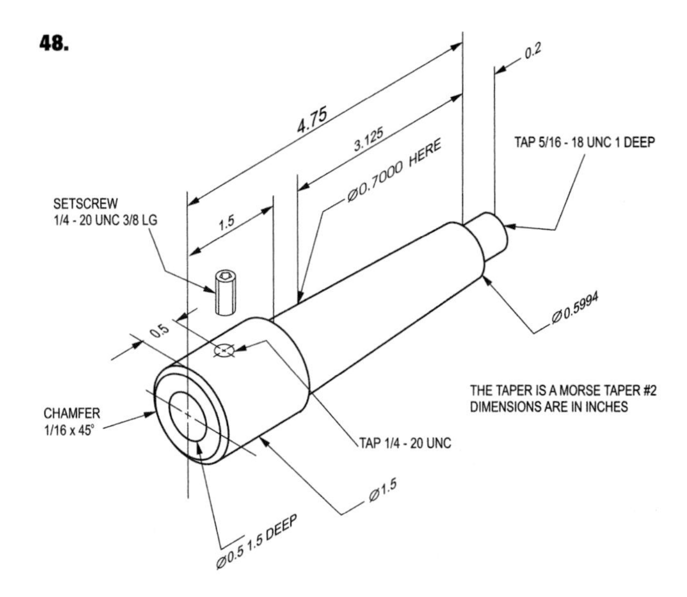

49.

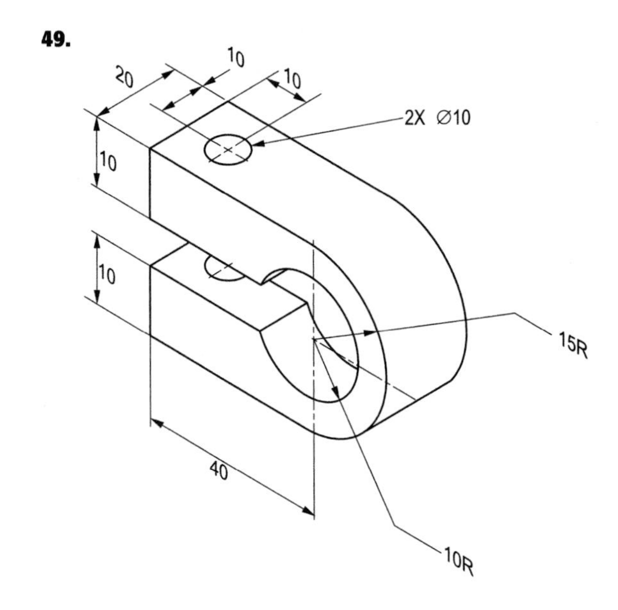

50.

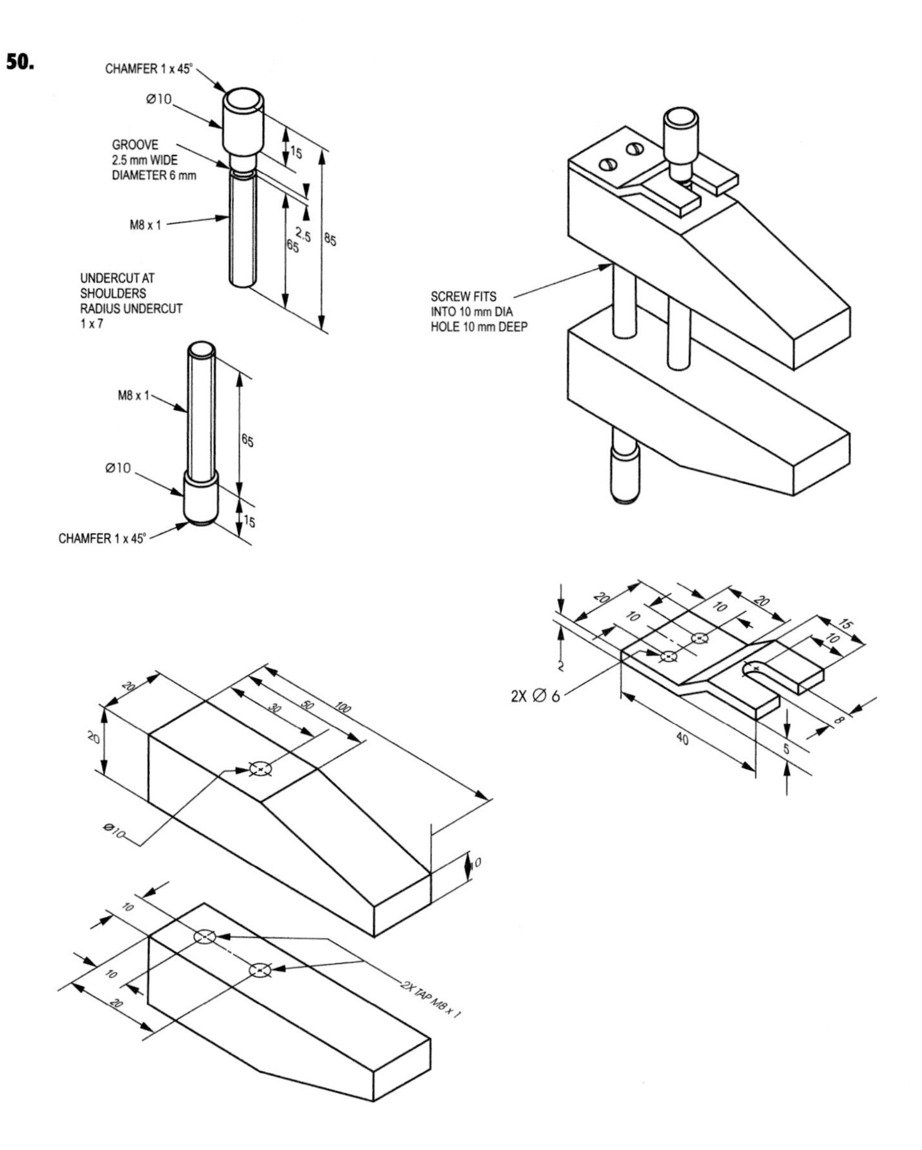
CHAMFER 1 x 45°
Ø10
GROOVE
2.5 mm WIDE
DIAMETER 6 mm
15
M8 x 1
2.5
65
85
UNDERCUT AT
SHOULDERS
RADIUS UNDERCUT
1 x 7
M8 x 1
65
Ø10
15
CHAMFER 1 x 45°
SCREW FITS
INTO 10 mm DIA
HOLE 10 mm DEEP
20
10
10
20
15
10
2X Ø 6
8
40
5
20
20
30
50
100
Ø10
10
10
10
20
2X TAP M8 x 1

51.

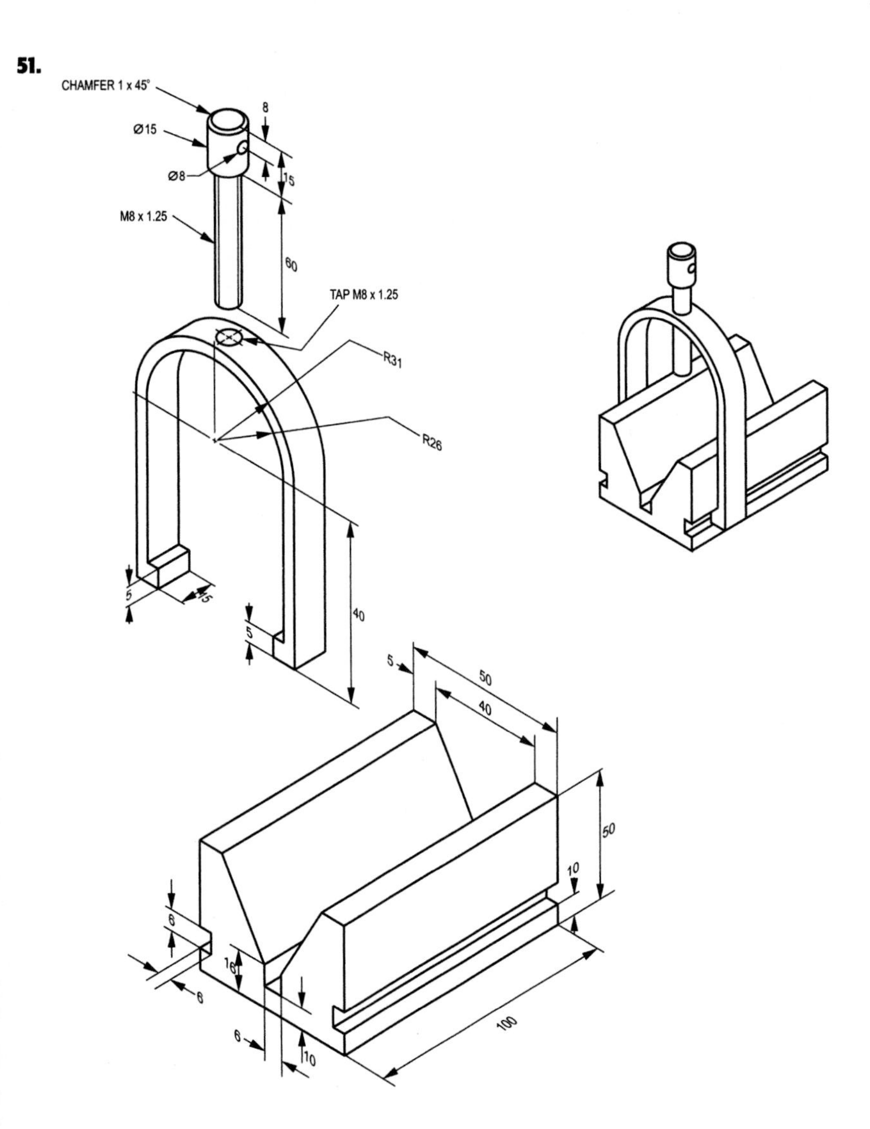
CHAMFER 1 x 45°
Ø15
Ø8
8
15
M8 x 1.25
60
TAP M8 x 1.25
R31
R26
5
5
5
40
5
50
40
50
10
6
16
6
6
10
100

52.

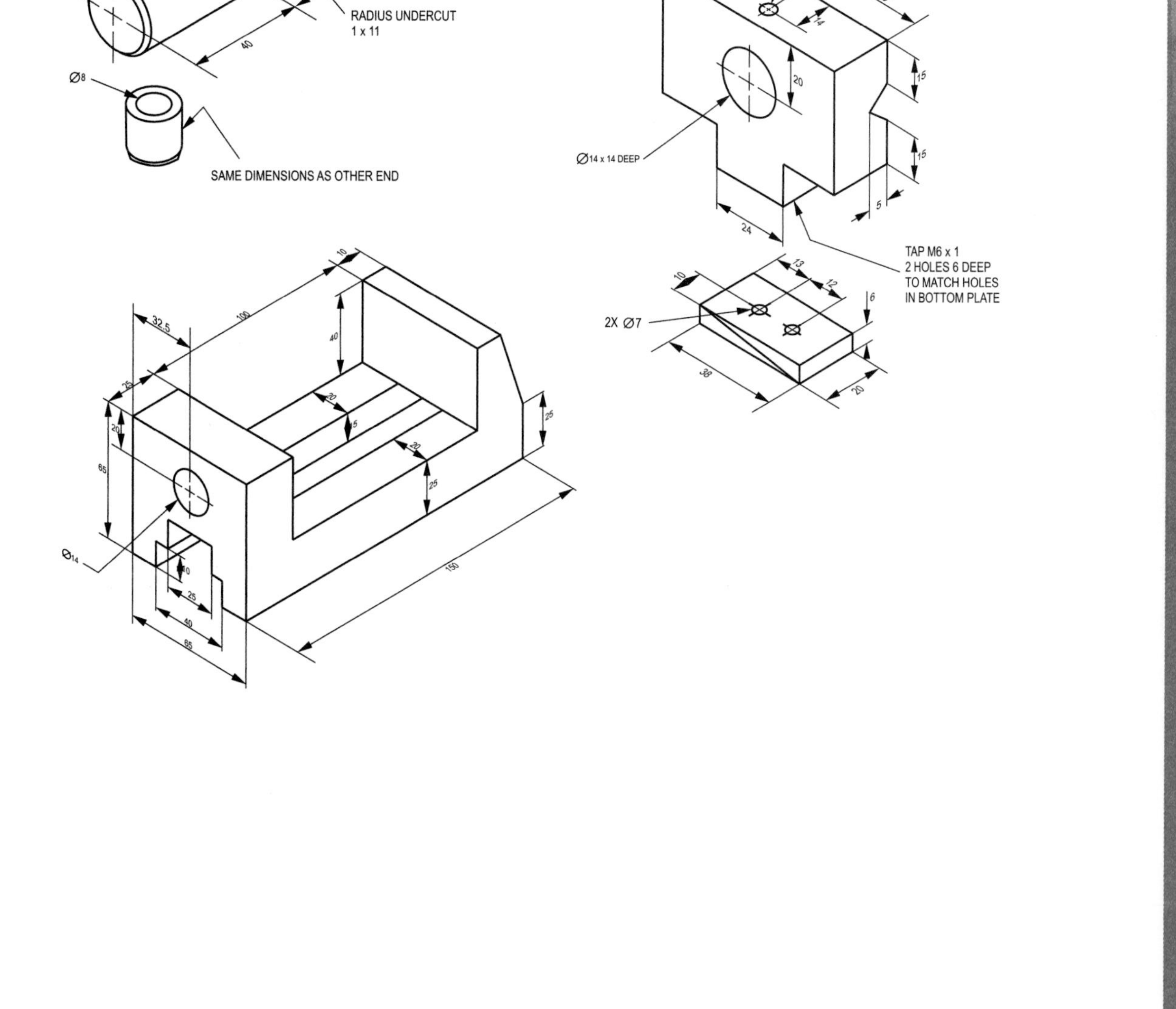
CHAMFER
1 x 45°
Ø12
Ø8
M12 x 1.75
GROOVE
7 mm WIDE
Ø6
Ø10
Ø20
CHAMFER
2 x 45°
RADIUS UNDERCUT
1 x 11
Ø8
SAME DIMENSIONS AS OTHER END
SETSCREW
M6 x 1 16 LONG
TAP M6 x 1
Ø14 x 14 DEEP
TAP M6 x 1
2 HOLES 6 DEEP
TO MATCH HOLES
IN BOTTOM PLATE
2X Ø7

53.

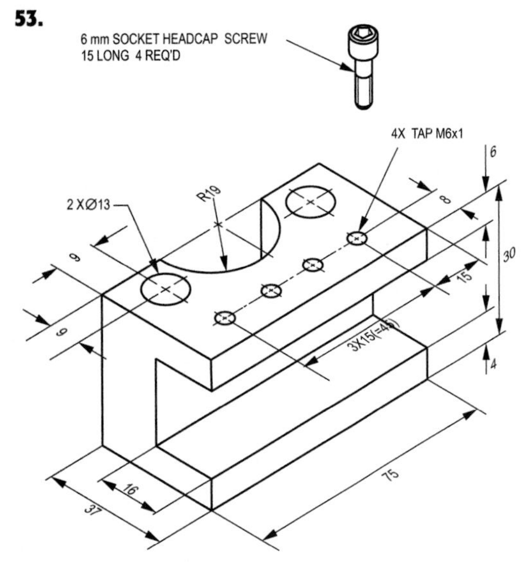
6 mm SOCKET HEADCAP SCREW
15 LONG 4 REQ'D
4X TAP M6x1
2 XØ13
R19
9
9
6
8
30
15
3X15(=45)
4
16
37
75

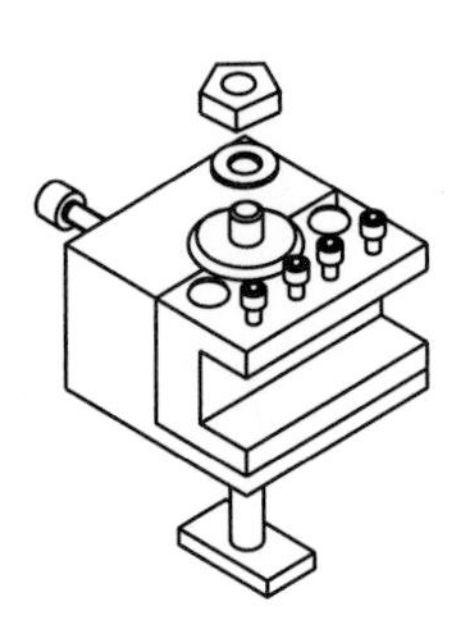

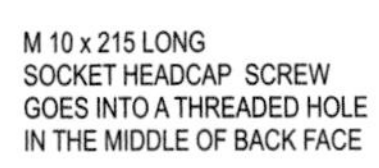
M 10 x 215 LONG
SOCKET HEADCAP SCREW
GOES INTO A THREADED HOLE
IN THE MIDDLE OF BACK FACE

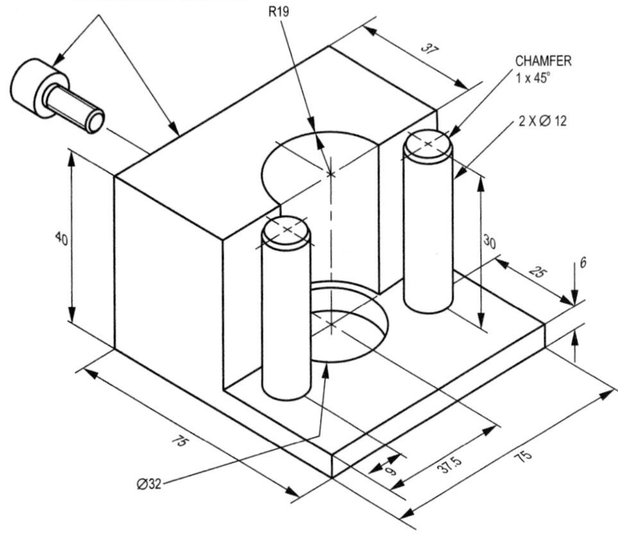
R19
37
CHAMFER
1 x 45°
2 X Ø 12
40
30
25
6
75
Ø32
9
37.5
75

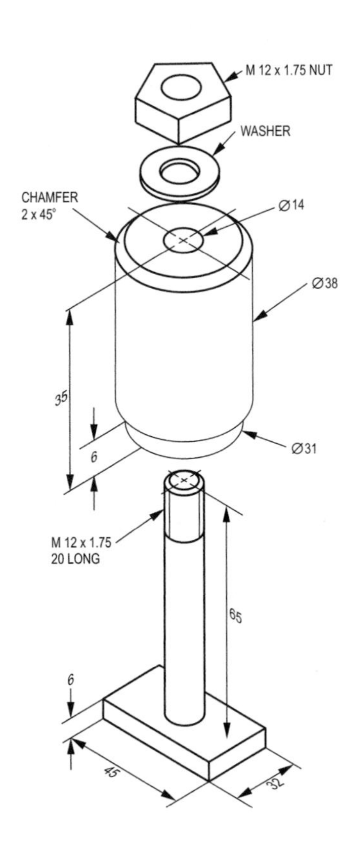
M 12 x 1.75 NUT
WASHER
CHAMFER
2 x 45°
Ø14
Ø38
35
6
Ø31
M 12 x 1.75
20 LONG
65
6
45
32

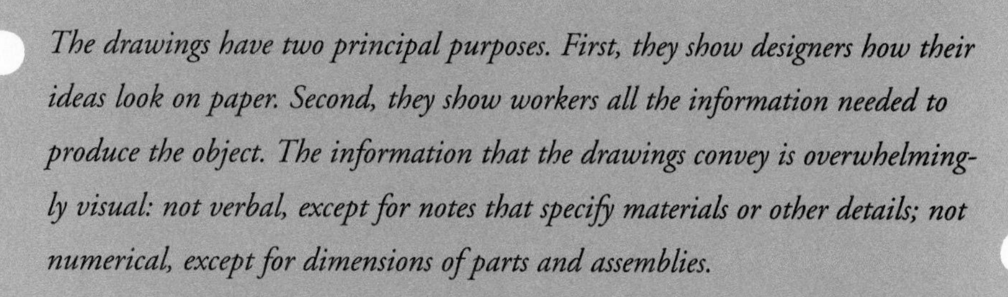

The drawings have two principal purposes. First, they show designers how their ideas look on paper. Second, they show workers all the information needed to produce the object. The information that the drawings convey is overwhelmingly visual: not verbal, except for notes that specify materials or other details; not numerical, except for dimensions of parts and assemblies.
Eugene S. Ferguson, Engineering in the Mind's Eye, *The MIT Press, 1992*

CHAPTER

5 Engineering Drawings

Engineering drawings, sometimes referred to as **technical drawings**, are used to communicate design, project, and production information, under the direction of an engineer.

Engineering covers many different disciplines; for example, mechanical, mining, metallurgical, structural, civil, environmental, geotechnical, electrical, electronic, industrial, petroleum, and nuclear, to name a few. Engineering drawings that are used to supply information and instructions for the manufacture of products or the construction of a structure, are called **working drawings** or **production drawings**.

WORKING DRAWINGS

Working drawings are a complete set of documents specifying all of the information required to make or build the design. Working drawings may be on one or more sheets depending on the size and complexity of the design and may include written specifications or instructions.

A working drawing is a legal document and should be drawn in accordance with the relevant national or international standards governing the particular discipline. Many companies adopt their own standards that are generally based on the national standards.

Examples of national standards are:

Canada	CSA B78
USA	ANSI Y14
UK	BS 308

National standards are internationally referenced by the International Organization for Standardization (ISO) in Geneva, Switzerland. The ISO standard for technical drawings comprises some 71 international standards grouped under: ISO/TC 10, *Technical Drawings, product definition and related documentation.*

In most cases, engineers do not create working drawings, although this could happen in a small enterprise. Usually, they leave the drafting to technicians or drafters. However, it is essential that engineers are familiar with the standards since they (the engineers) are ultimately responsible for the content and correctness of the documents, and generally stamp them accordingly. Engineers must therefore be able to interpret what is drawn and be able to give instructions for design changes (**engineering change orders**), if necessary.

Working, or engineering, drawings must be clear as to the designer's intent and must contain all of the information necessary to manufacture or build the design. Any misunderstanding or ambiguity when reading the drawings can cause expensive changes during production or building. This, in turn, wastes time and money. Because it is a legal document, the information on the working drawing must be accurate. If there is any failure to the design or structure, people will first look at the information on the working drawing from which the design was implemented.

Working drawings can take several different forms depending on the discipline; for example, in mechanical, civil, chemical, and electrical/electronic engineering, engineering drawings can be generalized for content-discussion purposes. First, we will discuss some items common to all working drawings.

Drawing Sheets

Whether the finished drawing is done by hand or plotted from a CAD (computer-aided design) system, working drawings should be produced on standard-sized sheets. The following table gives the standard sheet sizes for technical drawings:

Metric (millimetres)	Inch (standard)	Inch (civil/architectural)
A4 – 210 × 297	A – 8.5 × 11	9 × 12
A3 – 297 × 412	B – 11 × 17	12 × 18
A2 – 420 × 524	C – 17 × 22	18 × 24
A1 – 594 × 841	D – 22 × 34	24 × 36
A0 – 841 × 1189	E – 34 × 44	36 × 48

The drawing media used can make a difference to the quality of the output. While manufacturers or builders rarely see plotted or drawn "originals," they do see copies of the original, reproduced by either photocopying or the diazo whiteprint process or from a microfilm negative. In most cases, some of the original quality, that is, line sharpness or text legibility, is lost. The engineer "signing off" the drawings must ensure that the drawings are properly prepared and leave no room for misinterpretation.

The most popular forms of drawing media are opaque paper (bond), translucent paper (vellum), and polyester film (Mylar®). Film is the most expensive medium and the most stable for clarity. It is also water- and heat-resistant. It is recommended when original drawings are to be kept for a long time and handled or modified frequently. Opaque paper is normally used for master documents, such as maps and charts. Engineering production and construction drawings are plotted or drawn on translucent paper or film. Drawings on opaque paper are reproduced using an electrostatic process, such as photocopying. Drawings created on translucent paper and film can be reproduced using photocopying or the less expensive diazo process.

Drawing Sheet Format

Every production or construction drawing sheet will include an area on the right-hand side containing a title block and any general notes pertaining to the drawing. In addition, drawing sheets should include a grid reference zoning system to locate changes, and centring marks to facilitate positioning of the sheet when microfilming, as shown in Figure 5.1.

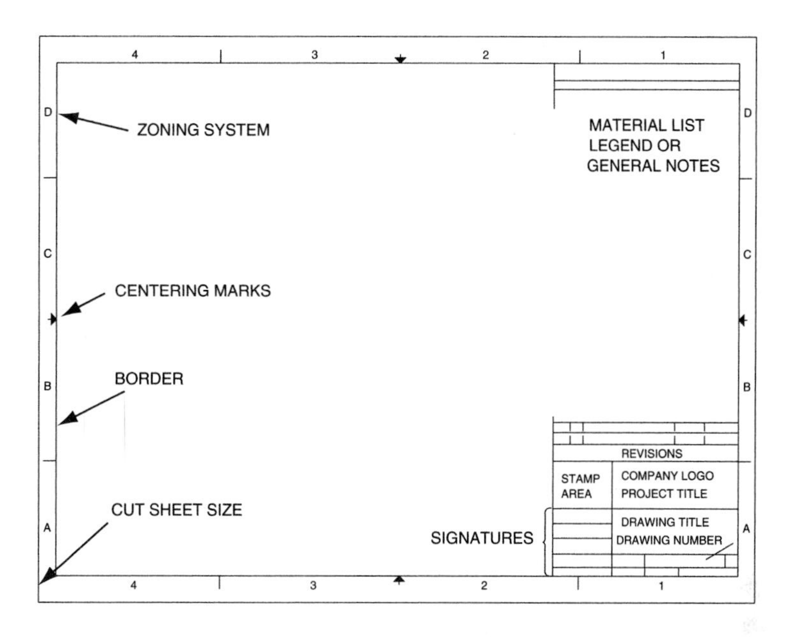

Figure 5.1 Drawing sheet

Title Blocks Title blocks vary in layout and contain general information relating to the drawing, such as the company's name, project title, drawing title, date drawn, scale used, and drawing number. Figure 5.2 is an example of a consultant engineering title block. Note that the client's name is included in addition to the consultant's name. Complex parts may require several sheets to fully describe the part or structure; therefore, sheet numbering is added to indicate the number of sheets in the drawing set; for example, sheet 1 of 2.

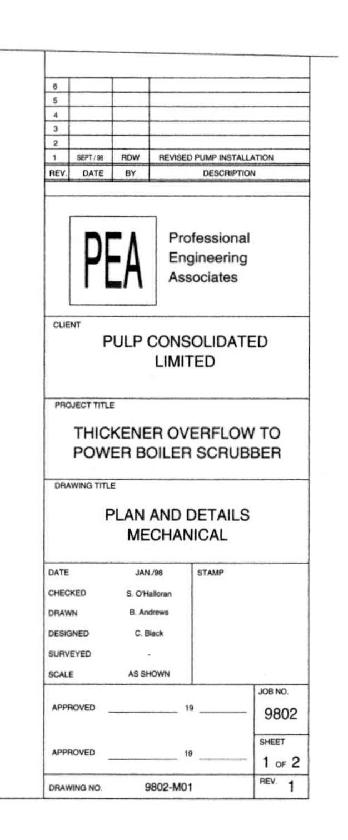

Figure 5.2 Title block

The project title is the general title of the design; for example, THICKENER OVERFLOW TO POWER BOILER SCRUBBER. The drawing title is for a particular drawing description; for example, PLAN AND DETAILS—MECHANICAL.

Most firms have their own title block already printed on cut sheets or stored in a CAD database for plotting. For presentation purposes, it is important that all drawing sheets in a set are consistent throughout; that is, the same size, the same title block text. Auxiliary drawing numbers may be added in the upper left corner, usually to read from the top-down, to aid hard copy (paper) filing systems.

Provision is usually made in a title block for approval signatures and professional stamps. Manufactured parts' information, such as general tolerances, material used, projection symbols, and drawing standards, may also be included within the title block.

Companies using CAD store predrawn title sheets and use them whenever they start a new drawing. CAD programs usually come with predrawn title sheets, as templates, which you are encouraged to use while doing the projects in this book.

Revision Column An important area of the title block is the **revision column** or **change table**. This area contains a brief description of any changes made to the drawing, along with the dates of the changes and approval signatures. Any changes to the drawing after it has been issued for manufacture or construction must be recorded and approved. The area containing the change can be listed using the grid reference or often a "cloud" is drawn around the area with the revision number indicated. Figure 5.3 shows a change to a dimension.

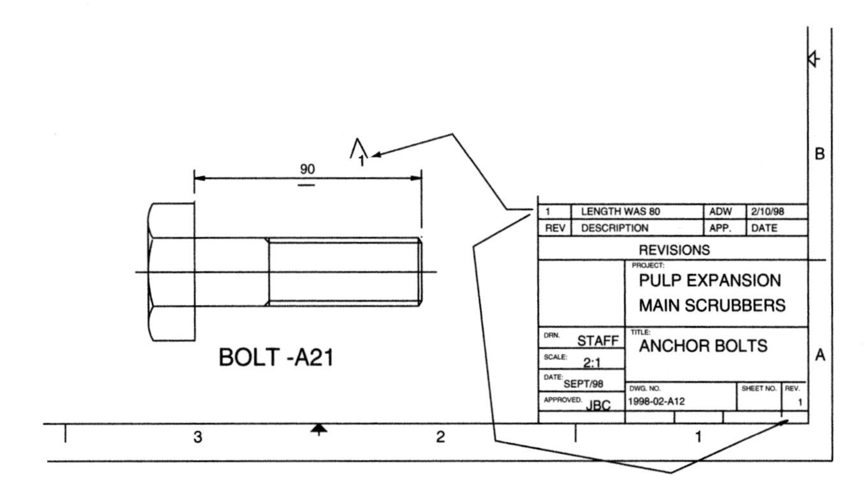

Figure 5.3 Revision column

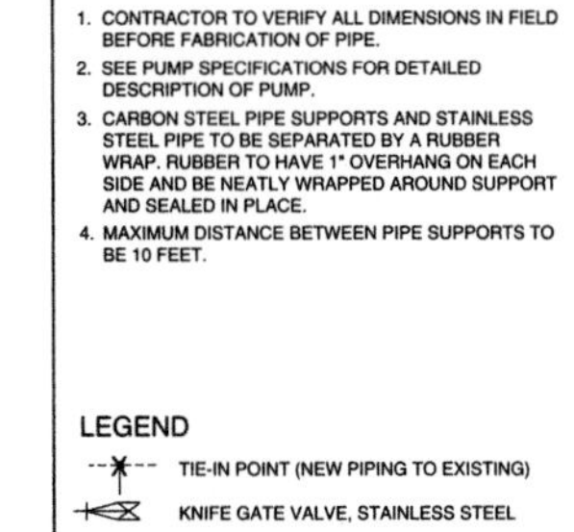

Figure 5.4 Material list/legend area showing general notes and legend

Material List/Legend Area The area above the title block should be reserved for listing parts or materials, general notes pertaining to the entire drawing, or a legend describing the symbols used on the drawing (see Figure 5.1 and Figure 5.4).

Drawing Sheet Layout

Drawing sheets should be laid out in accordance with good drafting practice. This includes the arrangement of views, angles of projection, and clarity of presentation. The engineer who signs off the drawing must ensure that it is correct and well presented. Remember, the quality of a drawing reflects on both the engineer and the company.

Drawing Views

As discussed in earlier chapters, the North American standard method of laying out multiple views in a two-dimensional (2-D) space is to use third-angle projection. This includes orthographic, section, partial, enlarged, and auxiliary views. As the intention of the drawing is to communicate, the drawing should be clear, with views spaced apart and dimensions and notes placed in appropriate locations. On an engineering drawing, the number of views should be limited to the minimum necessary to fully describe the design without ambiguity.

Let's look at the various types of views in more detail, starting with orthographic and section views.

Orthographic and Section Views The principles of orthographic projection have been discussed in Chapter 1. The third-angle projection system practised in North America places "flat" or projected views parallel to each other, as shown in Figure 5.5.

The number of views needed to describe objects or structures for production depends on the complexity of the objects. The two important characteristics of engineering drawings are that they:

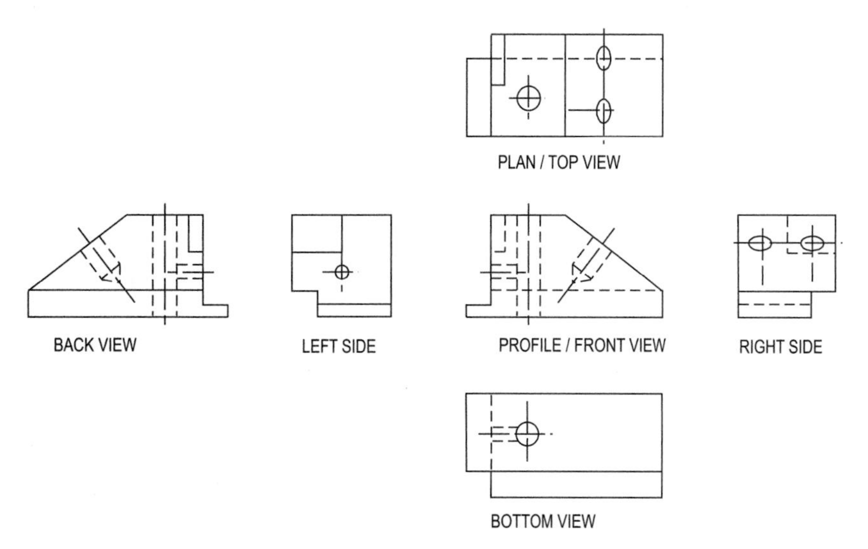

Figure 5.5 Orthographic views

1. provide sufficient "flat" views to fully describe features (holes, edges, cutouts) of the object; and
2. provide sufficient dimensions to make the object.

Bear in mind the placement of dimensions (discussed in Chapter 4). Edges that are hidden from view should not have dimensions. Place dimensions on the view that best describes the feature.

Section views are used to show interior detail (discussed in Chapter 3). Using section views in place of regular orthographic views can save drawing time and reduce the number of drawing sheets used. The shaft hanger shown in Figure 5.6 makes use of a full section, in place of the traditional front view, to illustrate the cutout and recess information and a revolved section to describe the rib profile.

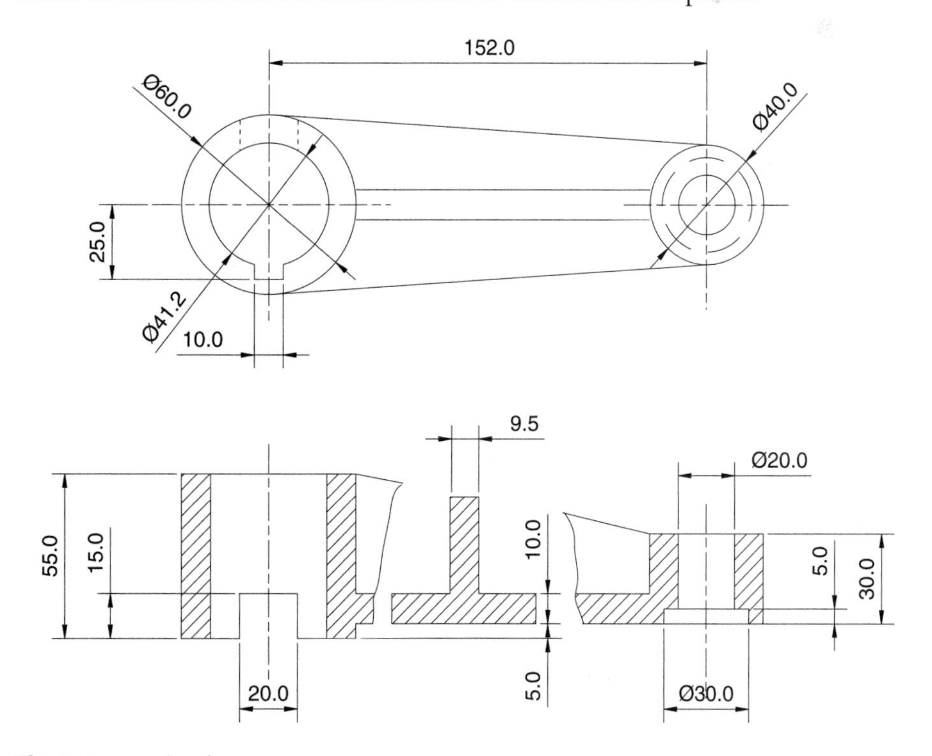

Figure 5.6 Section views

Partial and Enlarged Views Often, it is not necessary to show a complete view, when a partial view will suffice. A **partial view** is used to aid clarity and where drawing the complete view would not improve the information given. A symmetrical object can be adequately described by showing only one half or a quarter of the whole view (Figure 5.7). The line of symmetry is indicated by a centreline with two short parallel lines drawn across it. When using CAD, it is important to adhere to these guidelines, as it is often easier to give a complicated full view from the three-dimensional (3-D) model rather than the required simpler one.

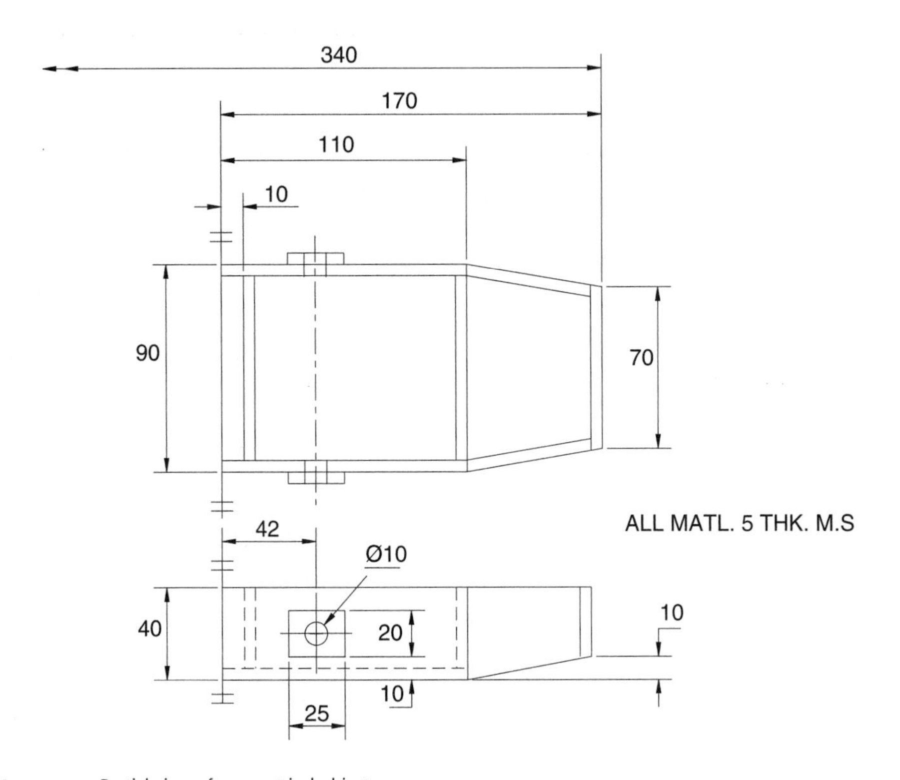

Figure 5.7 Partial view of symmetrical object

Sometimes, parts are so complicated, or structures so large, that full views or partial views have to be drawn on separate sheets. The arrows indicating the viewing or cutting plane will include the sheet number where it is drawn.

An **enlarged view** is used to show a partial view or feature detail in an enlarged scale. This is done to prevent overcrowding of dimensions or to simply save time and/or improve clarity. The view is titled and has the scale underneath. In CAD, the entire drawing is created full size in the model or work space, and the views scaled, cropped, and arranged using the paper layout space.

All views should be plotted to a standard scale (as described in Chapter 4). If the scale is the same for the entire drawing, it will be noted in the title block. If the drawing sheet contains partial views or details drawn at different scales, then the scale will be noted underneath the view title. Figure 5.8 gives an example of this.

Auxiliary Views Figure 5.5 illustrated the six principal orthographic views (discussed in Chapter 1). Note that the inclined surface appears shortened in both the top and the right-side views, leaving no place to correctly dimension the holes. Inclined surfaces are only perpendicular to one viewing plane, in this case the front plane.

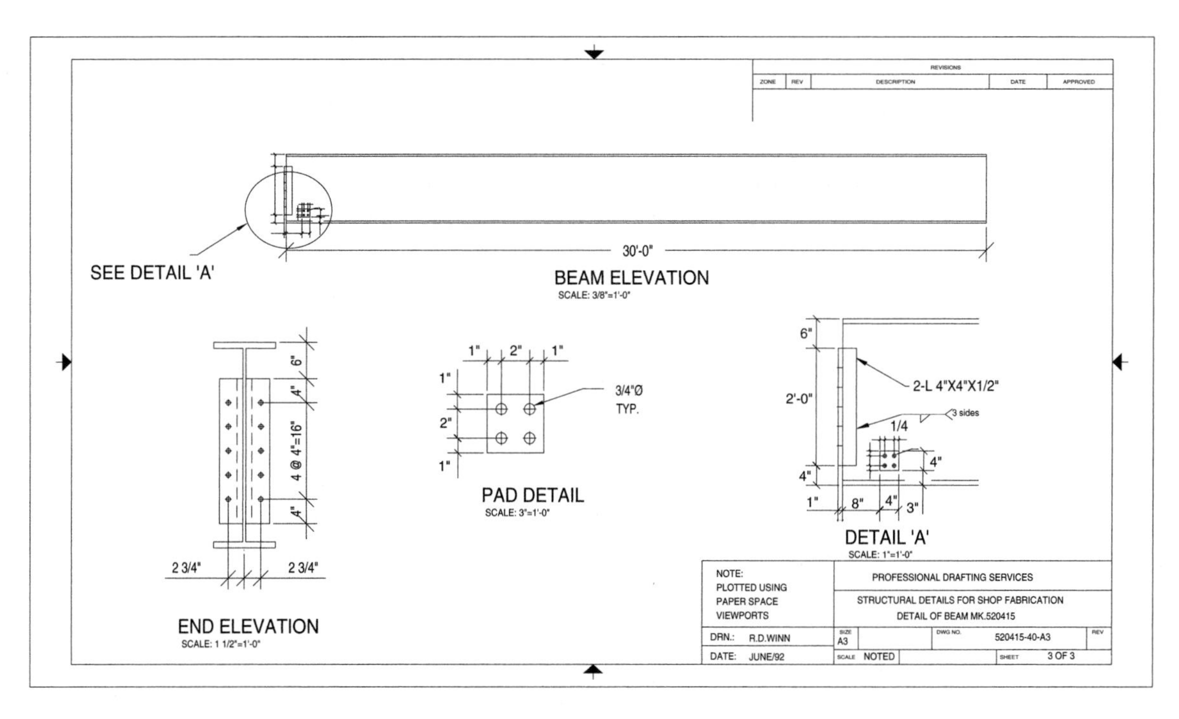

Figure 5.8 Multiple scaled plot sheets

When it is necessary to show the true size and shape of an inclined surface, you project a view parallel and perpendicular to the inclined surface. This view is called an **auxiliary view** and is normally drawn adjacent to the inclined surface, although it may be drawn anywhere as a removed view.

In Figure 5.9, an auxiliary view has been drawn, enabling the holes in the inclined surface to be dimensioned. Only the true size and shape of the inclined surface are shown, not the entire projection. When only part of a view is shown, it is known as a partial view. Partial views are often used for auxiliary views. A full auxiliary view would not help describe the features and should be avoided. The top view is also shown as a partial view, with a break line cutting off the unnecessary part of the view. The right-side view is also omitted as it serves no purpose.

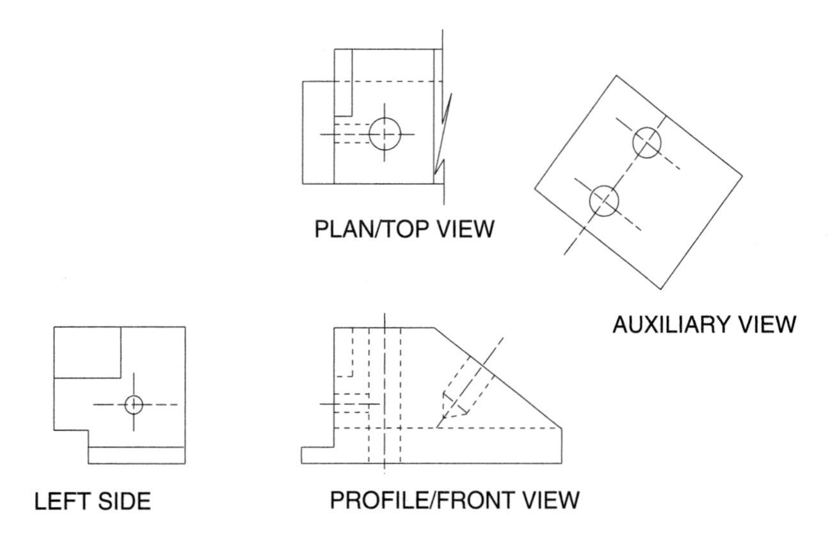

Figure 5.9 Auxiliary view

In the case of oblique surfaces—that is, surfaces that are not perpendicular to any of the three standard viewing planes—a second auxiliary view must be projected from the first, or primary, auxiliary view. Figure 5.10a shows a bracket that has an oblique surface. Figure 5.10b shows the fully dimensioned working drawing for the bracket. The secondary auxiliary view is projected parallel to the primary auxiliary view.

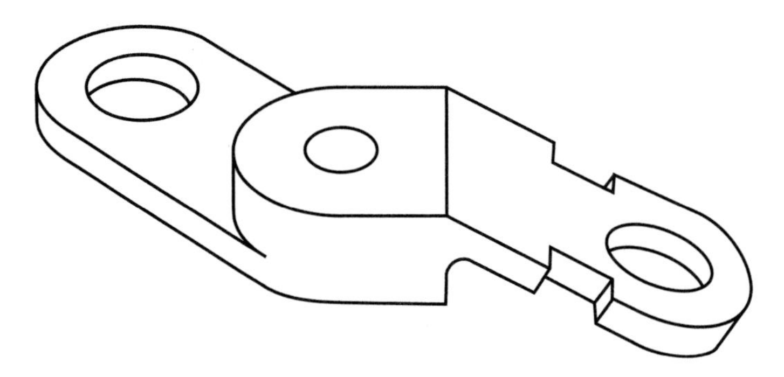

Figure 5.10a Bracket with oblique surface

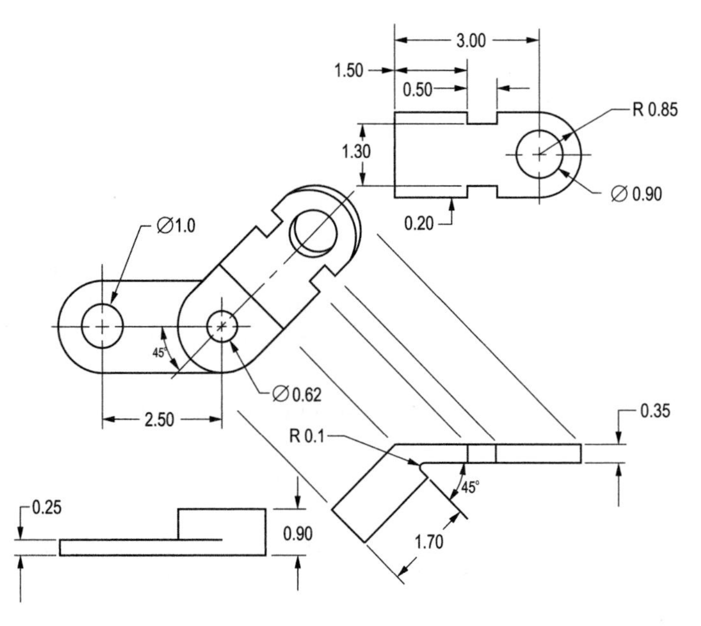

Figure 5.10b Secondary auxiliary view

Tabulated Drawings

Another time-saving tactic used while creating working drawings is to make a standard drawing and include a table of dimensions for similar parts or arrangements. This is called a **tabulated drawing**. In Figure 5.11 some of the dimensions have been replaced with letters. The values for these are found in the table next to the particular part variety. The drawing here is obviously not to scale for more than one part.

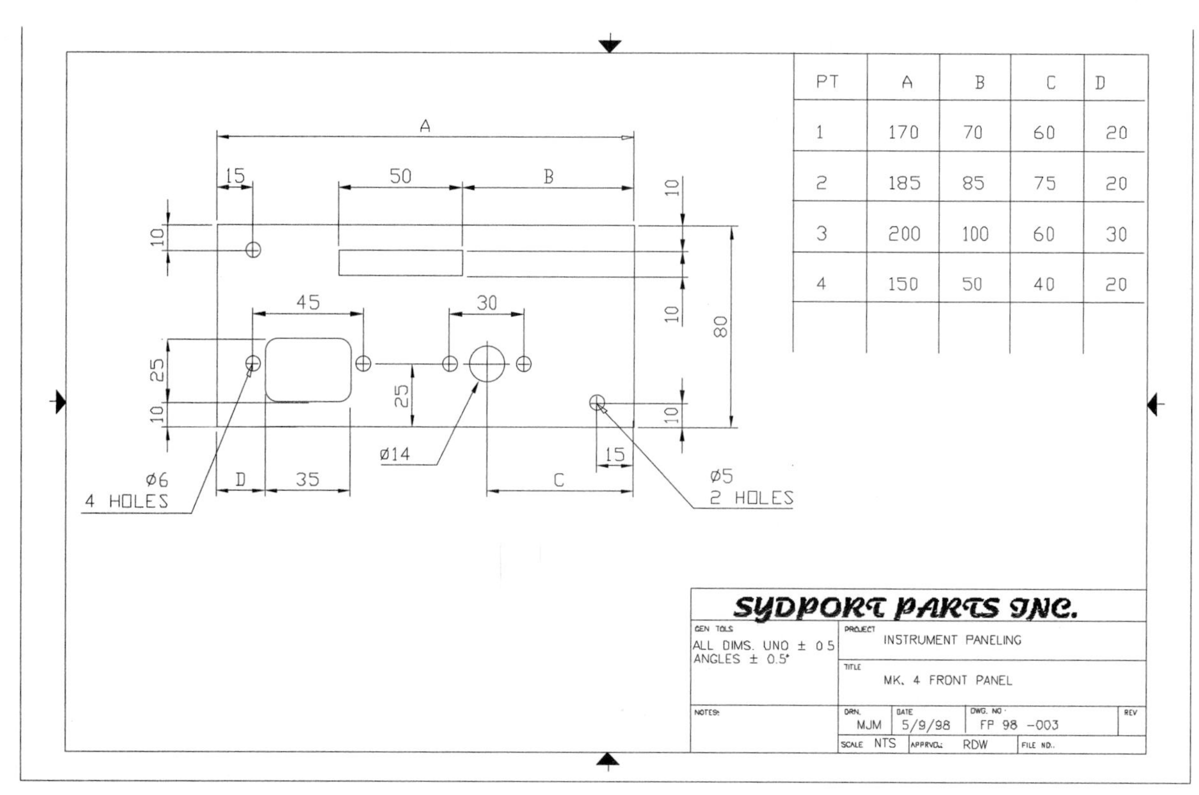

Figure 5.11 Tabulated drawing

MECHANICAL ENGINEERING DRAWINGS

Mechanical engineering covers a multitude of different types of engineering drawings. Different companies have different meanings for the term **mechanical**. For example, in manufacturing, mechanical engineers design mechanisms and parts, whereas in consulting engineering, the mechanical engineer is responsible for heating systems and ventilation, sizing and routing of ductwork, and plumbing systems. The following table lists the general headings along with a description of the working drawing contents.

Mechanical Engineering	Working Drawings Content
Manufacturing	Product parts and assemblies
Building services	Plumbing, heating, and air conditioning
Materials handling	Conveying, equipment, and supports
Industrial piping	Routing and supporting pipes
Fluid power	Hydraulic and pneumatic applications

Manufacturing

The concepts of engineering graphics are taught using examples from product manufacturing. Small parts can require all of the principal views, projection methods, and dimension methodology, without having to contend with the concepts of large-scale drawings.

Manufacturing working drawings are divided into two main types: **detail drawings**, which contain all of the necessary information to manufacture the parts, and **assembly drawings**, which provide information for the assembly of the parts.

Detail Drawings A detail drawing is a complete description of a part. This includes shape description, size description, and general specifications. The drawings used so far in this text for examples and problems are detail drawings.

Shape description refers to the views used to describe the part. Orthographic views, auxiliary views, sections, and pictorial views, as described elsewhere in this text, may be used.

Size description refers to dimensioning and tolerancing of the part features. To some extent, the manufacturing process and part function will determine the method of placing dimensions and assigning tolerances. The cost of the part can be greatly reduced by decreasing the number of digits to the right of the decimal point!

General specifications include such items as the material used, surface finish, heat treatment, type of finishing process required (e.g., chromium plate), or applicable standards to be used.

Detail drawings may be drawn as one part per sheet (Figure 5.12) or with several parts drawn on one sheet (see Figure 5.13 on page 154). One part per sheet is the preferred method for production drawings, as the parts may be used on several designs. For complicated parts, several sheets may be required to give a complete description for manufacture. The drawing must be complete and not leave anything open to misinterpretation. Figure 5.14 (page 155) shows the use of an enlarged detail view to describe features more clearly.

Standard Parts The number of detail drawings for a product can be reduced (and hence the time taken to get the design to production) by making use of standard parts wherever possible. Standard parts are purchased from a supplier and may include such items as gears, cams, bolts, screws, keys, and handles. These items are then listed in a materials list on an assembly drawing. The design engineer may need to see a purchase part drawing or shop drawing for bought-out parts. This overcomes the problem of the supplier making changes unknown to the user.

Fabrication Drawings Some component parts are fabricated from pieces of steel welded together, called a **weldment**. Automobile frames and bodies, machine tools, and mining equipment all use fabricated components as well as machined parts. These detail drawings will include information on the type and size of weld used, in addition to the items specified above. Often, each weldment has all of the separate pieces of steel listed on the right-hand side of the drawing in a material or item list, to aid procurement. Figure 5.15 (page 156) is an example of this.

Assembly Drawings

Most products and designs consist of several parts assembled together. The drawing that describes how they fit together is the **assembly drawing**, sometimes referred to as a **general arrangement** (or **GA**). The purpose of this drawing is two-fold: first, to give an overall picture of the completed design; and, second, to show the relationship of the various parts that make up the assembly. The parts are listed in a **bill of material** (BOM), **parts list**, or **item list**, usually found on the right-hand side of the assembly drawing sheet. Figure 5.16 (page 157) shows a parts list on an assembly drawing.

Figure 5.12 Detail drawings–One part per sheet

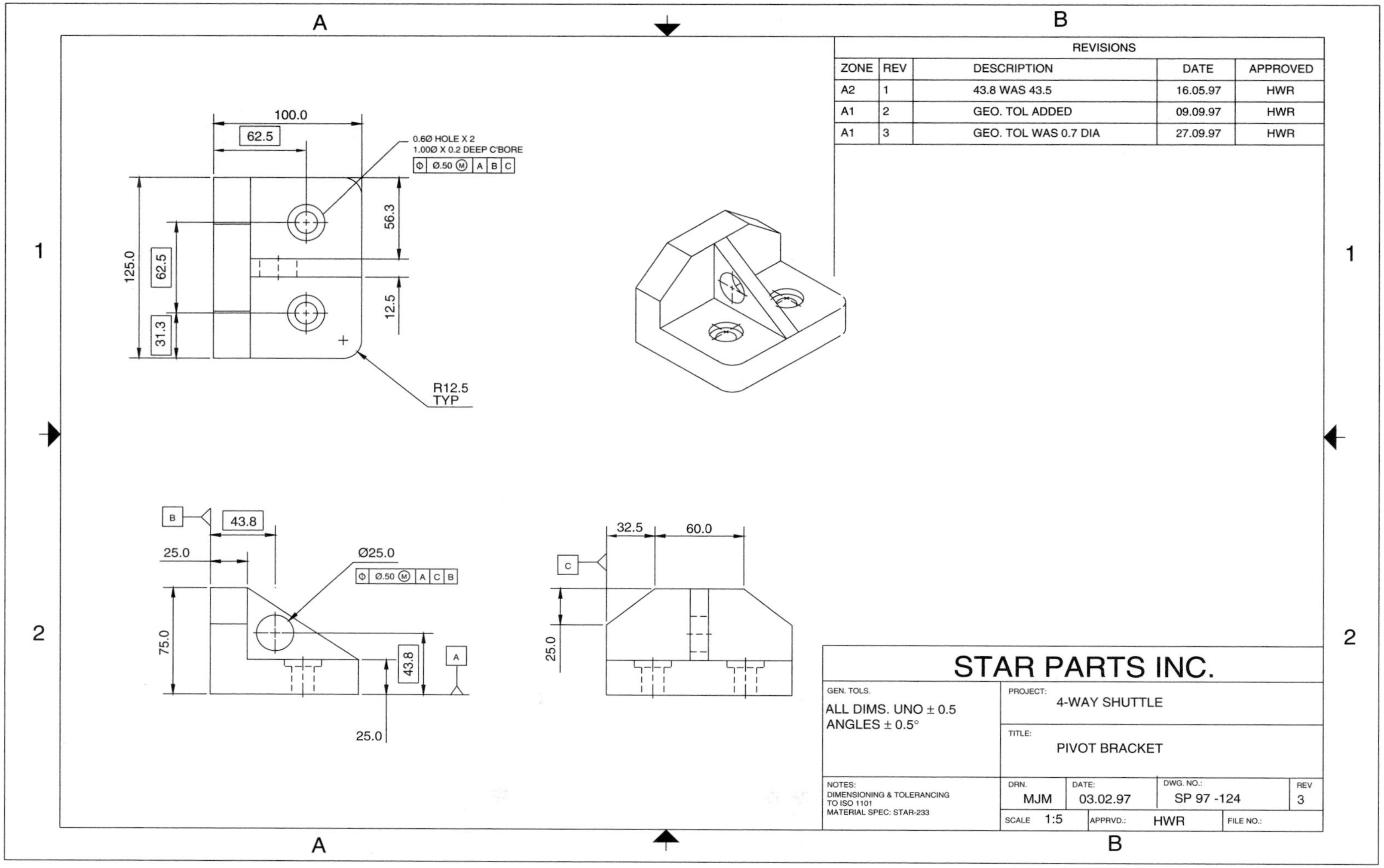

A B
1 2

REVISIONS				
ZONE	REV	DESCRIPTION	DATE	APPROVED
A2	1	43.8 WAS 43.5	16.05.97	HWR
A1	2	GEO. TOL ADDED	09.09.97	HWR
A1	3	GEO. TOL WAS 0.7 DIA	27.09.97	HWR

100.0
62.5
0.6Ø HOLE X 2
1.00Ø X 0.2 DEEP C'BORE
Ø.50 Ⓜ A B C
56.3
12.5
125.0
62.5
31.3
R12.5
TYP

B
43.8
25.0
Ø25.0
Ø.50 Ⓜ A C B
75.0
43.8
A
25.0

32.5
60.0
C
25.0

STAR PARTS INC.

GEN. TOLS.
ALL DIMS. UNO ± 0.5
ANGLES ± 0.5°

PROJECT: 4-WAY SHUTTLE

TITLE: PIVOT BRACKET

NOTES:
DIMENSIONING & TOLERANCING TO ISO 1101
MATERIAL SPEC: STAR-233

DRN.	DATE:	DWG. NO.:	REV
MJM	03.02.97	SP 97 -124	3
SCALE 1:5	APPRVD.: HWR	FILE NO.:	

Figure 5.13 Detail drawings—Several parts per sheet

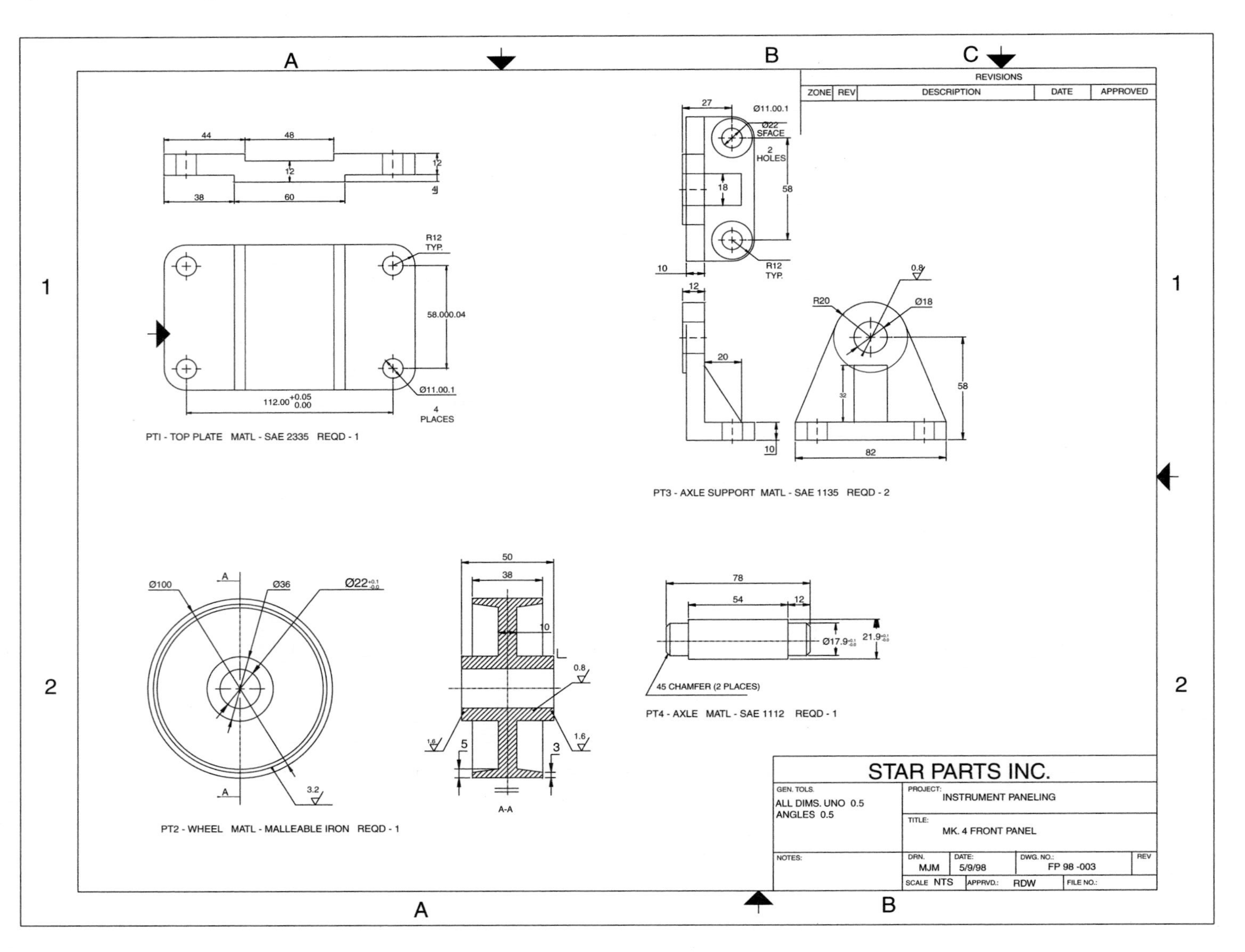

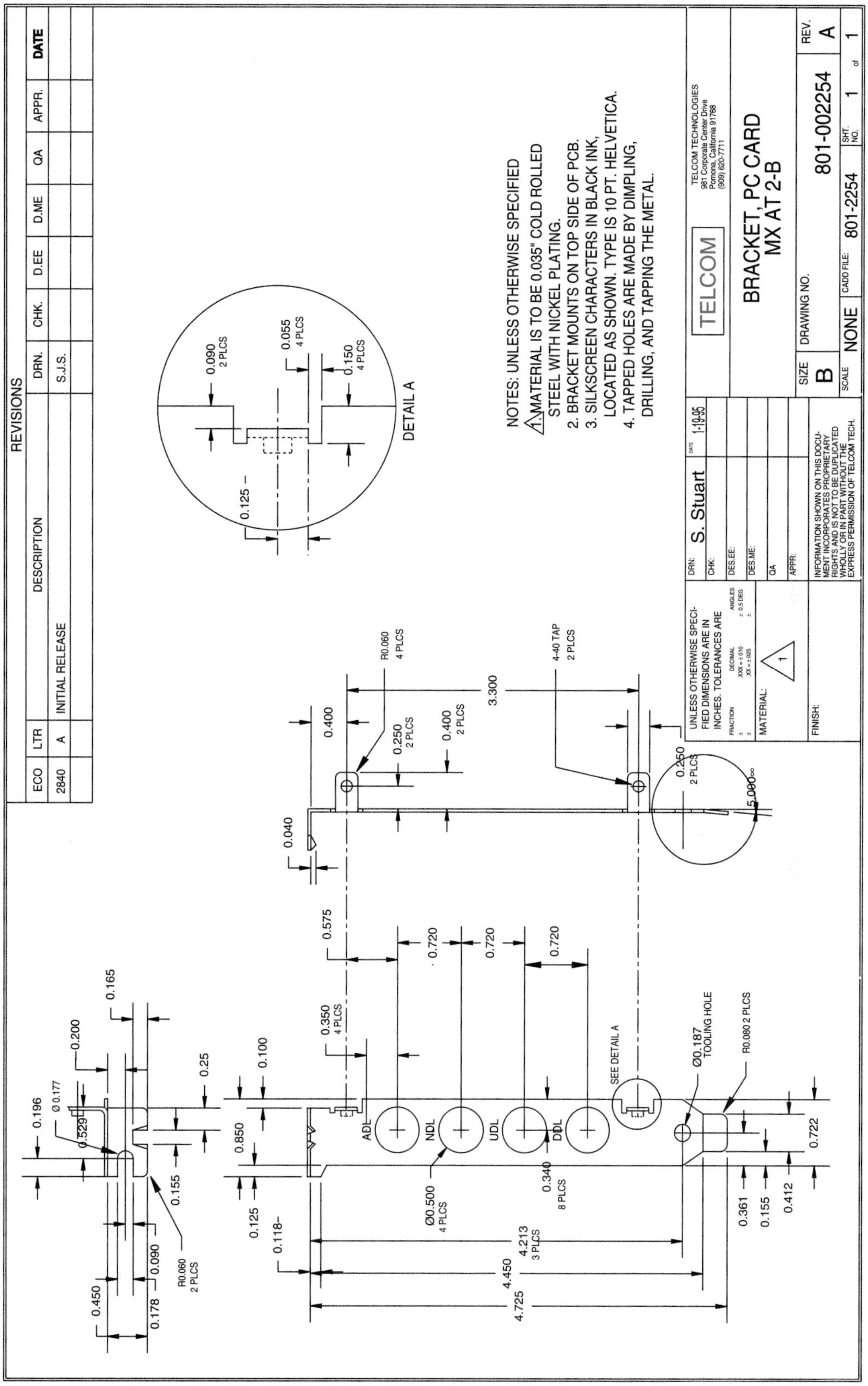

Figure 5.14 Enlarged views
(Courtesy: Telcom)

REVISIONS

ECO	LTR	DESCRIPTION	DRN.	CHK.	D.EE	D.ME	QA	APPR.	DATE
2840	A	INITIAL RELEASE	S.J.S.						

DETAIL A

NOTES: UNLESS OTHERWISE SPECIFIED
1. MATERIAL IS TO BE 0.035" COLD ROLLED STEEL WITH NICKEL PLATING.
2. BRACKET MOUNTS ON TOP SIDE OF PCB.
3. SILKSCREEN CHARACTERS IN BLACK INK, LOCATED AS SHOWN. TYPE IS 10 PT. HELVETICA.
4. TAPPED HOLES ARE MADE BY DIMPLING, DRILLING, AND TAPPING THE METAL.

UNLESS OTHERWISE SPECIFIED DIMENSIONS ARE IN INCHES. TOLERANCES ARE
FRACTION ± DECIMAL .XXX = ±.010 .XX = ±.025 ANGLES ± 0.5 DEG
MATERIAL: 1
FINISH:

DRN: S. Stuart DATE 1-19-95
CHK:
DES.EE:
DES.ME:
QA
APPR:

INFORMATION SHOWN ON THIS DOCUMENT INCORPORATES PROPRIETARY RIGHTS AND IS NOT TO BE DUPLICATED WHOLLY OR IN PART WITHOUT THE EXPRESS PERMISSION OF TELCOM TECH.

TELCOM
TELCOM TECHNOLOGIES
981 Corporate Center Drive
Pomona, California 91768
(909) 620-7711

BRACKET, PC CARD
MX AT 2-B

SIZE B | DRAWING NO. 801-002254 | REV. A
SCALE NONE | CADD FILE: 801-2254 | SHT. NO. 1 of 1

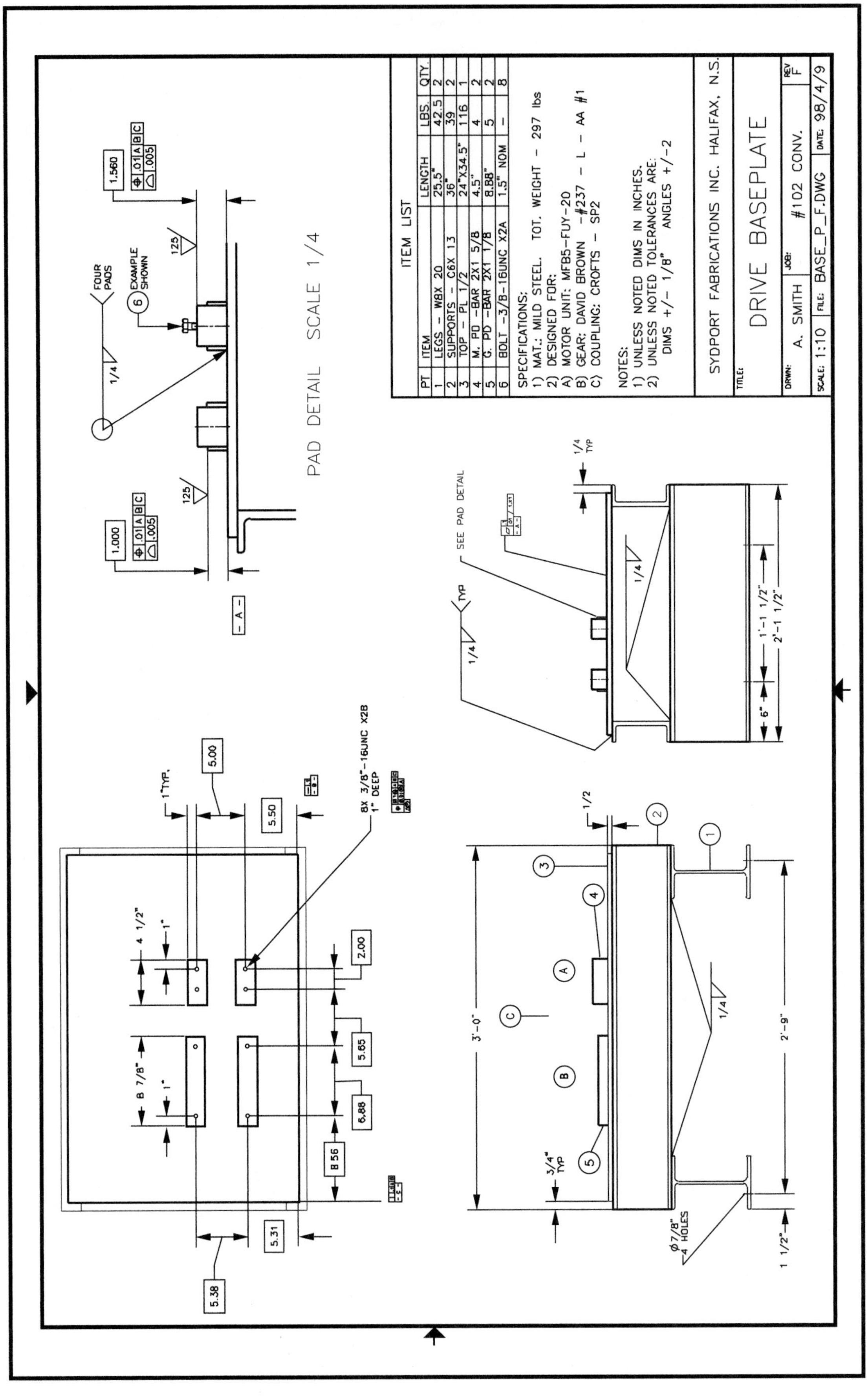

Figure 5.15 Fabrication detail

8 7/8"
1"
4 1/2"
1"
1"TYP.
5.38
5.31
5.00
5.50
8.56
6.88
5.65
2.00
8X 3/8"-16UNC X2B
1" DEEP

1.000
.01 A B C
.005
1/4
FOUR PADS
6 EXAMPLE SHOWN
125
1.560
.01 A B C
.005
- A -

PAD DETAIL SCALE 1/4

3'-0"
C
3/4" TYP
B
A
3
5
4
1/2
2
Ø7/8" 4 HOLES
1/4
1
1 1/2"
2'-9"

1/4
TYP
SEE PAD DETAIL
1/4 TYP
1/4
6"
1'-1 1/2"
2'-1 1/2"

ITEM LIST

PT	ITEM	LENGTH	LBS.	QTY.
1	LEGS - W8X 20	25.5"	42.5	2
2	SUPPORTS - C6X 13	36"	39	2
3	TOP - PL 1/2	24"X34.5"	116	1
4	M. PD -BAR 2X1 5/8	4.5"	4	2
5	G. PD -BAR 2X1 1/8	8.88"	5	2
6	BOLT -3/8-16UNC X2A	1.5" NOM	–	8

SPECIFICATIONS:
1) MAT.: MILD STEEL. TOT. WEIGHT - 297 lbs
2) DESIGNED FOR:
A) MOTOR UNIT: MFB5-FUY-20
B) GEAR: DAVID BROWN -#237 - L - AA #1
C) COUPLING: CROFTS - SP2

NOTES:
1) UNLESS NOTED DIMS IN INCHES.
2) UNLESS NOTED TOLERANCES ARE:
DIMS +/- 1/8" ANGLES +/-2

SYDPORT FABRICATIONS INC. HALIFAX, N.S.

TITLE: DRIVE BASEPLATE

DRWN: A. SMITH | JOB: #102 CONV. | REV F
SCALE: 1:10 | FILE: BASE_P_F.DWG | DATE: 98/4/9

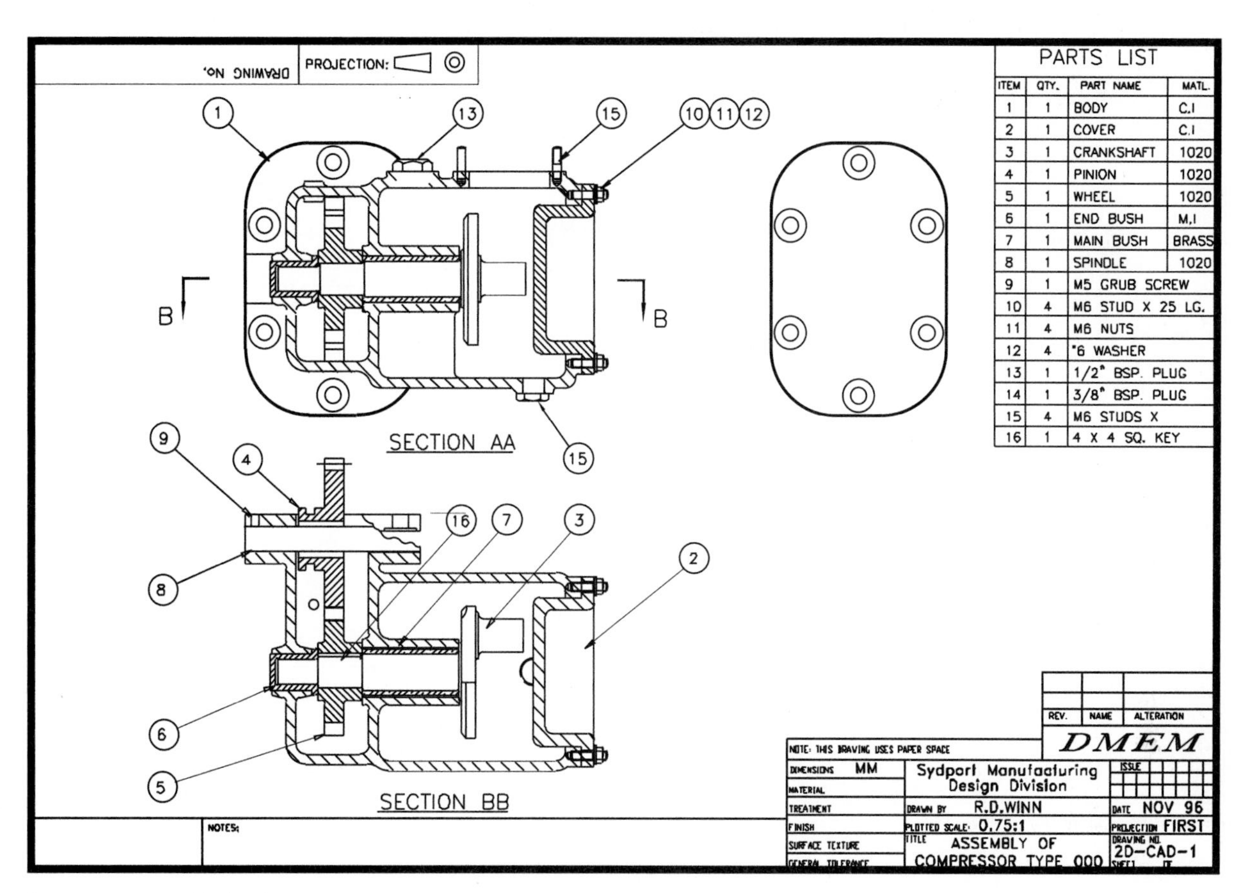

Figure 5.16 Assembly drawing

The assembly can be drawn orthographically or pictorially. Orthographic drawings are preferred for engineering assembly, whereas pictorial assemblies are used by people unskilled in drawing-projection methods. Typically, this includes part reordering and do-it-yourself kits.

Orthographic assemblies often make use of a section view to show interior parts. All parts are identified by inclined leader lines attached to balloons containing an item or identification number. Parts are listed in the item list along with the quantity per assembly and a description or drawing number (see Figure 5.17).

Dimensions in an assembly drawing are restricted to overall sizes, capacity dimensions, limits or extent of operation, and distances between parts. Other information may be design data and operating instructions. Figure 5.17 shows a dimensioned assembly.

Subassemblies **Subassemblies** are small assemblies that are part of a larger design. For example, for an automobile engine, subassemblies would include the oil pump, water pump, and transmission. The wheel assembly in Figure 5.17 is a subassembly of a conveying system. Each subassembly may be manufactured and assembled in separate departments and then the units combined to create the final assembly. Figure 5.18 shows a pictorial cut-away subassembly. A portion of this pictorial is an **exploded assembly**. The components are separated, and lines are added to show the relation between the parts.

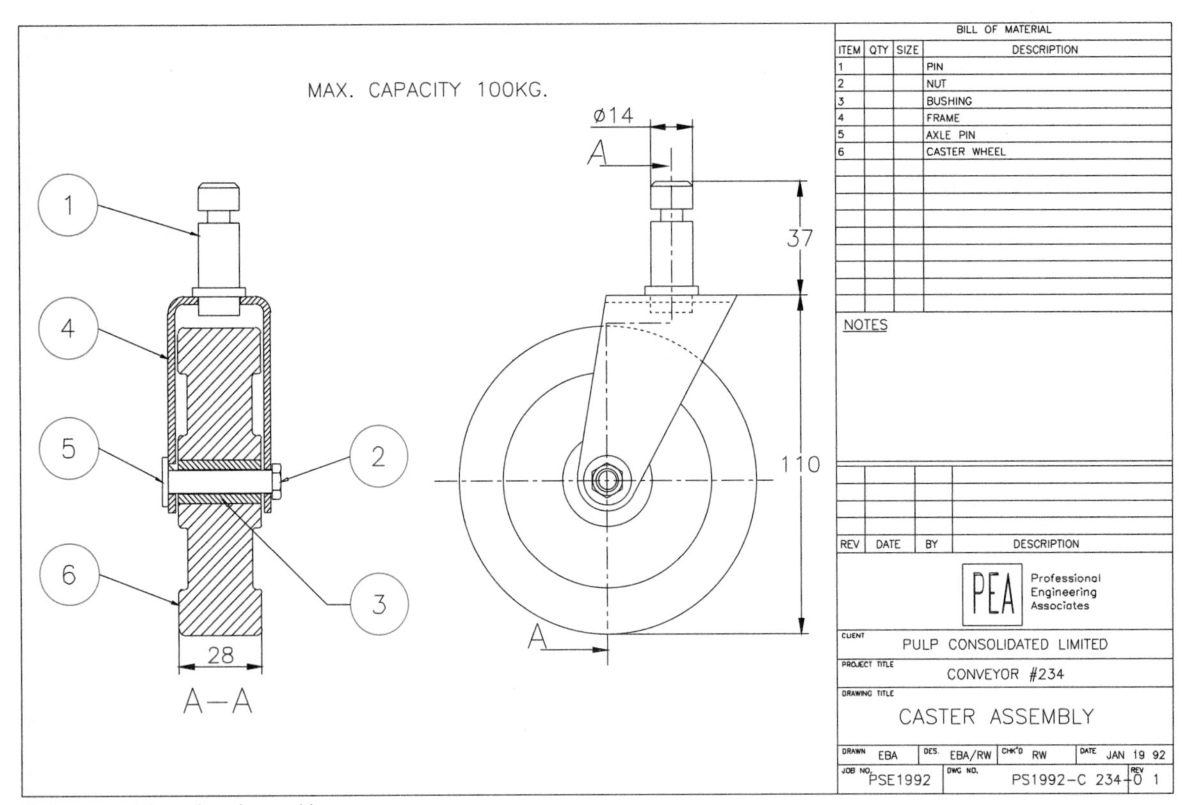

Figure 5.17 Dimensioned assembly

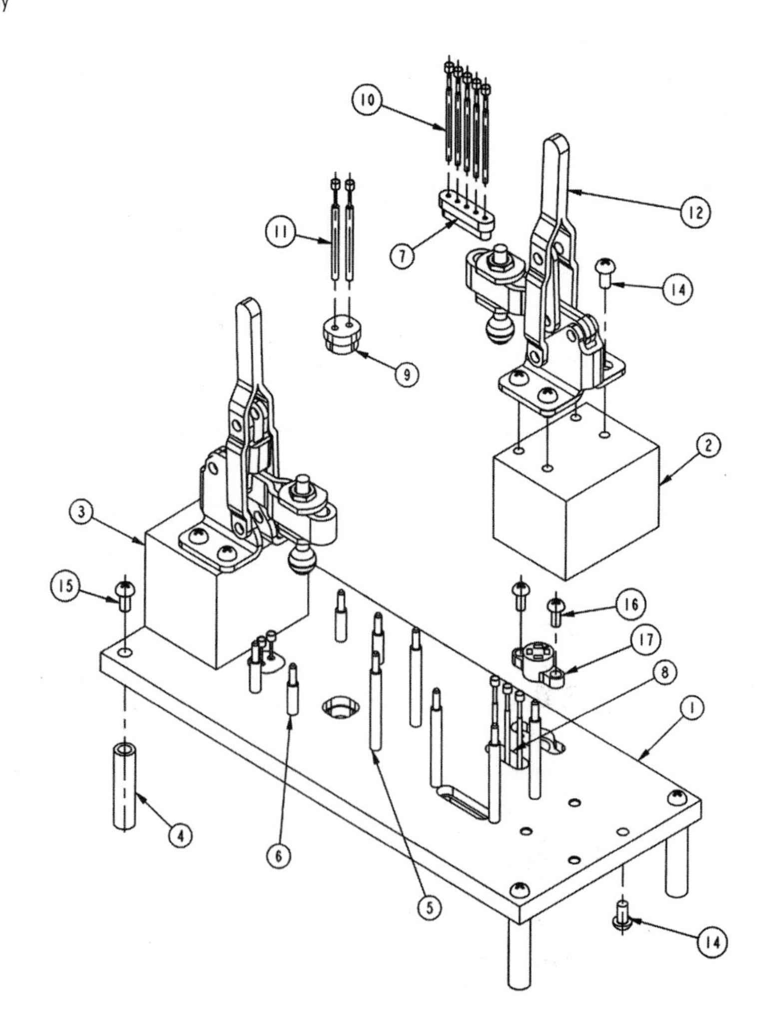

Figure 5.18 Pictorial subassembly
(Courtesy: Autodesk)

Building Services

Mechanical engineering drawings used to communicate the design of plumbing, heating, ventilation, and air conditioning systems in commercial buildings, schools, and hospitals are known collectively as **drawings for building services**. The design of the service is generally known as the **system**; for example, plumbing system, heating system, and so on, and the equipment, pipes, and ducts are the **elements** of the system.

A mechanical engineer is responsible for the design of the system, size of the elements within the system, and to some extent the routing and location of the elements. Drawings for building services make extensive use of symbols. It is in the best interests of the mechanical engineer to have a working knowledge of the content and nomenclature of these types of drawings.

Building Drawings

Building drawings are provided by an architect or consulting engineer. They are drawn using orthographic projection techniques, but because of the size, only one view per sheet is common. **Building drawings** comprise plans (similar to a top view), elevations (similar to a side view), and sections. Dimensions are given using the metric system (millimetres); however, the standard system of feet and inches is still commonly used. Mechanical engineers should familiarize themselves with the architect's scale in these cases.

Most of building drawing applications are of the general arrangement type or an assembly of equipment and connecting pieces. Detail drawings are made to clarify specific features of the system. Detail drawings are enlarged views of equipment installations or components, usually an elevation.

Plumbing Working with the building architect, the mechanical engineer designs the internal piping systems for the water and gas supply, waste pipes, and drainage. Both orthographic and isometric projection is used to explain the routing and size of pipes. Designs are based on the National Plumbing Code and include washbasins, WCs (water closets), and showers.

A large facility will require a fire prevention system, such as a sprinkler system, which requires a sophisticated valve distribution "station" detail.

Figure 5.19 illustrates a plumbing plan and schematic diagram for an office building. Figure 5.20 (page 161) shows a sprinkler distribution station for a public administration building.

Figure 5.21 (page 162) shows mechanical services details for a heating system (note the detail used on the pump). Figure 5.22 (page 163) is a plan view of the mechanical room.

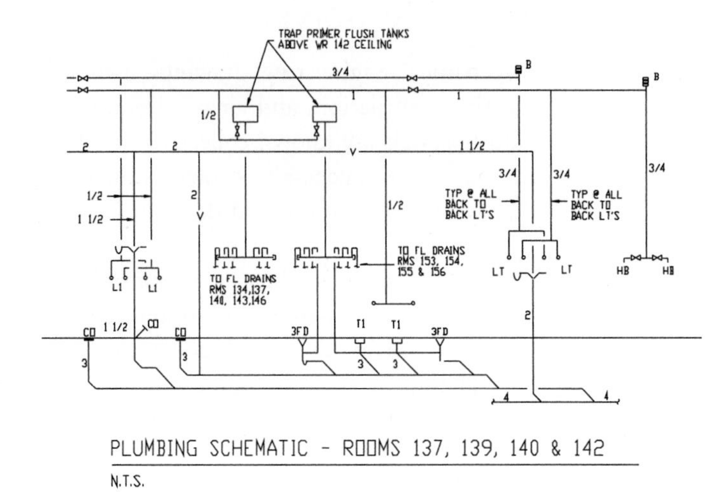

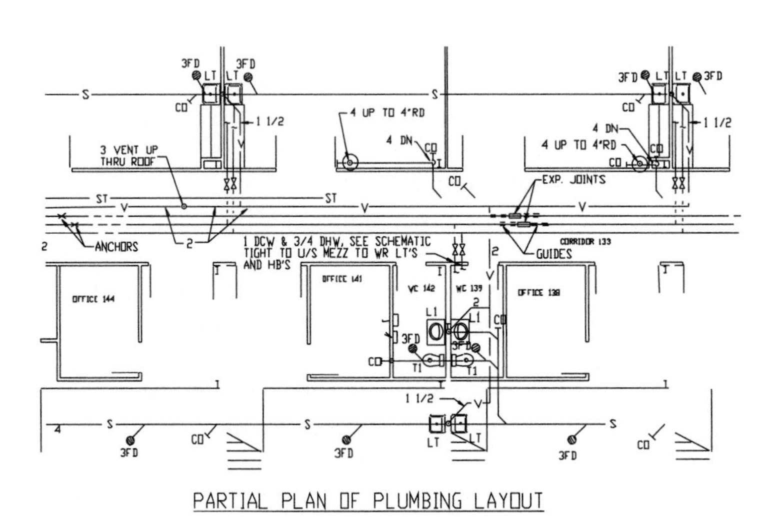

Figure 5.19 Plumbing plan and schematic diagram

Figure 5.20 Sprinkler layout

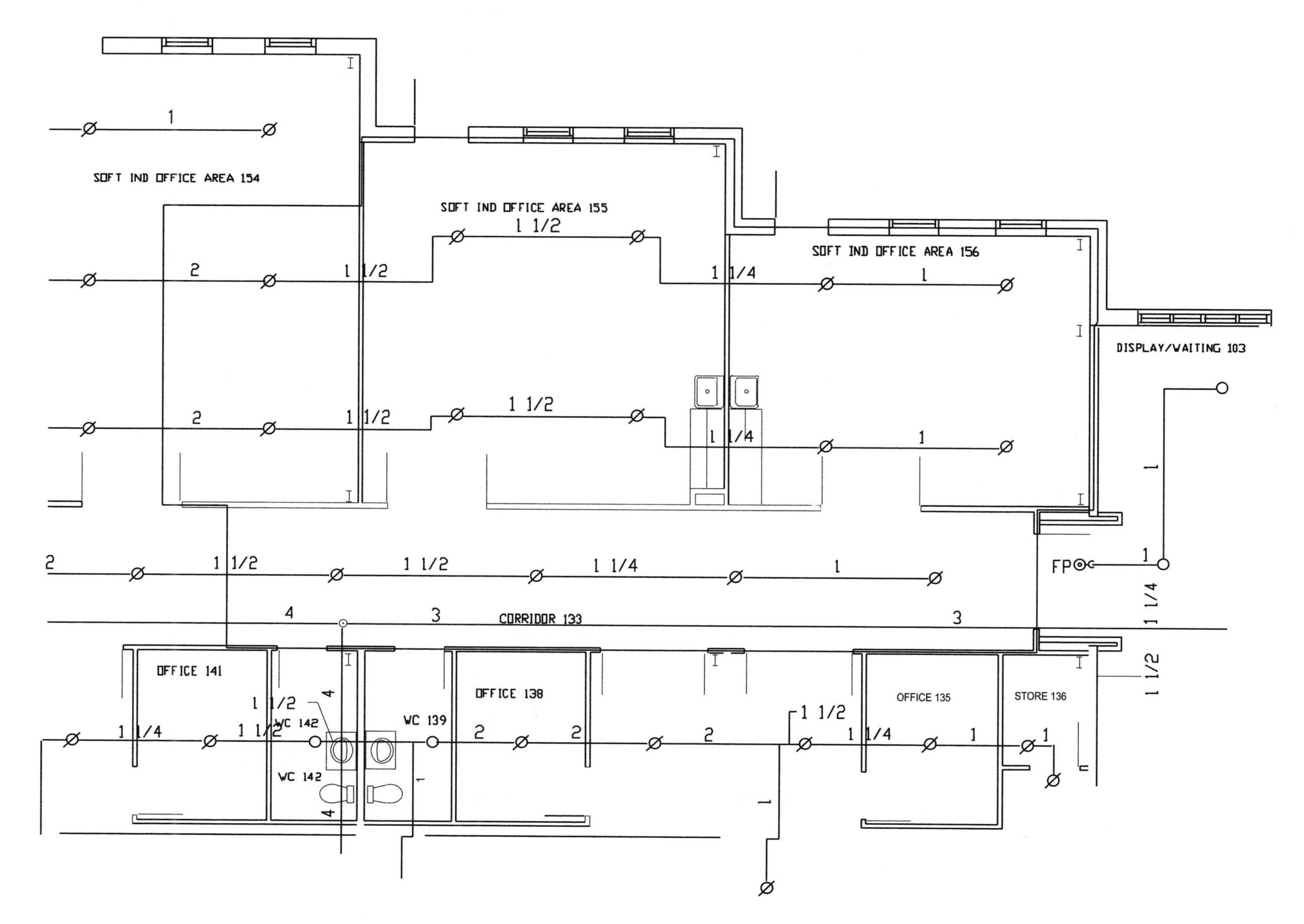

Figure 5.21 Heating equipment

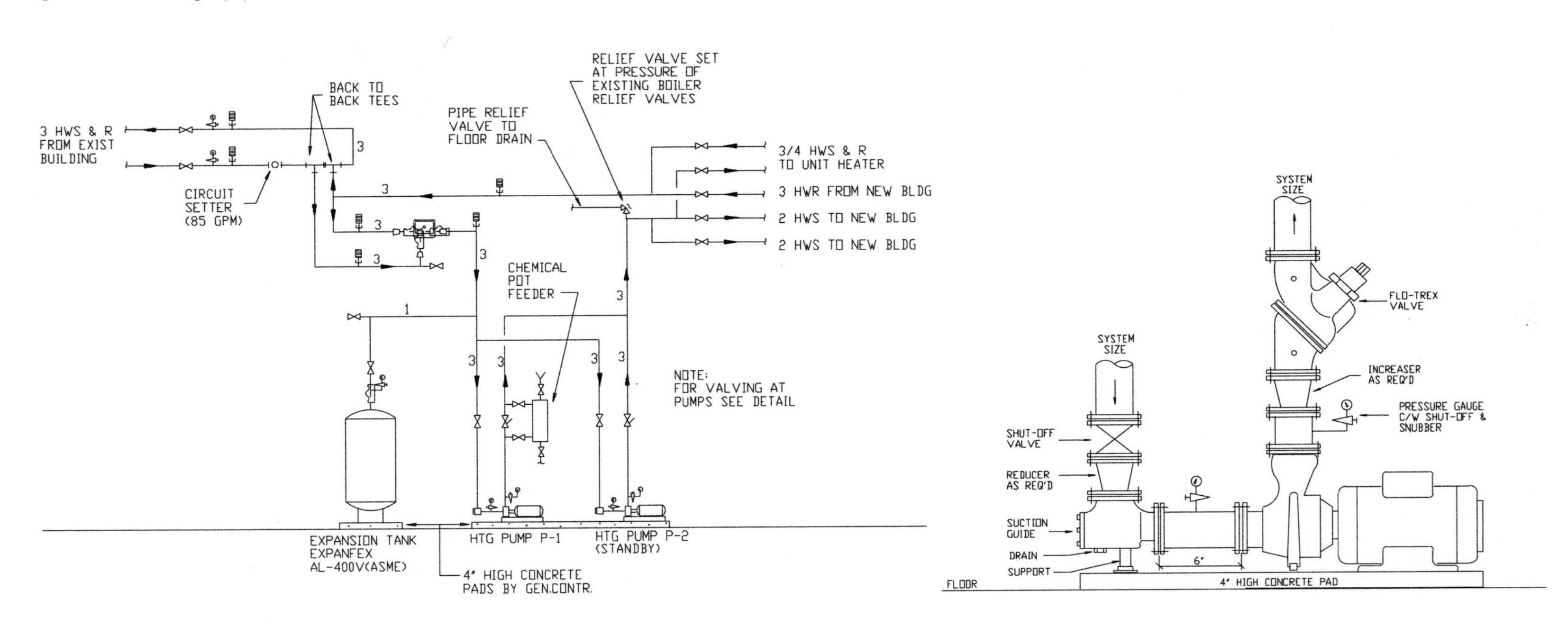

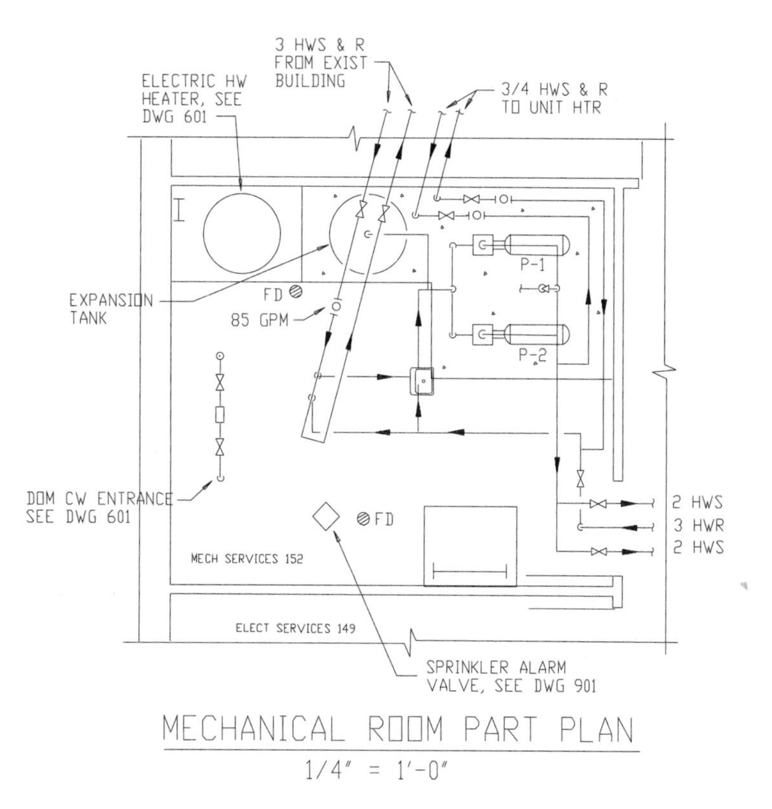

Figure 5.22 Mechanical equipment room plan

Heating, Ventilation, and Air Conditioning (HVAC) HVAC systems are designed by a mechanical engineer liaising with the building architect or engineering consultant. The purpose of heating and ventilating systems is to help maintain a normal comfort level within the building or to provide a purpose-designed environment (e.g., freezer unit). Ventilation systems remove excess heat, fumes, moisture, pollutants, and odour and supply fresh air.

HVAC systems consist of mechanical equipment—for example, air conditioners, heating units, dust collectors, louvers—that is interconnected by sheet-metal ductwork. Louvers are used to diffuse air into work areas. Detail drawings of individual pieces of ductwork are prepared for specialized or custom applications, making use of developments to create templates for forming the duct.

Generally, drawings for HVAC are similar to those for plumbing; that is, using architectural plans and elevations. The equipment is shown as an outline and the ductwork may be shown as a single line to scale as illustrated in Figure 5.23. The size of the ductwork is shown as the horizontal dimension followed by the vertical (depth) dimension. Elevation drawings and sections show the height (elevation) of the ducts from a baseline (ground) and the methods of supporting them.

Figure 5.23 shows part of an HVAC plan of an industrial building. Figure 5.24 (page 165) shows cross-sections (as indicated on the plan) that illustrate the ducting. Note that the principles you learned in earlier chapters of this text are used throughout these drawings. In Chapter 7, we will cover the principles of development used in the fabrication of chute and ductwork.

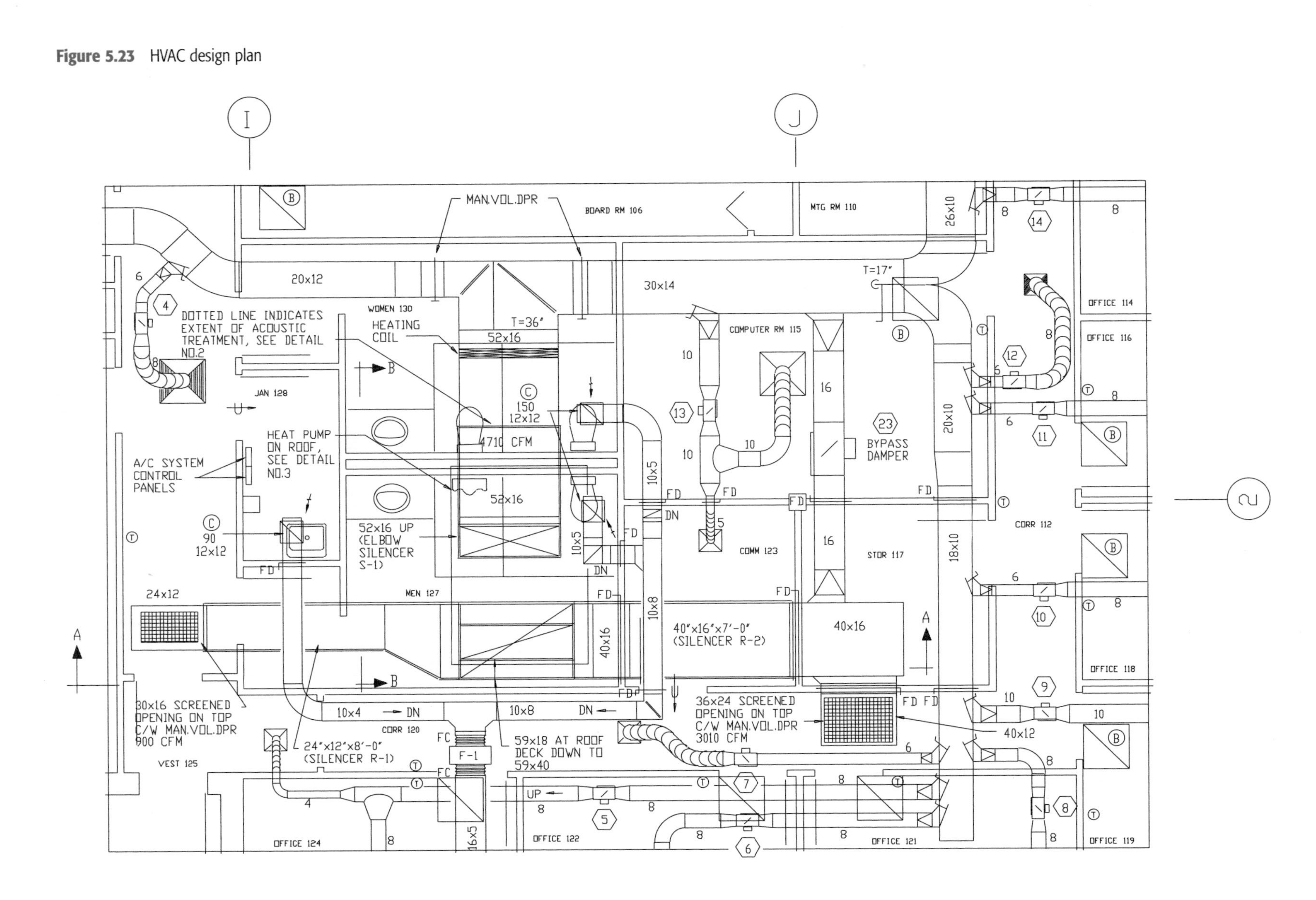

Figure 5.23 HVAC design plan

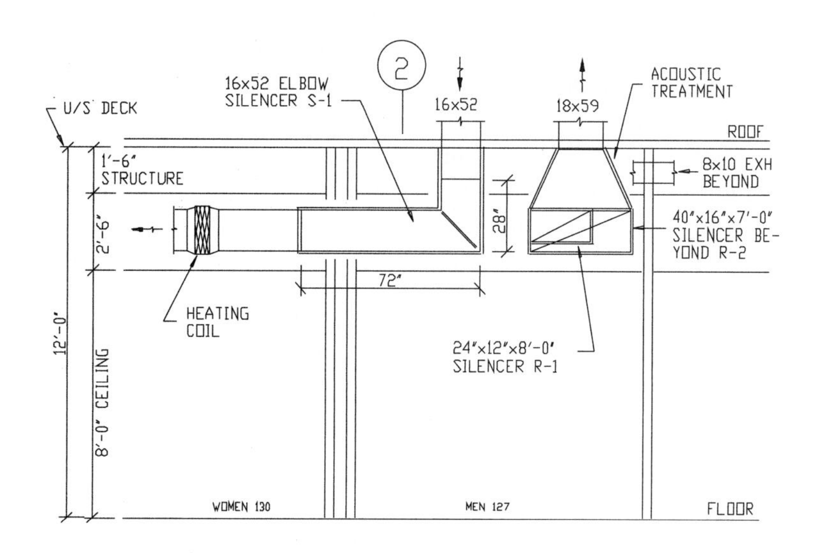

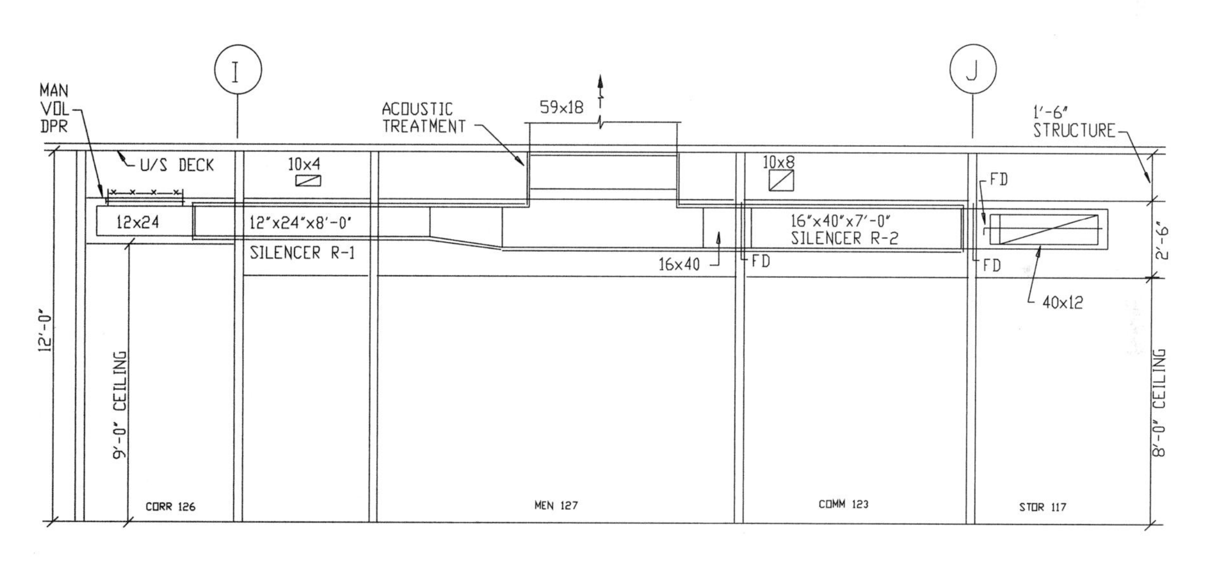

Figure 5.24 HVAC section drawings

Materials Handling

Engineering drawings used in materials handling generally make use of orthographic views and sections to describe equipment, storage vessels, conveyors, and associated steel structures. Perhaps along with the structural engineer (see civil engineering drawings), the mechanical engineer is responsible for the design and implementation of such systems. Again, resembling a general arrangement type of drawing, the equipment (e.g., rock crusher, weighing machine, cyclone washer, mine) is connected by conveyors and chutes. Other connections are to storage bins and loading areas for the material. Supporting structures are also designed and included in the arrangement.

Enlarged detail drawings are made to clarify specific features but generally do not include fabricating information. This is shown on **shop drawings** prepared by trained drafters or technicians from the engineering drawings. If the engineers are working for suppliers of material-handling systems, then they are also responsible for the detail design of the equipment and specifications. A working knowledge of structural steel properties and notations, conveying systems, and other materials-handling equipment is required in this field. Appendix F gives a table of notations used for some common structural steel shapes. When equipment specified for the system is to be supplied by others, detail drawings must be sent to the engineer in order that capacity, size, and connecting details can be verified. Building or site drawings used for the design of the system must always be the latest issue.

Figure 5.25 illustrates a conveyor and chute structure inside a building. The layout shows how the equipment is situated, clearances from other structures, and enough detail for suppliers of equipment to design the requirements.

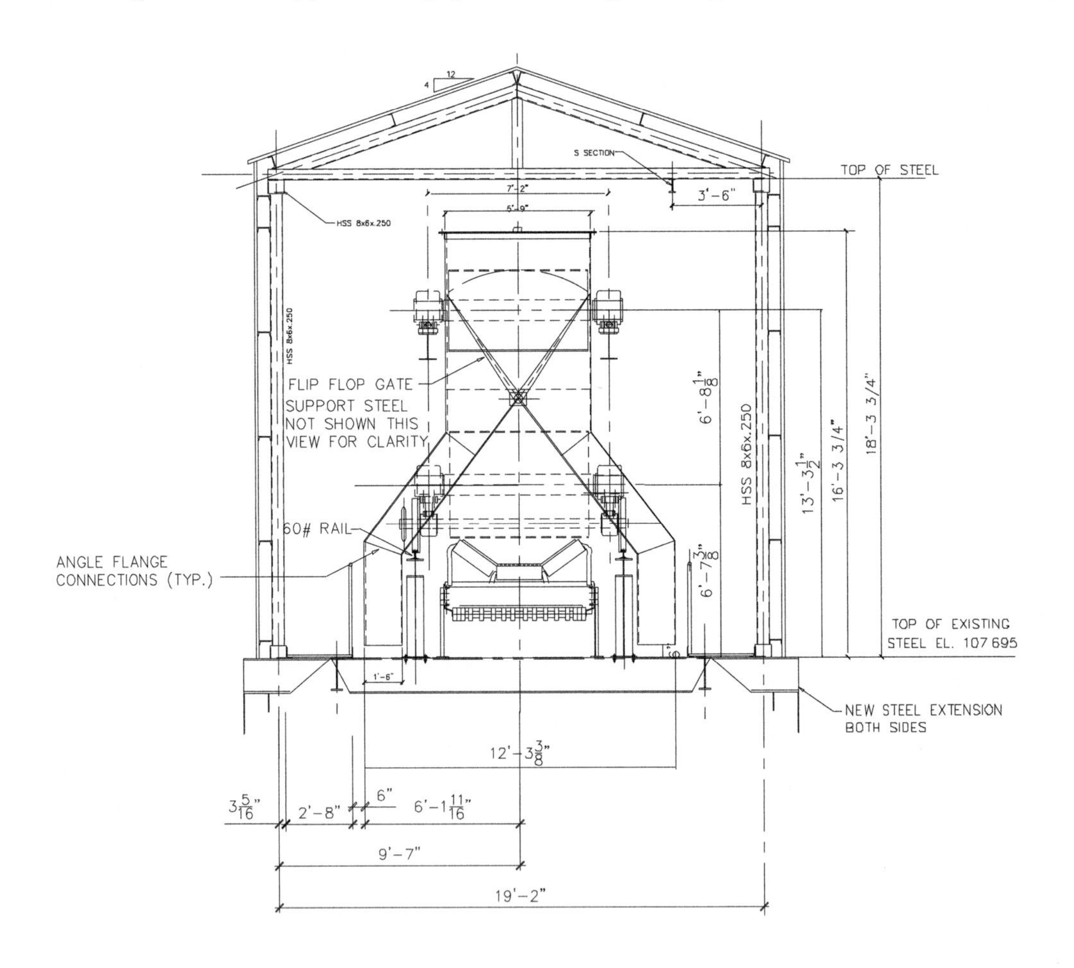

Figure 5.25 Conveyor housing

A general arrangement of materials-handling equipment for use underground is shown in Figure 5.26. Note that the dimensions show the equipment reach and capacity.

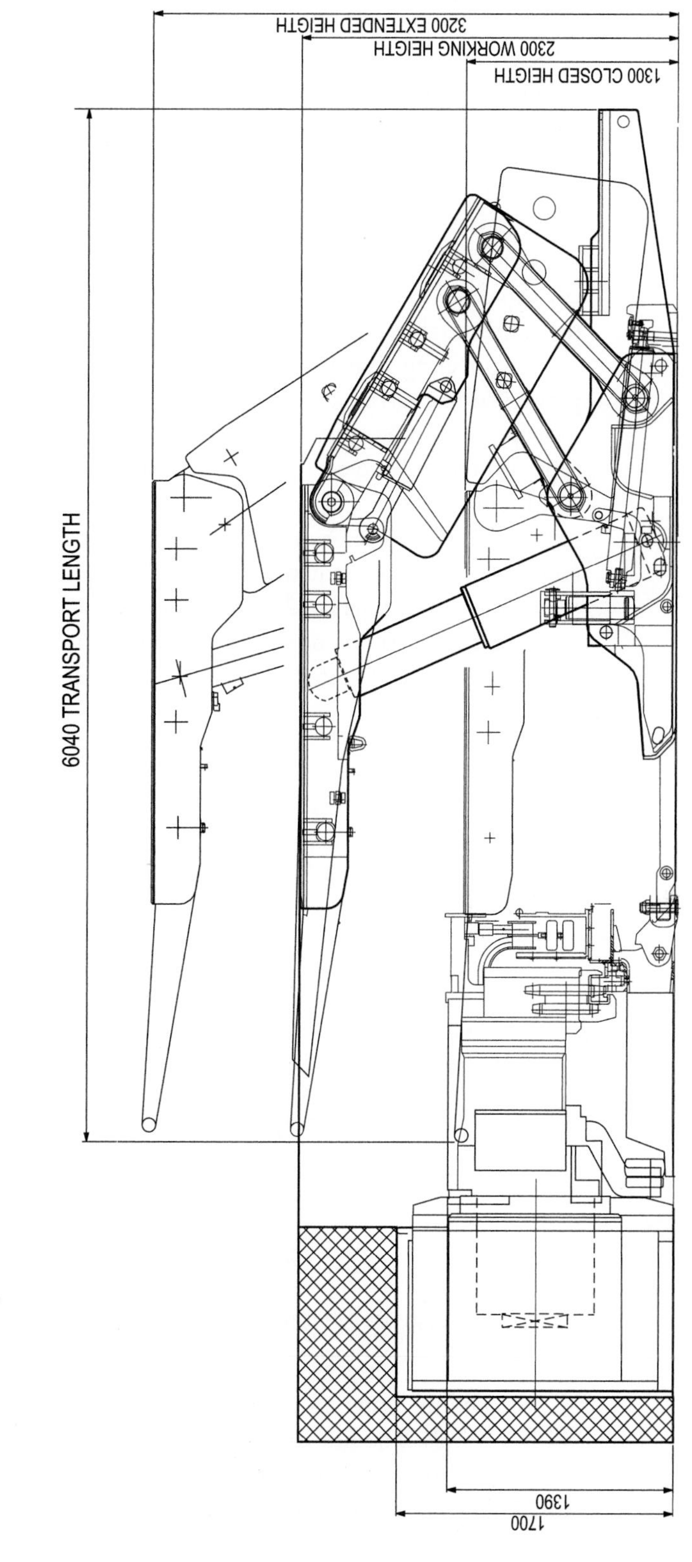

Figure 5.26 General arrangement

Industrial Piping

Industrial piping drawings are used in the design and construction of power stations, aviation fuelling stations, steel plants, coal wash plants, and other applications. Specialized mechanical engineers design systems for transporting such fluids as water as well as steam, and air through pipes and tubes. Fluids are carried by pipes to and from equipment for processing and storage within industrial facilities. Some crossover occurs as both the mechanical engineer and the chemical engineer design and select equipment for the transportation of fluids.

Pipes are made of steel, iron, plastic, or copper in standard sizes. They are connected to equipment and each other by fittings that can be welded, screwed, or bolted. Valves are used in piping systems to regulate or stop the flow of a fluid being transported through a pipe.

Engineering drawings for industrial piping show the size and location of the pipes, fittings, valves, and process equipment. To simplify the preparation of working drawings of piping systems, a set of symbols has been developed to represent various pipe fittings and valves. Appendix G lists common pipe sizes.

Piping systems can be so complicated that a 3-D model has to be constructed, either in a workshop or on a computer, to check clearances or interferences of pipes, equipment, support structures, or HVAC equipment. As with all engineering drawings, piping drawings are 2-D representations of the piping model.

The two methods of drawing piping systems in use are single-line and double-line drawings (see Figure 5.27a and 5.27b, respectively).

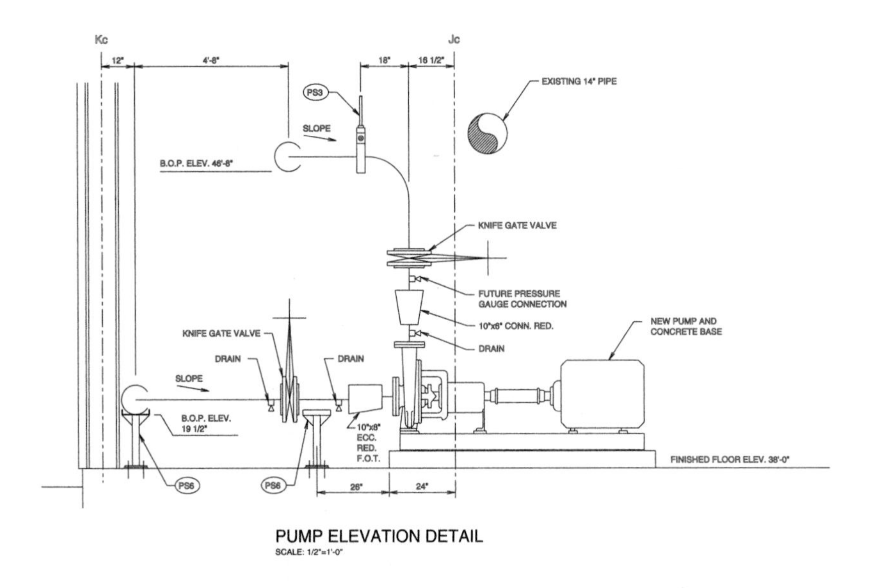

Figure 5.27a Single-line pipe drawing

Single-line drawings are a diagrammatic representation of the pipes and fittings using symbols. The line drawn represents the centreline of the pipe.

Single-line drawings are generally preferred over the more time-consuming double-line ones; however, many companies have their own standards which may, for example, use single-line piping up to 350 (14") diameter pipe and double-line drawings for pipes above that size. The majority of piping drawings are still prepared in feet and inches.

In orthographic drawings, a broken circle (the same diameter as the actual pipe) is used to represent a pipe turning away from the viewer and a solid circle for a pipe turn-

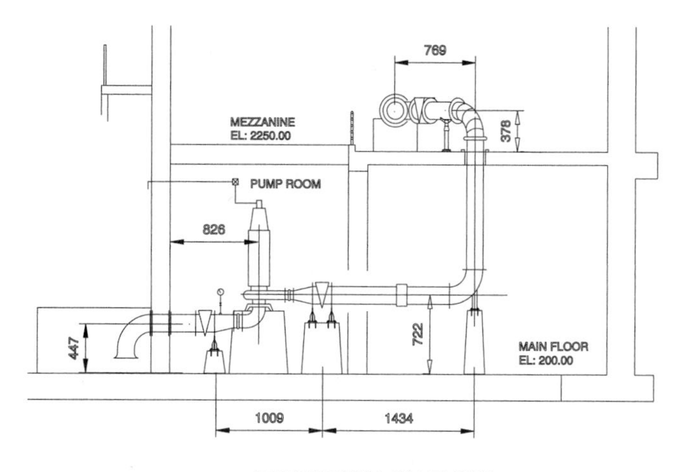

Figure 5.27b Double-line pipe drawing

ing toward the viewer. In Figure 5.28 an isometric pictorial sketch is converted into a two-view, orthographic detail drawing. Note how the different pipe locations are shown.

Horizontal dimensions for piping locations are given in the plan view, from centre to centre of pipe and to the outer face of a connecting flange (see Figure 5.29). Vertical locations are given as elevations from a base (e.g., site zero) or from the flange of an equipment connection (see Figure 5.30) in section views.

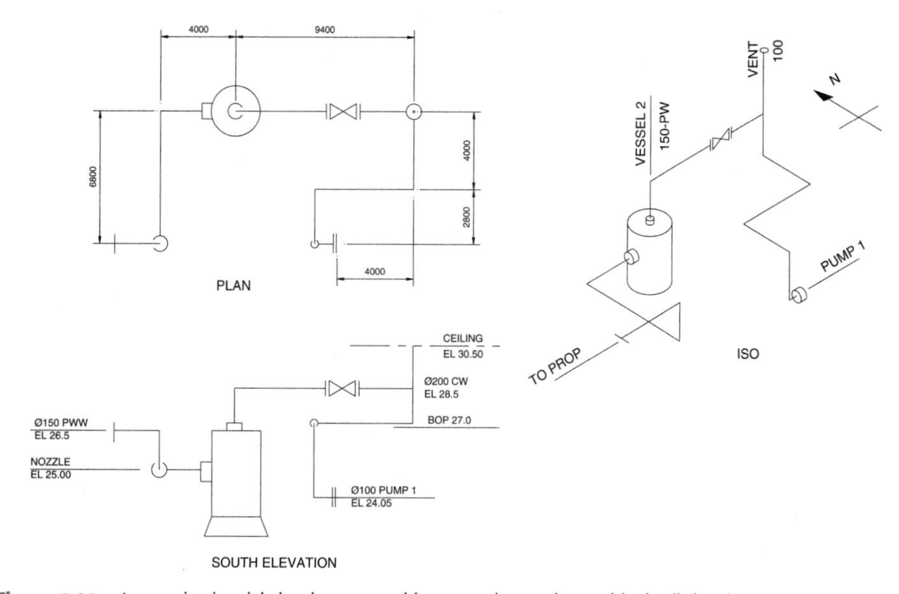

Figure 5.28 Isometric pictorial sketch converted into two-view, orthographic detail drawing

Equipment, supports, and vessels are generally shown as a thin line or a phantom outline. The location of these is given on the building layout drawings prepared by the civil engineer.

As industrial plants are generally large, they are divided into areas. The scale of engineering piping drawings is usually 1:50 or 1:20 metric (for inch units, a scale of 3/8" = 1' 0" is preferred). A key plan of the plant is drawn in the upper right-hand corner of the drawing sheet and the area shown on the sheet highlighted in the key plan. Drawing numbers for continuation of piping may be given on "match lines."

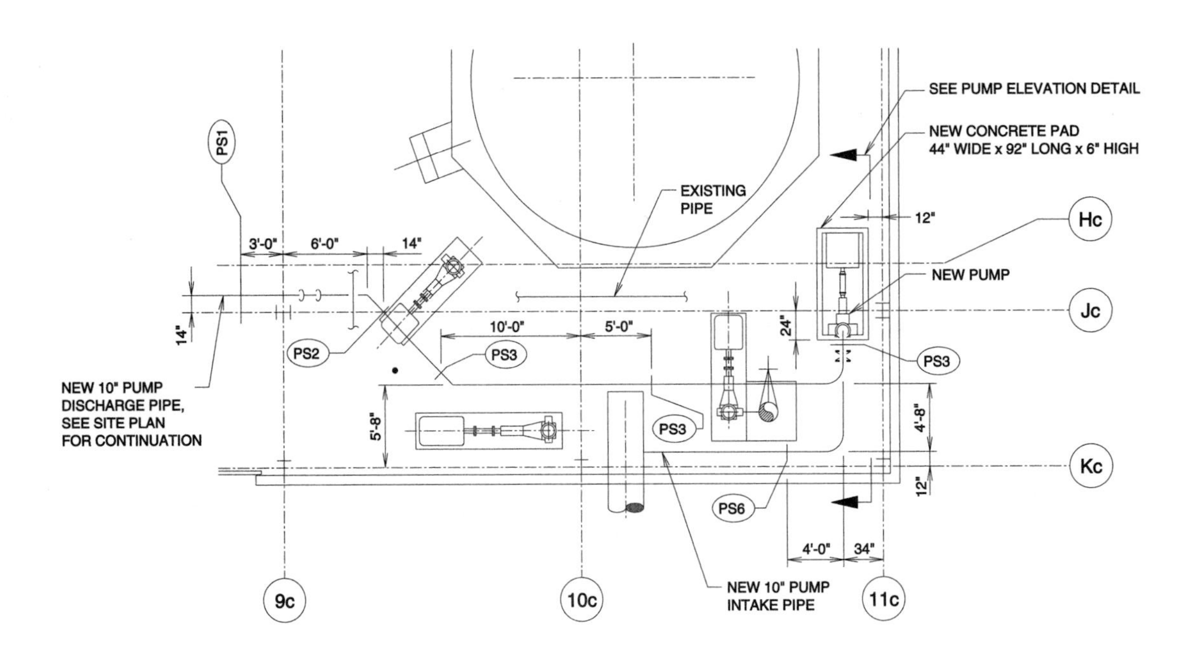

Figure 5.29 Piping design plan

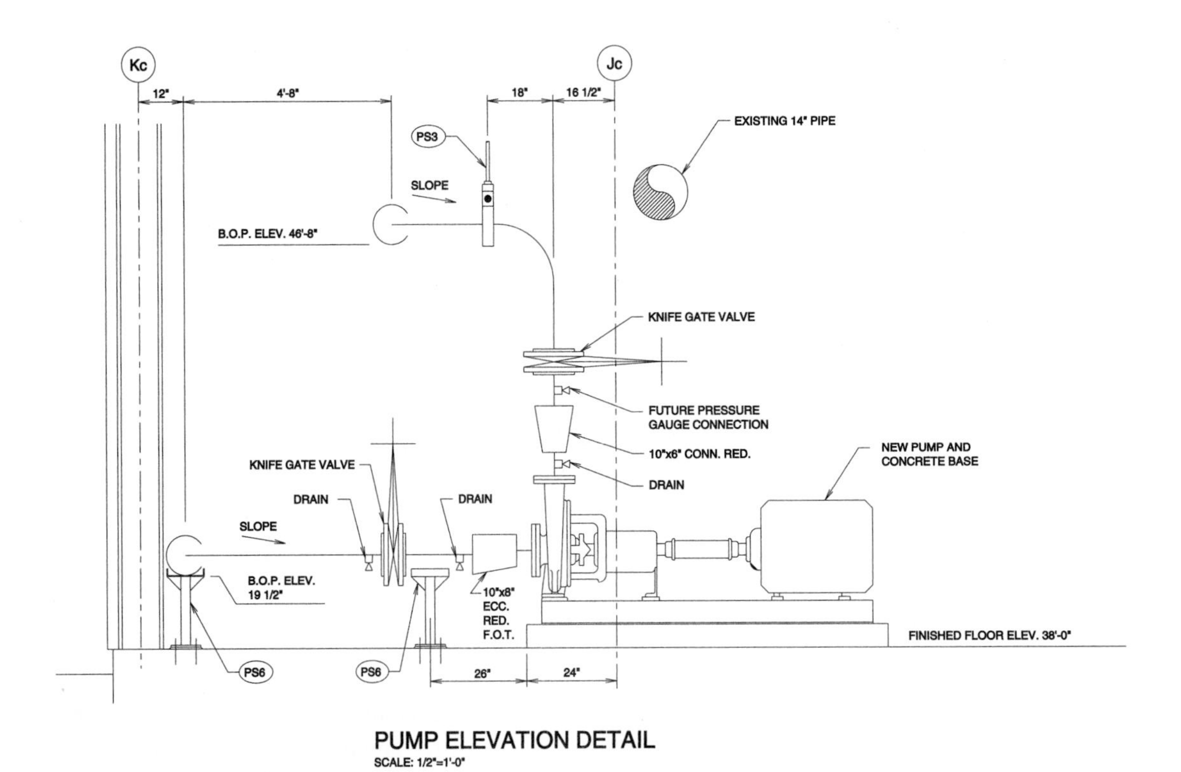

Figure 5.30 Piping design section from Figure 5.29

Figure 5.31 Spool detail of a section of pipe

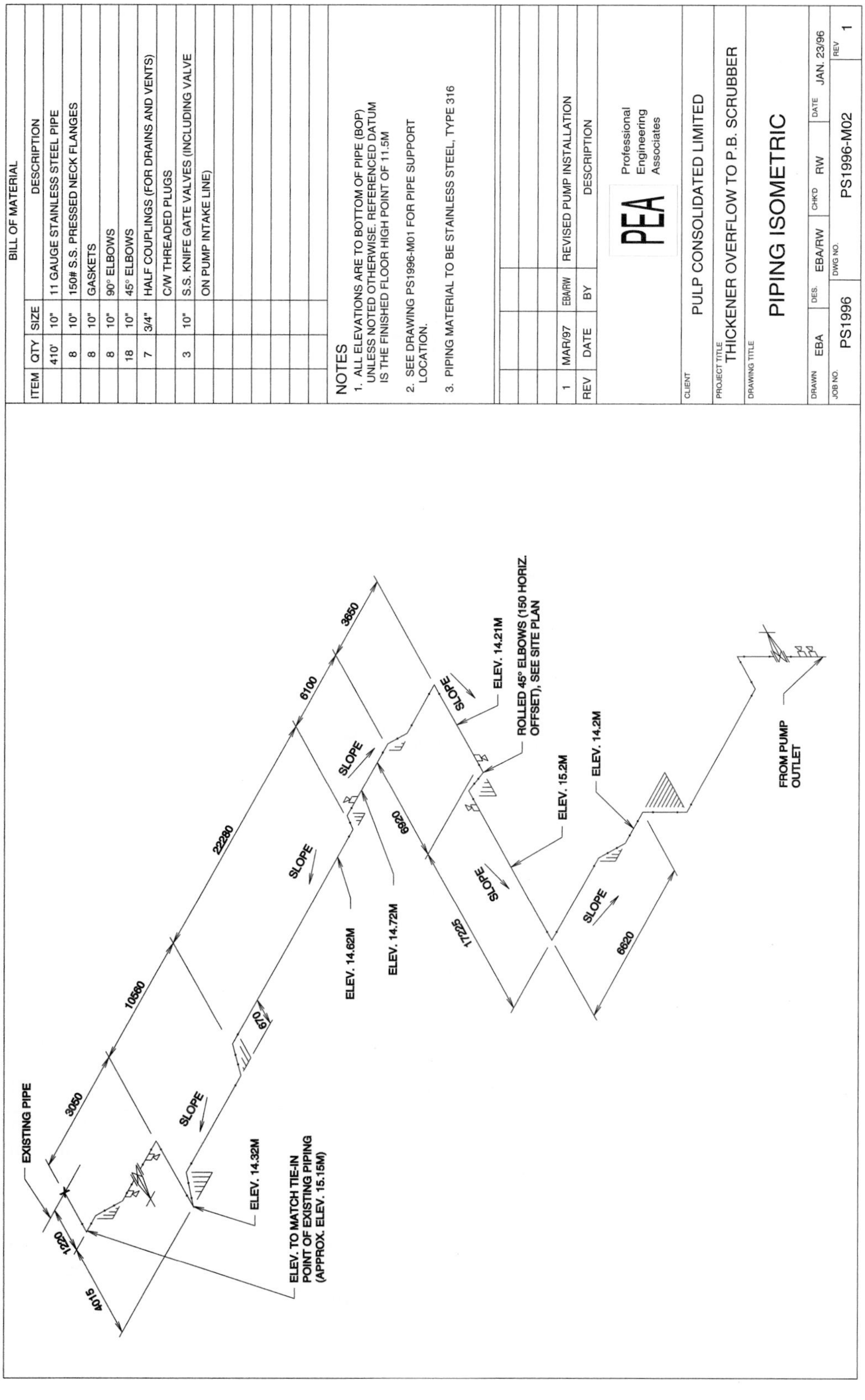

EXISTING PIPE
4015
1220
3050
10560
22280
6100
3650
670
6920
17225
6620
SLOPE
ELEV. 14.32M
ELEV. TO MATCH TIE-IN POINT OF EXISTING PIPING (APPROX. ELEV. 15.15M)
ELEV. 14.62M
ELEV. 14.72M
ELEV. 14.21M
ROLLED 45° ELBOWS (150 HORIZ. OFFSET), SEE SITE PLAN
ELEV. 15.2M
ELEV. 14.2M
FROM PUMP OUTLET

BILL OF MATERIAL

ITEM	QTY	SIZE	DESCRIPTION
	410'	10"	11 GAUGE STAINLESS STEEL PIPE
	8	10"	150# S.S. PRESSED NECK FLANGES
	8	10"	GASKETS
	8	10"	90° ELBOWS
	18	10"	45° ELBOWS
	7	3/4"	HALF COUPLINGS (FOR DRAINS AND VENTS)
			C/W THREADED PLUGS
	3	10"	S.S. KNIFE GATE VALVES (INCLUDING VALVE
			ON PUMP INTAKE LINE)

NOTES

1. ALL ELEVATIONS ARE TO BOTTOM OF PIPE (BOP) UNLESS NOTED OTHERWISE. REFERENCED DATUM IS THE FINISHED FLOOR HIGH POINT OF 11.5M
2. SEE DRAWING PS1996-M01 FOR PIPE SUPPORT LOCATION.
3. PIPING MATERIAL TO BE STAINLESS STEEL, TYPE 316

REV	DATE	BY	DESCRIPTION
1	MAR/97	EBA/RW	REVISED PUMP INSTALLATION

PEA Professional Engineering Associates

CLIENT: PULP CONSOLIDATED LIMITED

PROJECT TITLE: THICKENER OVERFLOW TO P.B. SCRUBBER

DRAWING TITLE: PIPING ISOMETRIC

DRAWN	DES.	CHK'D	DATE
EBA	EBA/RW	RW	JAN. 23/96

JOB NO.	DWG NO.	REV
PS1996	PS1996-M02	1

The fabrication drawings of industrial piping systems are called **spool drawings**. These can be drawn as isometric or orthographic views. Figure 5.31 (page 171) is a spool detail of a section of pipe from Figure 5.29.

Pipes have to be supported. Generally an engineer is assigned the task of designing piping support systems to withstand the load of the material, the pipe itself, and the shock loading from valving operations. Figures 5.29 and 5.30 showed plan and elevation views with the pipe supports (PS) noted. Further detail drawings will provide enough information for the fabricator to provide shop fabrication details of each support. A typical support detail from Figure 5.29 is shown in Figure 5.32.

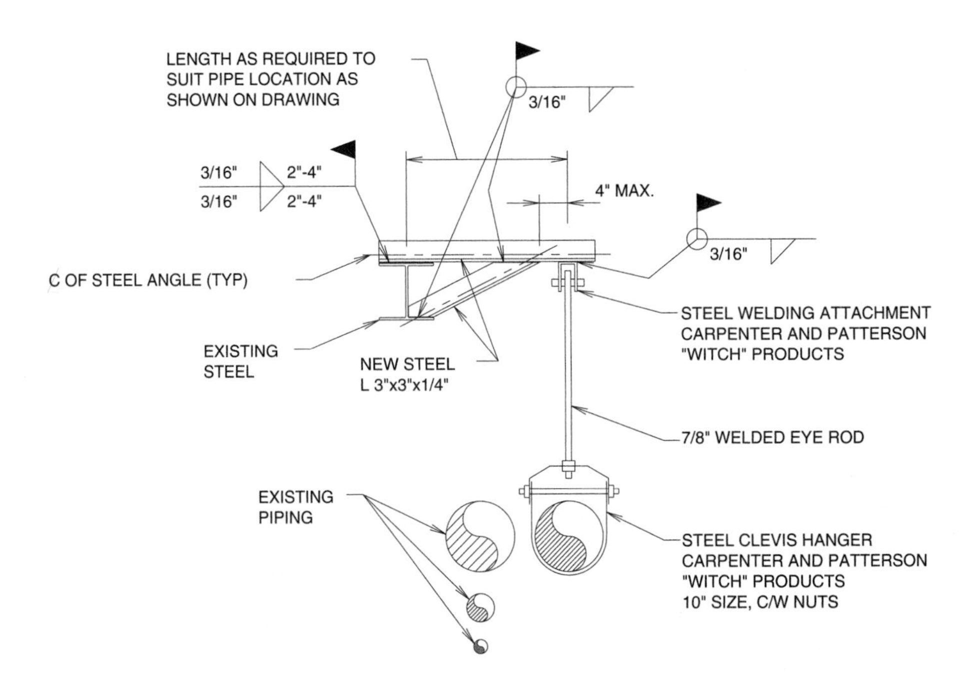

Figure 5.32 Piping supports

Fluid Power

Fluid power is a branch of mechanical engineering dealing with hydraulic and pneumatic operating systems. A **fluid** is defined as something that can flow and is able to move and change shape without separating, when under pressure. Fluid power includes both liquids and gases used extensively in the operation and control of automobile and aircraft systems, machine tools, heavy-duty equipment, and ships.

Usually, the pipes for fluid power are small diameter (bore) tubes of plastic, nylon, or copper. The engineering drawings generally consist of schematic diagrams drawn with symbols to illustrate the components of the system. The symbols are single-line graphics, as illustrated in ANSI T3.28.9-1989 or ISO 1219-1976. As these drawings are schematic, they are not drawn to scale (see Figure 5.33). The schematic diagram shown emphasizes the function of the circuit by using symbols to illustrate the operation of the components. A list of materials and components could be included as well as specifications for the components.

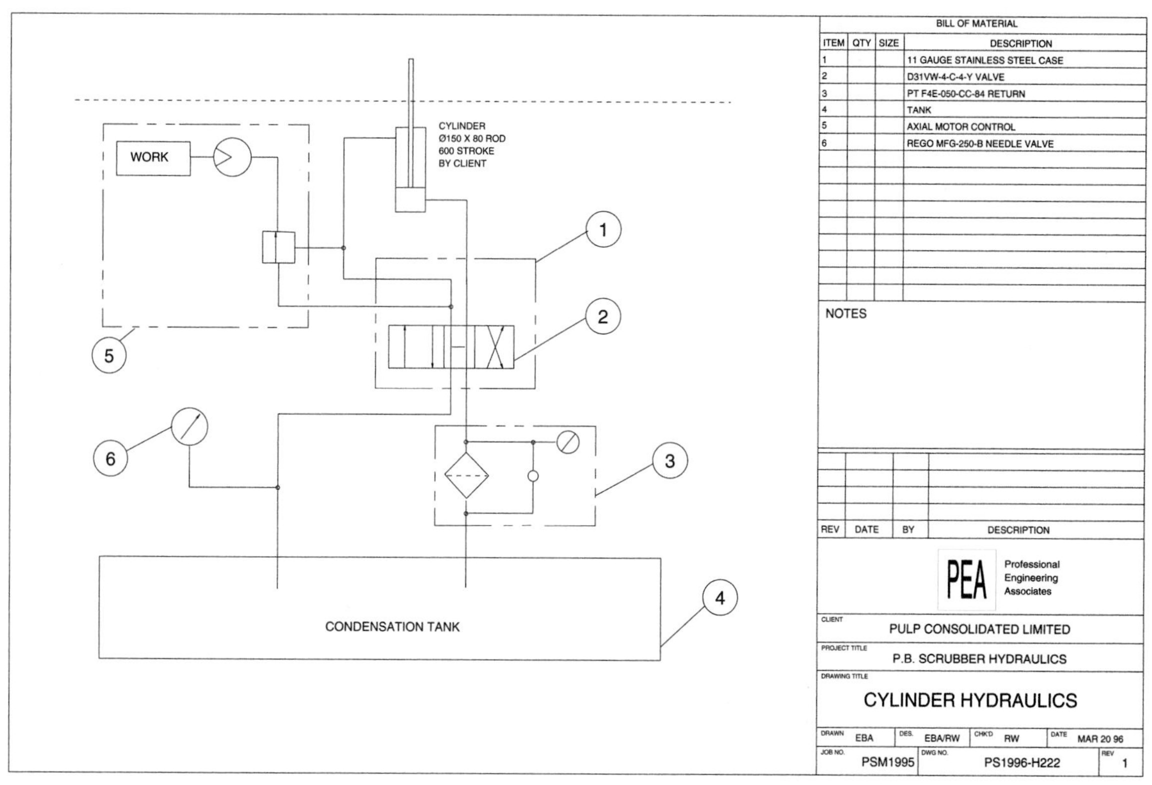

Figure 5.33 Fluid power drawing

CIVIL ENGINEERING DRAWINGS

Civil engineers also deal with many different types of engineering drawings. The following table lists some of the general areas, which cross over each other in many respects. Much of the work today uses the metric system of measurement—that is, the millimetre or metre. However, many agencies, especially in the United States, still use decimal feet as the unit of measure. Engineering drawings using these units are drawn to scale using the civil engineer's scale that uses 1 inch as a base; for example, 1" = 30' (1 inch on paper equals 30 feet on site).

Civil Engineering	**Working Drawings Content**
Site preparation and road design	Land development and preparation, survey information, GIS/GPS applications, plan and profile, cross-sections, and details
Structural steel, concrete, and wood	Structures: bridges, dams, and buildings. Equipment foundations, tanks and silos
Water and waste treatment	Disposal, treatment, and underground piping

Site Preparation and Road Design

Engineers use their knowledge of surveying, global positioning, and earthwork to design the site for an intended project. The site may encompass drainage, earth removal, and suitable material for foundations to rest on.

Engineering drawings for site preparation start with a topographical plan showing land contours and curves, location of buildings or structures, roads, water and utility lines, and the proposed project boundaries or traverse. **Contours** are curved lines that represent a certain elevation above a known point, usually sea level. The curves are drawn closer together to represent a steeper gradient or change in elevation. The elevation is indicated on the contour line. Contour lines are discussed in more detail in Chapter 6.

The location of the site boundaries and proposed structures are given by distances from a known point together with the angle. The angle can be given as either an **azimuth** (0–360 degrees) or a **bearing** (measured from a quadrant; for example, N45°W is the same as 315° as an azimuth). You will learn more about bearings and azimuths in Chapter 6.

A grid system is often used giving distances in a north or east direction. Figures 5.34 and 5.35 are examples of site drawings.

Highway design drawings start with a **key plan drawing** of the entire area involved in the project. This is divided into sections so that the drawing scale is manageable. Each section is drawn as a topographical plan drawing, showing the land layout. This includes contour lines expressing the altitude of the land and a grid system giving distances in a north direction and east direction. Such features as rivers, lakes, buildings, and railroads are included on the plan, which is drawn at a scale convenient for the area.

The proposed highway is plotted onto the topographical plan, using survey information and design criteria. For identification purposes, the highway is divided into "stations" along the centreline. Typical scales for plan drawings are 1:1000 or 1:500 metric (1" = 100' or 1" = 50' standard).

Plan and Profile Drawings

Below the plan, a full-section drawing is created—called a **profile drawing**. Generally, the profile is taken at the centreline of the highway and is drawn to an exaggerated scale to describe differences in elevation more clearly. Profile drawing scales are typically 10 times that of the plan—that is, 1:100 or 1:50 (1" = 10' or 1" = 5'). The surface of the natural ground is labelled "existing grade" and is shown as plotted from survey data. The centreline of the proposed highway sometimes cuts into the existing ground or requires filling to raise it to the design elevations.

The material to be cut or filled is shown on the profile drawing. The material that has to be removed or "borrowed" can be a decisive factor in the cost of the project, in relation to the number of transport operations, and hence the time and money required to prepare the site. Figure 5.36 (pages 177 and 178) is a plan and profile drawing of a proposed highway.

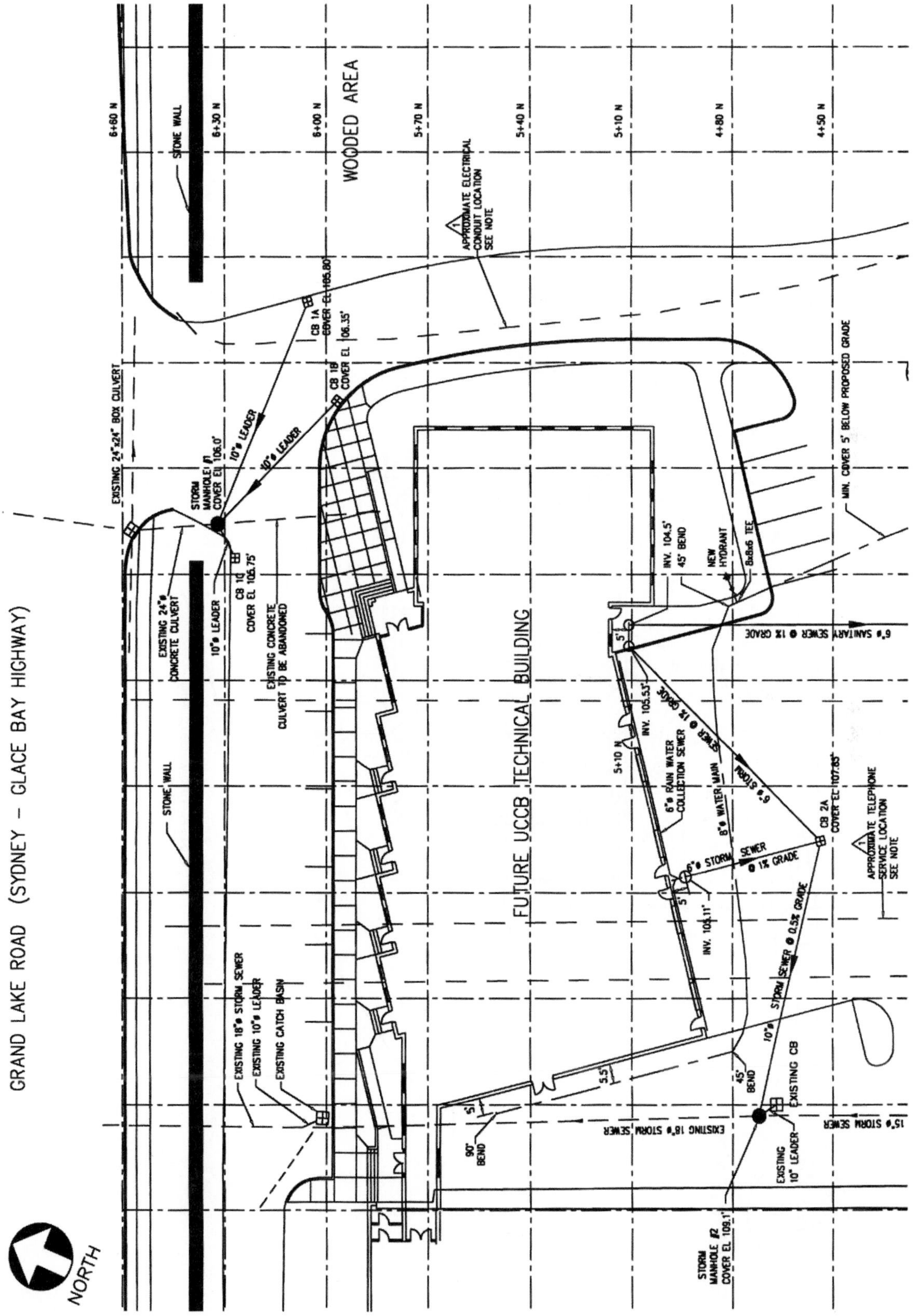

Figure 5.34 Site plan for a new building

GRAND LAKE ROAD (SYDNEY – GLACE BAY HIGHWAY)
NORTH
EXISTING 24"x24" BOX CULVERT
6+60 N
STONE WALL
EXISTING 24"ø CONCRETE CULVERT
STONE WALL
STORM MANHOLE #1 COVER EL 106.0'
10"ø LEADER
6+30 N
CB 1C COVER EL 105.75'
10"ø LEADER
EXISTING 18"ø STORM SEWER
EXISTING 10"ø LEADER
EXISTING CATCH BASIN
EXISTING CONCRETE CULVERT TO BE ABANDONED
10"ø LEADER
CB 1A COVER EL 105.80'
6+00 N
CB 1B COVER EL 106.35'
WOODED AREA
5+70 N
APPROXIMATE ELECTRICAL CONDUIT LOCATION SEE NOTE
90° BEND
5'
FUTURE UCCB TECHNICAL BUILDING
5+40 N
5.5'
5+10 N
5+10 N
5'
INV. 105.53'
6"ø RAIN WATER COLLECTION SEWER
6"ø SEWER @ 1% GRADE
INV. 104.5'
45° BEND
5'
INV. 105.11'
6"ø STORM SEWER @ 1% GRADE
8"ø WATER MAIN
6"ø STORM
NEW HYDRANT
8x8x6 TEE
4+80 N
STORM MANHOLE #2 COVER EL 109.1'
EXISTING 18"ø STORM SEWER
45° BEND
10"ø STORM SEWER @ 0.5% GRADE
EXISTING 10" LEADER
EXISTING CB
6"ø SANITARY SEWER @ 1% GRADE
CB 2A COVER EL 107.85'
4+50 N
MIN. COVER 5' BELOW PROPOSED GRADE
15"ø STORM SEWER
APPROXIMATE TELEPHONE SERVICE LOCATION SEE NOTE

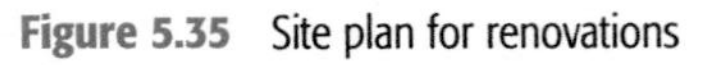
Figure 5.35 Site plan for renovations

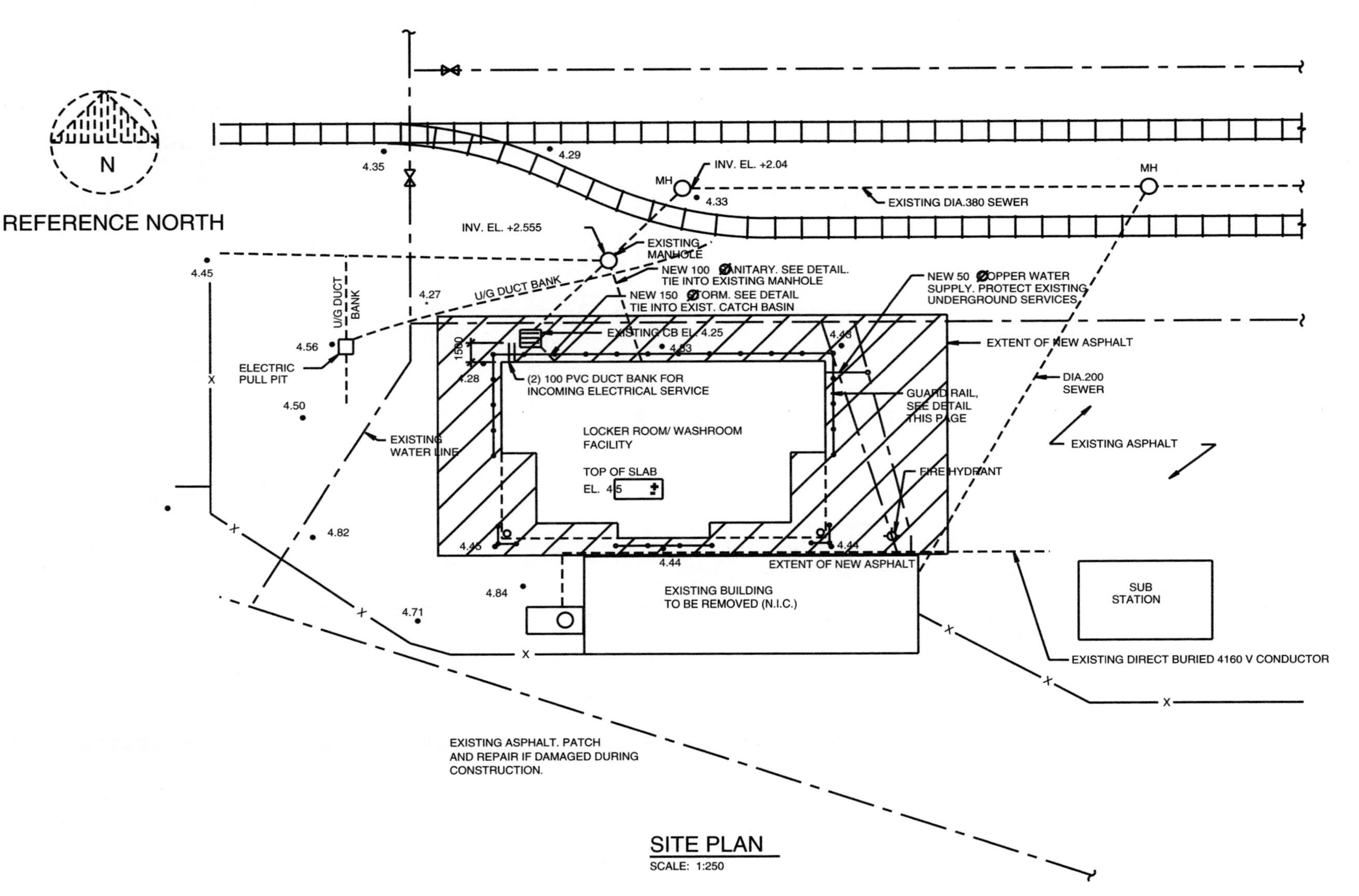

Figure 5.36a Plan of proposed highway interchange

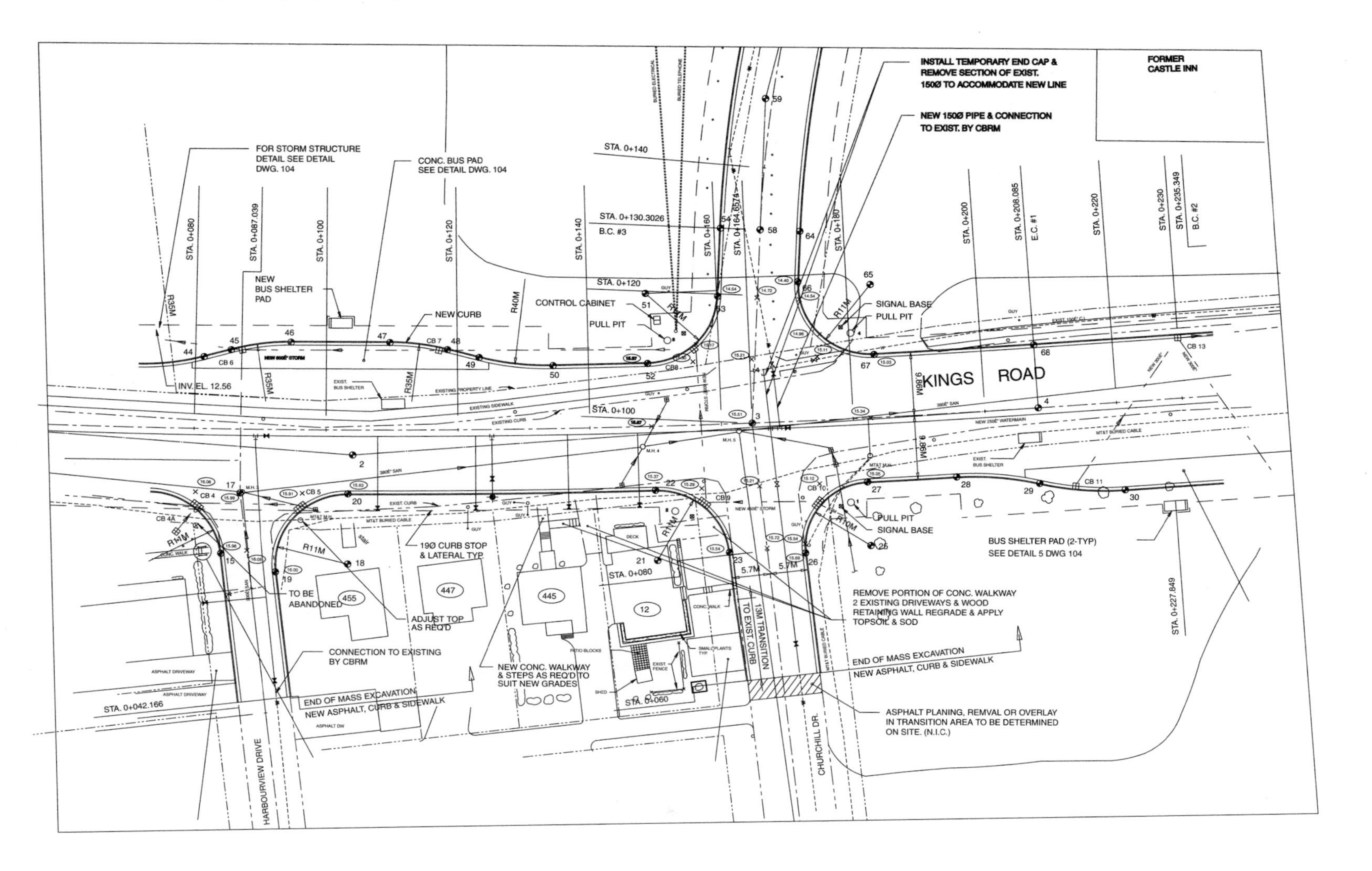

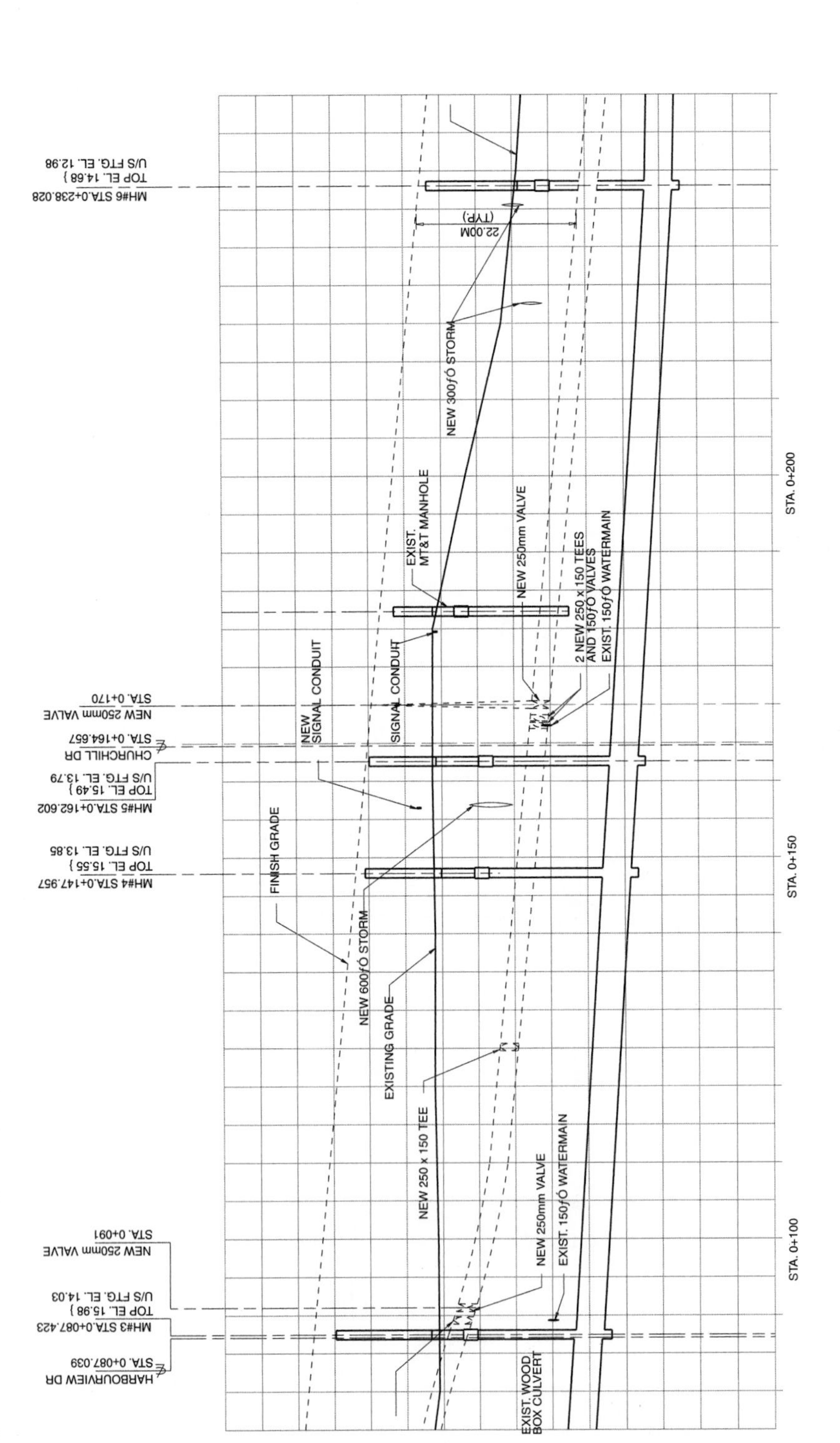

Figure 5.36b Profile drawing of a highway interchange

Cross-Section Drawings

As a profile drawing is taken along the longitudinal centreline of the road or site, **cross-section drawings** show the conditions at various points perpendicular to the centreline. In order to accurately determine the material to be removed or added, cross-sections are created at each station line. The roadway or structure is superimposed on each cross-section, which is again drawn at a larger scale than the plan to show detail. A typical road cross-section is shown in Figure 5.37.

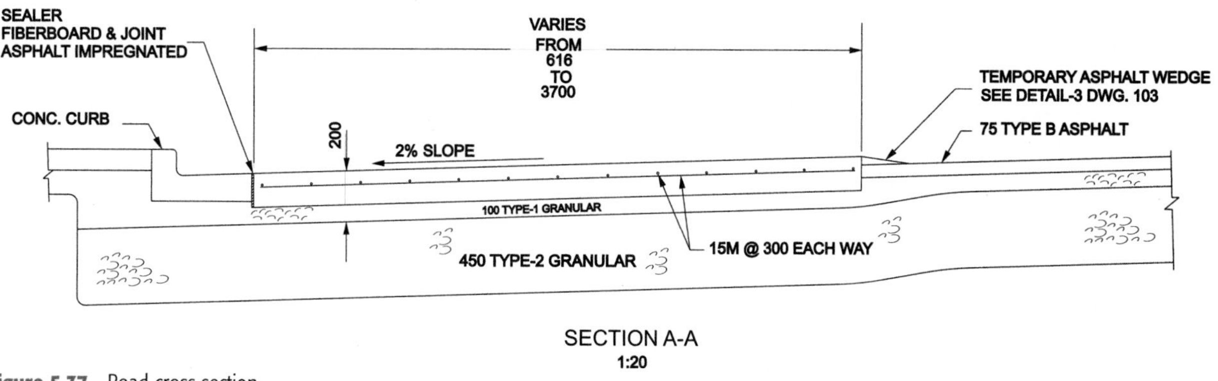

Figure 5.37 Road cross-section

In the case of highway construction, the cross-section drawings include such details as depth of material (subgrade), slope of the road for drainage, culverts or drains to remove water, super-elevation details for curves in the road, and often the location of underground pipes and conduits, curb details, and lighting.

Structural Steel, Concrete, and Wood

Structural engineering is generally associated with large steel, wood, or concrete structures and buildings. A structural engineer works with designers and architects to engineer structural components, supporting members, foundations, and geotechnical requirements of the building. Although grouped under the same heading, structural steel and concrete engineering drawings differ in the amount of content. Wood structures can be framed buildings or intricate, ornate, arched structures for public buildings.

Structural drawings generally fall into two categories: engineering design drawings and shop (manufacturing) drawings.

Engineering design drawings are set of working drawings that completely describe the project. Drawings include a site location plan, design layout plans and elevations of the locations of structural members, foundation layouts and concrete details, building elevation views, and typical wall sections. Figures 5.38 through 5.40 show several design working drawings for a small building.

Figure 5.38 Design working drawings–Sheet 1 of 3

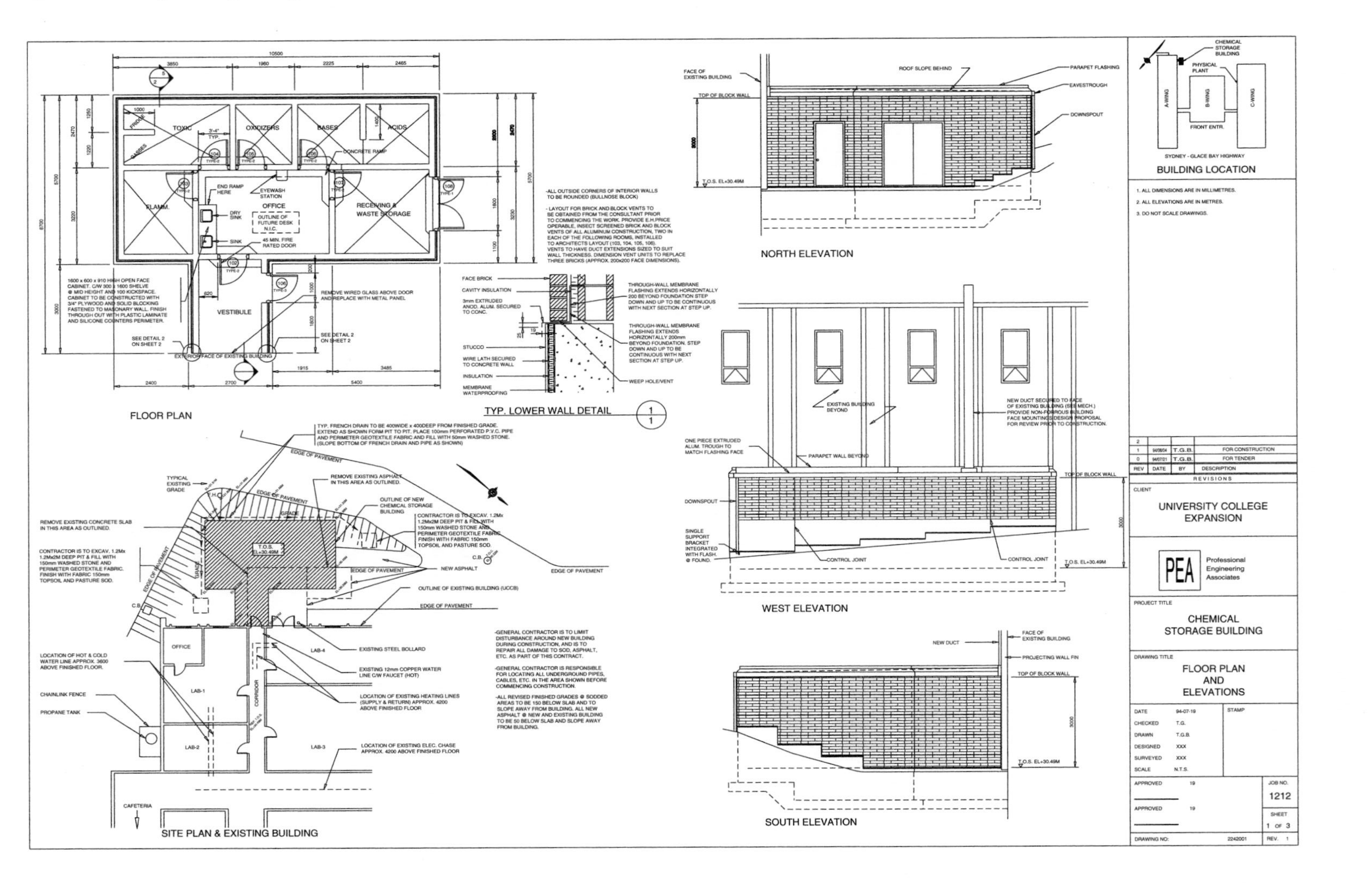

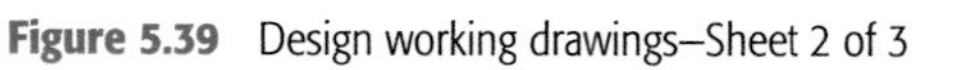
Figure 5.39 Design working drawings–Sheet 2 of 3

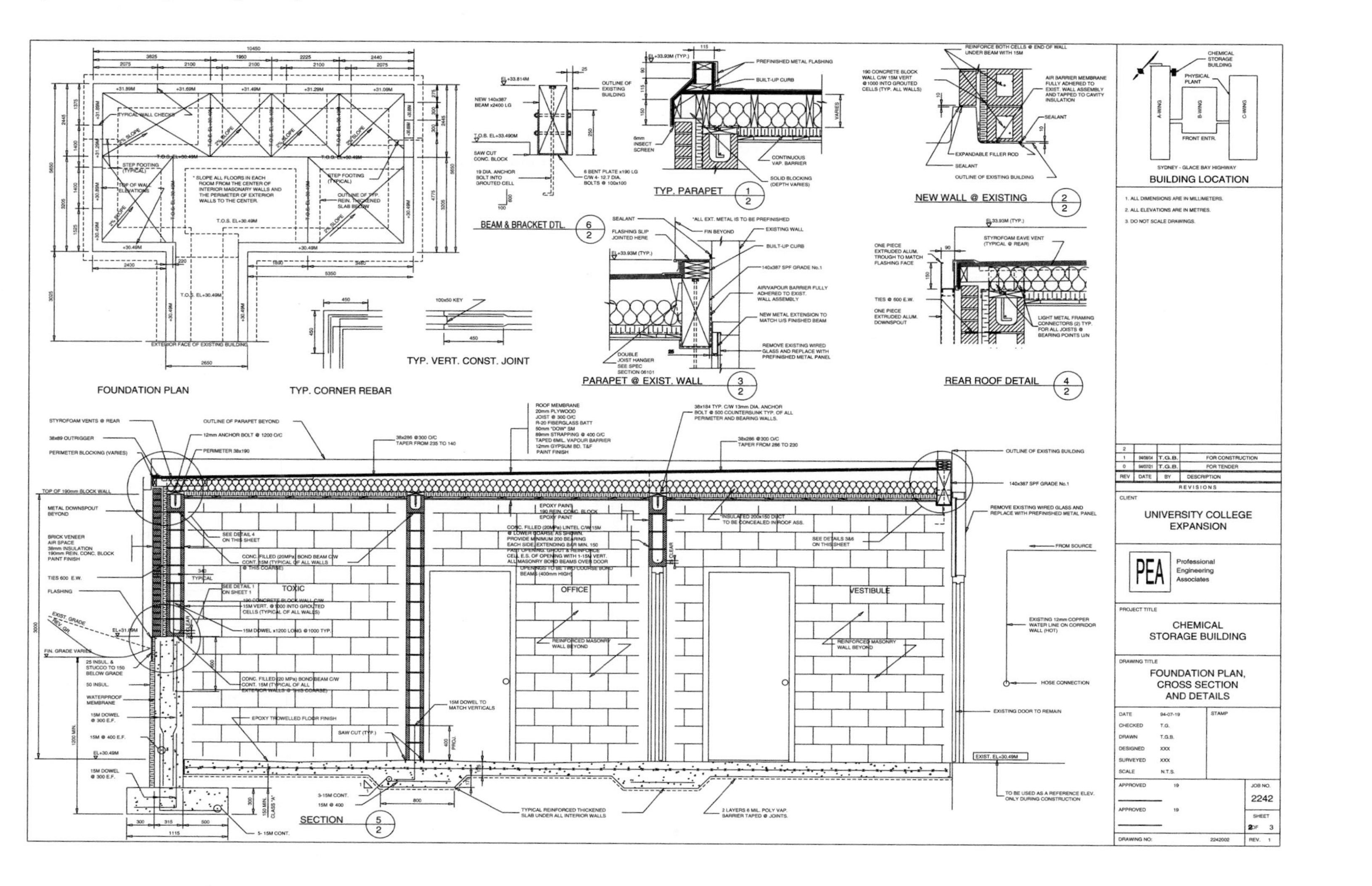

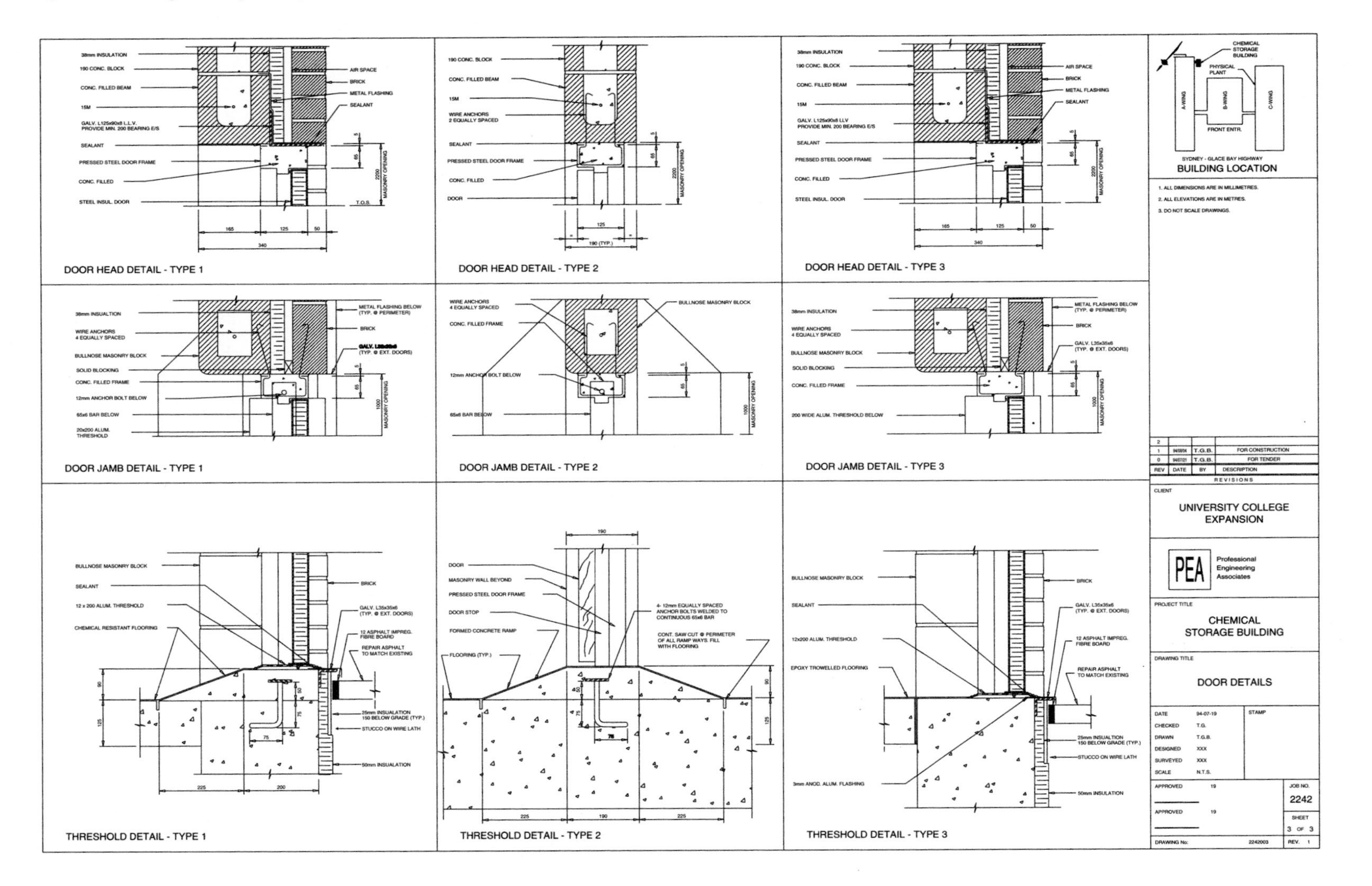

Figure 5.40 Design working drawings–Sheet 3 of 3

The engineer must become familiar with the terminology used on structural design drawings. The method of indicating structural steel members is given in ANSI and CSA standards (see following example). Structural members that are repeated on a plan are noted with "Do" meaning ditto (see Figure 5.41).

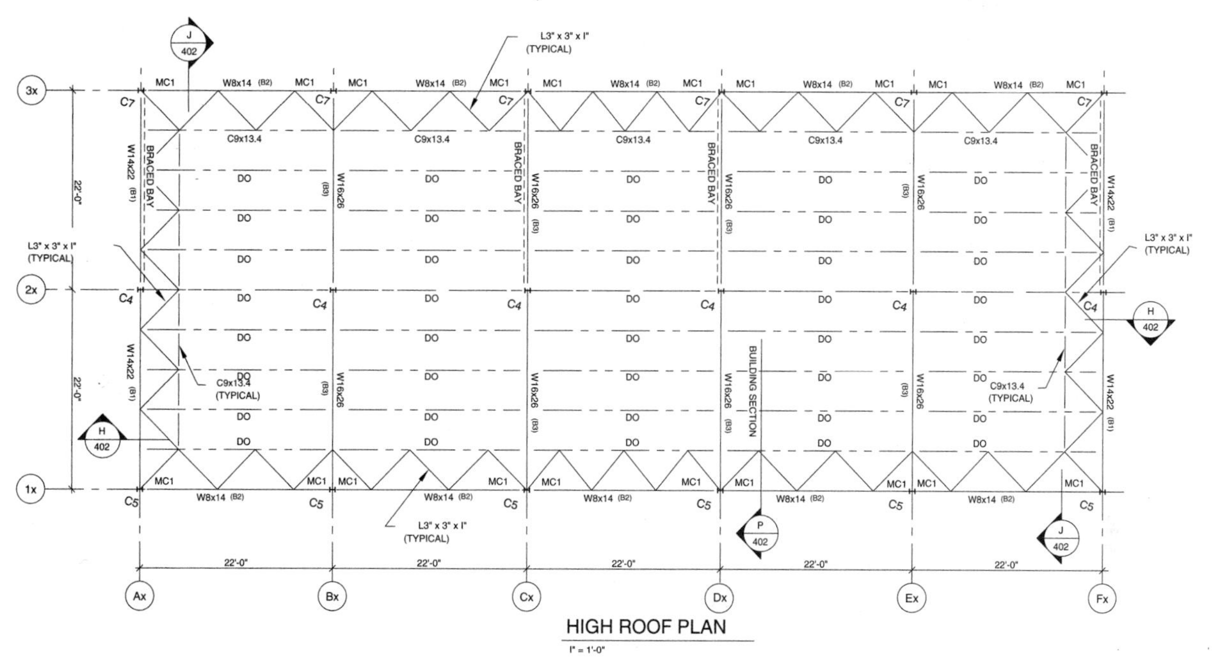

Figure 5.41 Structural roof plan

The design drawings for steel structures are primarily concerned with clearly showing the structural elements' size and location. Details are left to the fabricator who prepares shop drawings.

Shop (manufacturing) drawings for steel structures are generally made by the steel fabrication supplier.

Structural drawings illustrate the details and dimensions of all pieces of the structure and how they are to be fabricated (clearances, weld information, and so on). Each piece is identified by a marking system and listed in a bill of material (BOM). The fabricator provides an **erection drawing** or an **assembly drawing** showing the parts *in situ* (which is similar to the design drawing). The following figures show design drawings: Figure 5.41 shows a structural roof plan, and Figure 5.42 shows a section from the plan.

Drawings for concrete construction include all of the information necessary to install the concrete structure: size and placement of concrete, ground requirements,

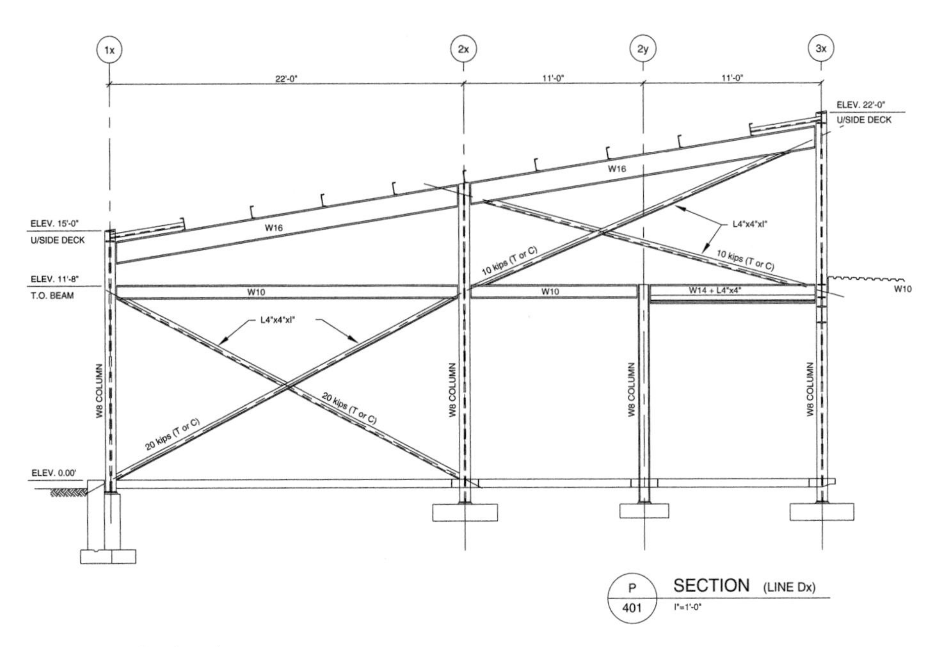

Figure 5.42 Steel section

reinforcing bar information, and connection details to steel or wood structures. Bar sizes are indicated as follows: **20M @ 200 E.W.** means 20 millimetre diameter bar spaced at 200 millimetres in both horizontal directions (Each Way).

Although concrete construction drawings follow standard engineering projections, they often appear cluttered. Figure 5.43 shows the concrete details for a steel-framed building.

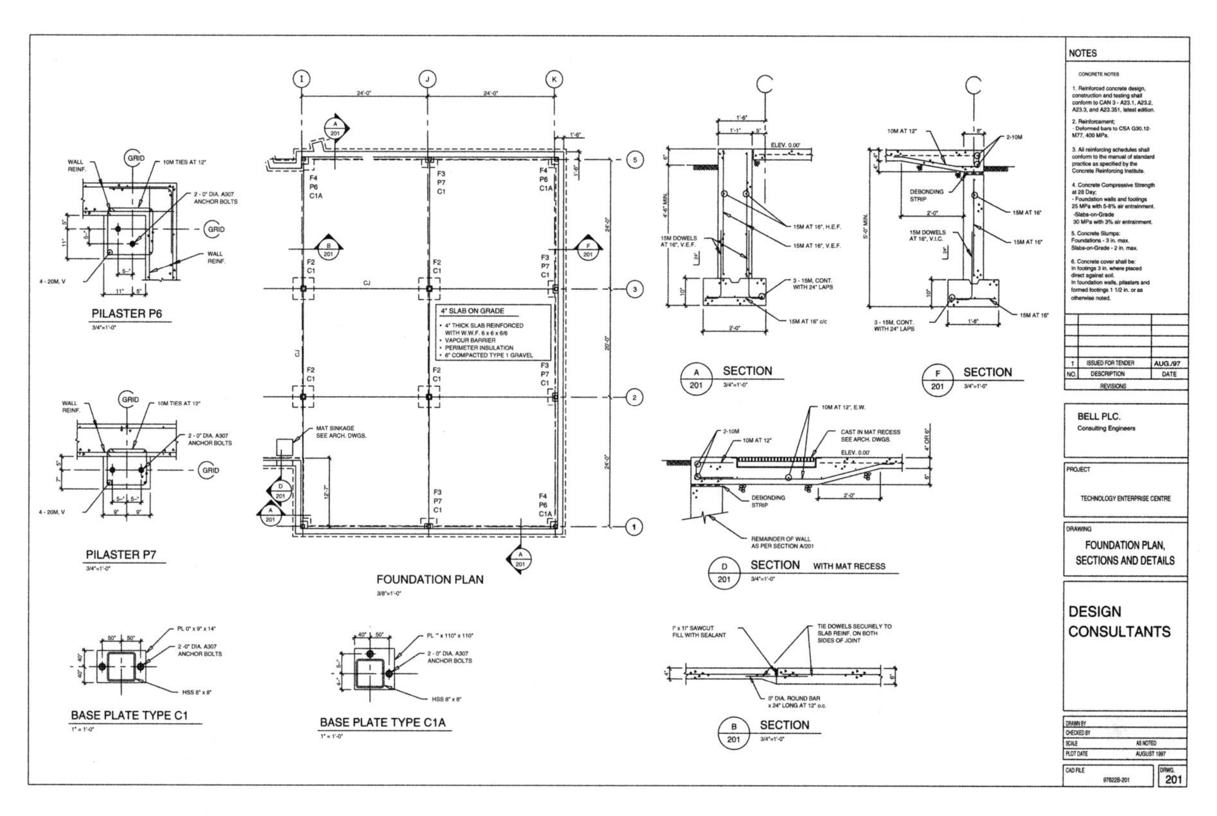

Figure 5.43 Concrete details

Water and Waste Treatment

Civil engineering incorporates the designs of fresh, waste, or sanitary water transport and of underground drainage. These drawings are termed **underground services** for the civil project.

The pipe and drainage information is shown on a plan and profile drawing, as discussed in the previous section. In addition, detail drawings of disposal beds and underground fixtures may be included. Figure 5.44 shows a plan and profile drawing of a proposed sewer and water line. Figure 5.45 (page 187) shows a typical detail sheet for underground fixtures.

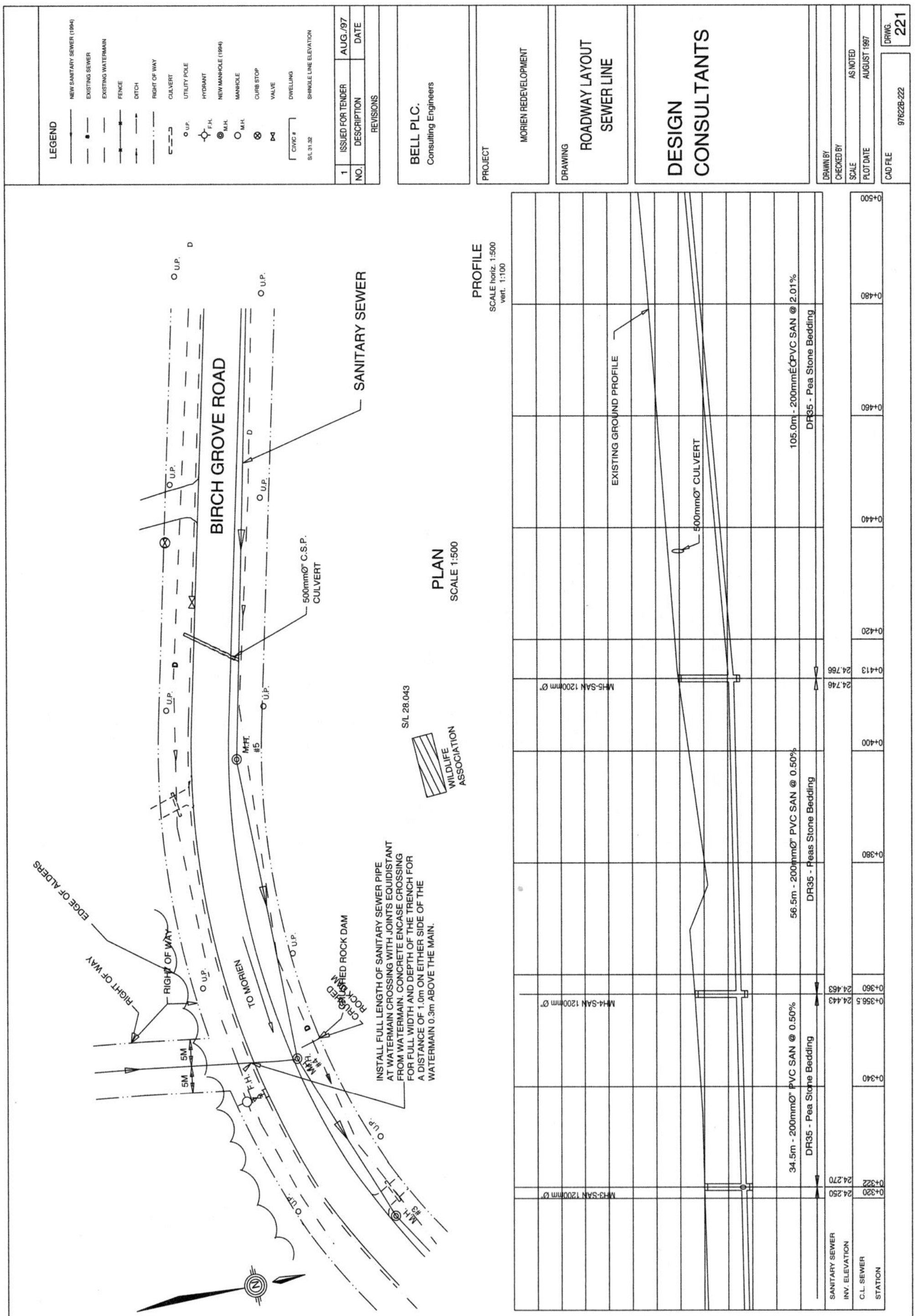

Figure 5.44 Plan and profile of proposed sewer and water line

RIGHT OF WAY
EDGE OF ALDERS
RIGHT OF WAY
5M 5M
F.H.
U.P.
TO MORIEN
M.H. #5
M.H. #4
D
CRUSHED ROCK DAM
M.H. #3
BIRCH GROVE ROAD
500mmØ" C.S.P. CULVERT
SANITARY SEWER
INSTALL FULL LENGTH OF SANITARY SEWER PIPE AT WATERMAIN CROSSING WITH JOINTS EQUIDISTANT FROM WATERMAIN. CONCRETE ENCASE CROSSING FOR FULL WIDTH AND DEPTH OF THE TRENCH FOR A DISTANCE OF 1.0m ON EITHER SIDE OF THE WATERMAIN 0.3m ABOVE THE MAIN.
S/L 28.043
WILDLIFE ASSOCIATION
PLAN
SCALE 1:500
PROFILE
SCALE horiz. 1:500
vert. 1:100

MH3-SAN 1200mm Ø
MH4-SAN 1200mm Ø
MH5-SAN 1200mm Ø
EXISTING GROUND PROFILE
500mmØ" CULVERT
34.5m - 200mmØ" PVC SAN @ 0.50%
DR35 - Pea Stone Bedding
56.5m - 200mmØ" PVC SAN @ 0.50%
DR35 - Peas Stone Bedding
105.0m - 200mmÉÓPVC SAN @ 2.01%
DR35 - Pea Stone Bedding
34 33 32 31 30 29 28 27 26 25 24 23 22 21

SANITARY SEWER
INV. ELEVATION
24.250 24.270 24.443 24.463 24.746 24.766
C.L. SEWER
STATION
0+320 0+322 0+340 0+356.5 0+360 0+380 0+400 0+413 0+420 0+440 0+460 0+480 0+500

LEGEND
NEW SANITARY SEWER (1994)
EXISTING SEWER
EXISTING WATERMAIN
FENCE
DITCH
RIGHT OF WAY
CULVERT
U.P. UTILITY POLE
F.H. HYDRANT
M.H. NEW MANHOLE (1994)
M.H. MANHOLE
CURB STOP
VALVE
CIVIC # DWELLING
S/L 31.32 SHINGLE LINE ELEVATION

NO.	DESCRIPTION	DATE
1	ISSUED FOR TENDER	AUG./97

REVISIONS

BELL PLC.
Consulting Engineers

PROJECT
MORIEN REDEVELOPMENT

DRAWING
ROADWAY LAYOUT
SEWER LINE

DESIGN
CONSULTANTS

DRAWN BY
CHECKED BY
SCALE AS NOTED
PLOT DATE AUGUST 1997
CAD FILE 97622B-222
DRWG. 221

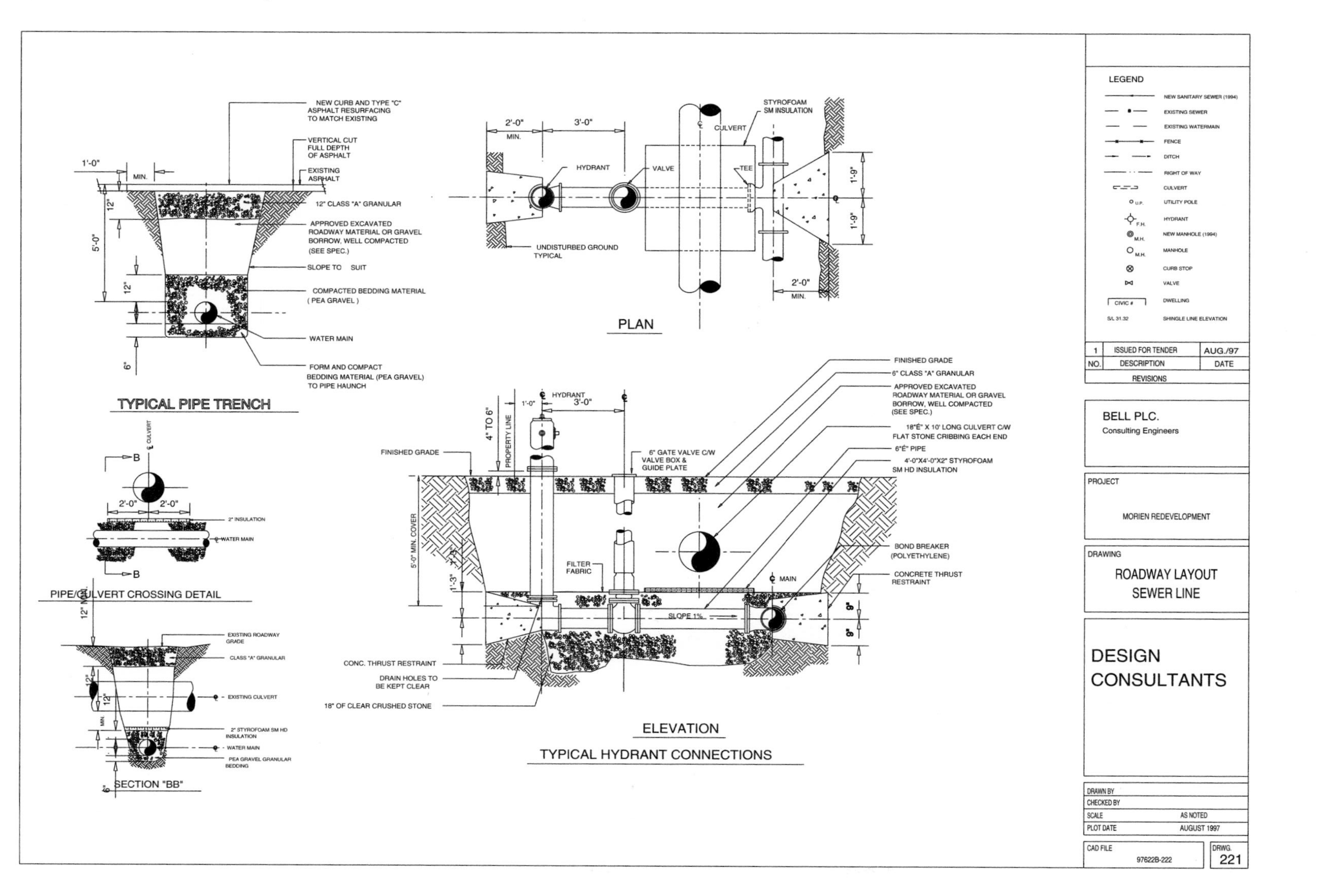

Figure 5.45 Detail sheet for underground fixtures

CHEMICAL ENGINEERING DRAWINGS

Chemical engineering involves equipment and processes for the transportation and manufacture of chemicals. Many of the drawings involved in chemical engineering are similar to those encountered in mechanical engineering—that is, piping, vessel design, and material-handling equipment.

The chemical engineer is more concerned with the process involved and the capabilities of the equipment. Therefore, the drawings initially used are schematic-type diagrams indicating the flow of material and the process steps. Because of the nature of industrial competition or for security reasons, most firms use codes to name chemicals or processes. Alternatively, the chemicals and processes may be omitted from the drawing entirely.

Process and Instrumentation Diagrams

The **process and instrumentation diagram** (P&ID) is the chemical engineer's road map of the system. It shows all of the major equipment and information relevant to the process in a schematic form, for example, equipment names and numbers, the sizes of pipes and line numbers, valves, capacities, pressure ratings, instrumentation, and any relevant data tables.

The drawing is not scaled. It is used to draw the design piping drawings in conjunction with the plant drawings. The P&ID should make use of standard symbols, such as those found in ISO 3511, and should be laid out consistently and evenly. A partial P&ID is illustrated in Figure 5.46.

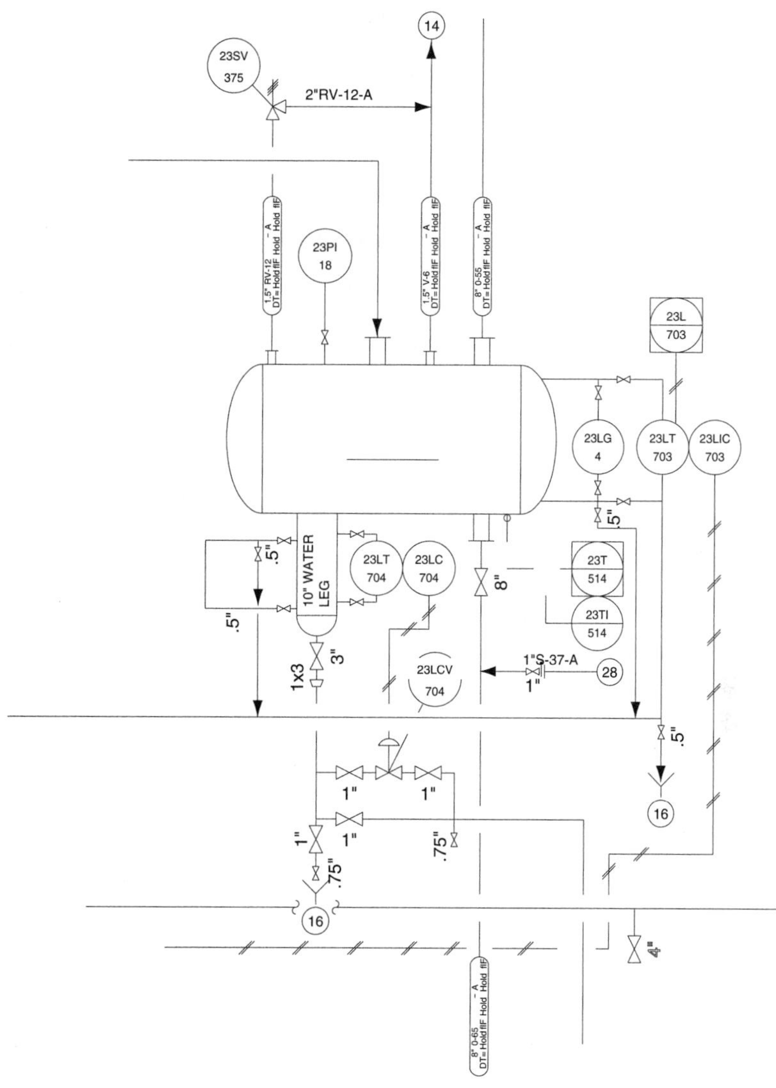

Figure 5.46 P&ID

ELECTRICAL AND ELECTRONIC ENGINEERING DRAWINGS

Electrical engineering deals mainly with power generation and distribution and is used in everything from industrial plants to computers. Control of electric motors, heating and lighting systems, and industrial processes all require electrical engineering drawings.

Electronic engineering is concerned with using small amounts of power to transfer electrons. Printed circuits, computers, audio/visual equipment, satellites, and instrumentation are all part of the electronic engineer's domain.

The types of drawings used by electrical/electronics engineers vary according to the particular field they are in. The majority of the drawings are schematic-type diagrams, using symbols to represent components, not drawn to scale. Their main purpose is to communicate how electrical/electronic devices work together. The major types of diagrams are the following:

- Schematic diagrams
- Connection and wiring diagrams
- Printed circuit board diagrams

Schematic Diagrams

Schematic diagrams are very common in electrical/electronic engineering. They show the functional relationship and connection of components used in an electrical/electronic circuit. Schematic diagrams show neither the physical size of the components nor the circuit. Often, components are rotated or inverted from their actual orientation to improve the drawing layout. In electrical engineering, single-line diagrams that show power distribution to equipment throughout a project are common. They are simplified schematic diagrams that show only the major components and the power feed to them. Figure 5.47 is an example of a single-line electrical schematic diagram used in the construction of a public health facility.

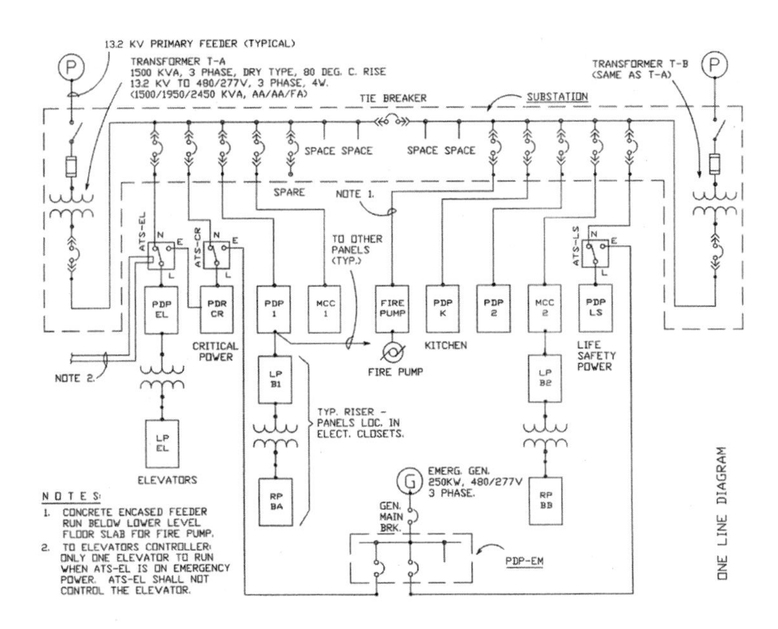

Figure 5.47 Single-line electrical schematic diagram

Electronic schematic diagrams use a component numbering system that can generate a bill of material list or be converted into a net list for use with other software. Typically, an electronic schematic diagram is the circuit design that is translated into a printed circuit board or "chip" layout for production. Figure 5.48 illustrates a partial schematic diagram for a communications satellite.

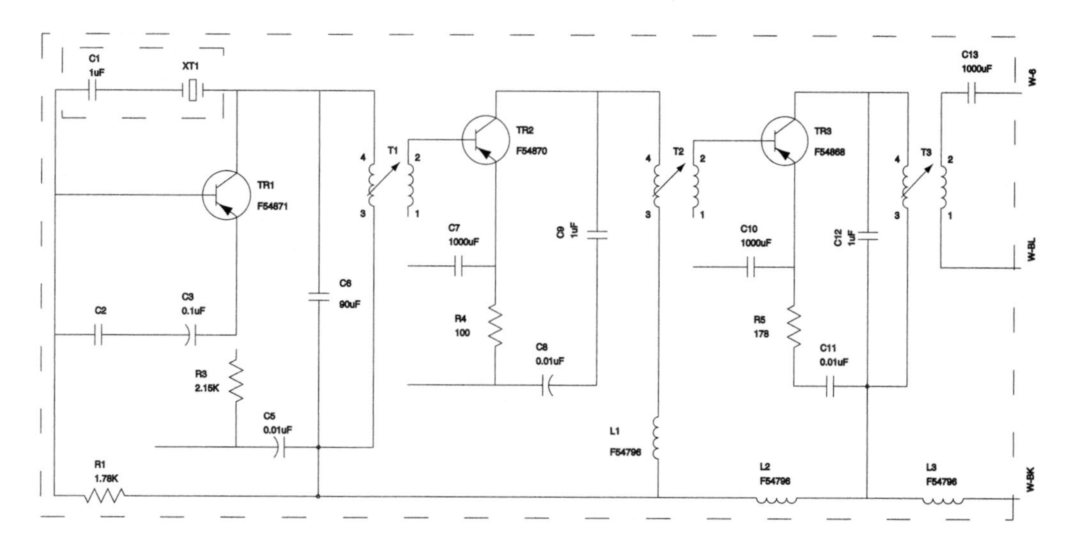

Figure 5.48 Partial schematic diagram for a communications satellite

Connection and Wiring Diagrams

Connection diagrams are used to show how components in an electrical assembly are connected. The types of connection diagrams include **point-to-point, highway, baseline**, and **harness diagrams**. Figure 5.49 shows a highway connection diagram. Each terminal of each component is numbered. Connecting wires often use a colour-coding scheme to lessen the chance of error during assembly. The components are drawn to represent the shape of the component and are placed in relative positions to each other unlike the symbols used in schematic diagrams.

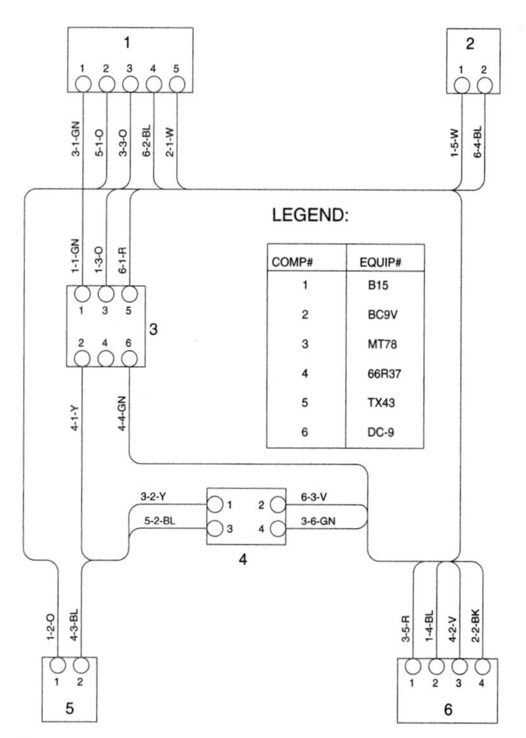

Figure 5.49 Highway diagram

The term **wiring diagram** is used in industrial and commercial wiring applications. One common method is to show the connections on a *ladder*. The **ladder diagram method** is used in motor control circuits, alarm circuits, and programmable logic controllers. A typical motor control circuit is shown in Figure 5.50. Note that the power supply is shown at the top of the diagram, and then a transformer to reduce the voltage to the control circuit is shown below. Each line is usually numbered. Again, the symbols are described in the appropriate standards.

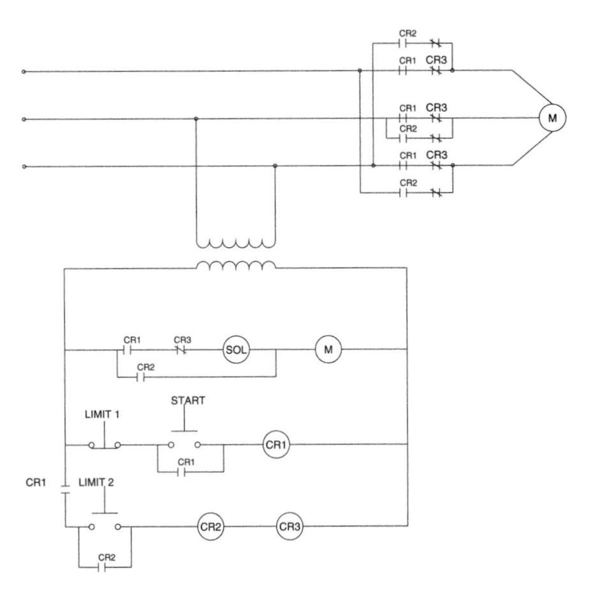

Figure 5.50 Ladder diagram

Printed Circuit Board (PCB) Diagrams

Most electronic devices use a circuit board to mount components and wiring connections between components. The board is made of a laminated insulating material (e.g., Fiberglass®) and the connections are made by a thin layer of conductive material (e.g., copper) etched into the board. These copper **traces** act as wires to connect the components as designed in the schematic diagram. The traces are generally 0.2 to 0.5 mm wide and 0.05 mm thick, depending on the current carried and the number of components on the board. The components can be resistors, capacitors, transistors, and so on, and integrated circuits (IC) or chips. **Chips** are miniaturized printed circuits made of silicon for generating very low current.

The components can be mounted on one or both sides of the board with holes that are plated through to carry current from one side to the other. Traces on both sides are noted by using different colours or line types (e.g., dashed for the other side).

Drawings to manufacture PCBs are done in different stages and are more mechanical in nature than electronic. The traces and component layout are done on an **artwork drawing**. A **master drawing** shows the board geometry, connector pattern, mounting holes for the components, and manufacturing dimensions. The board is designed to fit in its appropriate space within the equipment. The components have to be mounted in predrilled holes located with precision on a **drilling drawing**. An assembly drawing is used to show the components mounted in position. Figure 5.51 shows a PCB master drawing, used to develop the working drawings for production.

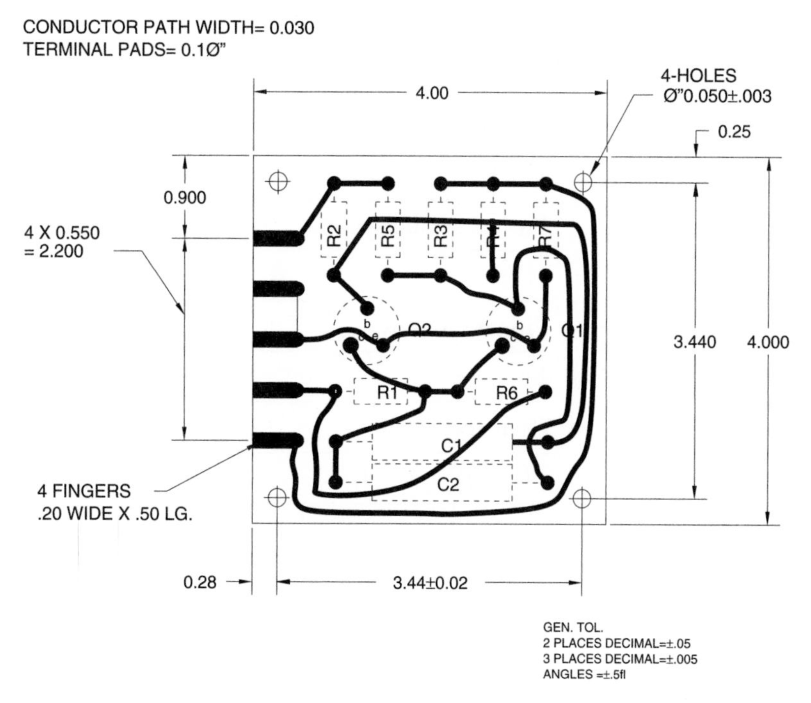

Figure 5.51 Printed circuit board master drawing

STANDARDS AND SPECIFICATIONS

Standards

Almost all engineering drawings are governed by standards of one form or another. **Standards** refer to a set of symbols or methods of illustration that have been accepted throughout the industry and adopted into a national or international document. Applying standards to engineering drawings saves time and money. Time is saved because the standards already exist and are recognized throughout the industry. There is no need to create or redesign parts and components that have been standardized.

Threaded fasteners, springs, gears, and many holding devices are examples of standardized mechanical parts. For economic reasons, standard shapes, sizes, and tolerances for steel and wood products are to be used, whenever possible. Electronic symbols, circuitry, and power supplies are all available as standardized symbols.

The way items are shown on an engineering drawing should be in accordance with the applicable standards to prevent misunderstanding. The following table lists the various standards pertaining to engineering drawings in the disciplines we have discussed.

General presentation	ANSI Y14.100	ISO 128
Dimensioning	ANSI Y14.5	ISO 129
Tolerancing	ANSI Y14.5	ISO 406 / 1101
Welds	AWS A2.4	ISO 2553
Process symbols	ANSI Y32.11	ISO 3511
HVAC	ANSI Y32.2.4	
Piping symbols	ANSI Z32	ISO 6412
Building		ISO 2594
Civil engineering		ISO 3766 / 7084
Electrical symbols	ANSI Y32.2d	

Specifications

Specifications include any standards the designer wants incorporated or adhered to; for example, steel specifications may relate to an ASTM (American Society for Testing Materials) standard for the composition and abilities of steel.

Drawings must convey the designer's intent as far as the geometric and dimensional characteristics are concerned. Specifications are written so that the manufacturer will make or build the objects on the drawing in accordance with the designer's requirements for material, finish, standards to be adhered to, and any particular methodology to be followed.

Specifications are sometimes added to the drawings, in note form, on the right-hand side. On large projects, the specifications are usually produced as a separate document, issued with the drawings. If the information given in the specifications differs from that given in the dimensional content of the drawing, generally, the drawing overrules.

Writing specifications is often the job of a senior engineer, as the wording has legal implications; however, many companies use juniors to prepare the "specs" and add them to the drawings or related documents.

For civil engineering projects in Canada, a National Master Specification has been written. It can be used by editing and adding in the appropriate wording.

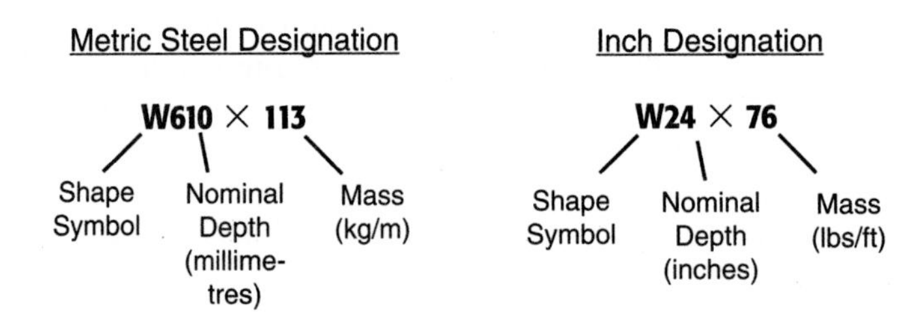

Problems

Problems for this chapter are at the discretion of your instructor, as each of you will eventually complete your engineering instruction in a specific discipline area. Your instructor may ask you to prepare working drawings based on the sample drawings in this and other chapters.

Complete all problems on a title sheet designed for your course, and lay them out in accordance with engineering graphics practices.

1. Create sufficient views to fully describe the part shown below. Include an auxiliary view. You may use partial views if sanctioned by your instructor. Add dimensions to the views according to standard practices. Tolerances may be added at the discretion of your instructor.

2. From the details given below, create an assembly drawing. The axle goes through the wheel with an axle support on each end, facing out. The axle supports bolt onto the top plate. The drawing should have at least one full cross-section view and one regular view. If directed by your instructor, include a bill of materials (parts list) and details of bolts, nuts, and so on.

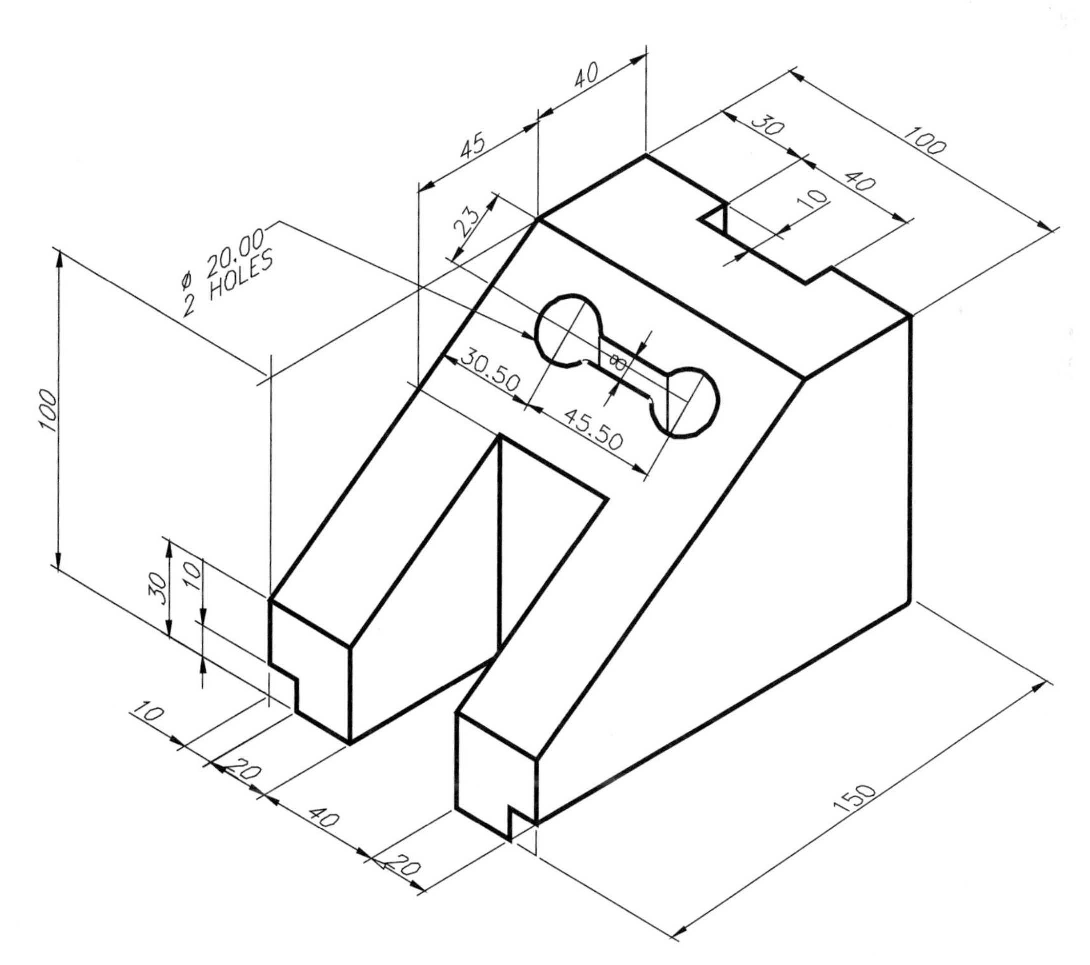

Figure 5.52

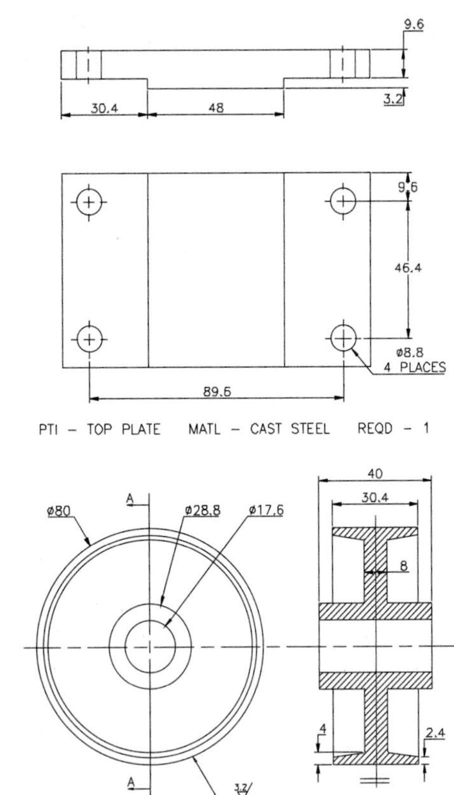

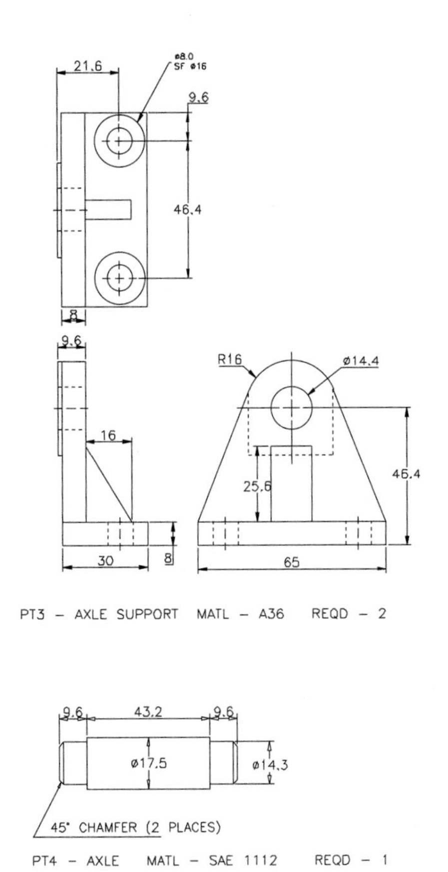

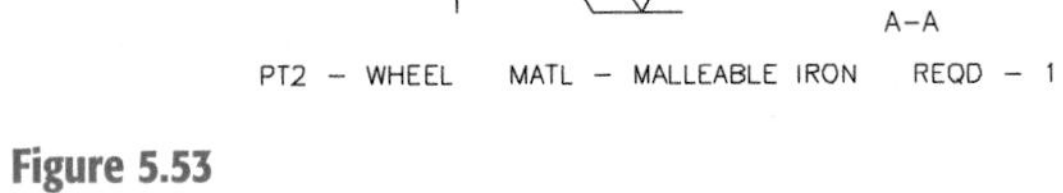

Figure 5.53

Problems

Space is without limit; all parts of space are alike; there is nothing characteristic about any particular part so it can serve as a term of reference for indicating the position of a particular point.

Thus, to define the position of a point in space it is necessary to refer this position to those of other objects which are of known position in some distinctive part of space, the number of objects being as many as are required to define the point; and for the process to be amenable for easy and daily use, it is necessary that these objects should be the simplest possible so that their position can be easily imagined.

Gaspard Monge, Geometrie descriptive, *1795*

CHAPTER

6 Visualization

As noted in Chapter 1, although we live in a three-dimensional (3-D) world, we must represent it on two-dimensional (2-D) paper or on a computer screen.

Isometric and perspective drawings are two ways of representing three dimensions on a 2-D surface. Mathematically, three dimensions can be represented by Cartesian coordinates, x, y, z, referenced to some origin and a set of axes. Position on the face of the earth can be defined by latitude, longitude, and elevation, above or below some datum, usually sea level. Latitude, longitude, and elevation are used for surveying and navigating. We also need a method of finding qualitative information on 3-D space on a 2-D surface. This is called **descriptive geometry**.

What an object looks like depends on its orientation and the viewpoint from which it is seen. To see how close one object is from another, you must choose the correct viewpoint. The procedure is similar whether you use paper or a computer. With paper, you create views (called **auxiliary views**) in addition to the regular front, top, and side views. With a computer, you set the point from which the objects are seen. The two methods are covered in separate sections of this chapter.

Let's start off by looking at some common visualization terminology.

VISUALIZATION TERMINOLOGY

Three common visualization terms are *point*, *line*, and *plane*.

Point

The **point** is the basic building block for an object in space. It has location but no dimensions. It can be located by assigning Cartesian coordinates, x, y, z, from an origin or by a distance from an arbitrary datum.

A point is represented on a 2-D surface by a small cross or by the intersection of two lines. Figure 6.1 shows front, top, and side views of a point, A.

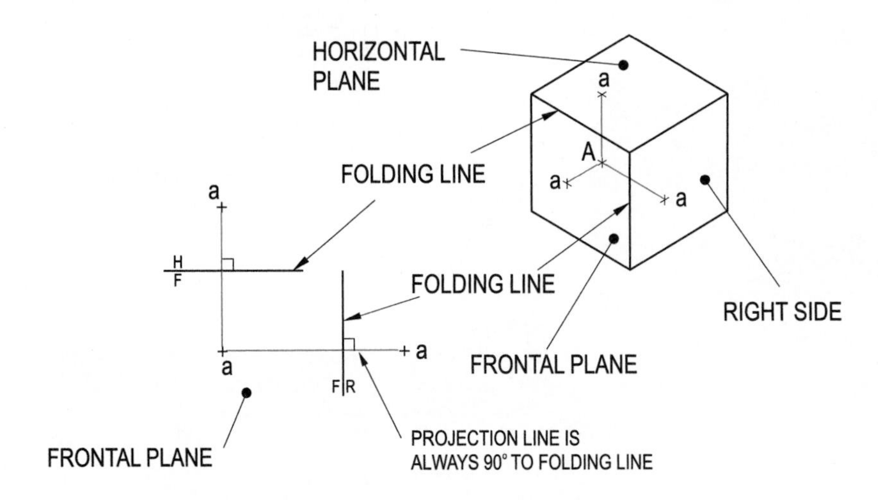

Figure 6.1 Three views of a point showing how it is projected from the front to the top

The folding lines in the orthographic views correspond to the folds between planes of an enclosing box. The location of the point is specified with reference to these folding lines. Since (with third-angle projection) the picture plane is always between the viewpoint and the point, the point is always behind the plane. The distance from the picture plane is used as a reference distance, and the datum is the folding line. Figure 6.2 shows where this reference distance appears on orthographic views. The three planes are identified by the letters on the folding line—F, frontal, H, horizontal, and R, right side.

The distance behind a plane, the frontal plane in this example, is seen in any and all views adjacent to the frontal plane. If the distance from the folding line is known in one adjacent view, it can be used to locate the point in any other adjacent view. The plan view and the right-side view are both adjacent to the front view. This same distance is used to locate the point in a left-side view.

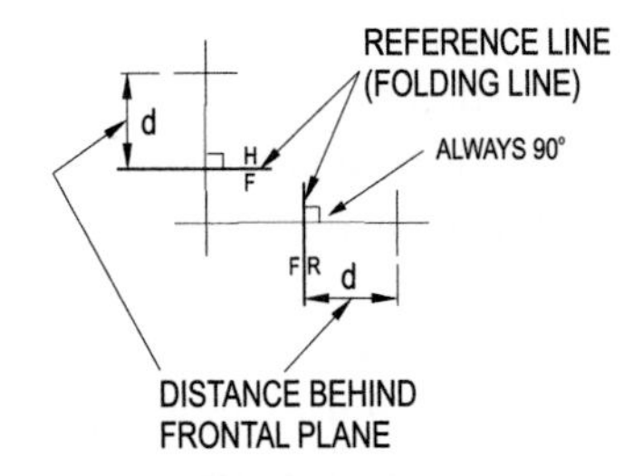

Figure 6.2 The distance behind the frontal plane appears in the views adjacent to the front view

Line

A straight line is made up of two points. Figure 6.3 shows three views of a line, AB. The frontal plane (F), top (horizontal) plane (H) and right side plane (R) are identified.

The ends of the line are identified by crosses.

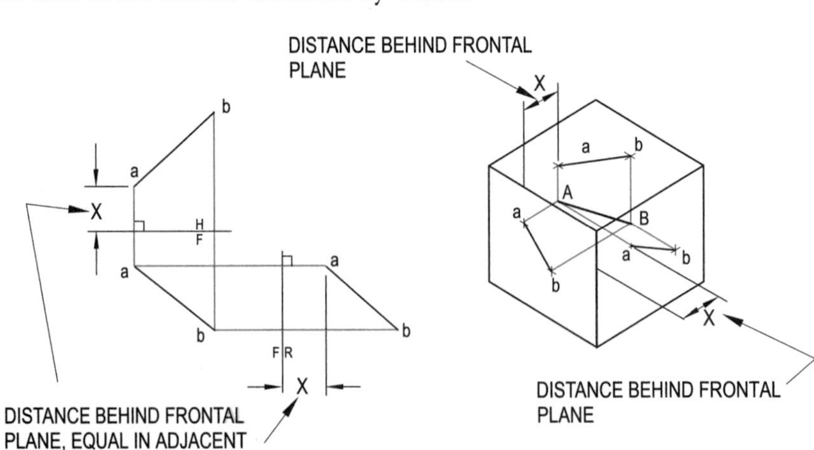

Figure 6.3 Three views of a line showing how end points are located in the top and side views

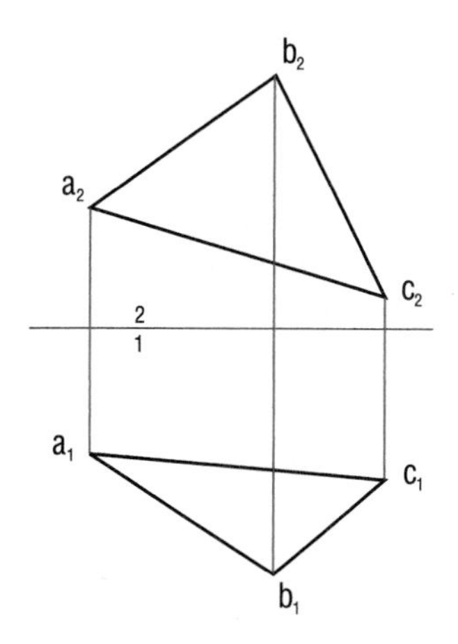

Figure 6.4 Front (called plane 1) and top view (called plane 2) of a plane

Plane

A **plane** is a flat surface such that a straight line joining any two points lies in the plane. The orientation, but not necessarily the limits, is defined by a minimum of three points. These points can be anywhere in the plane; they do not have to be at the edges (although they often are). Figure 6.4 shows two views—front and top—of plane ABC, made up of three points and three lines.

A subscript will be used to designate the plane on which the point is projected; for example, a_1, b_1, c_1, instead of calling them frontal, top, and side. We will be using many other planes besides the three principal ones. The frontal plane is arbitrarily designated as plane 1, and the top view is designated as plane 2. When there are several auxiliary planes, they should be identified with numbers (3, 4…). The folding line is also identified by the numbers assigned to each plane. Most problems, however, do not require more than two auxiliary planes.

PROPERTIES OF LINES AND PLANES

Let's look at the properties of lines and planes.

Properties of Lines

Four pieces of information are needed to define a line:

1. the location of one point on the line (which we have already discussed),
2. the direction,
3. the angle the line makes with the horizontal (the **slope**), and
4. the length.

Direction

Direction, given as a compass reading, is seen only in the horizontal plane (the plan view) because when you read a compass, you do not hold it in a vertical position. There are two ways to indicate a direction: bearing and azimuth. A **bearing** is measured from either north or south (Figure 6.5).

The angle is always less than 90°. A bearing of N 90°W would be given as west.

An arrow indicating north must be shown in the plan view. The "north arrow" usually points to the top of the page. A "north arrow" is seen only in the plan view.

An **azimuth** is a compass direction measured from north and stated as an angle from 0 to 360°. Figure 6.6 shows how an azimuth is measured.

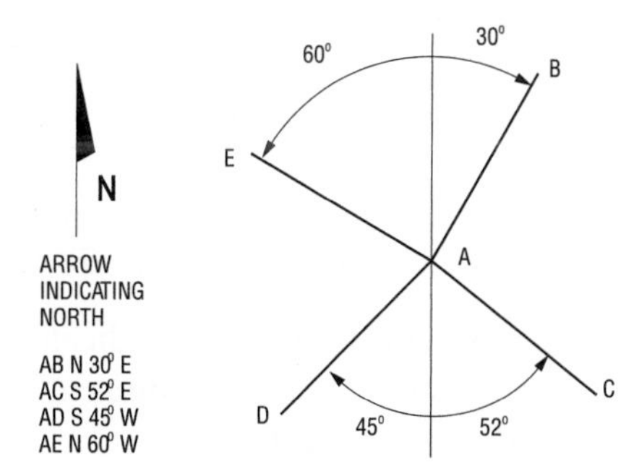

Figure 6.5 The bearing of a line is seen only in the horizontal plane, the plan view

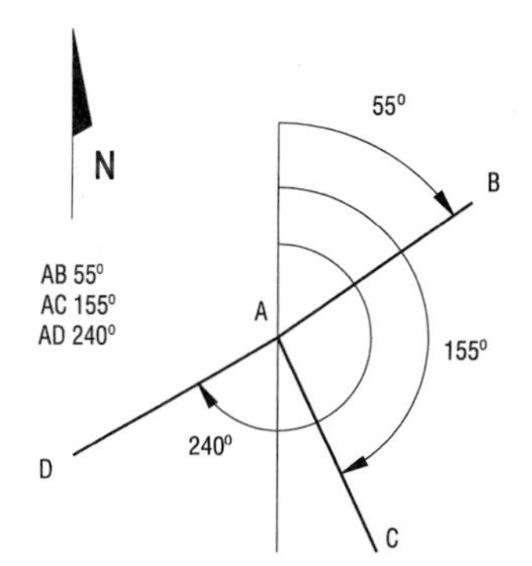

Figure 6.6 Azimuth is another way of specifying direction and is seen only in the plan view

Since an azimuth is always measured from north, there is no need to specify a compass direction, but sometimes a compass direction is given.

Slope

A slope indicates whether a line rises or falls from one end. You have probably used "rise over run" to describe slope. **Slope** is defined as the angle between a line and the horizontal plane (Figure 6.7). It can be seen and measured only in an elevation view where the line appears as true length. An elevation view is any view in which a vertical distance can be measured. (The front view is an elevation view.) In order to measure slope, you must be able to find the true length of a line in an elevation view (which will probably not be the front view). If slope is measured by "rise/run," both rise and run must be seen in true length.

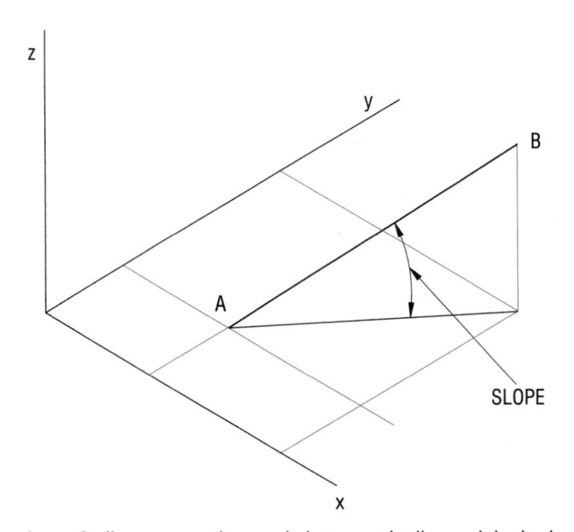

Figure 6.7 The slope of a line expressed an angle between the line and the horizontal

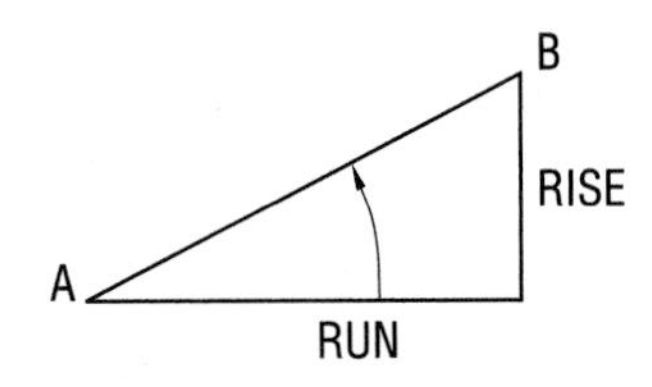

$$\text{GRADE} = \frac{\text{RISE}}{\text{RUN}} \times 100\%$$

Figure 6.8 Slope can be given as grade, the tangent of the line expressed as a percent

Civil engineering applications give slope as a grade. **Grade** is the tangent of the slope expressed as a percentage (Figure 6.8). A 100-percent grade is equivalent to a slope of 45°.

Length of a Line

A line can be viewed from any direction. Imagine walking around a pole that is in the ground at an angle. It will appear as a different length depending on how it is viewed. Length can be measured, but only if the line is first seen in true length. A line can be seen in true length only if you look in a direction perpendicular to it. When viewed from any other direction, it will appear as some length less than the true length. We must determine how to get a view that shows true length.

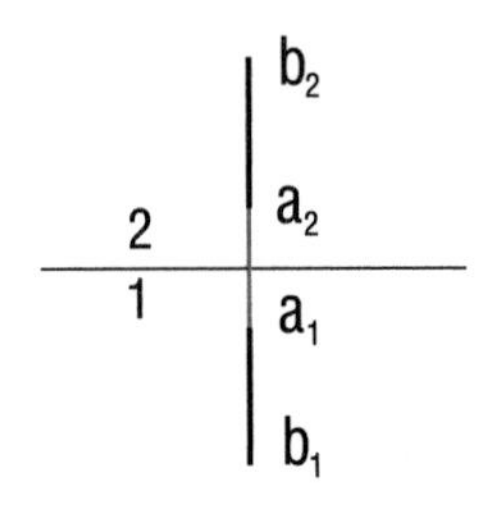

Figure 6.9 Front and plan views of a sloping line

Finding the True Length of a Line Figure 6.9 shows front and plan views of a line, AB. It is not true length in either view, as we will see later. To find the true length of line AB:

1. The line must be viewed from a perpendicular direction to see true length. Choose a view that satisfies this condition. A side view (either left or right) satisfies this condition for this line.

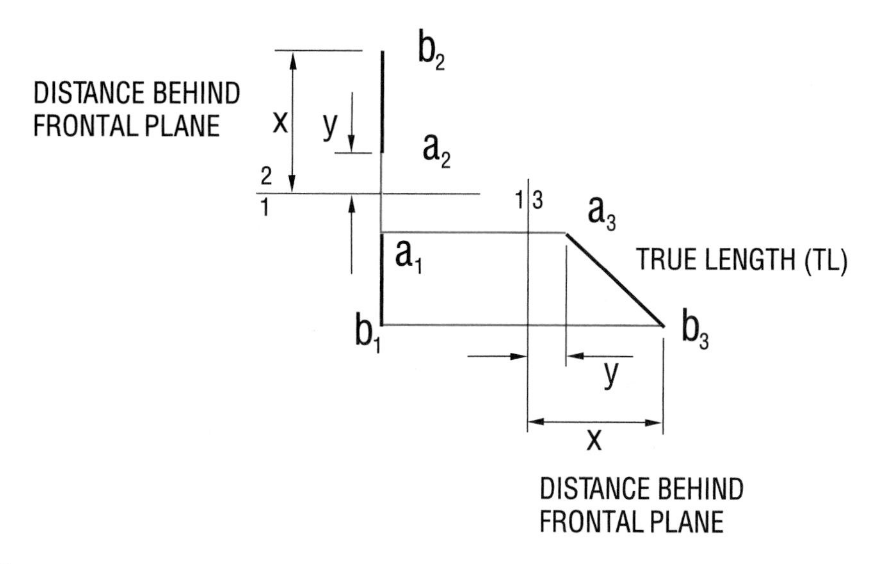

Figure 6.10 Points are located in the side view by taking distances from the plan view (adjacent to the front view)

2. Draw a folding line on the right side of the frontal plane so that the end points can be projected onto the right-side plane. The location of the folding line is arbitrary as long as it goes in the correct direction.

3. Project points A and B into the right-side view (designated plane 3 in this problem). Figure 6.10 shows how the end points are projected. The distance from the folding line in the plan view is used to locate the end points in the side view.

There is no need to measure this distance. It can be transferred with dividers. The true length of AB is seen and can be measured in the right-side view.

This example is a special case in which the true length appears in one of the principal orthographic views. This is because the line is parallel to one of the principal planes.

The same process is used to find the true length of the line shown in Figure 6.11. This line is not parallel to any principal plane, and so an auxiliary plane must be drawn.

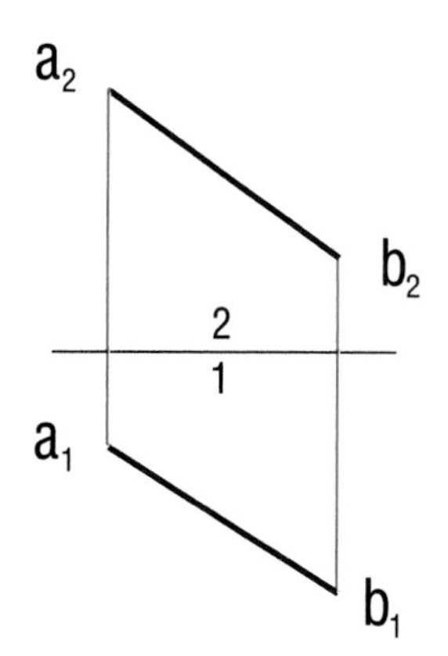

Figure 6.11 Front and plan views of a line

1. Locate an auxiliary plane such that the line of sight is perpendicular to AB. The folding line representing this auxiliary plane (called plane 3) is located parallel to line AB. (The auxiliary view can be taken from the plan view or the front view. The plan view is used here.) Since lines of sight are always perpendicular to picture planes, they will be perpendicular to AB (Figure 6.12).

   Figure 6.13 is an isometric drawing showing the picture plane for the auxiliary view relative to AB.

2. Project the end points onto the auxiliary plane (called plane 3). Distances from the folding line on the auxiliary plane are taken from the frontal plane, as shown in Figure 6.14. The true length (TL) can be measured in plane 3.

   An auxiliary plane was taken off the plan view in this example but it could have been taken off the front view. In some cases, an auxiliary view must be taken off a particular view, but in this example, it does not matter. If possible, locate the auxiliary plane so that it does not run off the paper.

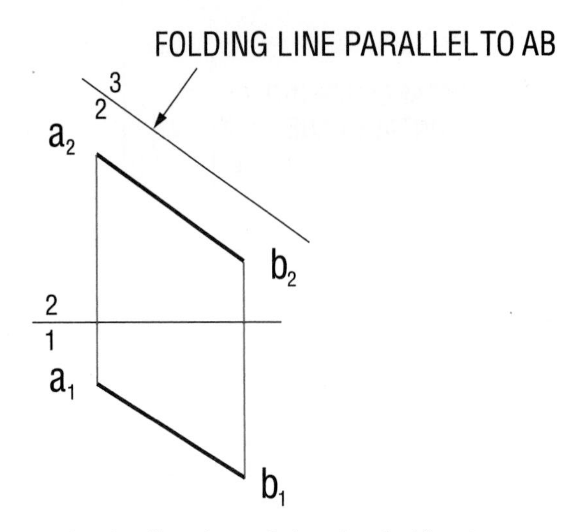

Figure 6.12 Location of auxiliary plane to find true length of line AB

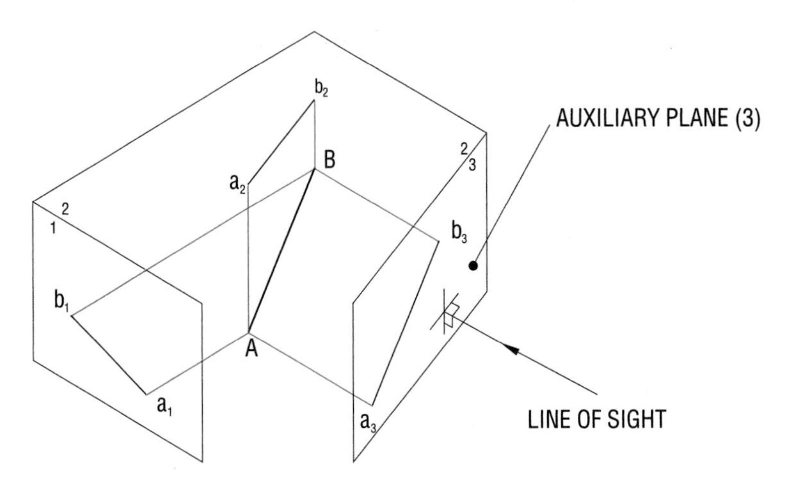

Figure 6.13 Locating an auxiliary plane parallel to a line to find true length

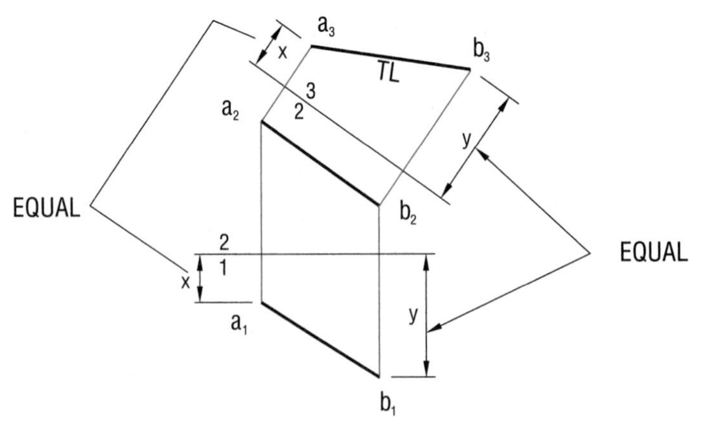

Figure 6.14 Projecting points onto an auxiliary plane

Point View of a Line

A point view is required to solve some line problems.

In order to see a line as a point view, you must "look along" the line. The line must be in true length before you can do this, and you must find a view showing the line in true length.

Finding the Point View of a Line As you may recall, we found the true length of the line in Figure 6.14. This figure is repeated as Figure 6.15.

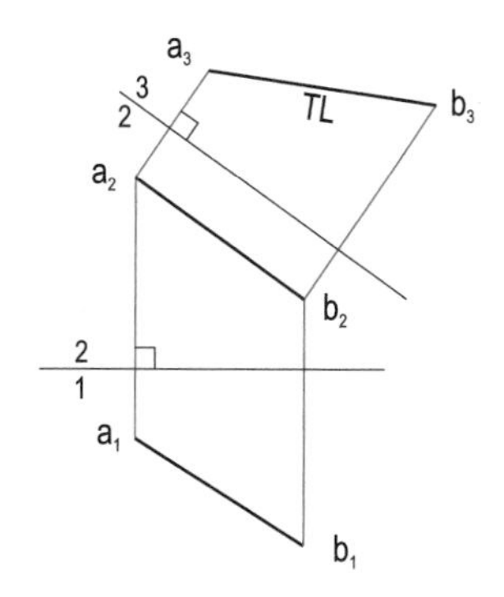

Figure 6.15 View showing a true length (part of Figure 6.14 repeated)

1. After finding the true length, locate an auxiliary plane (plane 4) so as to "look along" the line. The folding line is drawn perpendicular to the true length. Either end can be used.
2. The point view is positioned by taking distances from the plane adjacent to the true length view (the plan view in this example). The point view is seen in Figure 6.16.

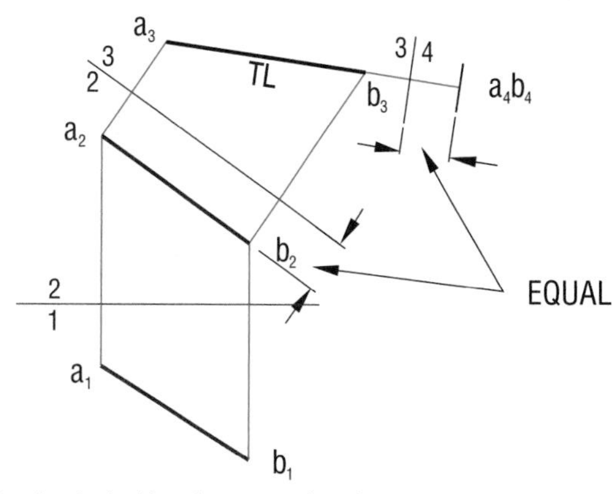

Figure 6.16 Find a point view by looking along a true length

In plane 4, the distance must be the same for each end of the line. If they are not, it cannot be a point view; there is an error.

Slope of a Line

Slope can only be seen in an elevation view showing the line in true length. This is so that the rise and run are seen correctly.

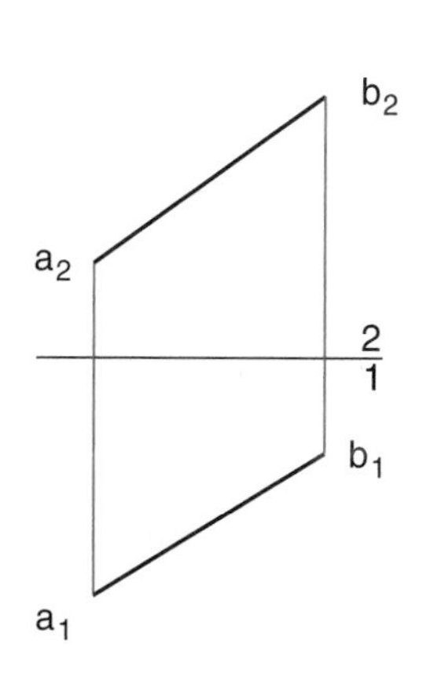

Figure 6.17 Front and plan views of a line

Finding the Slope of a Line Figure 6.17 shows two views of line AB. The slope is to be drawn and measured. To do this:

1. Place an auxiliary elevation view so that AB appears as true length. The folding line is drawn parallel to AB in the plan view. It can be located on either side of the line.
2. Draw the line in the auxiliary view. The distances from the folding line are taken from the front view (adjacent to the plan view). Figure 6.18 shows the true length.
3. Slope is the angle between the line and a horizontal line (in the auxiliary plane). Any line parallel to the folding line in this auxiliary view is a hori-

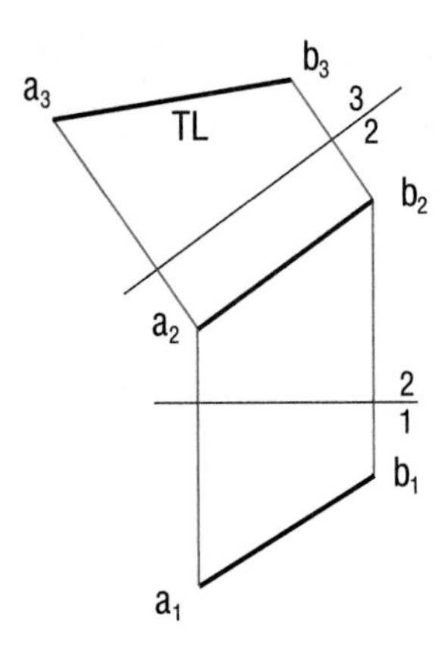

Figure 6.18 An elevation view showing true length is folded off the plan view

zontal line. (The folding line between the plan and auxiliary views is also a horizontal line.) Figure 6.19 shows the slope.

A line has a positive slope if it gets closer to the folding line and a negative slope if it gets further from the folding line when going from one end to the other. The slope is positive from A to B and negative from B to A.

The slope in Figure 6.20 is shown as a grade.

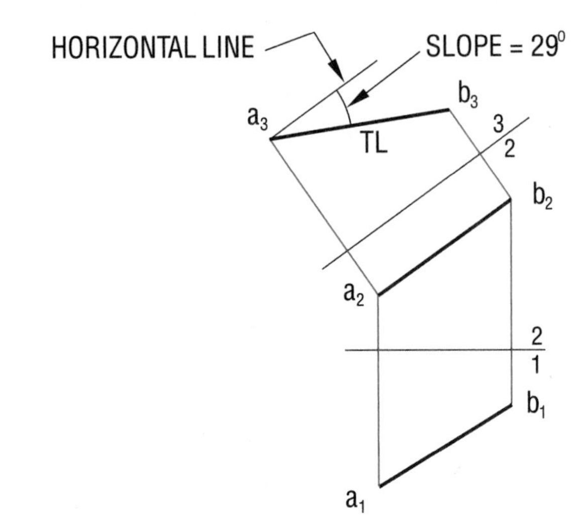

Figure 6.19 Slope of the line is measured between the horizontal and the true length

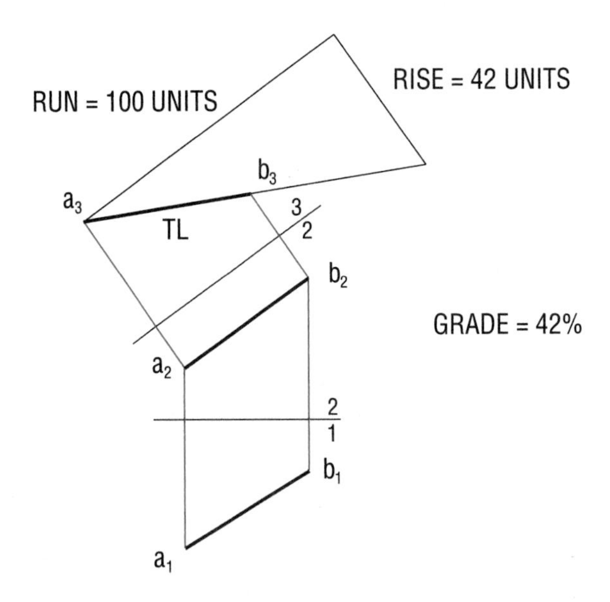

Figure 6.20 Slope can be given as grade

The grade can be read directly from a scale by laying out a horizontal run of 100 units (any convenient scale can be used) and measuring the vertical rise to the same scale (42 units in our example).

As grade = rise/run × 100, in our example the grade is
42/100 × 100 = 42%

To help you understand the practical application of what you have just learned, take a few minutes to look at the following examples.

Example 6.1

A wire rope supporting a broadcast antenna is anchored to the ground at elevation 400 m and to the antenna 1 m below the top. The antenna is 50 m tall, and the base is at an elevation of 390 m. The wire is anchored on the ground 14 m east, 20 m north of the mast. Determine the length and slope of the wire. The scale is 1:1000.

What is known?

- Location of the anchor, relative to the mast
- Elevation of the anchor
- Elevation of the attachment near the top of the mast
- Elevation of the base of the mast and the height of the mast.

1. Determine what view is necessary to find the required information–true length and slope. Both can be measured on an elevation view showing the true length of the line.
2. Lay out a plan view showing the location of the anchor and the mast. **Figure 6.21** shows the plan view, drawn to scale. We know this is a plan view because there is a north arrow.
3. Lay out an elevation view to show the elevations of the anchor, mast, and wire. Draw an elevation view that will show the true length and slope of the wire. (This is not the front view.) The folding line is located parallel to the wire to show true length in the auxiliary elevation view (plane 3). The ground elevation 400 m can be located anywhere. **Figure 6.22** shows these elevations and the plan view.
4. Draw the wire from the anchor to a point 1 m below the top of the mast. Measure the length and slope of the wire. **Figure 6.23** shows the length and slope.

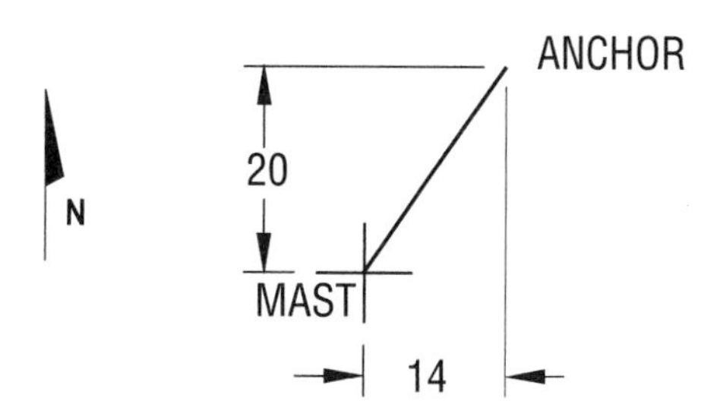

Figure 6.21 A plan view showing the layout of the mast and anchor

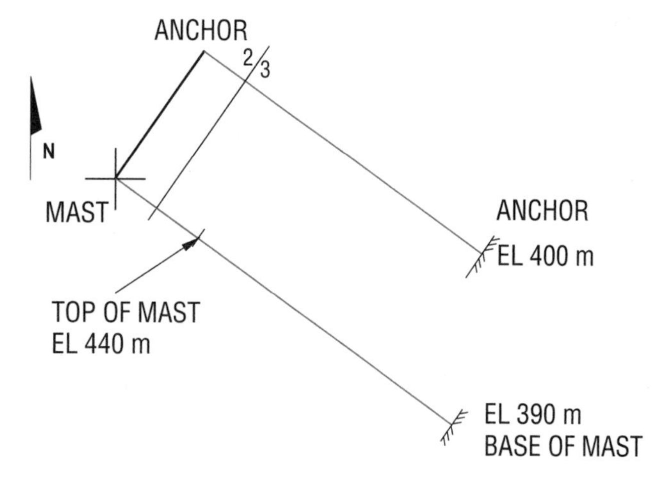

Figure 6.22 Locate an elevation view that will show the true length and slope

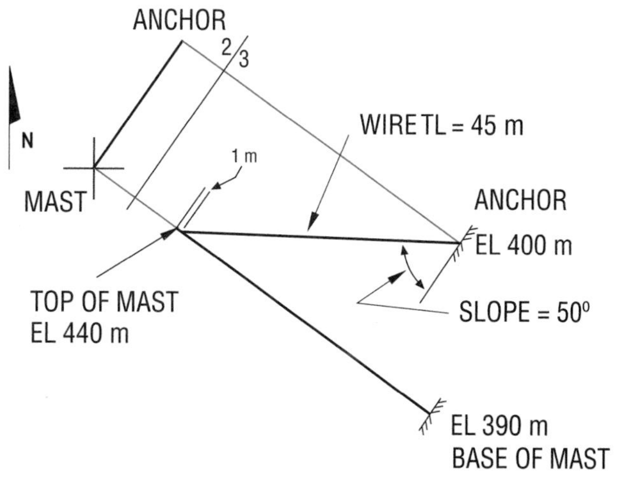

Figure 6.23 Length and slope of guy wire can be measured in the elevation view

Example 6.2

Two power lines pass through points A and B, as shown in **Figure 6.24**. The line passing through A bears N 30° E and slopes upward at 20°. The line through B bears N 80° W and slopes upward at 25°. A and B are at the same elevation and 7 m apart.

Determine the minimum distance between the two power lines. The scale is 1:200.

What is known?

- A starting point (A and B) on each of the lines
- Bearing and slope of each line (seen in the plan view)

1. Determine the view necessary to find the required information–the shortest distance between the two lines. The minimum distance will be seen by looking along one of the lines. The view required to measure the minimum distance is one that shows one line as a point view.
2. Show the given information on the plan view. A and B and the bearing are seen. **Figure 6.25** shows the plan view.

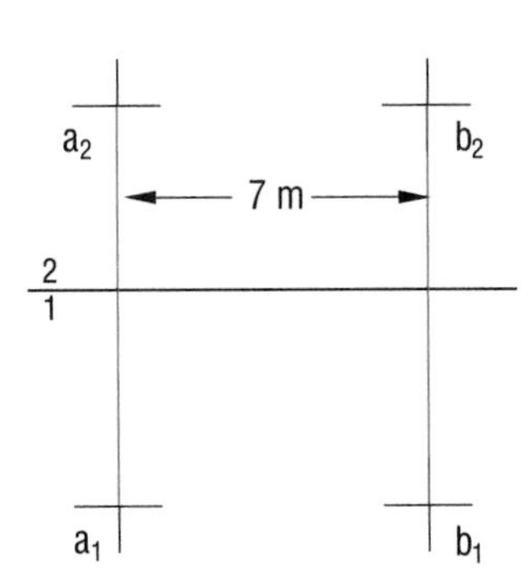

Figure 6.24 Plan and front views of a point on each power line

3. Use the slope of line A to construct a view showing the true length. An auxiliary view, 3, is located parallel to line A in the plan view. Line A will be true length in this auxiliary view. The actual length is not important. **Figure 6.26** shows the true length of line A.
4. Use the slope of B to draw the true length of the line from B. An auxiliary plane, 4, is located parallel to line B in the plan view. The true length of B will appear on this auxiliary. Line B cannot be projected from the auxiliary plane until you have established another point on it. Any point on the line, call it C, can be used. **Figure 6.27** shows the true length of line B, with point C located.

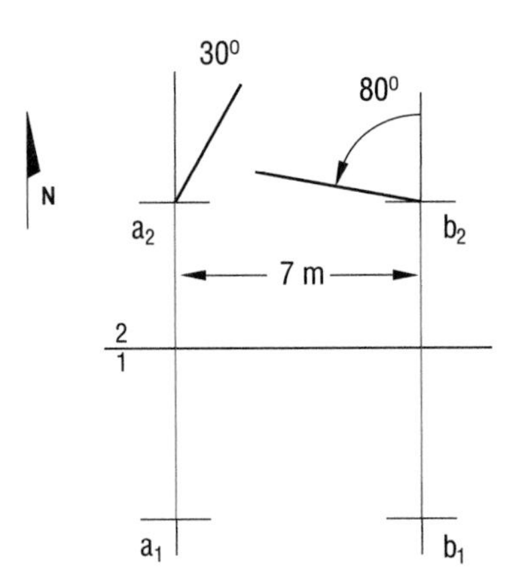

Figure 6.25 The bearing of each line is seen only in the plan view

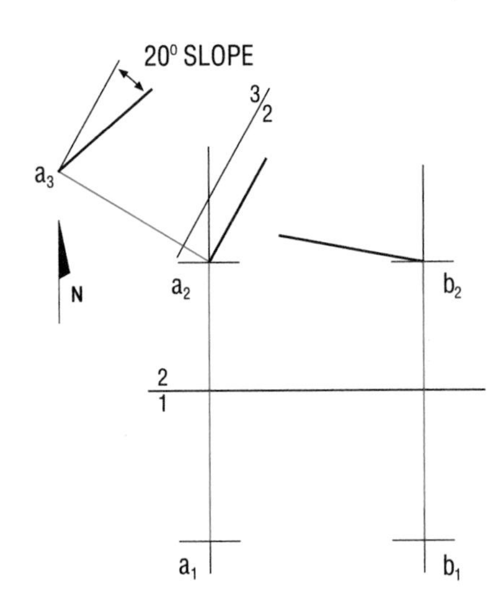

Figure 6.26 The slope and true length of line A are seen in elevation view 3

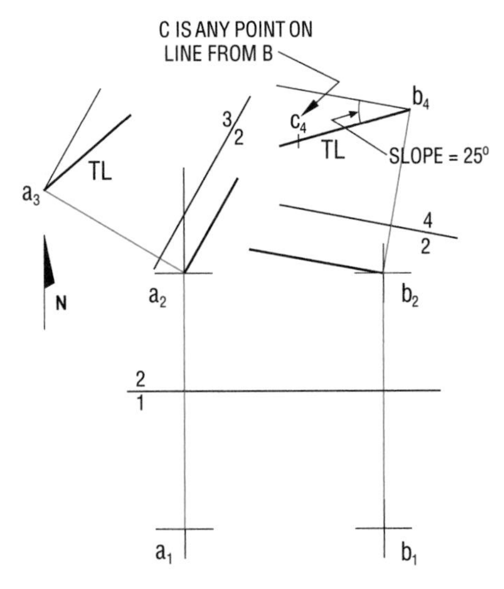

Figure 6.27 The slope and true length of line B are seen in elevation view 4

5. Project C to the plan view, and then project BC to the auxiliary view showing the true length of A. **Figure 6.28** shows the two lines on plane 3.

6. Locate an auxiliary view to show the point view of A. The auxiliary plane, 5, is perpendicular to the true length of A. Project A onto auxiliary plane 5.

   The shortest distance between A and B is the perpendicular distance between the point view of A and line B. **Figure 6.29** shows this distance.

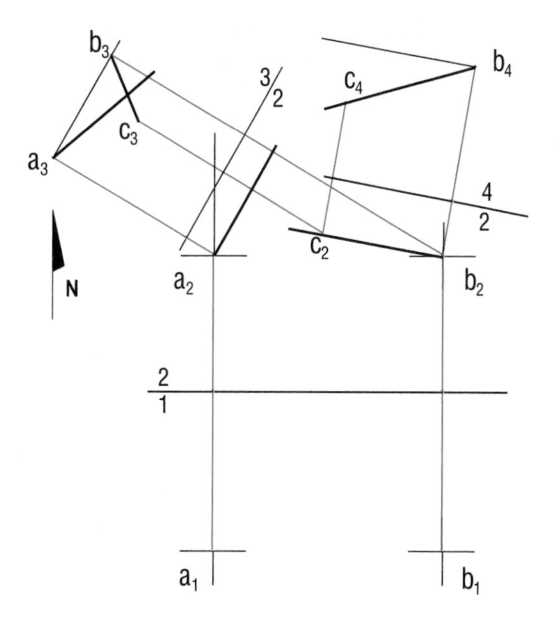

Figure 6.28 Line B is projected onto plane 3 using points B and C

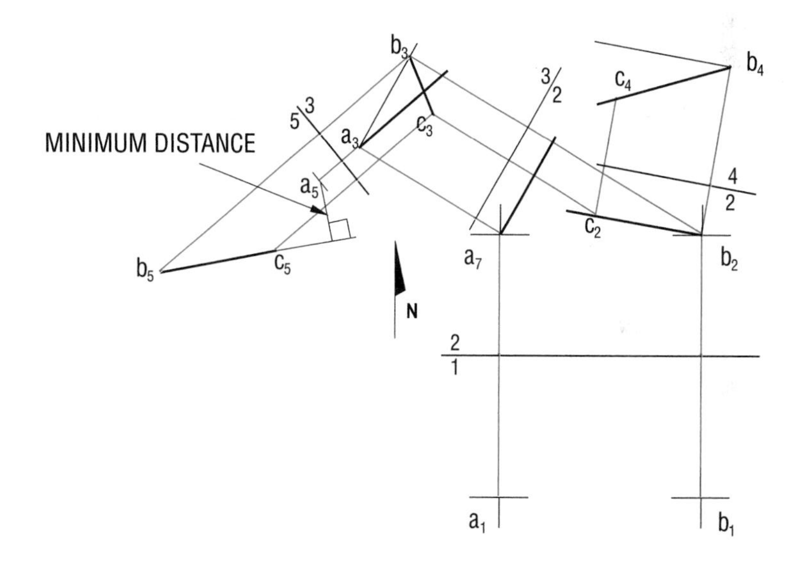

Figure 6.29 Distance between line A and line B is seen when one line appears as a point view

Example 6.2

There are three things you need to know to solve problems involving lines. These are summarized on the next page.

1. Find the true length of a line.

A line is seen in true length when viewed perpendicular to the line.

Required: The true length of a line.
To do this: Place an auxiliary view parallel to the line and project the end points, or any two points on the line, onto the auxiliary view.

See Figures 6.12 and 6.14 (page 202).

2. Find the point view of a line.

A line is seen as a point when viewed along the line.

Required: A true length must be found before a point view can be found.
To do this: Locate an auxiliary view perpendicular to a true length, and project the end points onto this plane.

See Figures 6.15 and 6.16 (page 203).

3. Find the slope of a line.

The slope is the angle a line makes with the horizontal. This is seen only in an elevation view where the line appears in true length.

Required: A true length of the line and an edge view of the horizontal.
To do this: Find an auxiliary view, folded off the plan, that will show the true length.

Measure the angle between the true length and the folding line, or a line parallel to the folding line.

See Figures 6.18 and 6.19 (page 204).

Things to Remember about Lines

- If a line is parallel to a folding line, it will appear in true length in the adjacent view.
- Bearing and azimuth are seen only in the plan view. See Figures 6.5 and 6.6 (page 199).
- Parallel lines always appear parallel, no matter how they are viewed.
- If two lines intersect, the intersection point will correspond in all views.
- The shortest distance between two lines is seen where one of the lines appears in point view.
- Perpendicular lines appear perpendicular in any view in which one, or both, of the lines appear in true length.
- A line will appear in true length in any view folded off a point view of the line.

Now that you know a little more about the properties of lines, try the following problems.

Problems

1. Draw a true length of line AB (**Figure 6.30**) by projecting (a) from the plan view and (b) from the front view, and measure the true length. The scale is 1:1.
   A is 14 mm behind the frontal plane and 8 mm below the horizontal plane.
   B is 22 mm behind the frontal plane and 20 mm below the horizontal plane.
2. Draw the point view of line CD (**Figure 6.31**). The scale is 1:1.
   C is 15 mm behind the frontal plane.
   D is 2 mm behind the frontal plane.
   Both points are 10 mm below the horizontal plane.
3. Draw a point view of line AB (**Figure 6.32**).
   A is 10 mm behind the frontal plane and 20 mm below the horizontal plane.
   B is 22 mm behind the frontal plane and 8 mm below the horizontal plane.

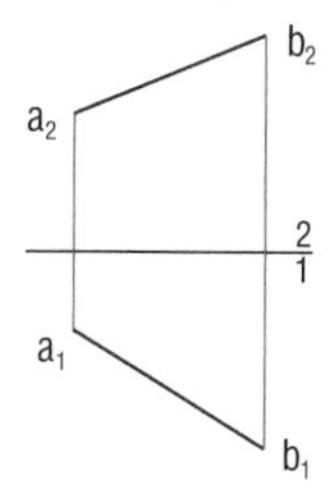

Figure 6.30

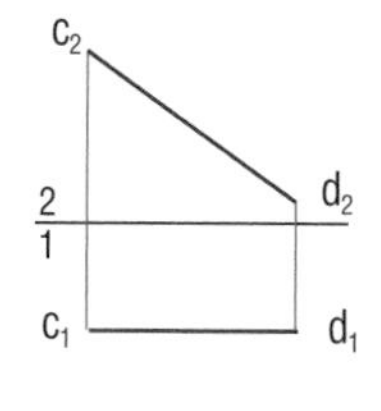

Figure 6.31

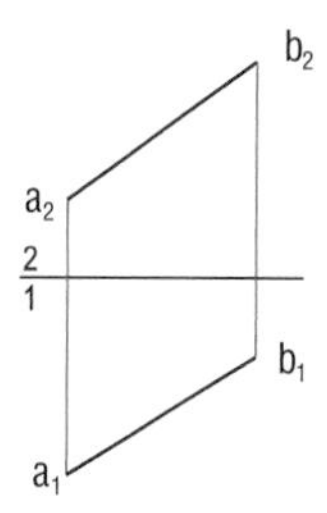

Figure 6.32

4. A line, AB, is 7 m long, bears N 40° E, and slopes down from A at 30°. Draw plan and front views of AB. Choose a suitable scale.
5. A line (CD) starting at any point, C, is 8 m long, bears N 35° W, and slopes downward from C to D at 40°. Draw the point view of CD. The scale is 1:200.
6. A line, BC, bears S 20° W from B and slopes downward at 30°. One end, B, is 700 mm higher than the other. Draw plan and front elevation views, and determine the length of the line. The scale is 1:50.
7. Point B is 300 m south and 200 m west of C and is at an elevation of 1200 m (above sea level). It slopes down from B at 25°. Draw the plan and north elevation (an elevation view looking north) views of the line and determine the elevation of C. The scale is 1:10 000.
8. Find the slope and the length of line AB.
   A is 200 mm behind the frontal plane and 120 mm below the horizontal plane.
   B is 50 mm behind the frontal plane and 200 mm below the horizontal plane.
   The scale is 1:10.
9. Point A must be connected to line BC by the shortest connector. How long is the required connector? Show the connector in the plan and front views.
   B is 150 mm behind the frontal plane and 150 mm below the horizontal. C is 150 mm north, 300 mm east, and at the same elevation as B.
   A is 230 mm east and 100 mm below B. The scale is 1:10.
10. Two lines, AB and CD, are shown in **Figure 6.33**. Determine the length, bearing, and slope of a connector that joins the midpoints of these lines.
    C is 15 mm east of A, and D is 40 mm east of A. Distances from the folding line are shown in Figure 6.33.

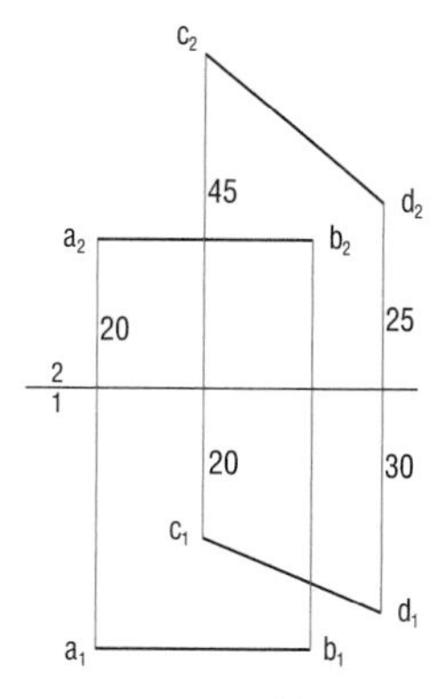

Figure 6.33

11. Line BC is 1.5 m long and has an azimuth of 25°. A point view (PV) is 25 mm behind the folding line on the auxiliary view, which shows the line as a point. Draw the missing views of line BC. The layout is shown in **Figure 6.34**. The scale is 1:50.

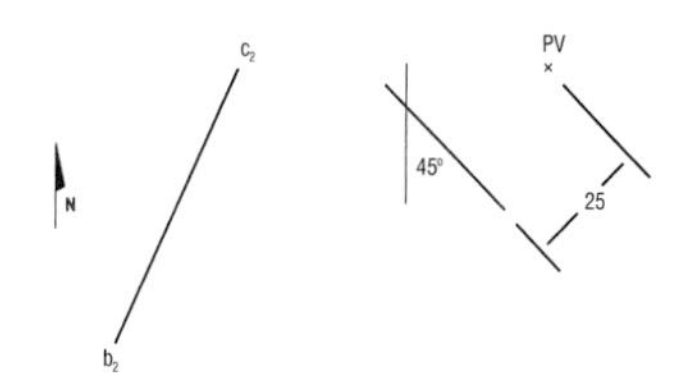

Figure 6.34

Now that you are familiar with the properties of lines, let's move on to look at the properties of planes.

Properties of Planes

You must know how to find the two limiting cases of a plane: the true shape and the edge view. These views, used in combination, can be used to solve any problems involving planes. A minimum of three lines are required to define a plane, but more lines can be added, if necessary. The only requirement is that any lines added must be in the plane. The edge view must be found before a true shape can be found.

Edge View of a Plane

A plane will be seen in edge view *if any line* in it is seen as a point. It is often easier to add a line than to use an existing line.

Finding the Edge View of a Plane Figure 6.35 shows plan and front views of plane ABC. The requirement is to find the edge view of this plane.

The plane can be seen as an edge *if any line* in it is seen as a point view. As you know, a line must be seen in true length before it can be seen in point view. Any line—AB, BC, or CA—can be seen in true length, and then a point view can be found. The plane will then be seen in edge view. This requires two auxiliary views; however, there is a better way.

None of the given lines, AB, BC, or CA, is true length, since none is parallel to the folding line in either view. If there is a true length line in the plane, it could be used to find a point view and thus an edge view of the plane. Since none of the given lines is true length, we will construct a true length line in the plane and then find the point view of this new line. Only one auxiliary view will be required to find an edge view.

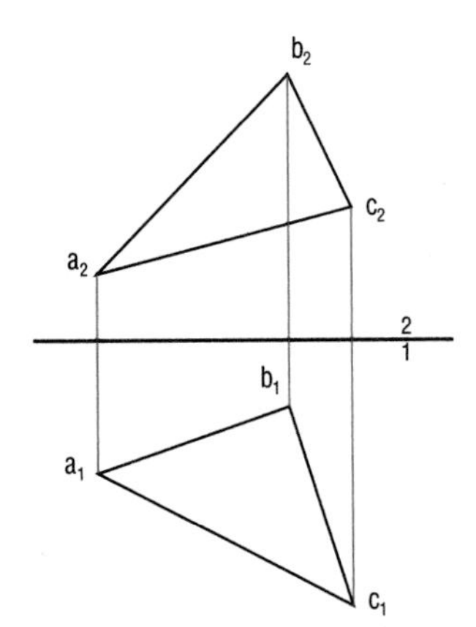

Figure 6.35 Front and plan views of a plane

1. Draw a line, in the plane, parallel to the folding line in the front view. This line will be true length when projected onto the plan view. Figure 6.36 shows the line added to the front view. The new line is a horizontal, or level, line.

   Start the new line at a given point in the plane (A). Call the other end X. If the line starts at a given point, only one end has to be projected onto the other view.

2. Project point X onto the plan view. It will be on BC. Label the line as true length (TL).

3. Find the point view of AX. Locate a folding line perpendicular to AX and project AX onto the new plane (plane 3). Figure 6.37 shows the point view of AX projected onto plane 3.

4. Project B and C onto plane 3, and join the points to see the edge view of ABC. Figure 6.38 shows the edge view.

   A true length line could be created in the front view by drawing a line parallel to the folding line in the plan view. The auxiliary plane showing the edge view can then be taken off the front view.

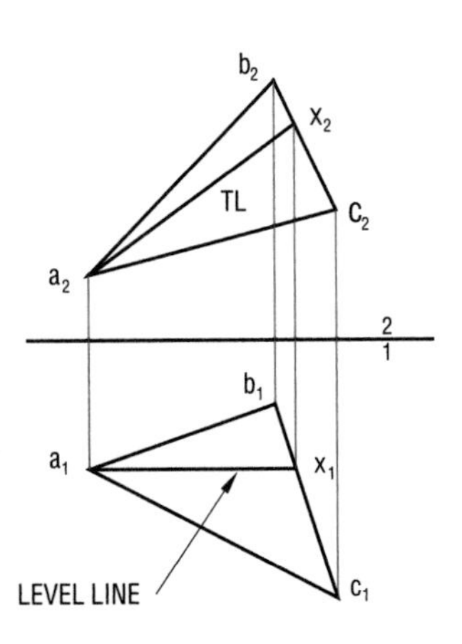

Figure 6.36 Create a true length line in a plan view by drawing a level line in the front view

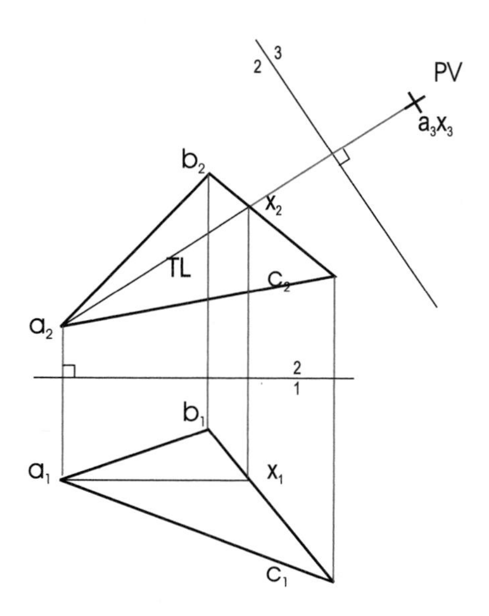

Figure 6.37 A point view of any line in the plane can be found

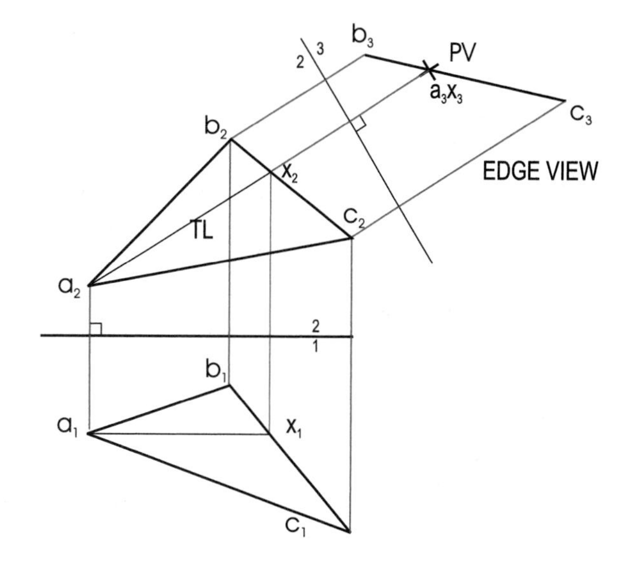

Figure 6.38 Edge view of a plane appears where any line in the plane is seen as a point

True Shape of a Plane

The true shape will be seen when viewed from a direction perpendicular to the plane. An edge view must be found before the true shape can be drawn. An auxiliary view is then positioned so that the line of sight is perpendicular to the edge view. The true shape will be seen in this auxiliary view.

Finding the True Shape of a Plane

1. Draw the plane in edge view (Figure 6.39). This is a repeat of Figure 6.38.
2. Place a folding line parallel to the edge view. It can be on either side of the edge view. Figure 6.40 shows the location of the new auxiliary plane.
3. Project all points onto auxiliary plane 4, and join them. Distance from the folding line is found from the view adjacent to the edge view, plane 2, in this problem. Figure 6.41 shows the true shape of the plane.

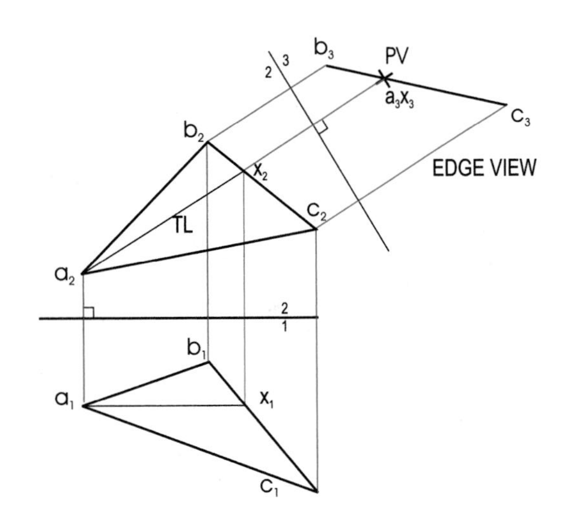

Figure 6.39 An edge view of a plane is seen when any line in the plane appears as a point

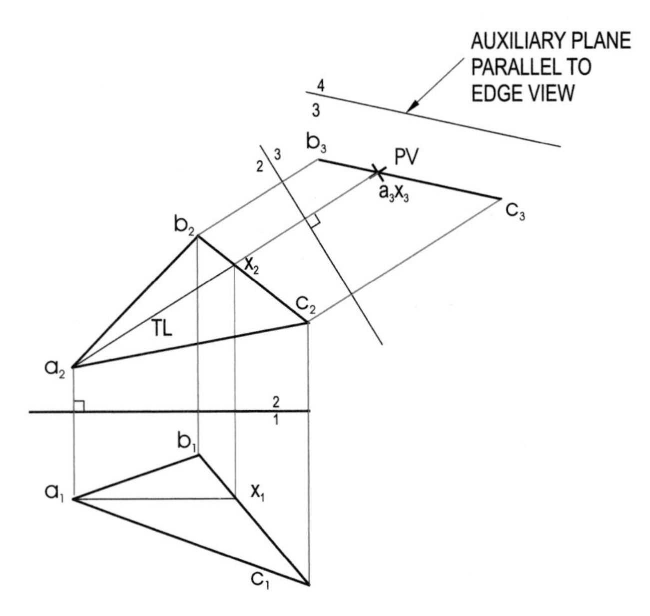

Figure 6.40 An auxiliary view is located parallel to the edge view to see a true shape

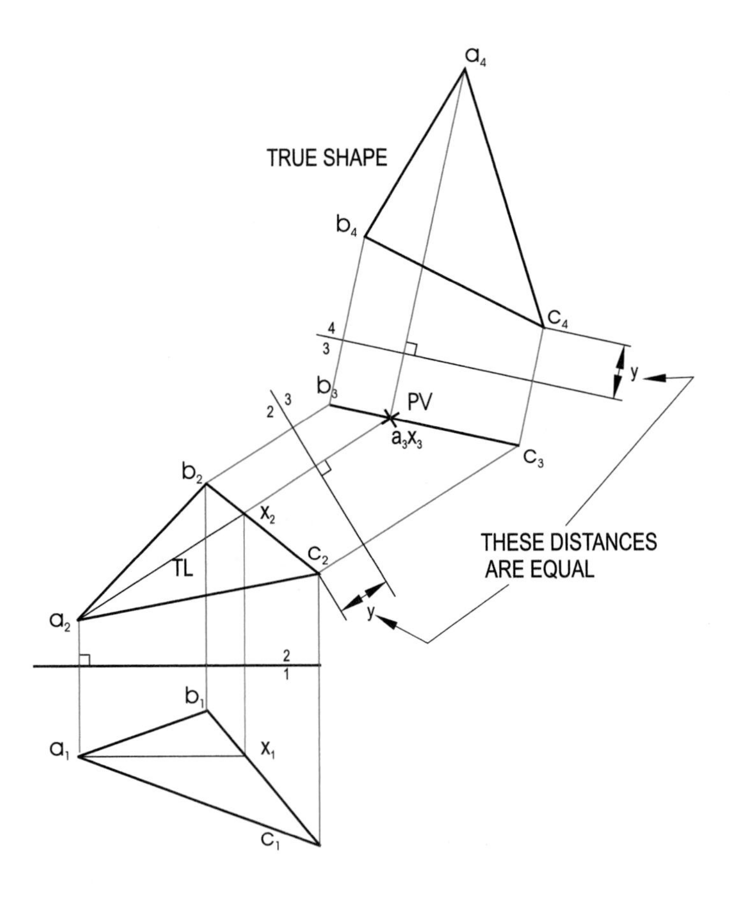

Figure 6.41 True shape of the plane is seen by projecting from an edge view

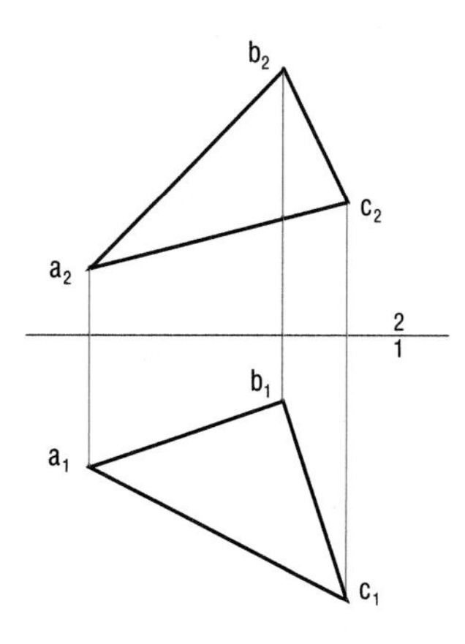

Figure 6.42 Front and plan views of a plane

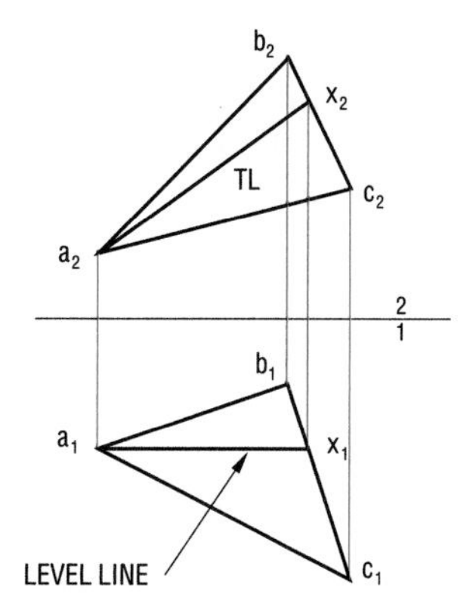

Figure 6.43 A true length line must be in plan view to find the slope of a plane

Slope of a Plane

The slope of a plane is the angle it makes with the horizontal. There can be only one slope. Slope is seen only in an elevation view where the plane is seen in edge view. An elevation view *must* be used because a horizontal plane (with which slope is measured) is seen as an edge only in an elevation view. The true length line used to find the edge view *must* be in the plan view for the edge view to appear in an elevation view.

Finding the Slope of a Plane Figure 6.42 shows two views of a plane ABC. The requirement is to find the slope of the plane.

1. Create a true length line in the plan view by drawing a horizontal line, AX, in the front view. Project AX onto the plan view where it will be true length. Figure 6.43 shows the true length line in the plan view.
2. Draw the edge view of ABC, and measure the angle it makes with the horizontal, as shown in Figure 6.44. Any line parallel to the folding line is a horizontal line.

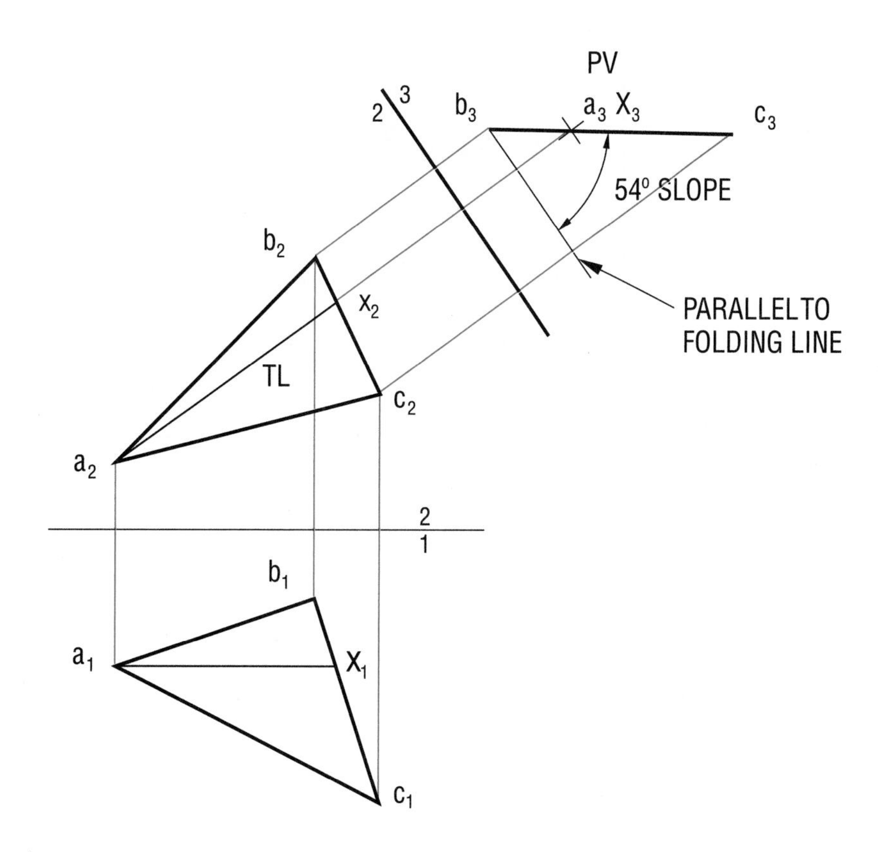

Figure 6.44 The slope of a plane is the angle it makes with the horizontal

The auxiliary view showing the edge view of the plane can be positioned on either side of the plan view.

There are two views of a plane, and you need to know how to find them. These are summarized on the next page.

Now that you know a little more about the properties of planes, try the following problems.

1. Find the edge view of a plane.

A plane will appear as an edge if any line in the plane appears in point view. The problem becomes one of finding the point view of a line.

Required: To find the edge view of a plane.
To do this: Draw a line in the plane parallel to a folding line, and project this line into the adjacent view where it will be in true length. Find a point view of this line and project the other points in the plane. The plane will appear as an edge. See Figures 6.36, 6.37, and 6.38 (pages 210 and 211).

2. Find the true shape of a plane.

A plane will be seen in true shape when viewed in a direction perpendicular to the plane. An edge view must be found before this can be done.

Required: To find the true shape of a plane.
To do this: Find the edge view of the plane and locate an auxiliary view parallel to the edge view. Project the point defining the plane onto this auxiliary view. The plane will be in true shape. See Figures 6.39, 6.40, and 6.41 (page 212).

Things to Remember about Planes

- The slope of a plane is seen only in an elevation view where the plane appears as an edge. See Figure 6.44 (page 213).
- An angle, or a length, in a plane can be measured in a view where the plane appears in true shape.
- The distance between a point and a plane is seen in a view where the plane is seen as an edge.

Problems

12. Draw the edge view of plane CDE by projecting from (a) the plan view, and (b) from the front view. A true length line has been drawn in each view. Layout dimensions are given in **Figure 6.45**. The scale is 1:1.

13. Draw the true shape of plane FGH (**Figure 6.46**). The edge view, which must be found first, can be projected from either the plan or the front view. Distances from the folding line are given in Figure 6.46.

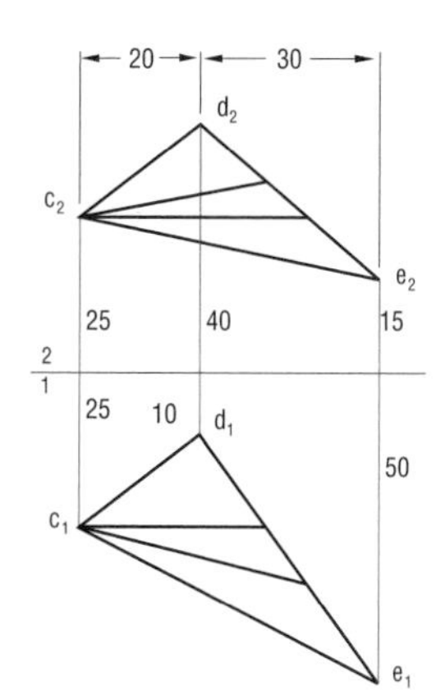

Figure 6.45

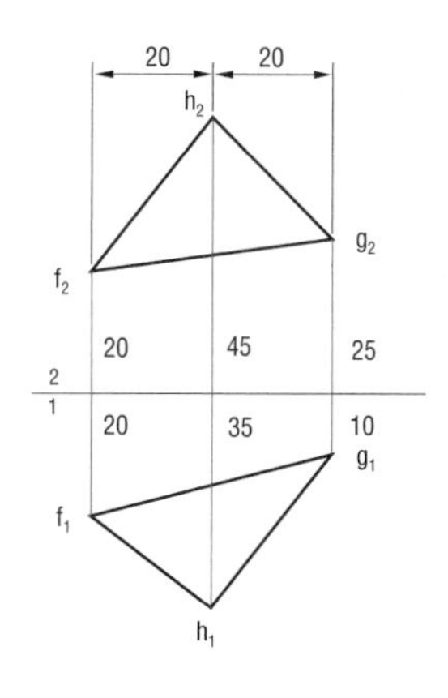

Figure 6.46

14. Determine the distance between point A and plane BCD. All points are referenced from the lowest point, D. The scale is 1:10.

B is 350 mm west, 100 mm north, and 200 mm above D.
C is 100 mm west, 250 mm north, and 300 mm above D.
A is 250 mm west and 300 mm above D.

15. Determine the slope of plane CDF. All points are referenced from the lowest point. The scale is 1:20.

D is 700 mm west, 600 mm north, and 450 mm above F.
E is 1000 mm west, 100 mm south, and 220 mm above F.

16. What is the area of the plane in Problem 14?

17. Line AX is 7 m long, bears N 32° E, and is in plane ABC. Determine the vertical distance between X and point B. The scale is 1:100.

B is 3 m north, 2.5 m east, and 2 m below A.
C is 1 m north, 4.5 m east, and 1 m above A.

18. The drawing in **Figure 6.47** shows a plane, ABC, and a line, AD. Line AD bears N 70° E and is in the plane ABC.

B is 100 m east and 150 m north of A.
C is 300 m east and 70 m south of A.
D is 300 m east of A.

The elevations of these points are: A 545 m, B 430 m, and C 670 m.

a. What is the slope of plane ABC?
b. What is the elevation of D?

The scale is 1:5000.

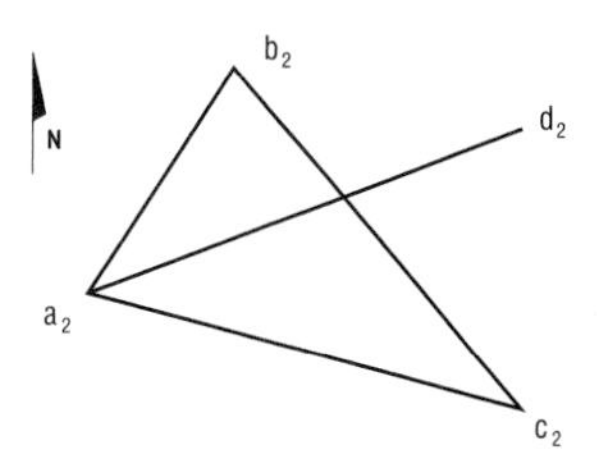

Figure 6.47

The following problems involve lines and planes and are more difficult.

19. A mining tunnel starts from A, bears S 65° E, and slopes down at 15°. Another tunnel starts at B, which is 250 m south and 160 m east of A. The elevation of A is 500 m. This tunnel bears N 55° E and slopes down at 20°. It meets the tunnel starting at A at C. A third tunnel starts at D, which is 500 m east of A and at an elevation of 250 m. This tunnel bears S 30° W and slopes upward at 18°. Point B must be connected to this third tunnel by a combination of a tunnel and a vertical shaft. The shortest possible tunnel is to be used.

What is the cost of connecting the tunnels in this way if the cost of drilling tunnels and shafts is $20 000 per metre?

At what elevation does the shaft meet the tunnel from B?

Figure 6.48 shows a layout of the three tunnels. The scale is 1:5000, vertical and horizontal.

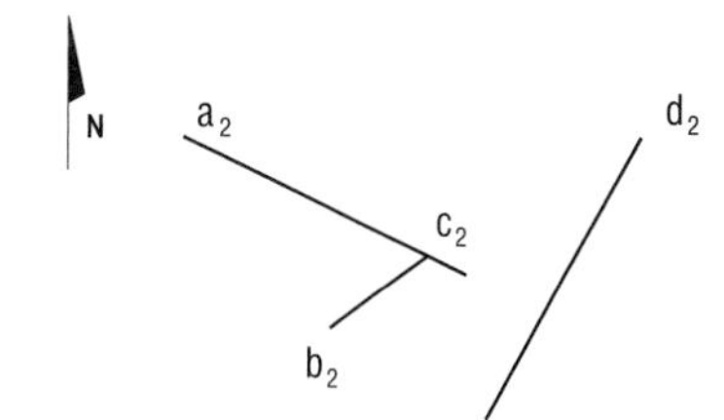

Figure 6.48

20. Plan and elevation views of two power cables are shown in **Figure 6.49**. The minimum distance between the cables must be 1 m. The only end that can be moved is B, and it can only be lowered. What is the minimum distance that B must be lowered to achieve the minimum clearance? The scale is 1:100.

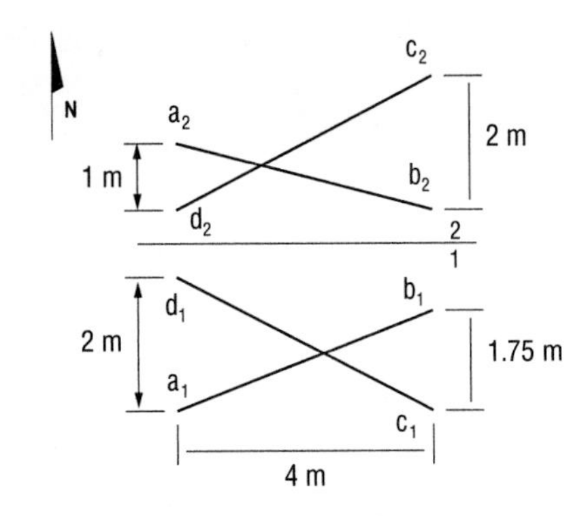

Figure 6.49

21. A plan view showing a 4 m mast near a power line, AB, is illustrated in **Figure 6.50**. The mast is supported by guy wires, one of which is anchored on the ground at X. The wires are attached to the mast 1 m below the top. The power line slopes down from B at 24°. There must be a minimum clearance of 1 m between the guy wire and the power line. The mast can be moved in a north–south direction to achieve the clearance. How far must the mast be moved to achieve the minimum clearance? The scale is 1:100.

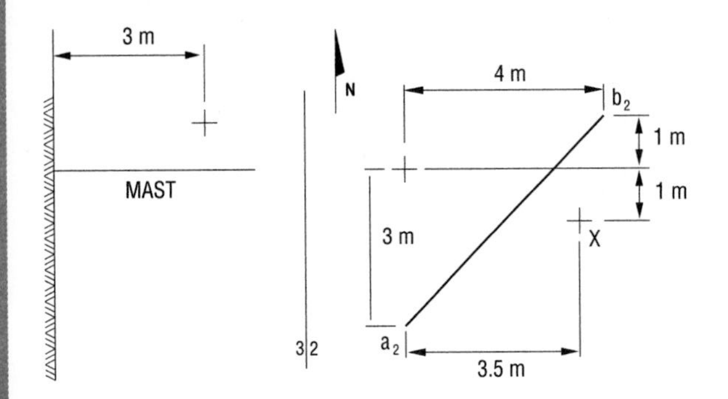

Figure 6.50

22. In the search for minerals and oil deposits, it is usual to drill a hole into the ground and determine where the hole intersects the deposit. By drilling a series of holes, the extent of the deposit can be estimated. A vertical hole drilled into level ground at A (**Figure 6.51**) hits the top surface of an ore vein at a depth of 40 m and the bottom surface at a depth of 46 m. Another hole drilled at B, on a bearing of N 30° E and a slope of 65°, hits the top surface of the ore vein at a depth of 30 m and the bottom surface at a depth of 60 m (measured along the drill hole). The top and bottom surfaces of the ore vein can be considered parallel.

Determine the thickness of the ore vein in this region from these drilling results. The scale is 1:1000.

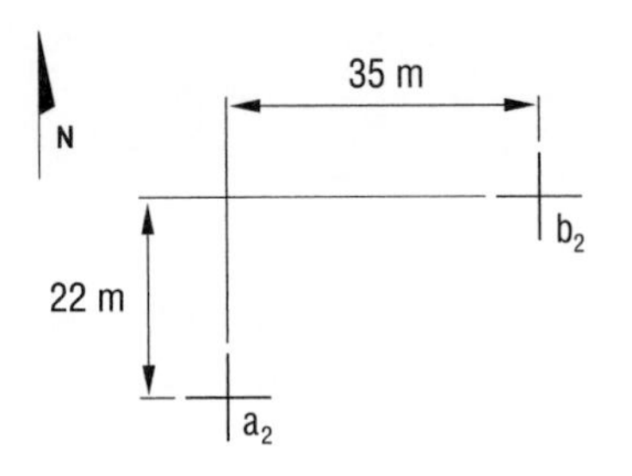

Figure 6.51

23. Two tunnels, AB and CD, are shown in **Figure 6.52**. They must be connected by a level tunnel bearing N 40° W. What is the length of the level connector? The scale is 1:200.

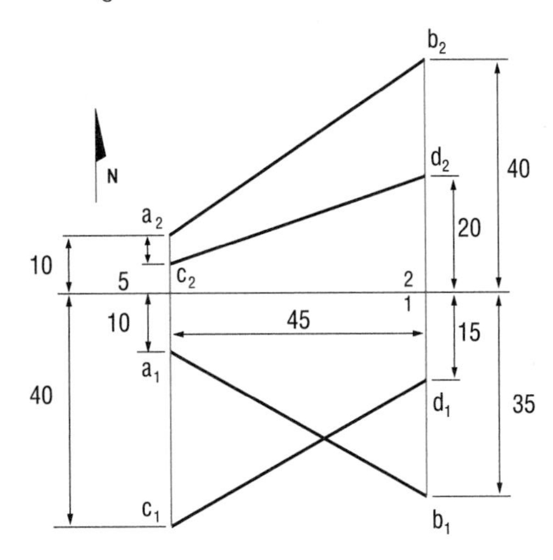

Figure 6.52

24. An ore vein meets the surface of ground at point A. This is called an **outcrop**. Holes (called boreholes) will be drilled in the ground in order to determine the location of the ore. An inclined borehole is drilled at a location, M, 50 m west, 40 m south, and 10 m below the elevation of A. This borehole bears S 60° E and slopes at –45°. The ore vein is hit after drilling 40 m, measured along the hole. Another inclined borehole is drilled at N, located 50 m east, 25 m south, and 8 m below A. This hole bears S 45° W and slopes at 60°. The ore vein is hit after drilling 60 m measured along the hole.

Scale: 1:1000 vertical and horizontal

Determine the slope of the ore vein. The slope of an ore vein is called the "dip."

Now let's move on to look at another visualization representation, contours.

CONTOURS

A **contour** is a line that indicates a profile on a surface. The most common application of contours is to show lines of constant elevation on the surface of the ground. Figure 6.53 shows contour lines on the surface of a human foot. These contours were determined from photographs and were used as inputs to a numerically controlled machine to make a model of the foot. Each line represents a constant distance above some datum.

Contour lines on a topographical map are lines of constant elevation measured above sea level. If you walked along a contour line, you would neither rise nor fall in elevation. Contour lines on a map show what the ground surface looks like without the use of an elevation view. Figure 6.54 shows a portion of a topographical map. The contour interval and scale must be indicated. In this example, the contour interval is 10 m. Moving from one contour line to the next requires a change in elevation of 10 m.

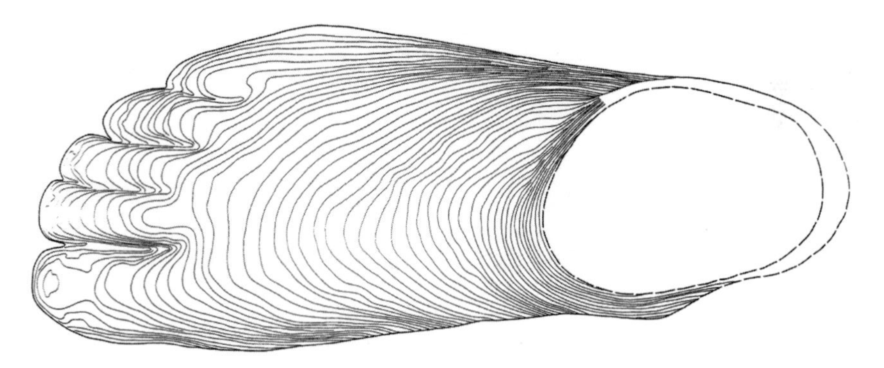

Figure 6.53 Contour lines on the surface of a foot
(Courtesy: Prof. James P. Duncan, Retired Department Head, Mechanical Engineering, University of British Columbia)

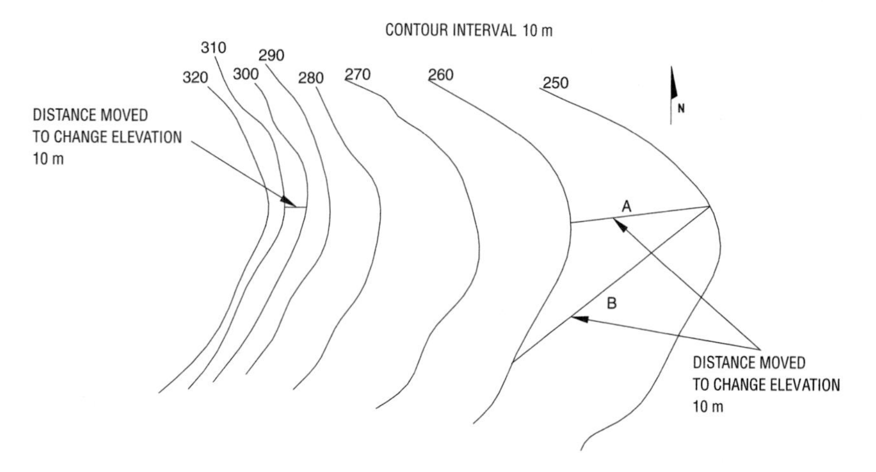

Figure 6.54 Horizontal distance travelled to rise 10 m depends on the direction

If contour lines are close together, as they are on the left side of the map, the slope of the ground is steep. A short run is required to rise or fall 10 m in this region. The wider spacing on the right shows a small slope, since the horizontal distance that must be moved to rise or fall 10 m is longer. If route B is followed, the horizontal distance is even longer. If you were in good physical condition, you might take route A. The run for route A is shorter than that for B, indicating that route A is steeper.

Contour lines represent the line of intersection of a horizontal plane with the ground surface. An elevation view showing these planes can be used to show the shape of the ground surface at any location. The process is equivalent to taking a section through the ground. The resulting outline is called a **profile**.

Finding a Profile

Figure 6.55 shows a topographical map.

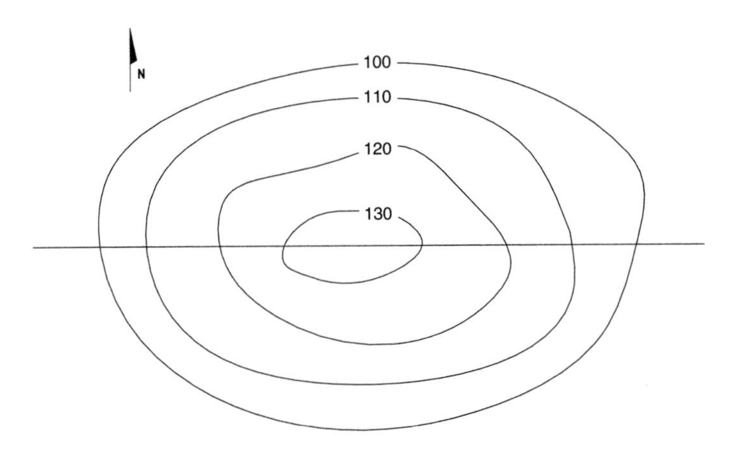

Figure 6.55 A topographical map shows a plan view

1. Draw a line indicating where the profile is taken. The east–west line in Figure 6.55 can be thought of as a vertical plane that cuts through the hill creating a section.

2. Draw an elevation view on which to see the profile. There is no need to draw a folding line. Draw horizontal lines corresponding to the contour interval in the elevation view. Vertical spacing is drawn to scale. Figure 6.56 shows the elevation view and horizontal lines representing the contour intervals.

3. Project the points where the east–west line intersects each contour line in the plan view to the corresponding contour in the elevation view (see left-hand side of Figure 6.57). These points are on the ground surface along the east-west line.

4. Join the points with a freehand line representing the ground surface. The symbol for "earth" is also drawn freehand. Be sure the symbol is on the correct side of the line. Figure 6.57 shows the completed profile.

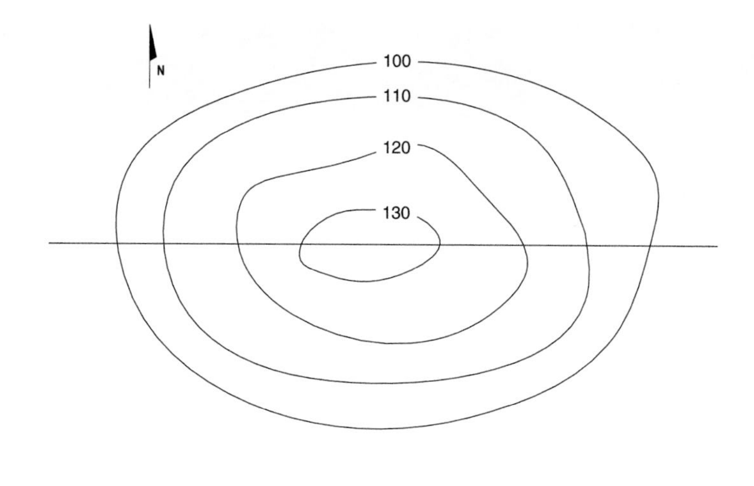

Figure 6.56 Elevation view looking north, showing contour intervals

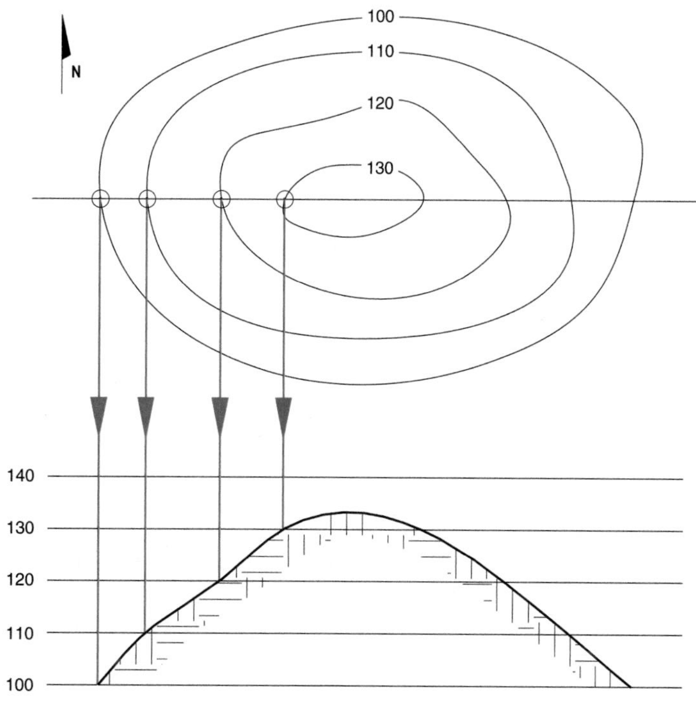

Figure 6.57 Ground profile in an east–west direction

To help you understand the practical application of your knowledge, take a few minutes to look at the following example.

Example 6.3

The topographical map in **Figure 6.58** shows a portion of a lake in a mountainous region. A float plane is on the water at the location shown. The airplane must take off on a bearing of N 60° E. The elevation of the lake is 275 m. In the worst case, the plane will lift off the water 20 m from the shore and climb at a rate of 250 m per 1000 m. Determine whether the plane should take off. The contour interval is 5 m. The scale is 1:1000, vertical and horizontal.

Obviously, the flight path of the airplane must not intersect the side of the mountain. If it clears the mountain, the take-off is safe. The path of the airplane must be compared with the profile of the mountain to determine whether they intersect.

What is known?

- Where the airplane leaves the water
- The rate at which the airplane can climb
- The direction in which the airplane takes off.

1. Draw the flight path on the map to show where the profile is being taken. Indicate where the plane leaves the water (20 m from the shore). **Figure 6.59** shows this information on the map.

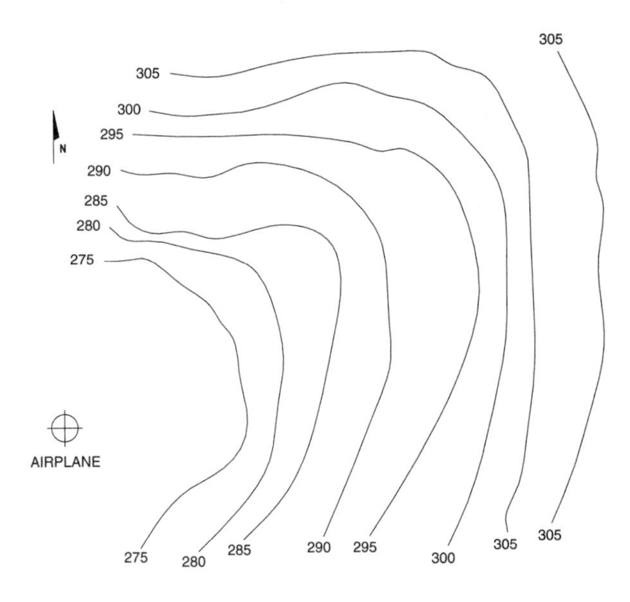

Figure 6.58 Topographical map showing the airplane relative to the shore

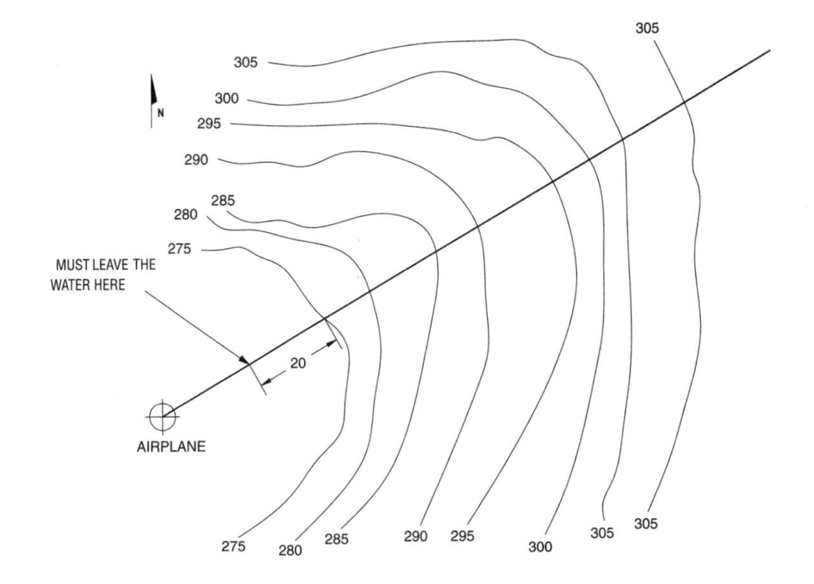

Figure 6.59 Bearing of the flight path of the airplane is shown

2. Draw an elevation view to show the profile and the flight path. The elevation view is located parallel to the direction in which the airplane flies. Show horizontal lines representing contour intervals on the elevation view (**Figure 6.60**).
3. Project the intersection of the flight path and the contour lines in the plan view to the corresponding elevation in the elevation view. Connect the points to show the profile (**Figure 6.61**)
4. Draw a line representing the rate of climb on the elevation view. **Figure 6.62** shows the flight path compared with the profile of the mountain.

The flight path does not intersect the mountain side, but it is very close (clearance is about 1 m). It would not be safe to take off under these conditions. The profile (drawn freehand) could not be expected to be accurate to within 1 m, and elevations do not take trees into account.

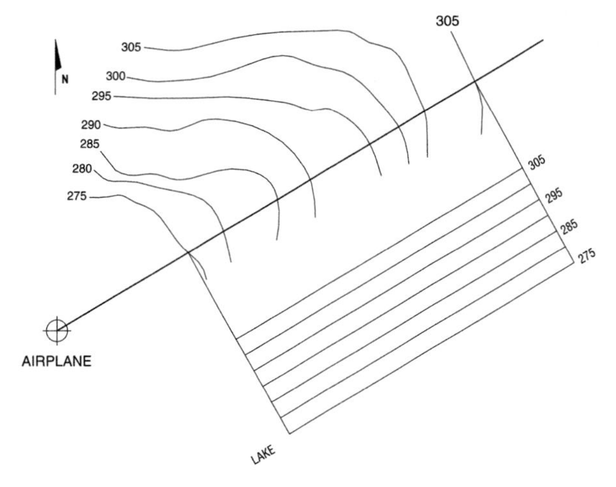

Figure 6.60 Elevation view located to show profile

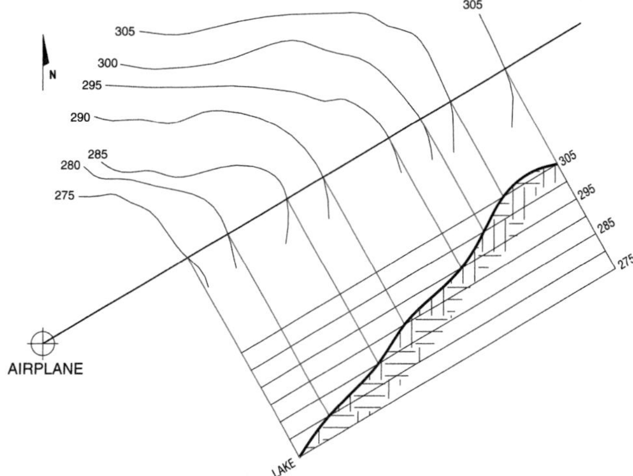

Figure 6.61 Profile of the mountain in the direction of the flight path

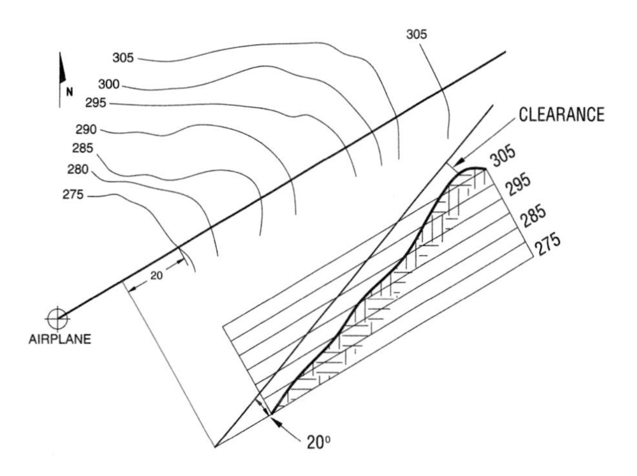

Figure 6.62 Flight path of airplane compared with mountain profile

Cut and Fill

A road or railway line must have a specific grade, and hills must be cut away or depressions filled to achieve this. The excavation and filling are called **cuts and fills**. Material excavated can be used to fill a depression if there is enough, or material can be brought in if there is not. In either case, the volume required to fill the depression and the volume excavated must be found. A contour map can be used to do this. Figure 6.63 shows a level road across a region where the ground is not level.

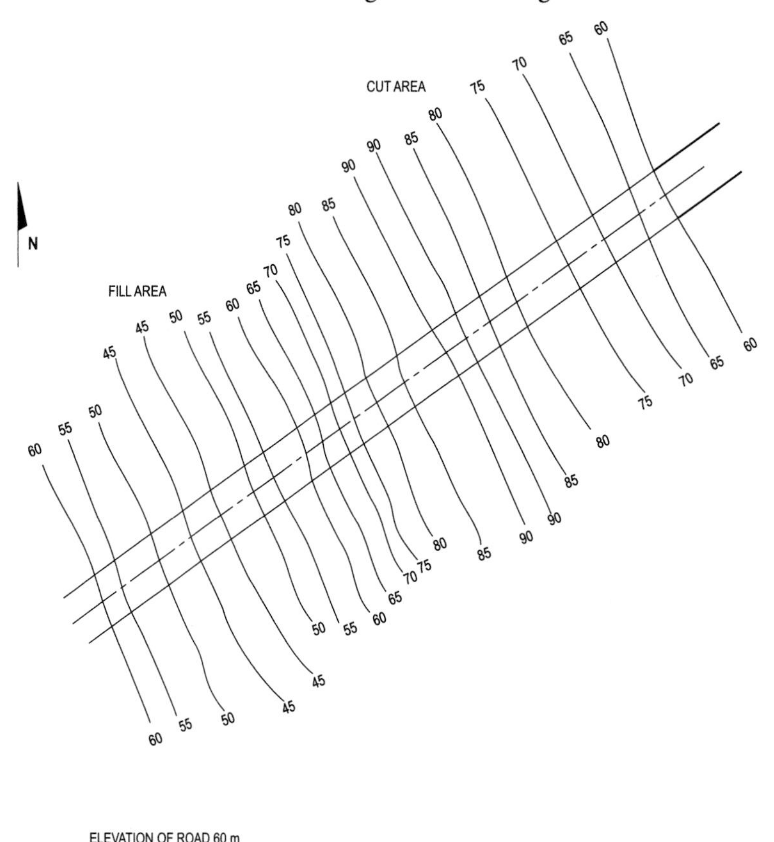

Figure 6.63 Topographical map showing the route of a road.

The depression must be filled, and the hill must be cut down to the level of the road. The volumes can be found by finding the outline of the fill and cut on the contour map. The slope of the sides of the fill and cut must be specified in order to do this. A profile showing the ground surface along the route of the road is drawn, and then sections showing the cut and fill are created. The outline of the cut and fill can be found and volumes estimated. Figure 6.64 shows a profile of the ground along the route of the road found by projecting from contours in the plan view.

Since the road is level, an elevation view looking along the road will show the sides of the cut and fill. If the road is not level, this cannot be done, since the centreline of the road is not seen in true length in the plan view. Figure 6.65 (page 224) shows auxiliary elevation views at each end of the road. The cut is shown at the east end and the fill at the west end. The sides of the cut and fill are drawn at the specified slope starting at the edge of the road clearance. Outlines of the cut and fill are created on the plan view by projecting the intersection of the cut and fill at each elevation. This is a point view of a line on the surface of the cut or fill. The points where these lines intersect contour lines are shown in Figure 6.65. The outlines of the cut and fill are shown in Figure 6.66 (page 225).

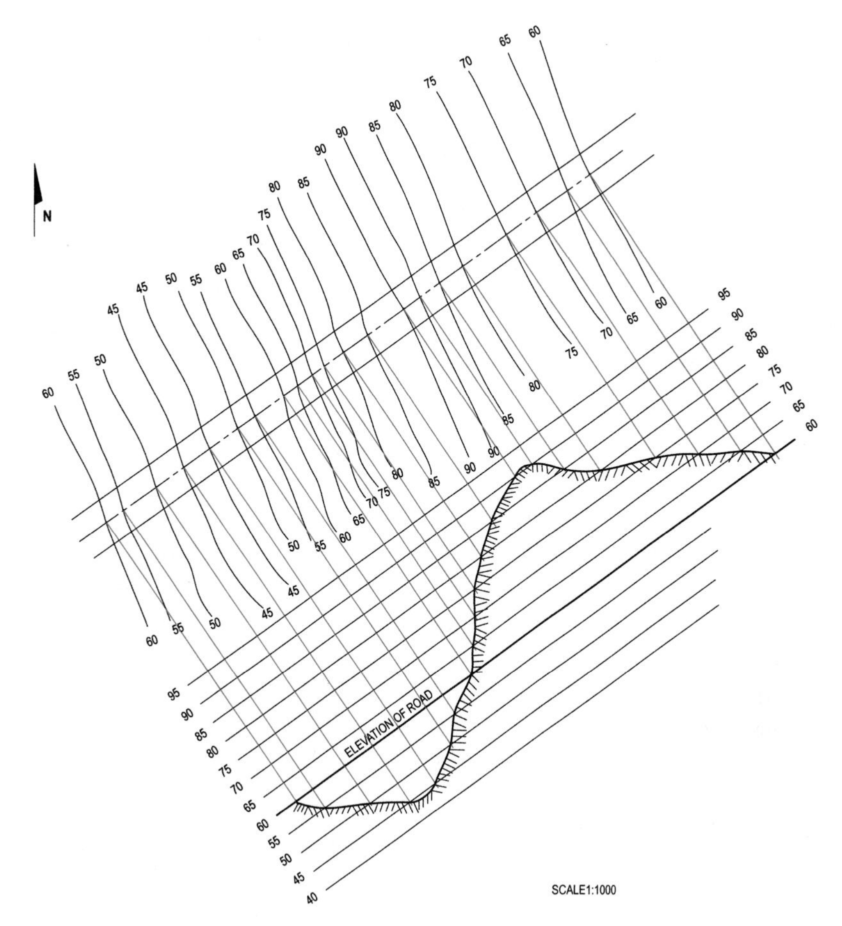

Figure 6.64 Ground profile created by projection from the plan view

When seen in the plan view, the line is parallel to the road and at a constant elevation and will go from one contour line to another at the same elevation. Lines on the side of the cut and fill will be parallel to the road only if it is level. It is not necessary to draw lines on the surface of the cut and fill, since only the intersection with the contour lines is necessary to draw the outline. They are drawn in Figure 6.66 to aid visualization. The outline is drawn by joining the points where the level lines intersect each contour line.

The points of intersection with the ground are found in Figure 6.66 by projecting from auxiliary views showing the edges on the cut and fill surface, but there is another way if the road is level. The horizontal distance between level contour lines on the cut and fill surfaces equals the run for these surfaces; thus, they can be located by measuring the horizontal distance between them (i.e., the run) staring at the edge of the road. This eliminates the need for auxiliary views.

There are various ways of finding the volume depending on how accurate you wish to be and how much time you want to spend on it. One way is to choose a cross-sectional area at a right angle to the road and multiply this by a length over which the chosen cross-section applies. This would be done for several lengths along the road, each with a different cross-section. The area of a cross-section can be estimated by breaking it into known shapes (rectangles and triangles) and summing the areas.

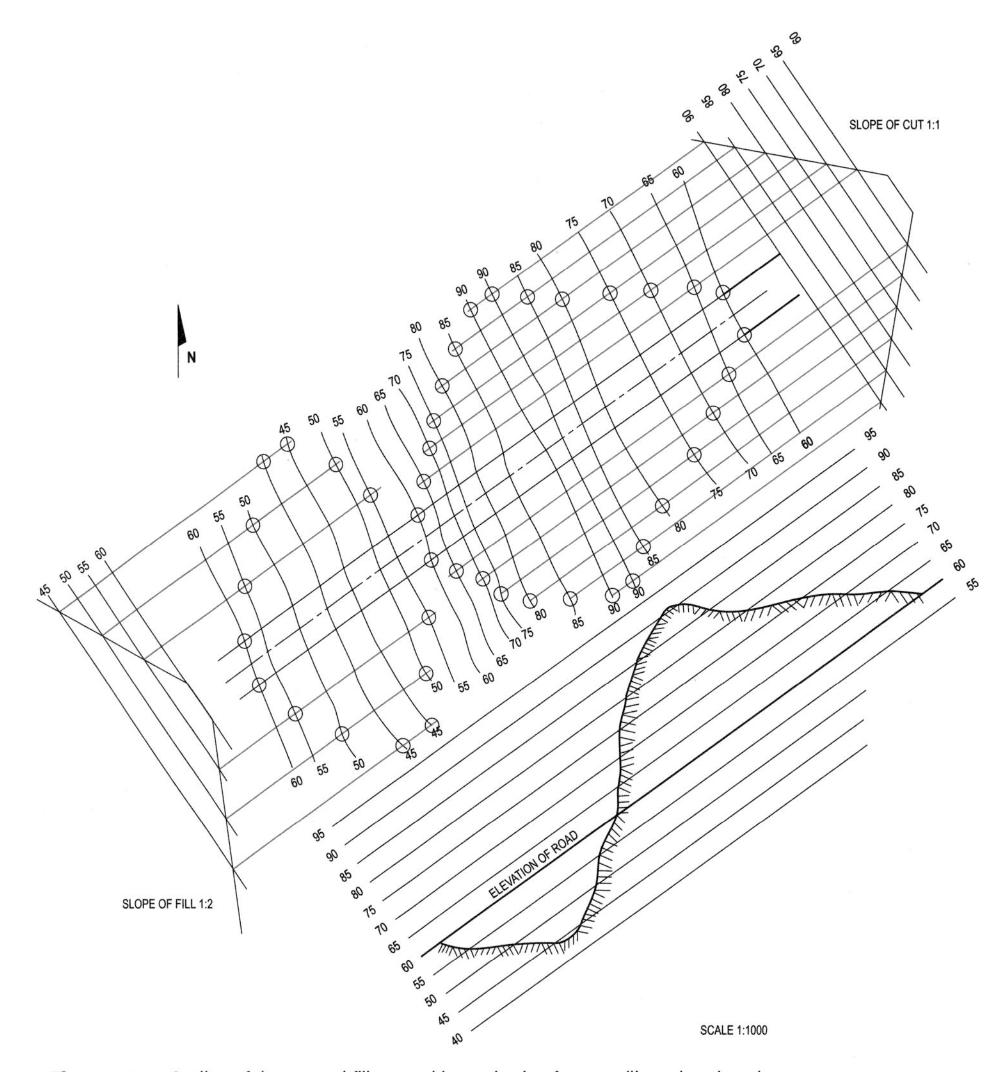

Figure 6.65 Outline of the cut and fill created by projecting from auxiliary elevation views

A more accurate variation is to divide the length along the road into segments of known length and find a representative cross-section for each segment. The volume of each segment is found by multiplying the cross-sectional area by the length of the segment. Total volume is found by summing the volume of each segment. The more segments you use, the more accurate the result will be.

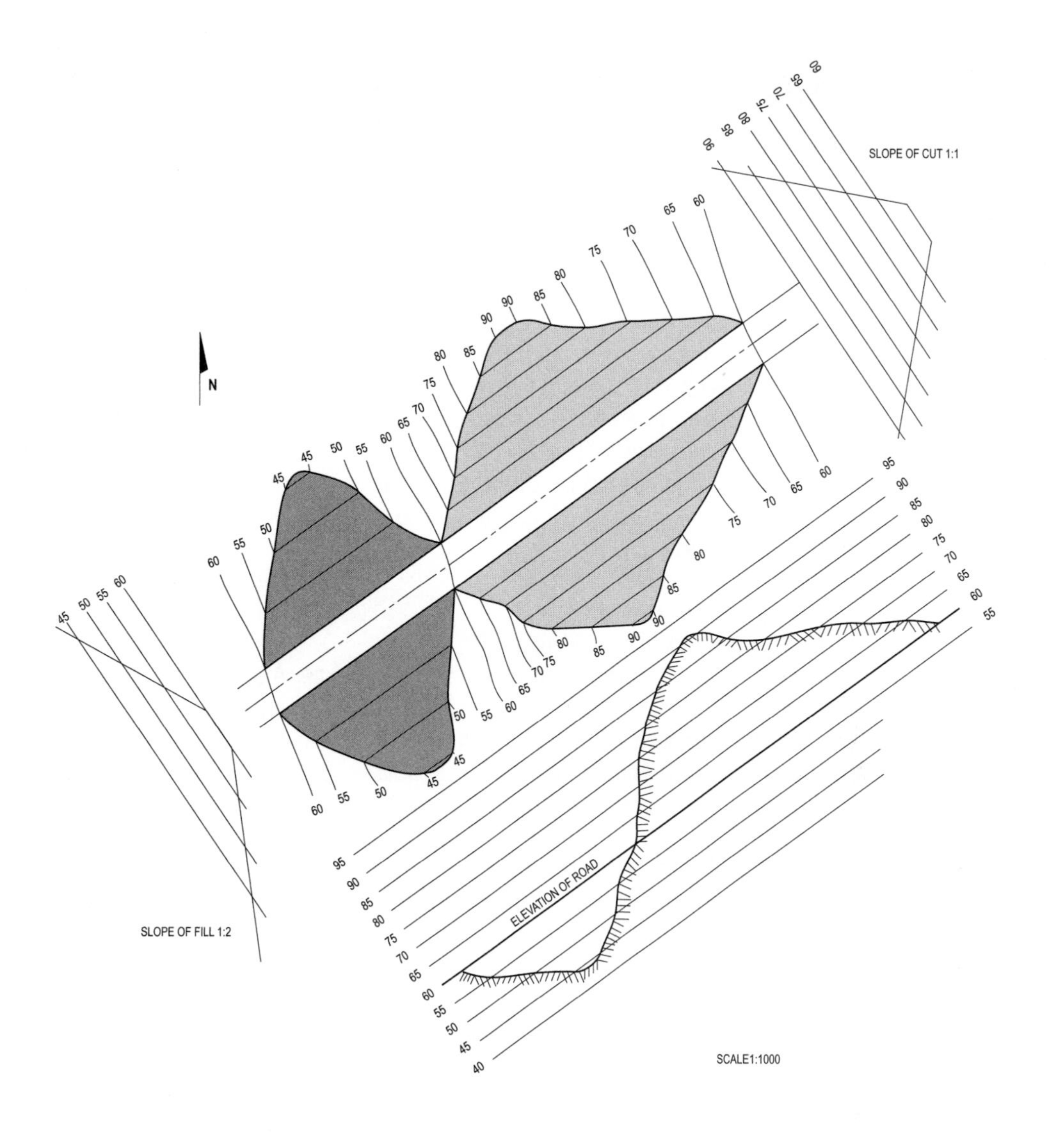

Figure 6.66 Surfaces of cut and fill showing contour lines

Problems

25. The topographical map in **Figure 6.67** shows the proposed location of a dam site at the convergence of two rivers (river A and river B). Water is carried from the reservoir behind the dam to the turbine in the powerhouse by a large pipe called a penstock. The penstock enters the powerhouse on a bearing of due east (point A) and an elevation of 1400 m at the centreline. This section of the penstock is 160 m long and is horizontal. From the west end of this horizontal section, the penstock bears N 20° W and slopes upward at 30°. The penstock has another turn so that it meets the dam at B, on a bearing of S 75° E and a slope of 30° (downward from B). Scale 1:10 000.

Determine:

- The total length of the penstock.
- The length the penstock that is underground.
- The elevation at which the penstock meets the dam (the elevation of B).

26. A ski tow is to go from elevation 470 m to point A at an elevation of 585 m (**Figure 6.68**). Support towers will be at elevations 470, 510, 530, 550, 570, and 580 m. Towers are vertical and are 10 m high.

Approximate the cable (at the top of each tower) as a series of straight lines and determine the length of cable required to go from the bottom tower to the top tower.

27. In a region where the ground has a uniform slope, the contour lines are straight parallel lines bearing N 15° W. The horizontal distance between contour lines is 50 m, as shown on the sketch in **Figure 6.69**.

A tunnel 300 m long and bearing N 60° E starts at the surface at an elevation of 550 m (point A in Figure 6.69). The tunnel slopes downward from A at 30°.

A new tunnel must be drilled from the ground surface to the bottom of this tunnel. The new tunnel will slope at 60° on a bearing of N 45° W.

Determine where the new tunnel must start in order to meet the end of the old tunnel. Specify the drilling location by giving the distances south and east of A. The scale is 1:5000, horizontal and vertical.

28. A contour map, **Figure 6.70**, shows a region where a vein of ore meets the surface of the ground. The top and bottom surfaces of the vein are parallel. A vertical hole at A (elevation 512 m) hits the top of the vein at an elevation of 502 m. A vertical hole at B, 34 m east of A, hits the top surface of the vein at 512 m and the bottom at 504 m. A point on the top surface of the vein is at the ground surface at C, 32.5 m north and 26 m east of A. The scale is 1:1000, vertical and horizontal.

Determine the thickness of the ore vein.

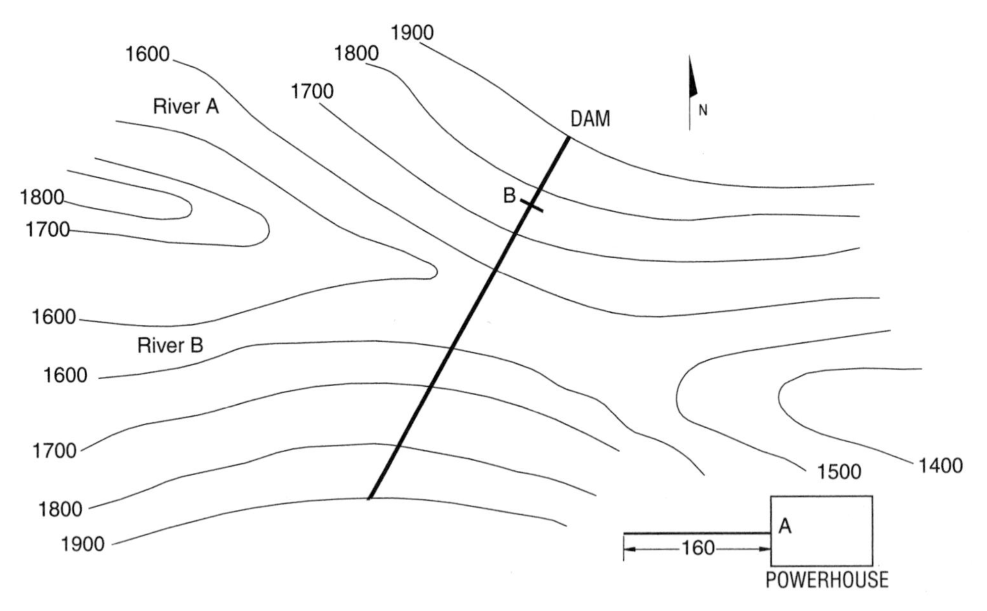

Figure 6.67

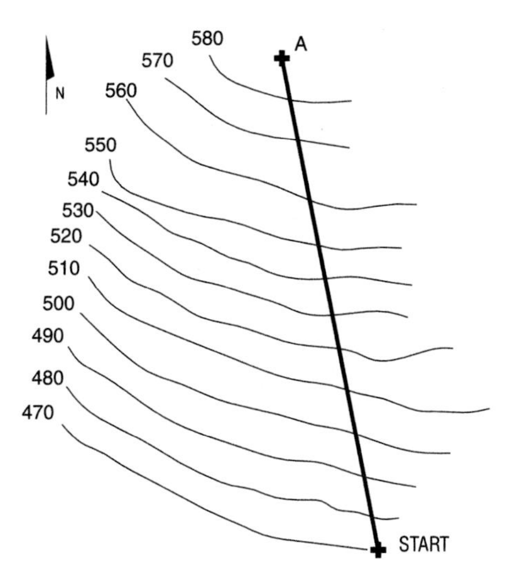

Figure 6.68

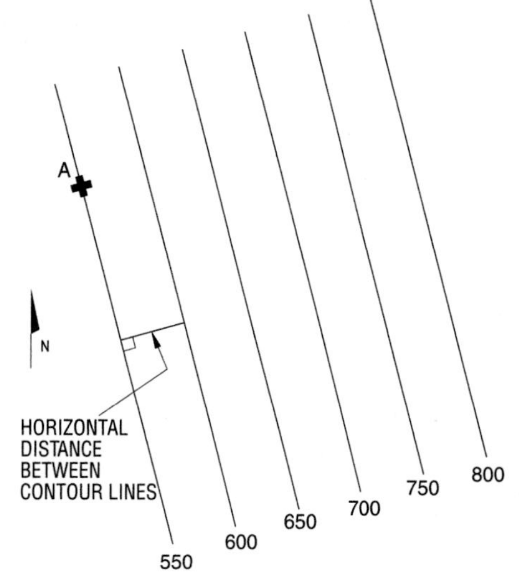

Figure 6.69

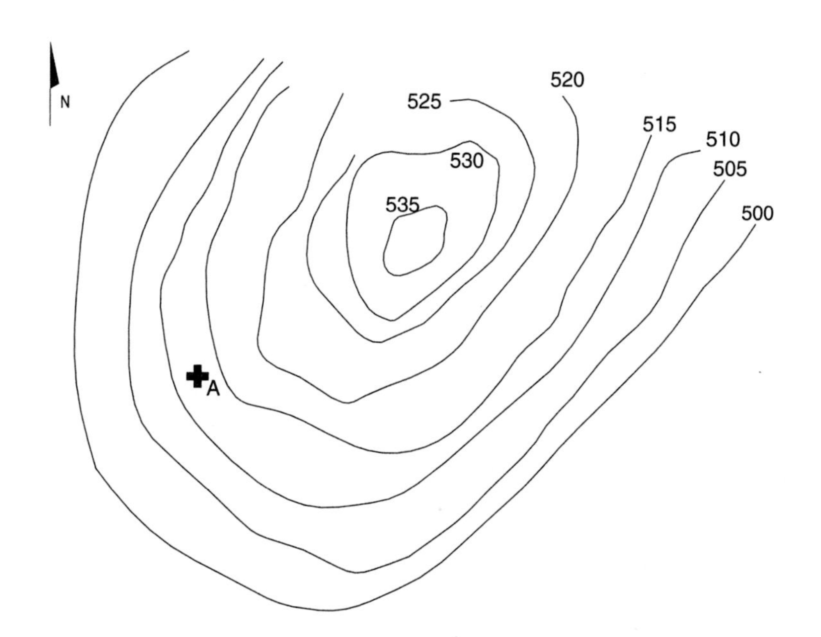

Figure 6.70

29. The contour map in **Figure 6.71** shows an area for a proposed ski resort. A ski tow is proposed from A, elevation 130 m, to B, elevation 202 m. The cable must be a minimum of 5 m, perpendicular distance from the ground. Approximate the cable by a series of straight lines and determine the length of cable required to go from A to B. The scale is 1:1000, vertical and horizontal.

A ski run is to start at the location shown and end at a location near A. The slope of the run is 20 percent. Show a possible run that meets these conditions. The ski run may have sharp turns.

30. **Figure 6.72** shows a topographical map with hills A, B, and C. Microwave towers are to be situated on the top of hills A and C. If there is no direct line of sight between A and C, another tower must be built on hill B. Using a profile from A to C, determine if a tower is required at B.

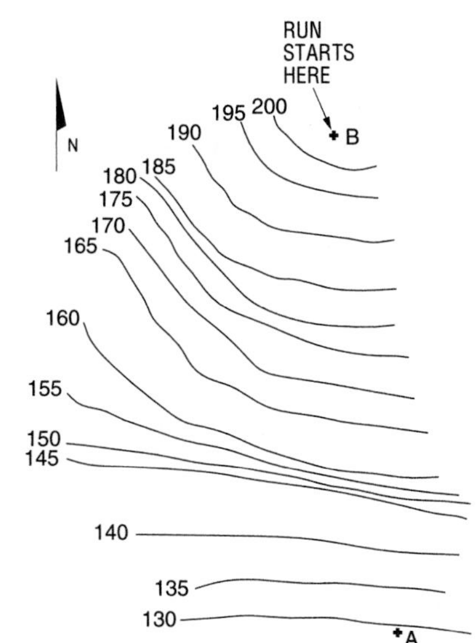

Figure 6.71

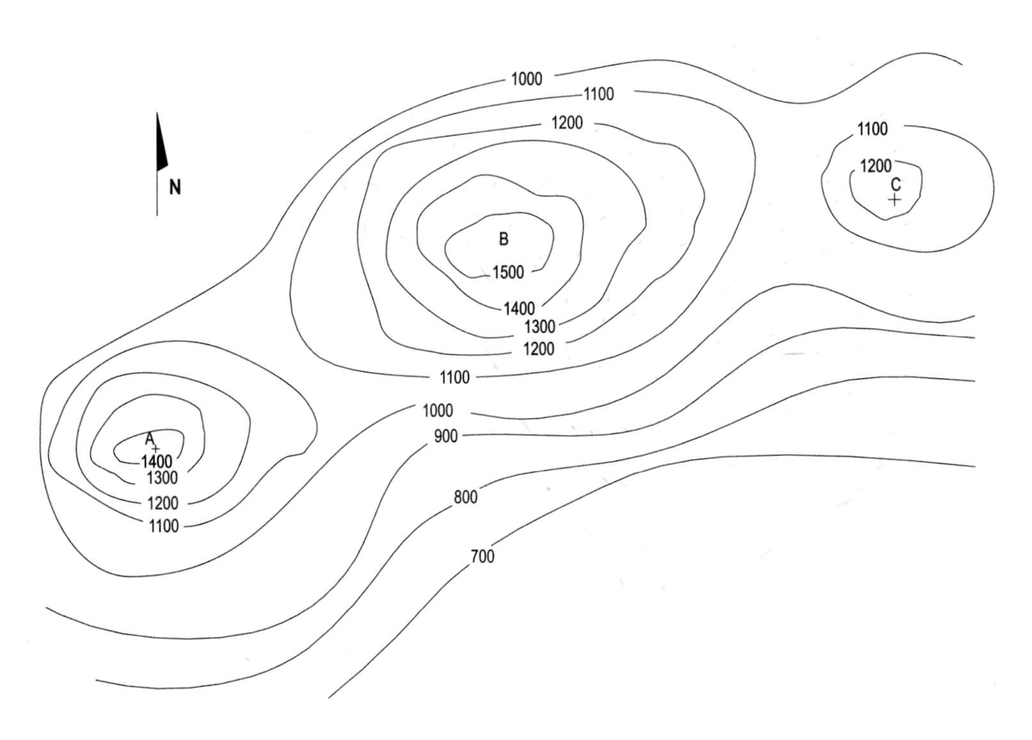

Figure 6.72

31. **Figure 6.73** shows the location of a small hydro project. Water from a lake (elevation 120 m) is taken to the powerhouse through a tunnel. The tunnel is not straight. The entrance to the tunnel is at an elevation of 118 m and the bottom of the tunnel is at elevation 88 m, where it meets the powerhouse. The tunnel bears N 30° E from the intake at a grade of 30 percent. The bottom end of the tunnel bears S 60° W from the location shown on the powerhouse. Determine the total length of the tunnel. At what elevation do the two sections meet? What is the maximum depth of the tunnel?

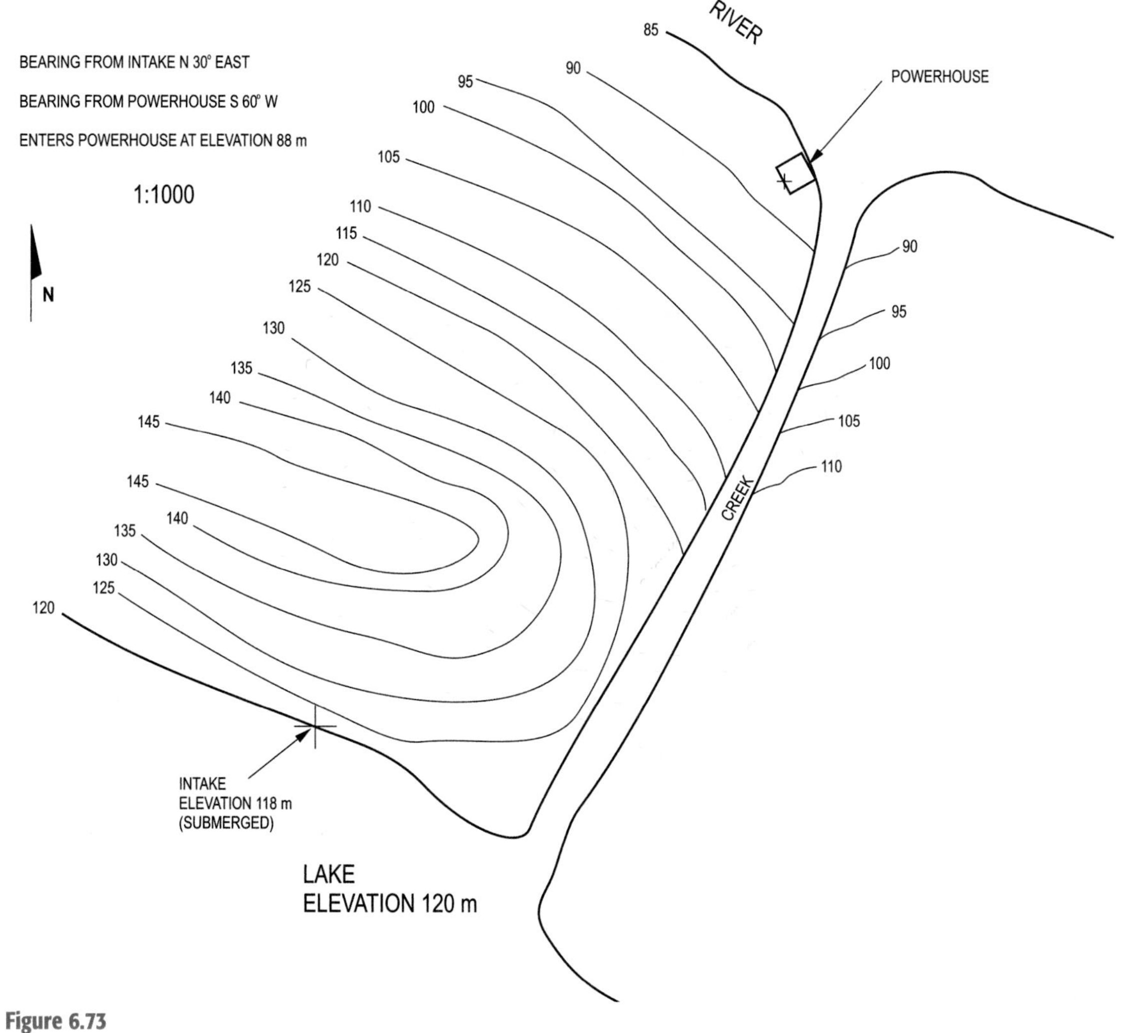

Figure 6.73

32. **Figure 6.74** shows the location of a road across a river valley. The road is 10 m wide and runs N 45° E. The hill must be cut so that the road can remain level at an elevation of 120 m. The slope of the sides of the cut is 1:1.

    Show the cut portion of the hill on the contour map, and estimate the volume of material that must be removed to make the cut.

33. A cableway is to go from elevation 740 m to point A at elevation 811 m on a bearing of N 75° E. (**Figure 6.75**) Support towers will be located a elevation 750, 760, 770, 780, 790, 800, 810, and 811 m. Towers are set perpendicular to the ground, except those at 740 m and 811 m, which are vertical. Towers are 15 m long measured along the tower. Scale 1:1000 vertical and horizontal.

    Approximate the cable as a series of straight lines and determine the total length of the cable.

    A construction road is to run from the starting point at 740 m to the 800 m level. The maximum grade is 20 percent, and the road can have sharp turns. Lay out a route for the road that meets these conditions. You will have to find the horizontal distance travelled to rise 10 m (one contour interval) with a 20-percent grade. Lay out this distance between the contour lines between 740 m and 800 m, and sketch the route.

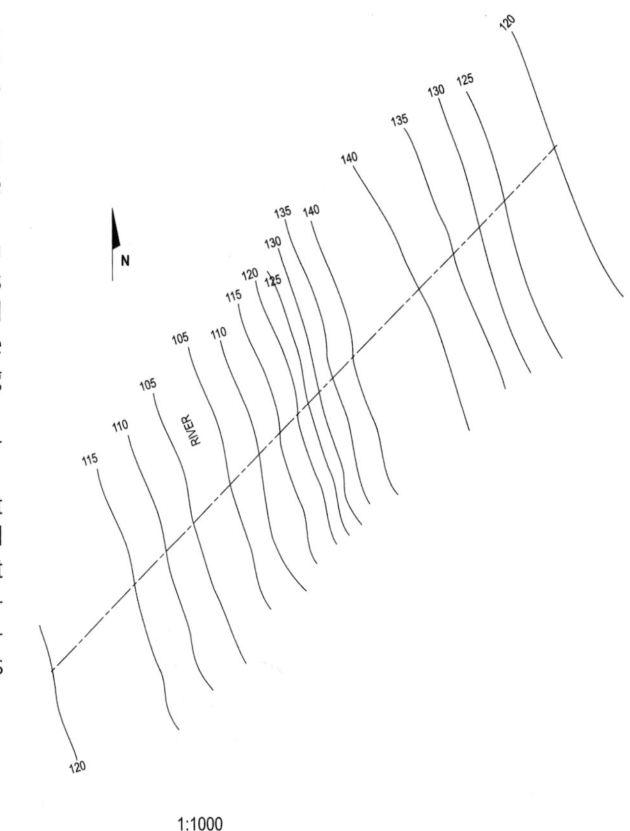

Figure 6.74

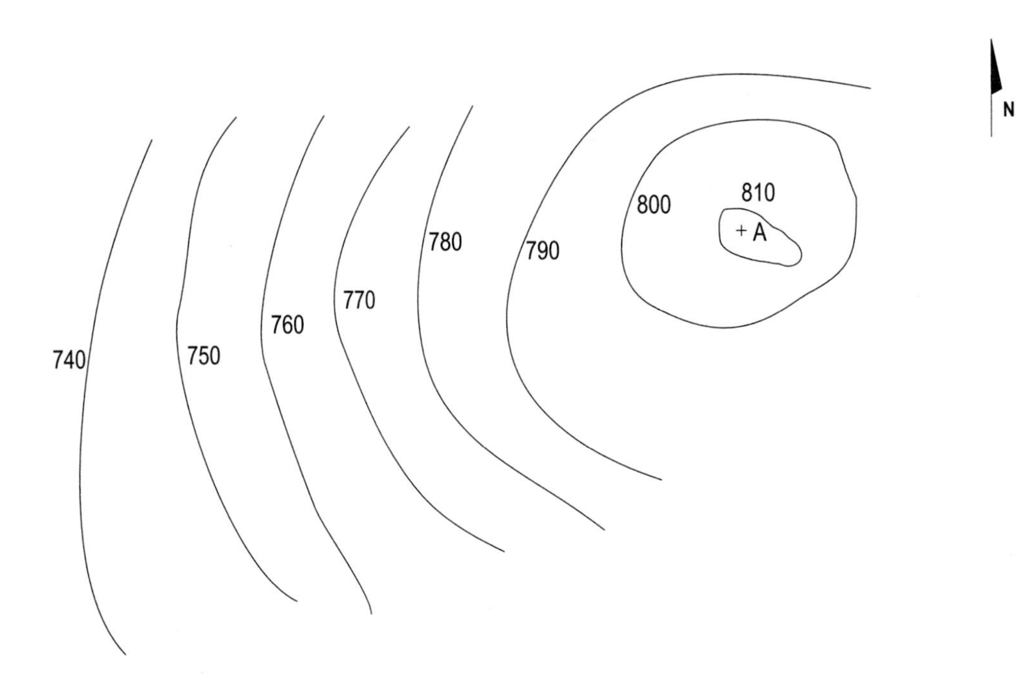

Figure 6.75

Now let's move on to look at some of the 3-D features available on computer programs.

3-D

The 3-D features of a computer program can be used to solve problems similar to those in the first section of this chapter, but instead of drawing auxiliary planes, the viewpoint is changed, and the computer screen is the auxiliary plane. The same views—true length, point view, and edge view—are used.

Because computer programs differ in details, this section takes a generic approach. The limiting cases—true length and point view for lines and the edge view for planes—are described.

Setting the Viewpoint

To set the viewpoint, first determine which view will provide the necessary information, and then specify the viewpoint. For example, if you want to measure the shortest distance between two lines, set the viewpoint so that one of the lines is seen as a point. The viewpoint can be set by specifying the angle *from* the *x-y* plane and *in* the *x-y* plane (Figure 6.76). The angle from the *x-y* plane is determined from the slope. The angle in the x-y plane is determined from the bearing or azimuth.

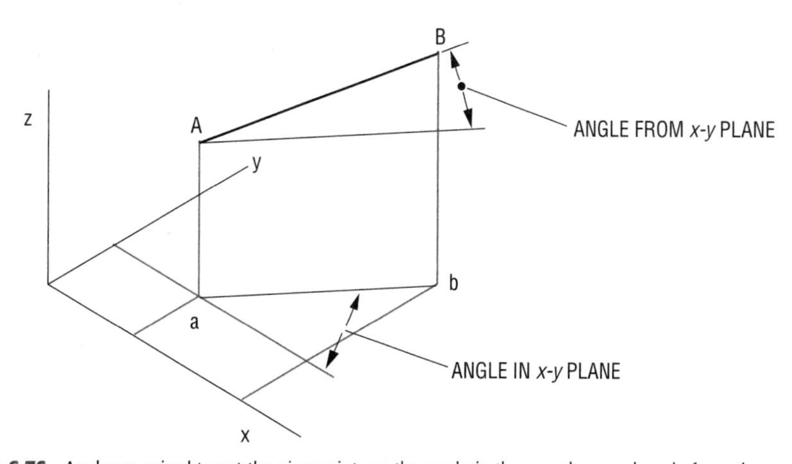

Figure 6.76 Angles required to set the viewpoint are the angle in the *x-y* plane and angle from the *x-y* plane

When you enter information on a line or other entity, it is stored as a vector. The program can use this information to determine length, so finding the length is not the problem. You want to set the viewpoint to see the true length on the screen. True length is seen when the line is viewed in a direction perpendicular to it. Figure 6.77 shows two of the many viewpoints that satisfy this condition.

These lines of sight are 90° to the bearing.

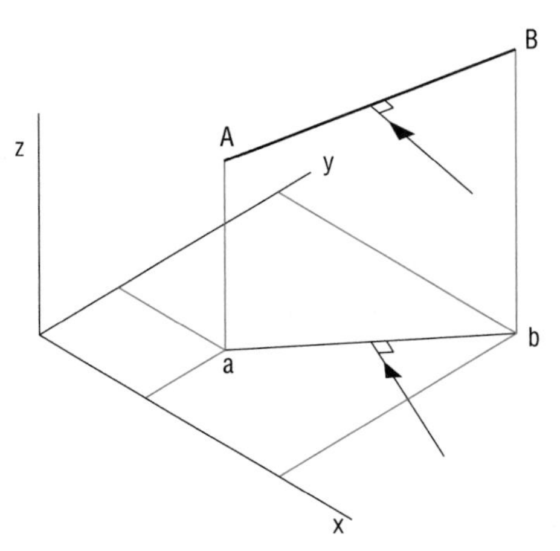

Figure 6.77 Line of sight must be perpendicular to the line to show true length

Showing the True Length of a Line

Figure 6.78 shows a line with end points A (2, 2, 6) and B (5, 6, 1). You want to specify the angle in the *x-y* plane so that the line is seen in true length.

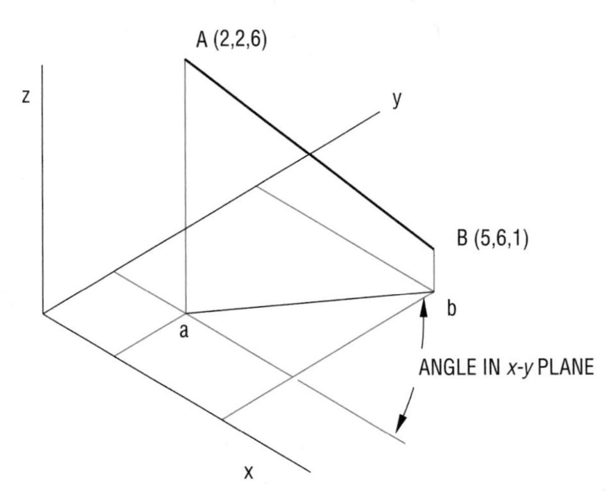

Figure 6.78 Line with end points A and B projected onto the *x-y* plane

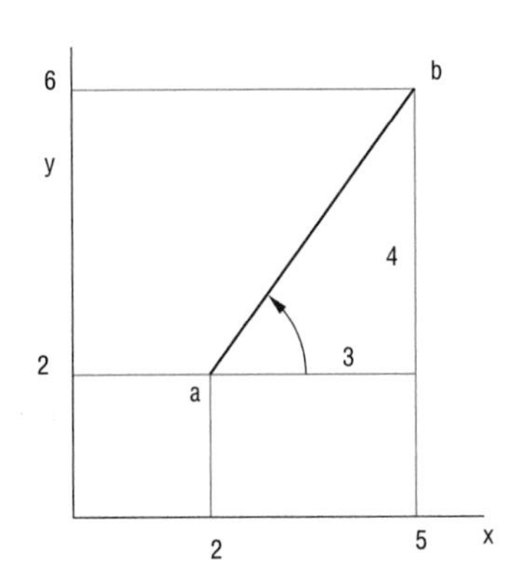

Figure 6.79 The bearing of the line (angle in the *x-y* plane) can be found if the coordinates are known

You can use your program to give the length and possibly other information. If this is not available, the angle in the *x-y* plane can be found from the coordinates. The line is 7.07 units long.

1. Determine the angle of the line in the *x-y* plane. Figure 6.79 shows how this is done using the coordinates. The tangent of the angle in the *x-y* plane is 4/3 and the angle is 53°. (The bearing is N 37° E.)

2. Determine the angle of the line of sight in the *x-y* plane. The line of sight is perpendicular to the line. This corresponds to an angle in the *x-y* plane of –37° or 143°. These angles are shown in Figure 6.80.

True length can be seen from many other viewpoints, but these are the easiest to set, since only the angle in the *x-y* plane must be determined. The angle from the *x-y* plane is 0°.

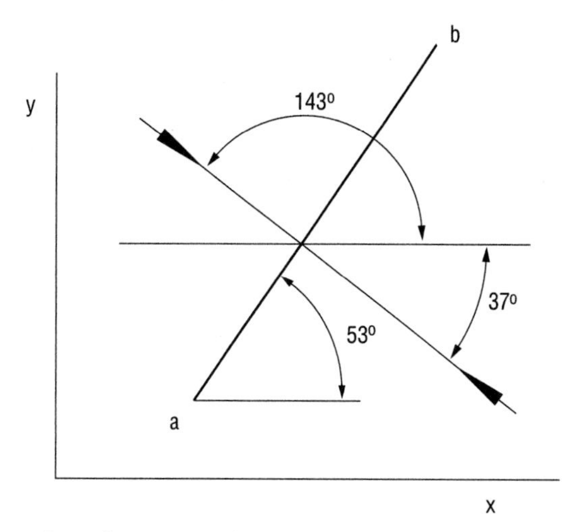

Figure 6.80 Viewpoint angles to see true length are perpendicular to the bearing

Showing the Slope of a Line

Slope can only be seen when the line of sight is parallel to the *x-y* plane (the angle from the *x-y* plane is 0°) and the line is true length. Either of the viewpoints used in Figure 6.80 to show true length will show slope. Figure 6.81 shows the slope of AB.

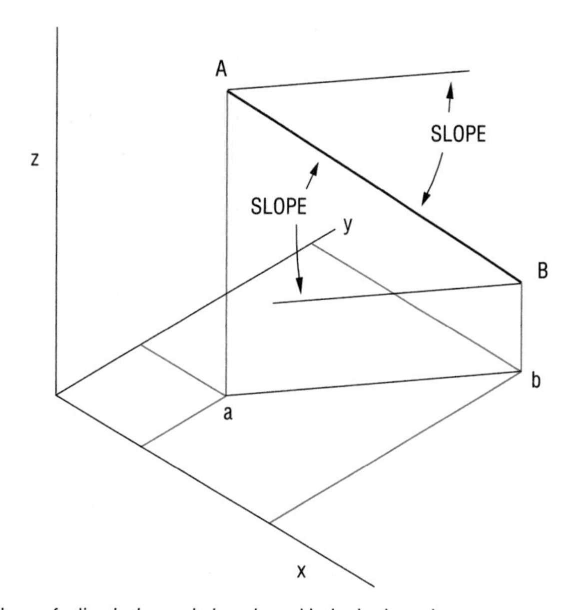

Figure 6.81 The slope of a line is the angle it makes with the horizontal

The method of finding slope is the same as that described for finding true length.

Showing a Line as a Point View

A point view can be seen by setting the viewpoint to look along the line. The angle in the *x-y* plane (direction) and the angle from the *x-y* plane (slope) must be specified to see a line as a point. Figure 6.82 shows how a point view is seen.

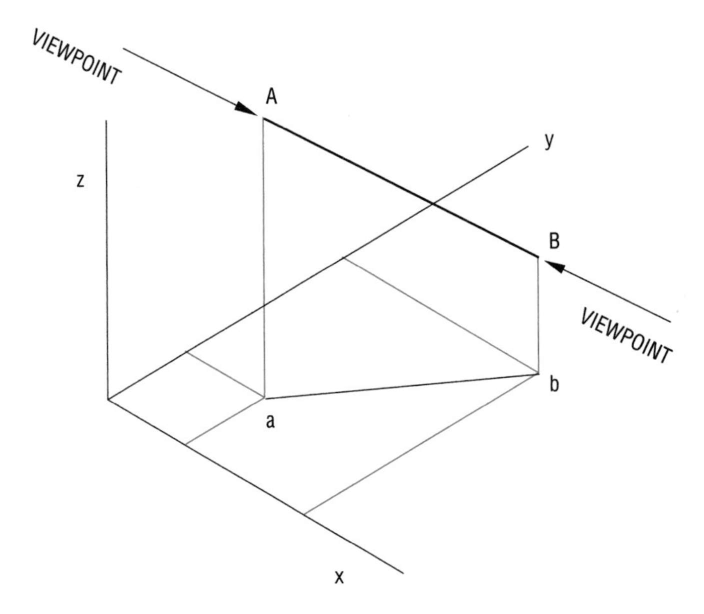

Figure 6.82 Viewpoints to see a point view can be from either end of the line

Figure 6.83 shows a line with end points A (2, 3, 3) and B (6, 6, 1). You want to specify the angle in the *x-y* plane and the angle from the *x-y* plane to see a point view of the line defined by end point A.

1. Find the angle of AB in the *x-y* plane. Your program may give this information, but you can also find it from the coordinates. The tangent of the angle in the *x-y* plane is 3/4 and the angle is 37°. Figure 6.84 shows how this is found.

2. Find the slope of AB. The line slopes down from A with a drop of 2 and a run of 5 (the length of AB in the *x-y* plane). The slope is –22° (down from A). This is shown in Figure 6.85.

3. A point view can be seen from either end of AB. Figure 6.86 shows both possible angles. The angle in the *x-y* plane is 37° when viewed from B and 217° when viewed from A.

4. Find the angle from the *x-y* plane. There are two possible angles depending on the direction in which the line is viewed. When viewed from the low end, B, the angle from the *x-y* plane is –22°, since the viewpoint is below the *x-y* plane. When viewed from the high end, A, the angle from the *x-y* plane is 22°, since the viewpoint is above the *x-y* plane. Figure 6.87 shows the angles required to set the viewpoint for each end.

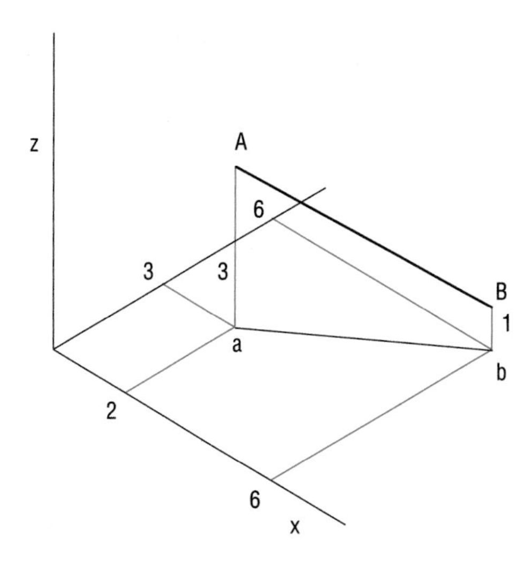

Figure 6.83 Line with end points A and B projected onto the *x*-*y* plane

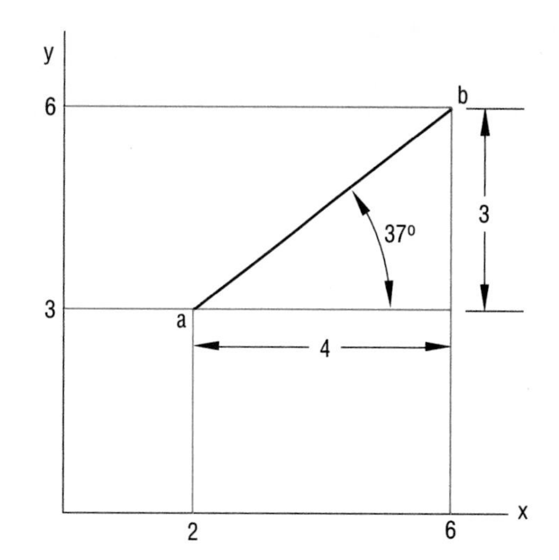

Figure 6.84 Finding the bearing (angle in the *x*-*y* plane) of AB

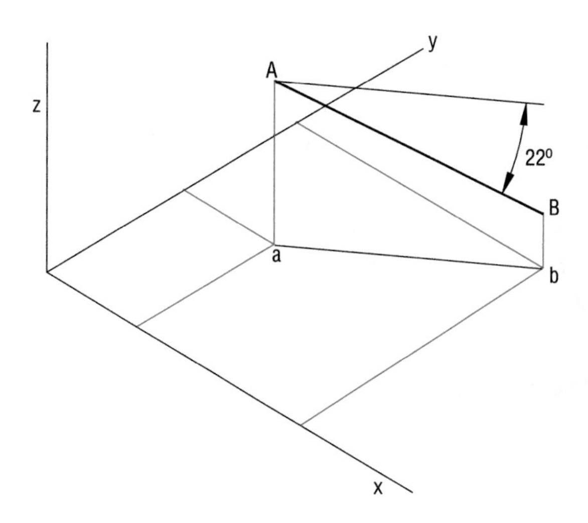

Figure 6.85 The slope of AB is the angle with the *x*-*y* plane

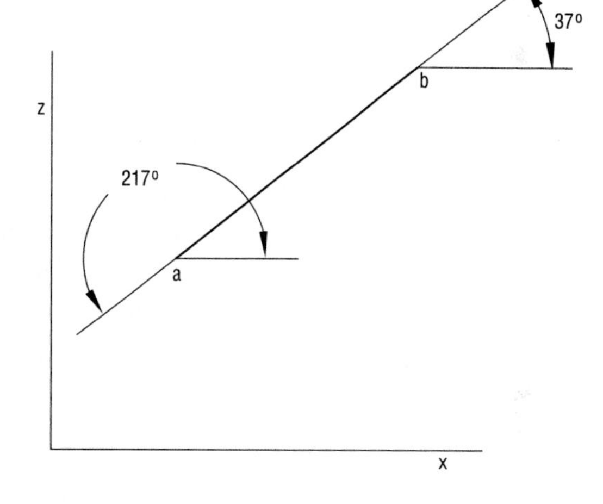

Figure 6.86 Viewpoint angles in the *x*-*y* plane

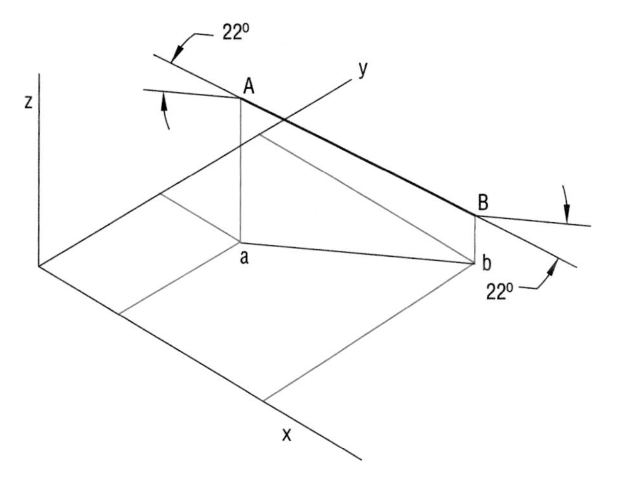

Figure 6.87 Viewpoint angles from the *x*-*y* plane to see a point view

Now that you know a little more about 3-D lines and viewpoints, try the following problems.

Problems

These problems can be solved with or without a computer.

34. Show the true length of a line AB, with end points A (6, 5, 7) and B (2, 3, 2).

35. Show the slope of the line with end points (1, 3, 3) and (9, 10, 8). Dimension the slope.

36. What is the bearing of a line AB with end points (4, 8, 6) and (10, 4, 3)?

37. Determine the length, slope, and bearing of the line with end points (2, 3, 6) and (8, 7, 1).

38. What is the shortest distance between the lines AB and CD, defined by:

A 6, 3, 0? B 1, 5, 6?
C 0, 2, 1? D 7, 8, 8?

Show and dimension this distance. The shortest distance is the perpendicular distance. Find one line as a point view and measure the perpendicular distance from the point view to the other line.

39. Determine the length of the shortest connection between AB and CD.

All points are referenced from A. The lines may be extended as required.

B is 330 mm east, 460 mm north, and 400 mm below A.

C is 300 mm north, 240 mm west, and the same elevation as A.

D is 100 mm east, 390 mm north, and 480 mm below A.

Now let's look at using your computer program's features to solve problems relating to planes.

SOLVING PLANES PROBLEMS USING A COMPUTER

You would want to view a plane as an edge if you wanted to measure the angle it made with another plane or to determine an intersection point. If any line in the plane is seen as a point view, the plane will be seen as an edge. All that is required is to find a point view of a convenient line in the plane. True shape is seen when a plane is viewed in a direction perpendicular to it. As with lines, both limiting cases are found by setting the viewpoint.

Showing an Edge View and Slope

Slope is seen only when the plane is seen as an edge view and the angle from the *x-y* plane is 0°. Figure 6.88 shows how slope is seen.

Measure the slope of plane ABC, defined by:

A	$x = 1$	$y = 1$	$z = 2$
B	$x = 6$	$y = 6$	$z = 6$
C	$x = 7$	$y = 2$	$z = 1$

The plane is shown in Figure 6.89.

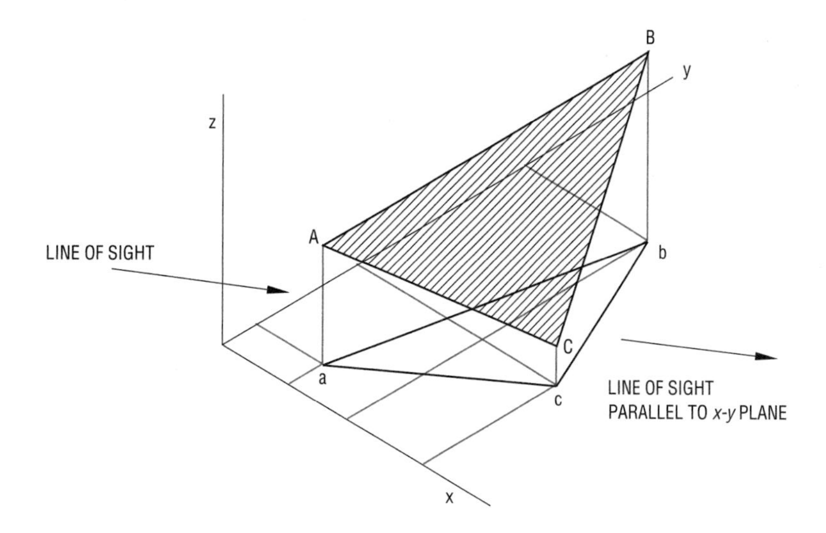

Figure 6.88 Line of sight to see the slope of a plane must be parallel to the *x-y* plane

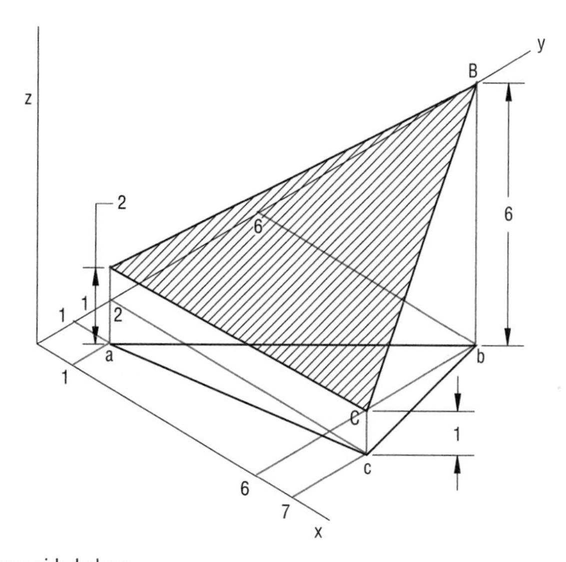

Figure 6.89 Three-sided plane

The 3-D commands in your computer program must be used to set the viewpoint (or rotate the object). There is no simple way of using coordinates to find the viewpoint as there is with lines. The procedure depends on which program you use and can only be described in general terms.

1. Decide how the plane must be viewed to show slope. Slope is the angle the plane makes with the horizontal plane so that slope can be seen only if the line of sight is horizontal. The angle from the *x-y* plane must be 0°, and the plane must be seen as an edge. There are only two viewpoints that satisfy these requirements.

2. Set the angle from the *x-y* plane to 0°, and move the viewpoint left or right until the plane appears as an edge. This is equivalent to rotating the plane about a vertical axis. There are two locations where the edge view appears: 17.25° and 197.25°. Accuracy is limited by how well you position the edge view by hand. Figure 6.90 shows these angles in the *x-y* plane.

3. Draw a horizontal line so that it passes through the plane and measure the angle the plane makes with this line. Figure 6.91 shows a horizontal line and the slope. How you measure the angle depends on the computer program you are using.

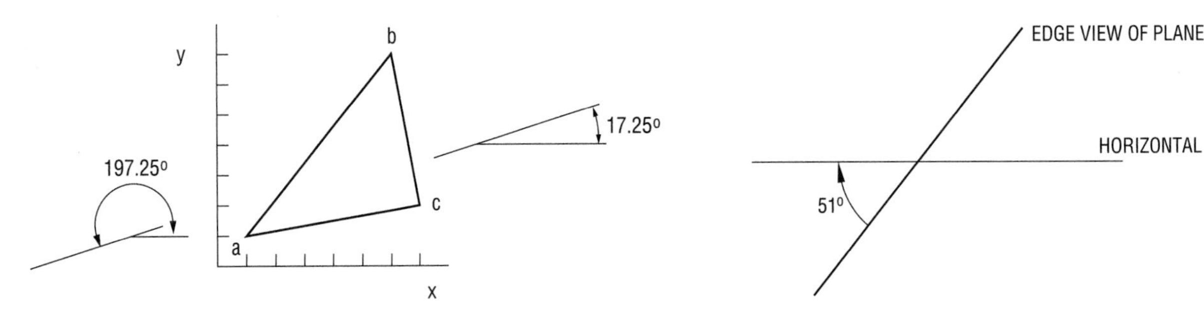

Figure 6.90 Angles in *x-y* plane to see an edge view

Figure 6.91 Slope of a plane is the angle it makes with the horizontal

Showing the True Shape of a Plane

A plane is seen in true shape when the line of sight is perpendicular to it. How this is done depends on the program you are using. AutoCAD, for example, allows you to align coordinates parallel to the plane, giving the equivalent of a plan view. You can then determine area and angles in the plane.

You must specify angles in, and from, the *x-y* plane to see true shape. The easiest way to find these angles is to first find the slope and the angle in the *x-y* plane necessary to see slope. The angle from the *x-y* plane is found from the slope. Figure 6.92 shows how the angle from the *x-y* plane is related to slope.

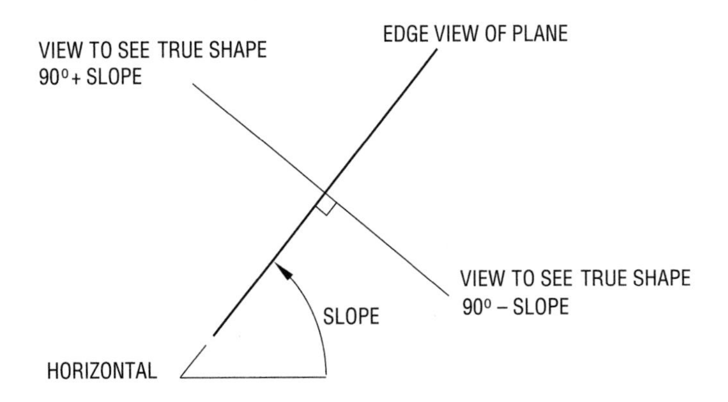

Figure 6.92 Angle from *x-y* plane to see true shape

The angle from the *x-y* plane is 90° plus slope or 90° minus slope depending on which side of the plane is seen.

You can find the angle in the *x-y* plane by knowing the viewpoint angle, α, necessary to see slope. Figure 6.93 shows how these angles are related.

The angle in the *x-y* plane is either $90° + \alpha$ or $90° - \alpha$ depending on which side of the plane is seen. Viewpoint angles to see the true shape of a plane are:

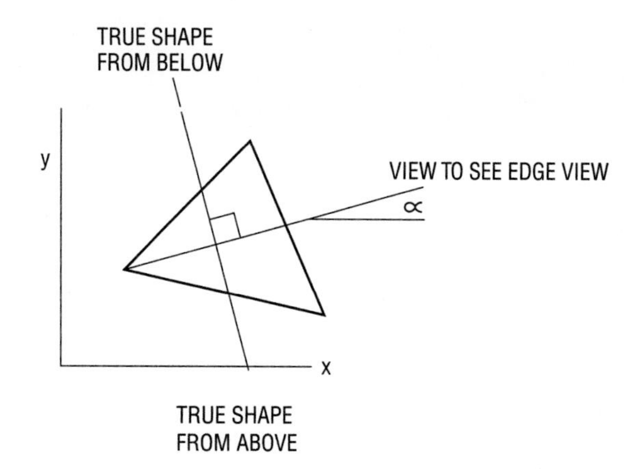

Figure 6.93 Angle in the x-y plane to see true shape

from above

in x-y plane	–(90° minus α)
from x-y plane	90° plus slope

from below

in x-y plane	90° plus α
from the x-y plane	–(90° minus slope)

Figure 6.94 shows the angles when viewed from above the plane.

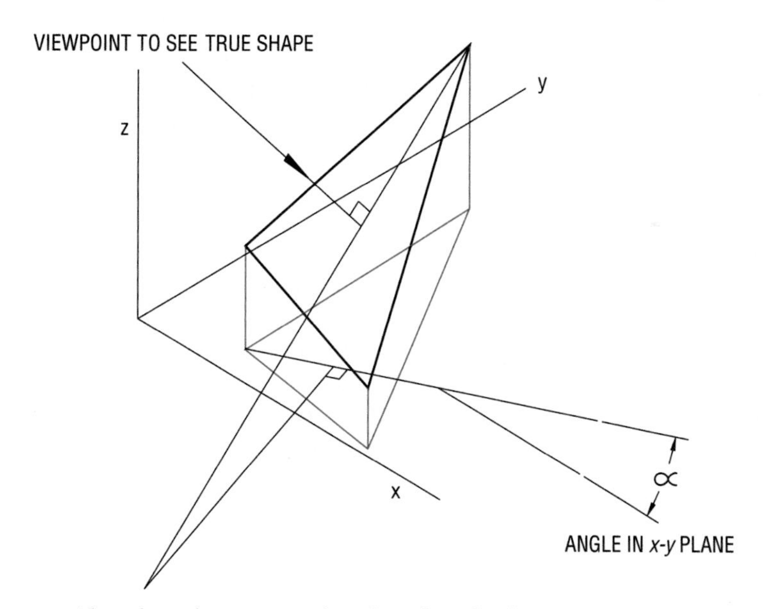

Figure 6.94 Viewpoint angles to see true shape from above the plane

The following problems can be solved with or without a computer.

Problems

40. What is the area of a plane, ABC, defined by A (1, 4, 7), B (7, 7, 4), and C (6, 4, 3)? The units are centimetres.

41. What is the slope of a plane with orientation defined by A (1, 4, 7), B (7, 7, 4), and C (4, 3, 2)? The units are centimetres.

42. Determine the shortest distance between a point A (3, 4, 0) and a plane defined by points B (1, 2, 7), C (1, 7, 5), and D (6, 4, 12). The shortest distance will be a perpendicular line from the plane through the point. The units are centimetres.

43. Determine the distance from the origin to a plane defined by A (3, 3, 10), B (5, 10, 8), and C (10, 5, 4). The units are centimetres.

44. Dimension the slope of the plane defined by A (4, 1, 8), B (1, 5, 4), and C (8, 9, 1). The units are centimetres.

45. Determine the area and slope of the plane defined by A (1, 9, 6), B (3, 2, 2), and C (7, 8, 4). The units are centimetres.

46. Dimension all angles in the plane defined by A (2, 2, 5), B (5, 9, 10), and C (10, 5, 2). The units are centimetres.

47. A plan view of a plane, ABC, and a line, XY, are shown in **Figure 6.95**. Locate the point at which the line intersects the plane. Specify the location of the intersection point with reference to the lowest point in the plane. All points are located from the lowest point, C.

A is 170 mm east, 350 mm south, and 250 mm above C.
B is 290 mm west, 200 mm south, and 80 mm above C.
X is 240 mm west, 120 mm south, and 220 mm above C.
Y is 350 mm south and 80 mm above C.

48. Referring to **Figure 6.96**, what angle does the sloping face make with the bottom surface of the block? The object is made from a cube with sides of 750 mm.

49. **Figure 6.97** shows an object with several plane surfaces. Determine the angle plane ABC makes with the horizontal.

50. Determine the area of the plane surface, A-to-H, shown in **Figure 6.98**. What is the slope of this surface?

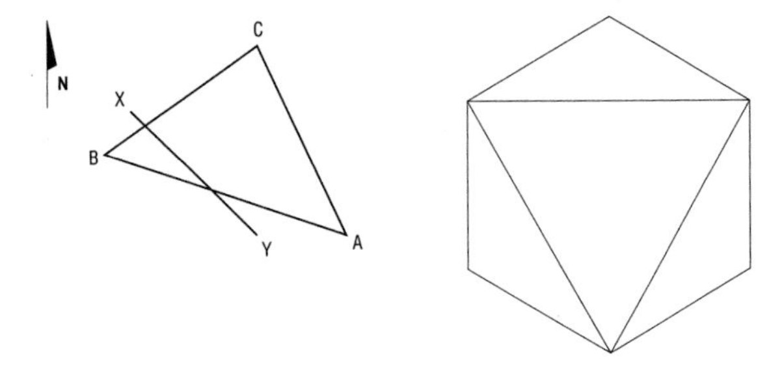

Figure 6.95

Figure 6.96

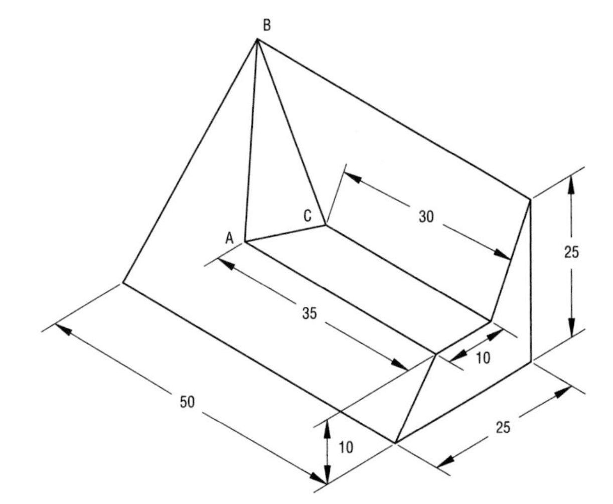

Figure 6.97

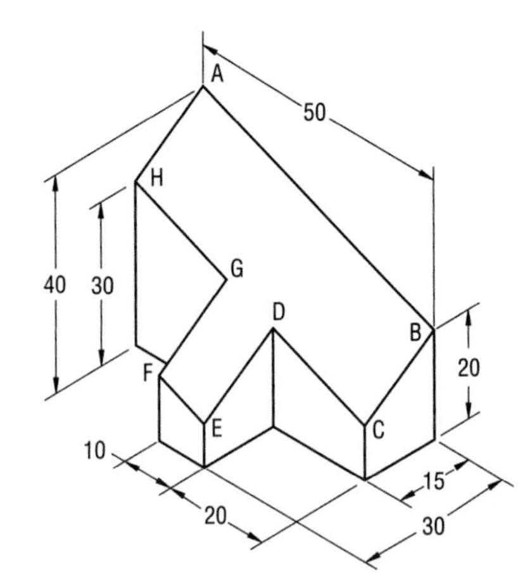

Figure 6.98

51. **Figure 6.99** shows a 25 × 25 × 50 mm block with one end cut at an angle. Determine the area and slope of surface ABCD.

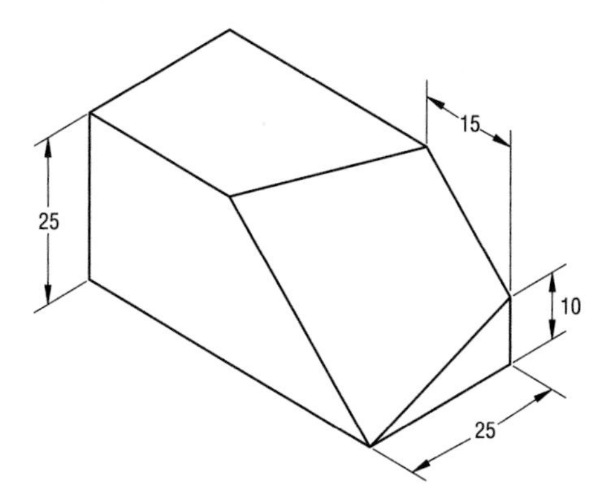

Figure 6.99

52. **Figure 6.100** shows plan and elevation views of a triangular steel frame, ABC. All points are in the same plane. A power line runs through point X on a bearing of S 45° W and slopes down at 23°. Determine the clearance between the line and the closest frame member. (The wire goes through the centre of the frame.)

B is 2500 mm east of A.

C is 5000 mm east and 3500 mm below A.

X is 5000 mm east 4000 mm north, of A.

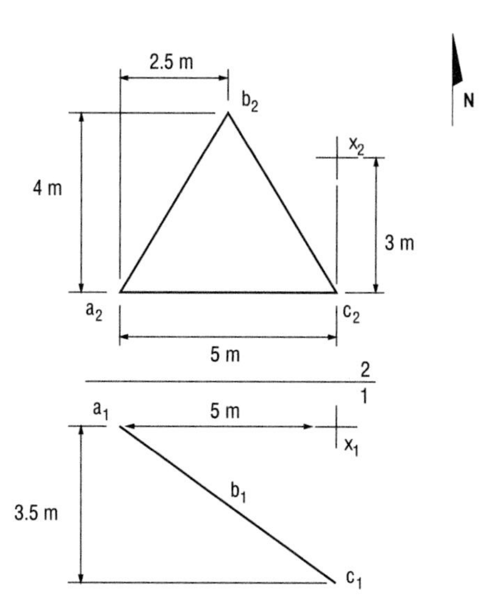

Figure 6.100

53. The centrelines of two pipes, AB and CD, are shown in **Figure 6.101**. What is the shortest distance between the two pipes?

B is 3300 mm east, 1400 mm north, 1000 mm higher than A.

C is 600 mm west, 2400 mm north, 1700 mm lower than A.

D is 3600 mm east, 400 mm north, and at the same elevation as A.

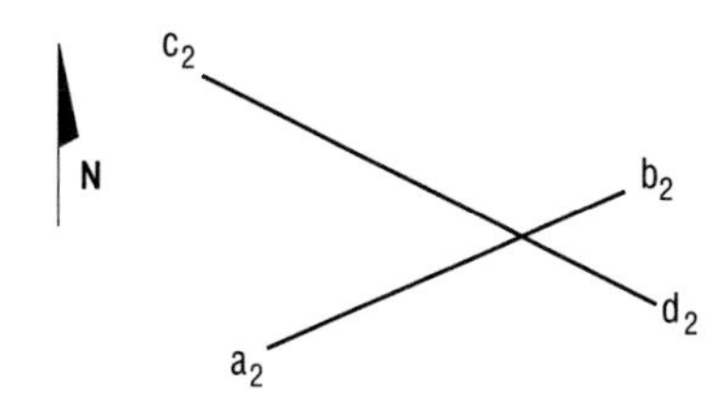

Figure 6.101

54. **Figure 6.102** shows an electrical wire near a vertical mast supported by guy wires (not shown). One of the guy wires is anchored to the ground at location X. The other end must be attached to the mast as close to the top as possible. There must be a minimum of 600 mm clearance between the guy wire and the electrical wire. At what height above the ground can the guy wire be attached to the mast under these conditions? The ground is level, so the base of the mast and the anchor (at X) are at the same elevation.

The anchor is 4000 mm east and 2000 mm south of the mast.

Point B (a point on the wire) is 5000 mm east, 1700 mm north of the mast, and 4100 mm above the ground.

The electrical wire bears S 35°W from B and slopes down (from B) at 22°.

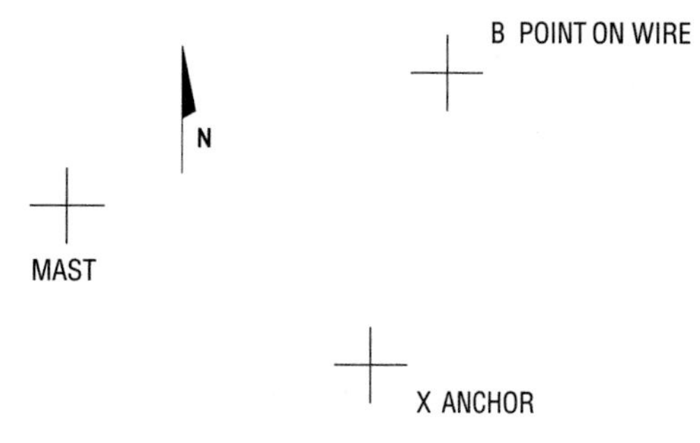

Figure 6.102

Problems

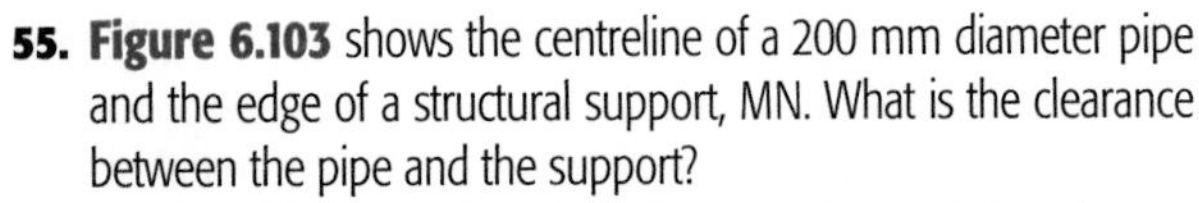

55. **Figure 6.103** shows the centreline of a 200 mm diameter pipe and the edge of a structural support, MN. What is the clearance between the pipe and the support?

    A and B are points on the pipe centreline and M and N are points on the edge of the structural support.
    B is 750 mm south, 300 mm east, and 150 mm above A.
    M is 120 mm south, 230 mm west, and 170 mm above A.
    N is 490 mm south, 150 mm east, and 220 mm above B.

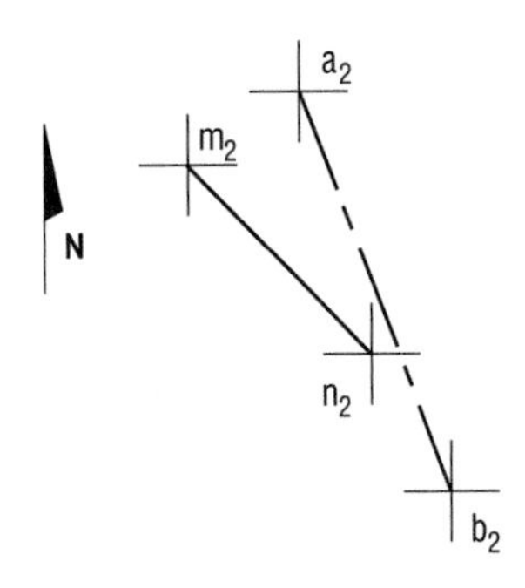

Figure 6.103

56. The drawing in **Figure 6.104** shows a connection between two rafters in a roof. (This type of roof is called a hip roof.) The sketch shows where these rafters are located in the roof. The rafter from the corner to the ridge is a hip rafter. The other is called a jack rafter. The end of the jack rafter must be cut so that it meets the hip rafter. Determine the angles on each face of the jack rafter so that there are no gaps at the connection. The dimensions of the rafters may be taken as 40 × 90 mm. The vertical face is 90 mm.

    The top of the roof is 2000 mm above the top of the wall. The end of the roof is 6 m, wide and the slope of the end panel is 1:2.

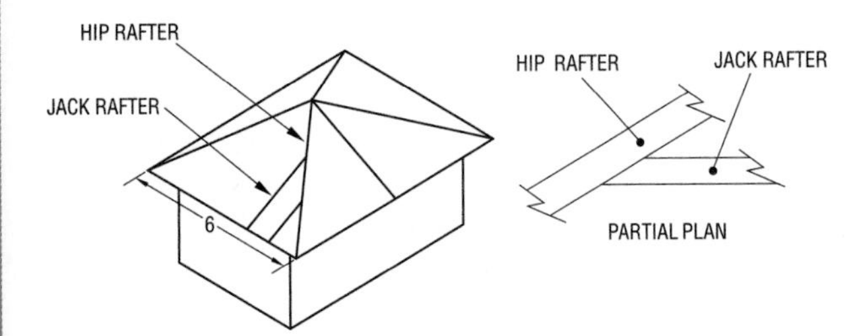

Figure 6.104

57. One of the components of an earthquake simulator is a vertical column made with an H-section. This must be braced from the floor with a hollow structural section welded to a steel plate on the floor. The brace has a square cross-section 75 × 75 mm. The brace will be welded to the column 2400 mm above the floor. The centre of the plate is 900 mm south of the attachment point on the column and 800 mm east. All dimensions are taken from the centreline of the brace. Determine the centreline length of the brace. **Figure 6.105** shows the layout.

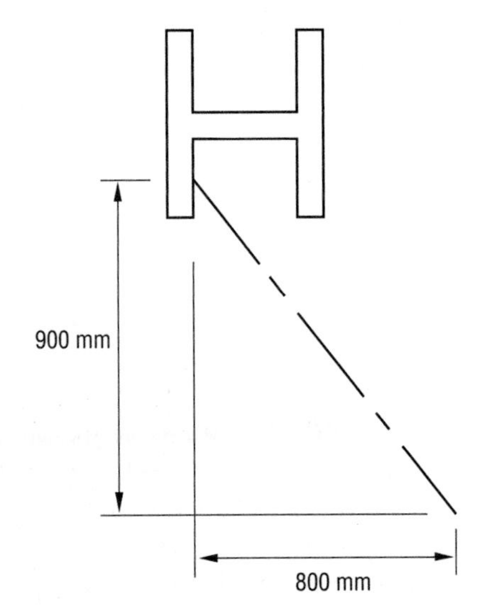

Figure 6.105

> *Every curved surface may be regarded as generated by the movement of a line, either constant in form while it changes its position, or variable at the same time both in form and in position space...*
>
> *Cylindrical surfaces may be generated in two principal ways, either by the movement of a straight line which always remains parallel to a given straight line, while it moves so as to pass through all the points of a given curve or by the movement of the curve which in the first method served as a guide for the straight line, in such a way that, one of its points moving along a given straight line, all the other points may describe lines parallel to this line. In both these methods the generating line, which in the first case was a straight line and in the second a curve, is constant in form and only changes its position.*
>
> *Gaspard Monge,* Geometrie descriptive, *1795*

CHAPTER 7

Intersection and Development

INTERSECTION

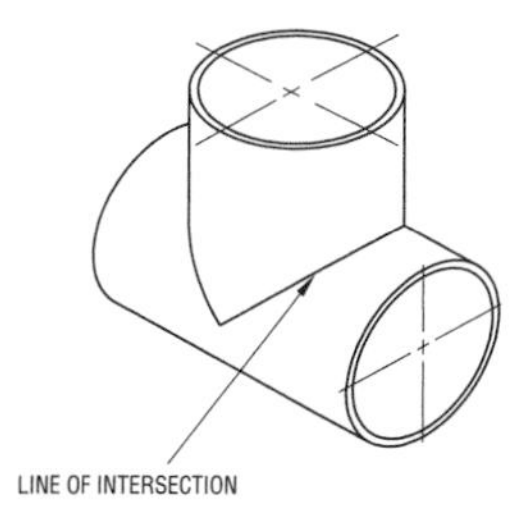

Figure 7.1 Intersection of two cylinders showing the line of intersection

Whenever two entities—lines, planes, or solids—meet, a point or line of **intersection** is formed.

When two lines meet, the intersection is a point; two planes meet on a straight line; and the intersection of curved surfaces produces a curved line.

Before pipes can be welded together to make a joint for a roof frame, for instance, the ends must be cut so that they can be joined. Before the ends can be cut, the line of intersection must be determined. When flat surfaces intersect, the angle created must be found. Figure 7.1 shows two pipes meeting at 90°. The end of the vertical pipe is cut so that it fits over the horizontal one.

A point or line of intersection is found using the same two-dimensional (2-D) methods that are used to find the true length and point view of a line. An intersection point can also be found using the three-dimensional (3-D) methods described previously. Computer methods will not be dealt with here.

Intersection of Two Lines

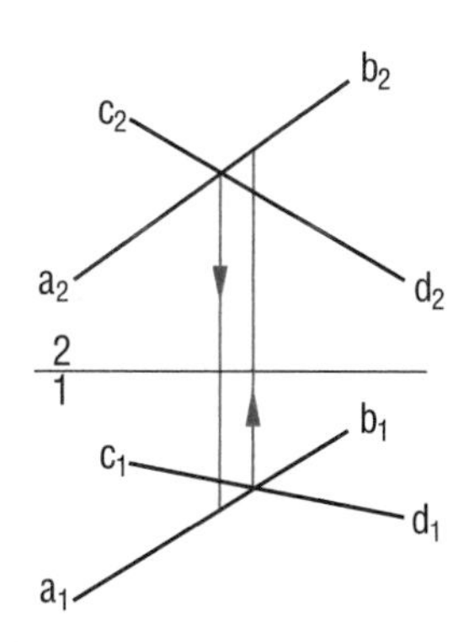

Figure 7.2 If lines do not intersect, the intersection point does not correspond in the two views

The basis for solving all intersection problems is the application of operations with a line. If you can find the point of intersection of two lines, you can find the line of intersection between two solids. Before you attempt to find a point of intersection, determine whether the lines in question do, in fact, intersect. Two views are needed to ensure that two lines do actually intersect. Figure 7.2 shows two lines, AB and CD, that appear to intersect in the plan view, but, in fact, they do not.

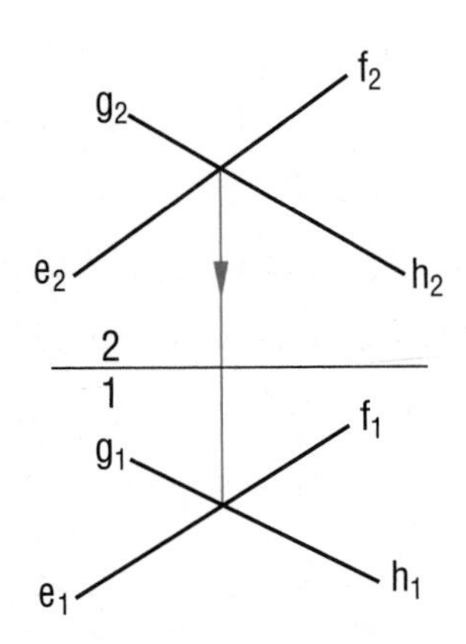

Figure 7.3 When lines intersect, the intersection point corresponds in all views

If these lines actually intersect, the intersection point, X, will be at the same point on the line in both views. There can be only one intersection between two straight lines. Projecting point X to the front view shows that it does not correspond to the apparent intersection point in the front view, and so lines AB and CD do not intersect. Lines EF and GH in Figure 7.3 *do* intersect; the intersection point is at the same point on each line in both views.

Intersection of a Line and a Plane

To find where a line intersects a plane, only an edge view of the plane is needed. The intersection can be found using a point view of the line, but more views are required, and the intersection point cannot be projected to the line without extra construction.

Example 7.1

Figure 7.4a shows a line that appears to pass through a plane. Find the point at which XY intersects ABC.

1. Draw an edge view of ABC. (**Figure 7.4b**)
2. Project line XY to the auxiliary view. (**Figure 7.4c**)
3. Project the point of intersection from the auxiliary view to the plan and front views. (**Figure 7.4d**)

Projection lines are shown only in the view in which they are used.

If the location of the intersection is to be dimensioned on the plane, the true shape of the plane must be found.

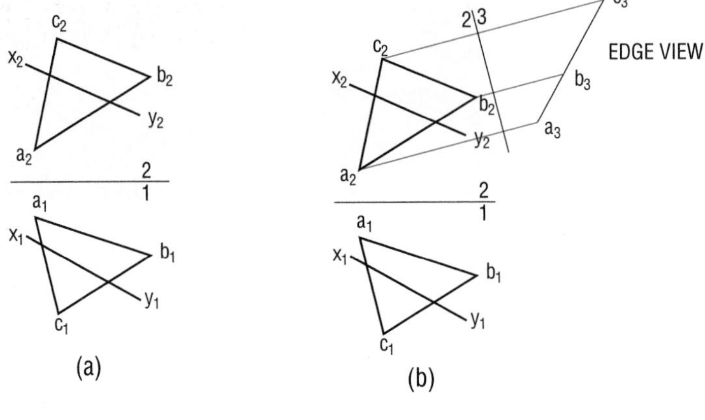

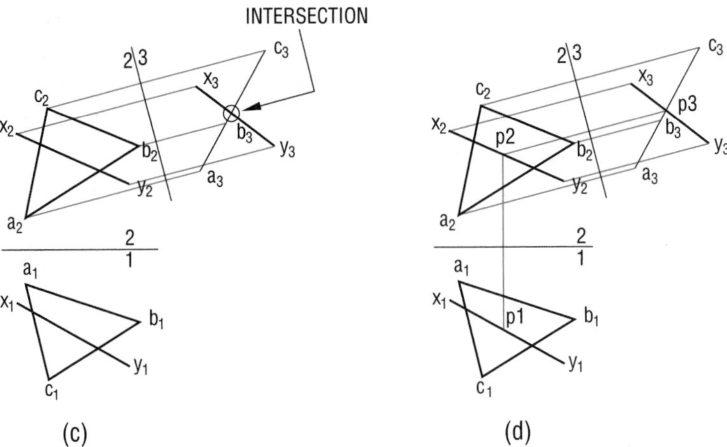

Figure 7.4 The intersection of a line and a plane shows where the plane appears as an edge

Intersection of Plane Surfaces

The line of intersection between two plane surfaces is a straight line.

Example 7.2

Figure 7.5 shows two planes and a line of intersection (BD). The angle between the two planes is called the **dihedral angle**.

The line of intersection must be found before the dihedral angle can be measured.

The line of intersection between two planes is in both planes. It can be found by viewing one plane in edge view. Two points on the line of intersection can be seen where this edge passes through the other plane.

1. Draw an auxiliary view showing the edge view of one plane (ABC is used here) and project the other plane (EFG) to this auxiliary (**Figure 7.6a**).
2. Two points, circled, on the line of intersection are located where the edge view of ABC intersects the sides of EFG. Project these points to the plan and front views (**Figure 7.6b**).

Projection lines are shown only in the view in which they are used.

Some parts of each plane will not be visible. However, we are not concerned with visibility; hidden lines are not used for the hidden parts of the planes.

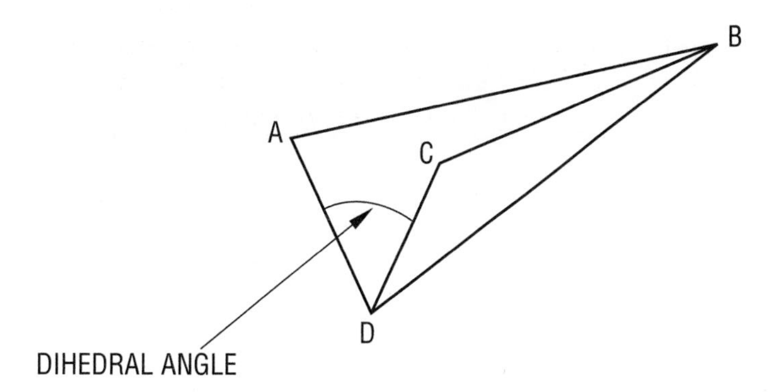

Figure 7.5 The dihedral angle is the angle between two planes

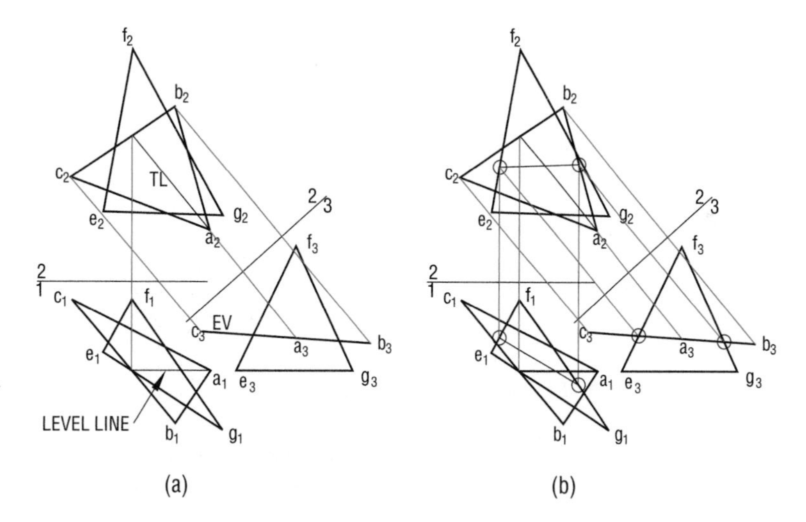

Figure 7.6 The line of intersection between planes is seen when one plane appears as an edge

Measuring Dihedral Angles

The angle between two planes can be measured when both planes are seen in edge view. We have observed that a plane is seen as an edge if any line in the plane is seen as a point. If the line chosen for point view is on both planes, as the line of intersection is, both will be seen in edge view, and the dihedral angle can be measured. The view required to measure the angle between two planes is a point view of the line of intersection.

Example 7.3

Figure 7.7 shows a roof that must be modified by adding the section containing the plane ABC. A steel support structure must be designed to support the addition, and the angle between panels ABC and ACDE must be known in order to design the support structure. Determine the angle between the existing roof, represented by ACDE, and the new panel, ABC.

The dihedral angle can be seen and measured when the line of intersection is seen in point view. The problem becomes one of finding a point view of line AC. A point view of any line can be found using the methods described in Chapter 6.

1. Find the true length of the line of intersection, AC. The true length must be found before a point view can be obtained.

   Figure 7.8 is an auxiliary view showing the true length of the line of intersection. Only one half of the roof is shown.

2. Find the point view of the line of intersection. Since a plane appears in edge view, if any line in it is seen as a point, both planes will be seen in edge view in this auxiliary view because the line of intersection is in both planes. **Figure 7.9** shows the point view of the line of intersection and edge views of both planes. The dihedral angle can now be measured.

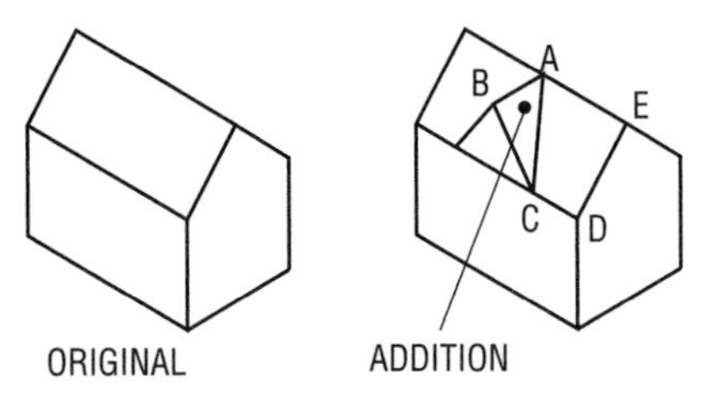

Figure 7.7 The angle between two roof panels must be found

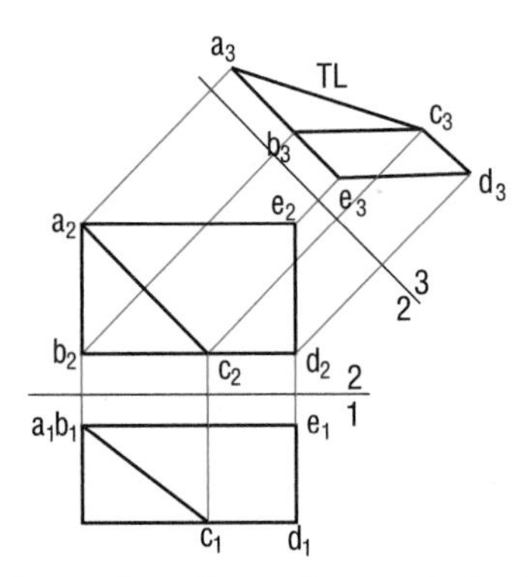

Figure 7.8 The true length of the line of intersection must be found before a point view can be found

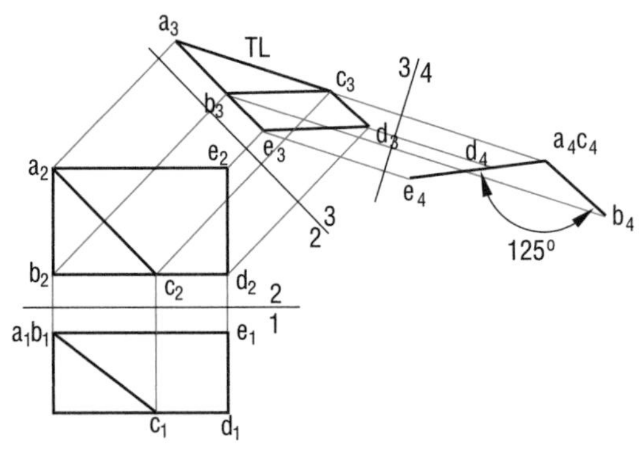

Figure 7.9 Dihedral angle can be measured when both planes appear as edge views

Here are some things you need to know when solving problems on the intersection of lines and planes.

- If two lines intersect, the intersection point corresponds in all views (see Figure 7.3).
- The intersection point between a line and a plane is seen where the plane appears as an edge (see Figure 7.4 on page 244).
- The line of intersection between two plane surfaces is seen where one plane is seen as an edge (see Figure 7.6 on page 245).
- The angle between two planes (dihedral angle) can be measured where both planes appear in edge view. Both planes will appear in the edge view where the line of intersection appears in point view (see Figure 7.9 on page 246).

Now that you know a little more about intersections of lines and planes and how to measure dihedral angles, try the following problems.

Problems

1. A line starting at A bears N 37° E and slopes upward at 27°. Another line starts at C, 800 mm west and 900 mm north of A, bears S 73° E, and slopes up at 13°. A and C are at the same elevation. Do these lines intersect?
2. Specify the bearing and slope of a line, starting at D, 1 m east and 2 m south, and 1.5 m above A, so that it will intersect plane ABC at the centroid (**Figure 7.10**). A **centroid** is like the centre of mass of a 2-D shape. The centroid of a triangle is at the intersection of the lines joining the vertices to the midpoints of the opposite sides. Scale is 1:100.
   B is 2.5 m north, 2.5 m east, and 1.5 m below A.
   C is 1.7 m south, 4.5 m east, and 2 m above A.
   D is 2.5 m north, 1 m east, and 1.5 m above A.
3. Refer to **Figure 7.11**. Line AB bears N 40° W and slopes down from A at 20°. Show the point at which this line intersects plane DEF in all views. Scale is 1:100.
   E is 2 m north, 3 m east, and 1.5 m above D.
   F is 2.3 m south, 2.3 m east, and 800 mm below D.
   A is 1 m south, 4 m east, and at the same elevation as E.
4. Refer to **Figure 7.12**. Determine the point on line XY where it intersects plane ABC. Scale is 1:100.
   B is 1.5 m north, 2.2 m east, and 2 m below A.
   C is 2.8 m north, 2 m west, and 1.7 m below A.
   X is 3.2 m north, 1 m west, and 200 mm below A.
   Y is 800 mm north, 1 m west, and 1.2 m below A.
5. A plan view of a plane, ABC, and an intersection line, XY, are shown in **Figure 7.13**. Locate the point at which the line intersects the plane. Specify the location of the intersection point with reference to the lowest point in the plane. Scale is 1:200.
   B is 9 m east, 9 m south, and 6 m below A.
   C is 3 m east, 6.4 m south, and 5 m below A.
   X is 8 mm east, 7 m south, and at the same elevation as A.
   Y is 4.4 m east, 2 m south, and 7 m below A.

   This defines the orientation of the plane but not the extent.

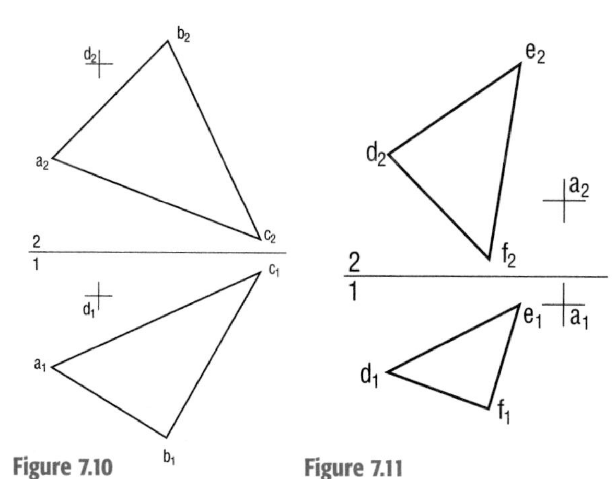

Figure 7.10

Figure 7.11

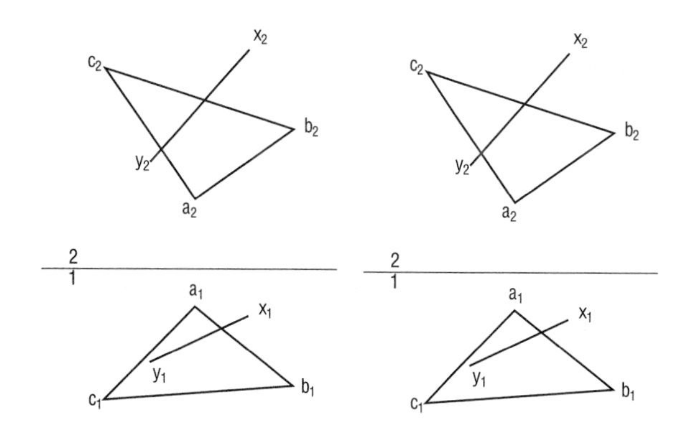

Figure 7.12

Figure 7.13

6. **Figure 7.14** shows two intersecting planes. What is the angle between the planes? Scale is 1:100.
   - A is 4 m north, 1 m west, and 2 m below C.
   - B is 2 m east, 1 m north, and at the same elevation as C.
   - D is 2.5 m west, 2 m north, and at the same elevation as C.

7. Details of the planes shown in **Figure 7.15** are given below. Determine the bearing and slope of the line of intersection. Scale is 1:100.
   - B is 4.7 m east, 500 mm south, and 3 m below A.
   - C is 2.3 m east, 2.6 m north, and 3.5 m below A.
   - D is 4 m east, 2.6 m north, and at the same elevation as A.
   - E is 6 m east, 1.6 m north, and 2 m below A.
   - F is 1.7 m east, 1 m south, and 4 m below A.

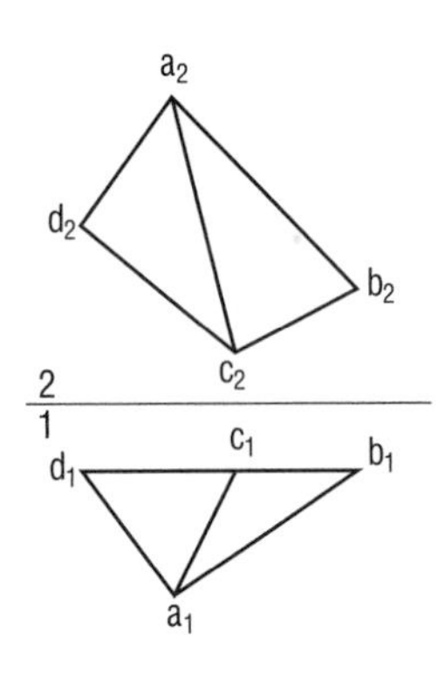

Figure 7.14

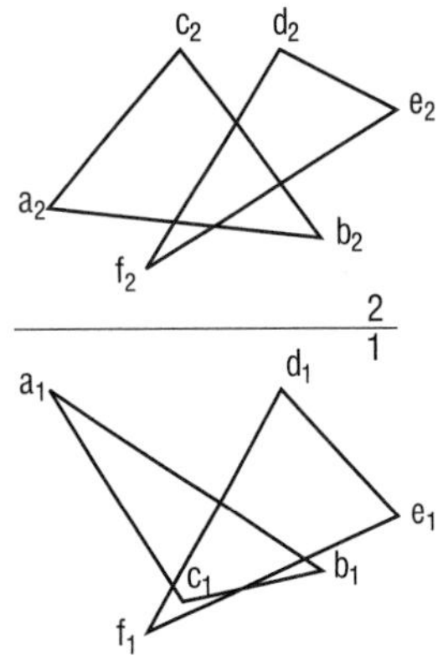

Figure 7.15

8. Determine the dihedral angle between planes ABD and DBC illustrated in **Figure 7.16**. Scale is 1:200.
   - B is 3.4 m north, 3 m east, and at the same elevation as A.
   - C is 4 m north, 10 m east, and at the same elevation as A.
   - D is 2.4 m south, 8 m east, and 4 m below A.

9. Two planes, ABC and DEF, are shown in **Figure 7.17**. These planes intersect, but the line of intersection is not shown. Determine the angle between the two planes. Scale is 1:5000.
   - B is 290 mm east, 40 mm north, and 70 mm above A.
   - C is 130 mm east, 130 mm north, and 100 mm below A.
   - D is 70 mm east, 100 mm north, and 100 mm above A.
   - E is 310 mm east, 130 mm north, and 100 mm below A.
   - F is 30 mm east, 50 mm south, and 70 mm below A.

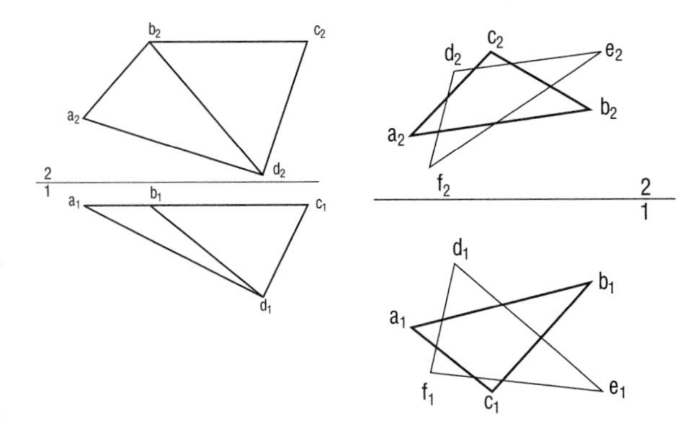

Figure 7.16

Figure 7.17

10. **Figure 7.18** shows two intersecting planes. What is the angle between the planes? Points B, C, and D are measured from A. Scale is 1:5.
    - C is 250 mm east, 100 mm south, 150 mm above A.
    - B is 50 mm east, 100 mm south, and 100 mm above A.
    - D is 50 mm west, 70 mm south, and 200 mm above A.

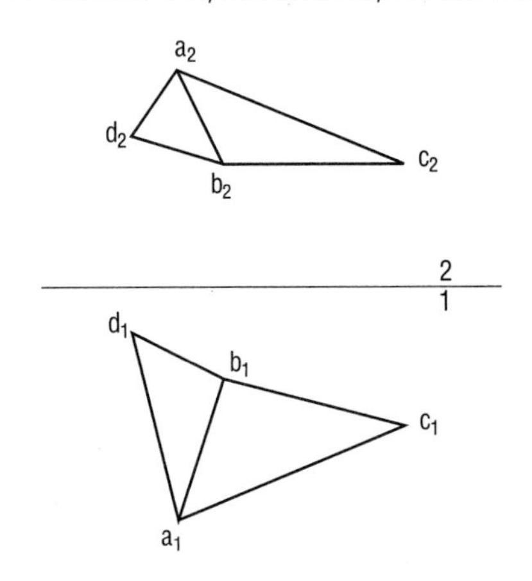

Figure 7.18

Now let's move on to look at the intersection of a plane and a solid.

Intersection of a Plane and a Solid

Conic sections are examples of the intersection of a plane and a solid. The line formed on the surface where the plane cuts through the cone is the line of intersection. It is important to remember that this line is on the surface of the cone. The shape of the resulting line depends on how the plane passes through the cone. A horizontal plane, parallel to the base, results in a circle; a vertical plane through the vertex results in a triangle; and a vertical plane that does not pass through the vertex results in a hyperbola. These are illustrated in Figure 7.19.

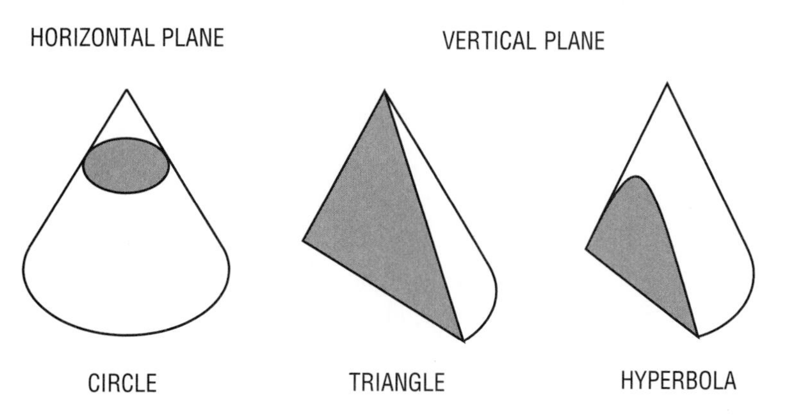

Figure 7.19 Conic sections are the shapes formed by planes passing through a cone

If a solid is intersected by a plane, some shape will be formed, and there will be a line of intersection on the surface of the solid. This line is in the plane and on the surface. The shape depends on how the plane passes through the solid.

The line of intersection formed when a plane intersects a rectangular prism is a straight line on each face of the prism. Figure 7.20 shows a plane intersecting a rectangular prism. The line of intersection forms a rectangle. An irregular shape with straight sides results when a plane passes through a rectangular prism at an arbitrary angle.

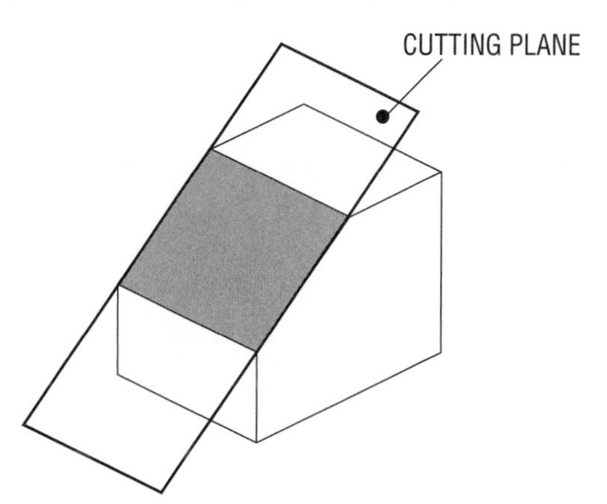

Figure 7.20 Intersection of a plane and a rectangular solid

When plane surfaces intersect, it is necessary to find only two points on the line of intersection. These points are usually the end points on each side. When a plane passes through a curved surface, more points must be found. Figure 7.21 shows a pipe passing through a wall (an example of a cylinder intersecting a plane). The shape enclosed by the line of intersection depends on the angle at which the pipe meets the

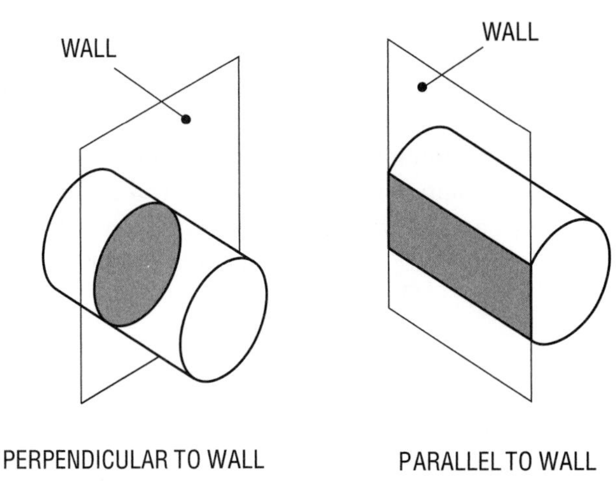

Figure 7.21 The shape formed by the intersection of a plane and a cylinder depends on the orientation of the plane

wall. If the pipe meets the wall at a right angle, it will be a circle; if it meets at an angle, it will be an ellipse; and if the plane is parallel to the axis of the cylinder, a rectangle is formed.

Finding a Line of Intersection

The following are practical examples of instances when a line of intersection must be found:

- If a hole is to be made in a wall, the size and shape must be known. A circle is an easier shape to cut on a wall than an ellipse (particularly if the wall is concrete), but it is not always possible to meet the wall at 90°.
- The angle between two panels on a roof must be found so that supports can be designed.
- If two pipes are joined by welding, the ends of one or both (depending on how they are joined) must be cut so that they meet correctly (assuming they are not joined end-to-end).

If a line on one surface intersects a line on another surface, the lines must intersect on the line of intersection between the two surfaces. A line of intersection between two solids can be determined by finding the intersection of a line on one surface with a corresponding line on the other surface. The problem becomes one of finding the intersection of two lines.

Lines can be created on the surface of a solid by passing cutting planes through it. Figure 7.22 shows a cylinder cut by a plane parallel to the axis. The plane intersects the top and underside of the cylinder and creates two lines on the surface. These lines appear as point views when viewed from the end of the cylinder. Any number of lines can be created on the surface by passing more cutting planes through it. The reason for using cutting planes to create lines on the surface will become apparent when dealing with the intersection of curved surfaces.

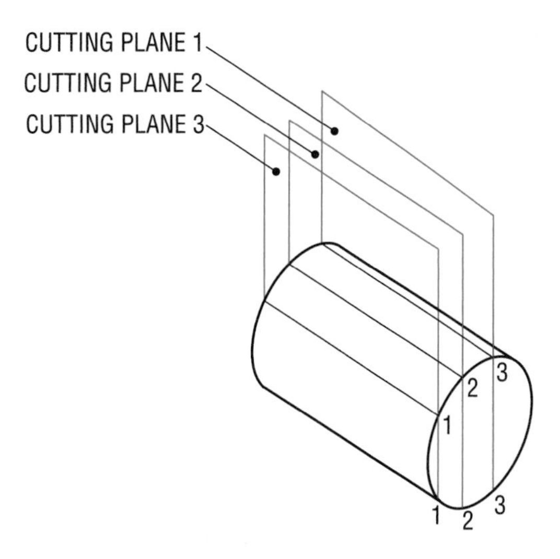

Figure 7.22 Creating lines on a surface using cutting planes parallel to the axis

Intersection of a Cylinder and a Plane

Example 7.4

Figure 7.23 shows a large duct passing through a wall. The line of intersection must be found so the hole can be cut. The hole will be an ellipse, since the duct meets the wall at an angle.

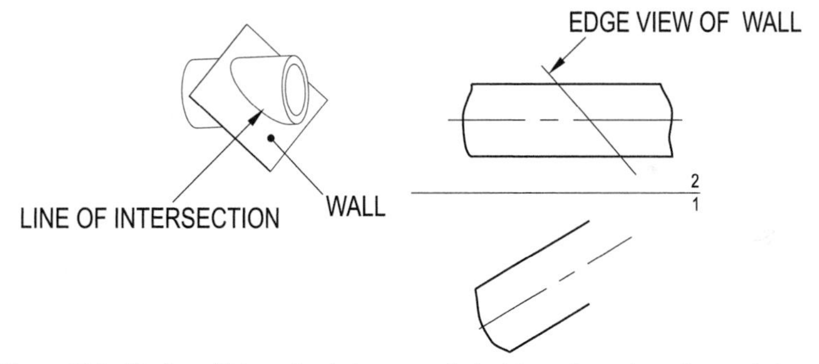

Figure 7.23 The line of intersection between a cylinder intersecting a plane at an angle is an ellipse

1. Create lines on the cylinder surface using cutting planes parallel to the cylinder axis. These planes create lines on the top and bottom surfaces, and these lines intersect the wall. In this instance, it does not matter how the cutting planes are spaced. **Figure 7.24** shows the duct with three cutting planes, CP1, CP2, and CP3, parallel to the axis. The points at which these lines intersect the wall are identified as 1, 2, and 3. These points will be projected to the front view, but before this can be done, there must be lines in the front view to project them to.

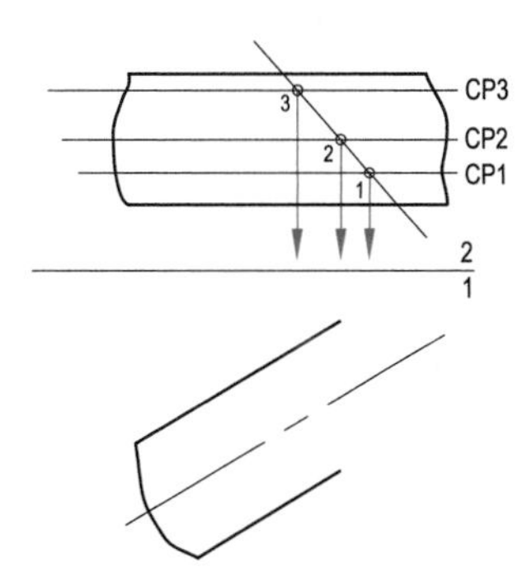

Figure 7.24 Lines on the top of the cylinder intersect the wall at a points 1, 2, and 3

2. Draw an auxiliary view looking along the axis of the duct. This will enable you to locate the lines around the circumference. **Figure 7.25** shows a point view of the duct axis and the duct (which appears as a circle). The auxiliary view is drawn following the procedure for finding the point view of a line outlined in Chapter 6.
3. Locate the cutting planes in the auxiliary view. They are located using distances from the folding line taken from the top view. The surface lines are seen as point views in the auxiliary view. (**Figure 7.26**). The numbers correspond to the cutting planes used to create the lines.
4. Project surface lines from the auxiliary view to the front view. **Figure 7.27** shows these lines in the front view.

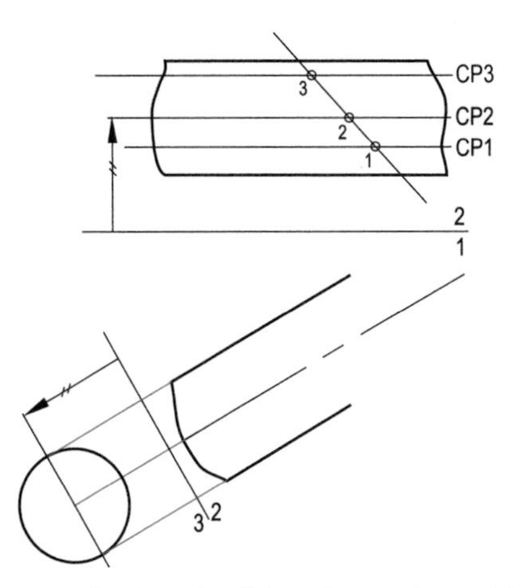

Figure 7.25 A point view of the duct axis will show the true shape of the cylinder

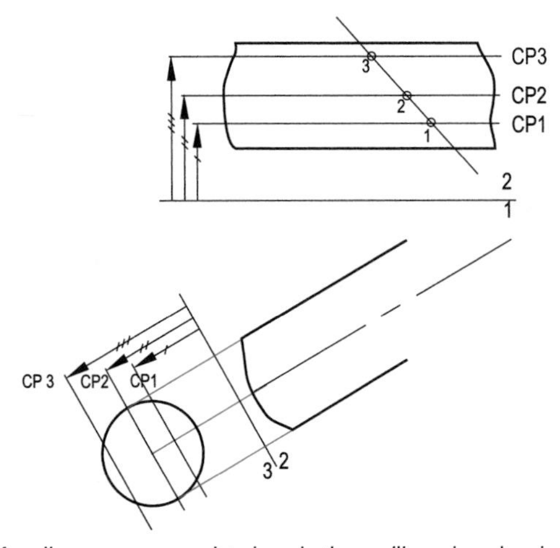

Figure 7.26 Surface lines appear as point views in the auxiliary view showing the axis as a point

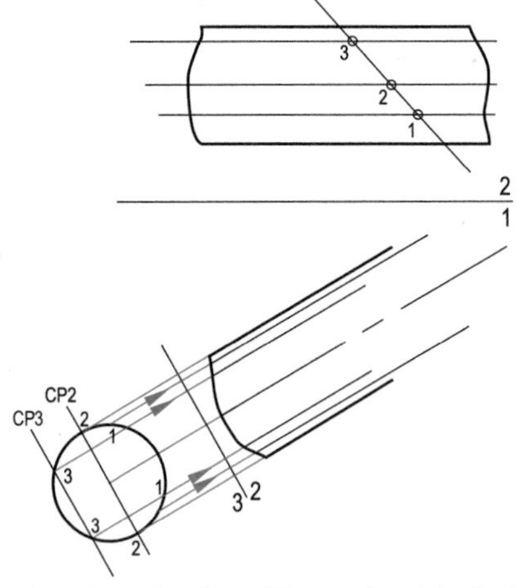

Figure 7.27 A point view of a surface line will be true length in the front view

Example 7.4

5. Project the points at which the surface lines intersect the wall to the corresponding line in the front view. **Figure 7.28** shows these points on the line of intersection. It is obvious that more points are required to define the line of intersection.

6. Create more surface lines around the circumference of the duct, and find the intersection points in the front view using the same procedure. Any number of surface lines can be created by using more planes. **Figure 7.29** shows the shape of the hole in the wall as it appears in the front view. This is not the true shape of the hole.

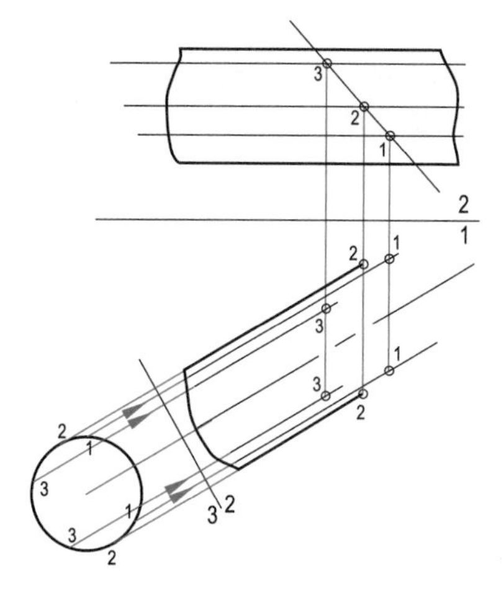

Figure 7.28 Intersection points projected to the corresponding line in the front view

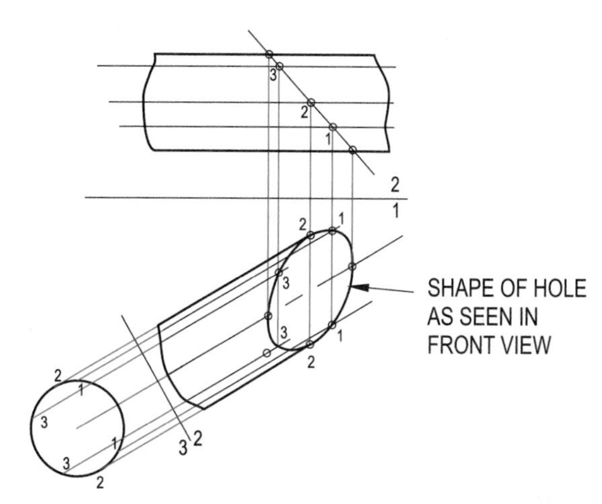

Figure 7.29 The shape of hole as it appears in the front view

Example 7.4

Here are some things you need to know when solving problems on the intersection of planes and solids.

- The line of intersection between a plane and a solid must be on the surface of the solid. See Figures 7.19, 7.20, and 7.21 (pages 249 and 250).
- Points on the line of intersection between a plane and a solid are found by determining where a lines on the surface intersects the plane. The intersection points can then be transferred to corresponding lines in other views. See Figures 7.23 to 7.29 (pages 251 to 253).

Now that you know a little more about intersections of planes and solids, try the following problems.

Problems

11. Find the area of the rectangle formed when a 100 mm diameter cylinder is cut by a plane, parallel to the cylinder axis and half-way from the axis to the outside (at half the radius). Find the area per unit length. Scale is 1:2.

12. The grain bin shown in **Figure 7.30** must be modified to allow all grain to flow out of the bin. Grain remaining in the bin attracts mice. By experiment it was determined that if the line of intersection between the end panel and the side panel(s) is 35°, grain will not "hang up" in the corner. (The line of intersection slopes at 35°, not the panel.) Determine the dihedral angle between the end panel and the side panel. The angle is the same on both sides.

The top of the bin is 1800 mm wide and 2500 mm long. The hole in the bottom is centred in the bin and is 600 × 600 mm. The top is 1400 mm above the bottom. Scale is 1:50.

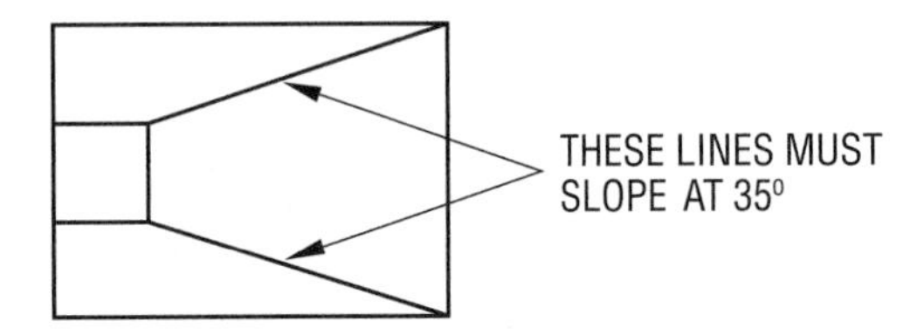

Figure 7.30

13. The roof of the building shown in **Figure 7.31** is made with six identical panels. The angle between the panels (dihedral angle) must be known in order to design the roof. The angle is the same for all panels.

Point A is used as a reference. Since the shape is a hexagon, the included angle between sides is 120°. Size is given knowing the diameter of an enclosing circle. Scale is 1:200.

The hexagonal roof is contained in a circle with a diameter of 14 m.

Point B is 5 m west and 3200 mm higher than A.
Points D and E are at the same elevation as A.
Points C and F are at the same elevation as B.

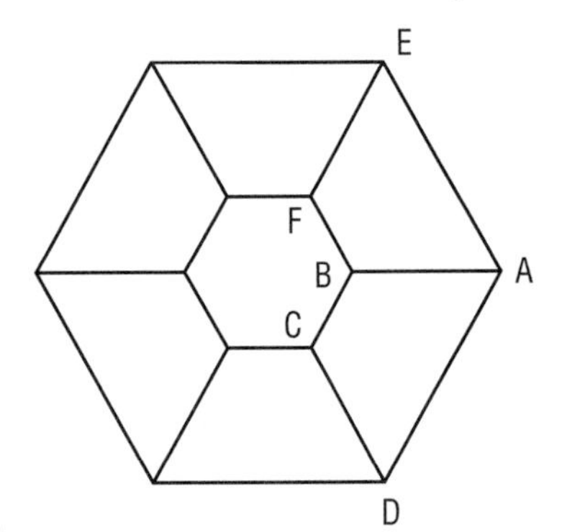

Figure 7.31

14. A vent line must go through a floor and then through a vertical wall. The hole in the floor is to be 400 mm diameter, and the line will pass through it at an angle. A hole must be cut in the wall to accommodate the cylinder. An oblique cylinder will be used because a right cylinder will not pass through a circular hole at an angle. An oblique cylinder has a circular cross-section parallel to the base but an elliptical shape when viewed along the axis. (An oblique cylinder is shown in Figure 1.9 on page 5.) Plan and elevation views are shown in **Figure 7.32**. Draw the shape of the hole required in the wall. The hole will be seen in true shape in the front view. (Find the point at which a line on the surface of the cylinder intersects the wall. Project the intersection point to the corresponding line in the front view.) Scale is 1:10.

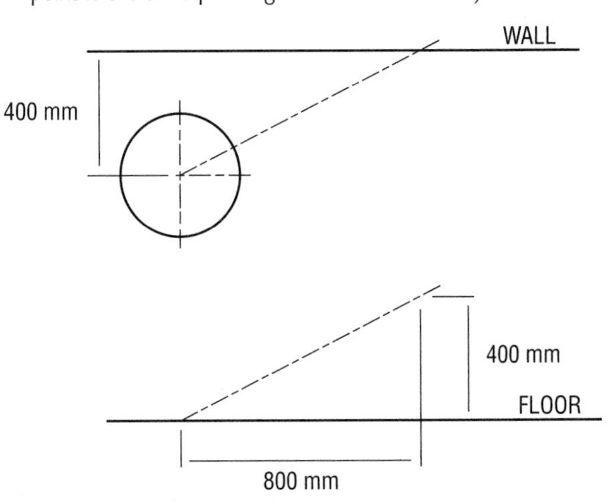

Figure 7.32

15. Find the true shape of the ellipse formed when a cone is cut by a plane passing through it at 30° to the horizontal. The diameter of the base of the cone is 100 mm, and the height is 100 mm. The plane passes through the axis of the cone half way between the base and the vertex. Scale is 1:2.

16. A plane passes through a cone such that it intersects the midpoint on the axis and the base at a distance of 25 mm from the axis (half the radius). Draw the true shape of the parabola formed. The cone is 100 mm high, and the base diameter is 100 mm. Scale is 1:2.

Now let's move on to look at the intersection of curved surfaces.

Intersection of Curved Surfaces

Lines created on both surfaces for the purpose of finding a line of intersection cannot be drawn randomly. You must be sure that they actually intersect. If two lines are in the same plane, they must intersect unless they are parallel. As you may recall, you use cutting planes to create surface lines to ensure that a line on one surface corresponds to a line on the other surface.

Figure 7.33 shows two intersecting cylinders. A cutting plane is used to create surface lines on both cylinders. Since the lines on the surface of both cylinders are in the same plane, we can be sure they intersect.

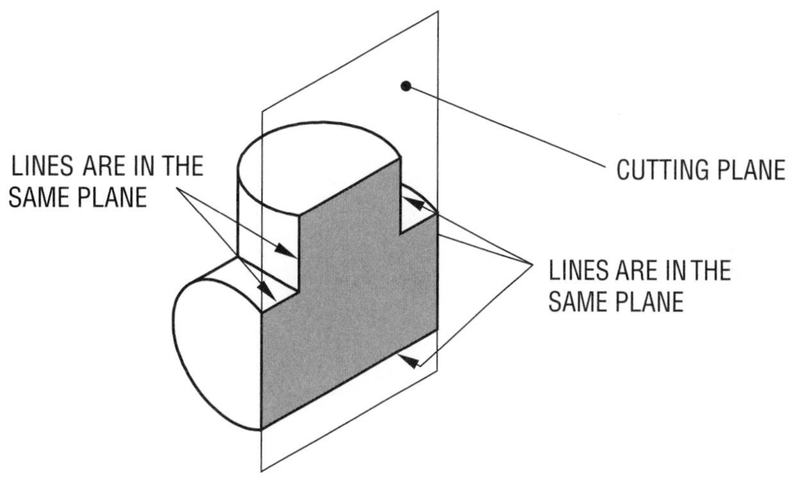

Figure 7.33 A vertical cutting plane is used to create intersecting lines on two cylinders

Example 7.5

Figure 7.34 shows two pipes that are to be welded together at 90°. The end of the vertical pipe (A) must be cut so it can be joined to the horizontal pipe (B). Determine the line of intersection on the vertical pipe.

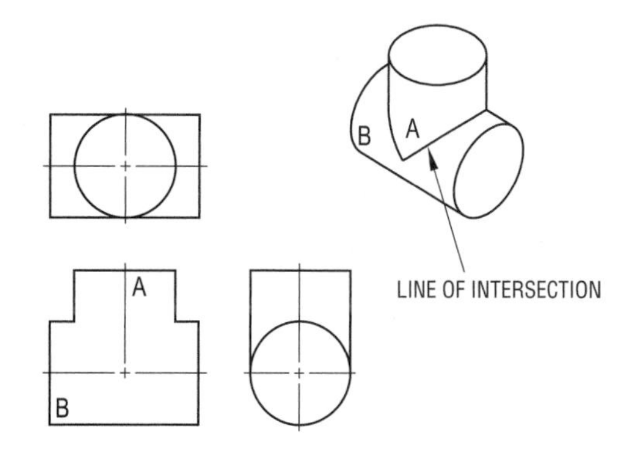

Figure 7.34 The end of the vertical pipe must be cut to fit exactly over the horizontal pipe

1. Create intersecting lines on the surface of each pipe by passing vertical planes through both pipes. The cutting planes used to create surface lines are seen as edge views in the plan view. **Figure 7.35** shows the cutting planes in the plan view. Only one cutting plane is shown on the isometric.
2. Project vertical lines on the surface of pipe A from the plan view to the front view (**Figure 7.36**). These lines are seen as point views in the plan view so that they will be true length in the front view. Now find corresponding lines on pipe B.
3. Use the distance from the folding line in the plan view to locate cutting planes in the right-side view. Surface lines on pipe B are seen as point views in this view (**Figure 7.37**). Surface lines are only required on the top of the horizontal pipe.

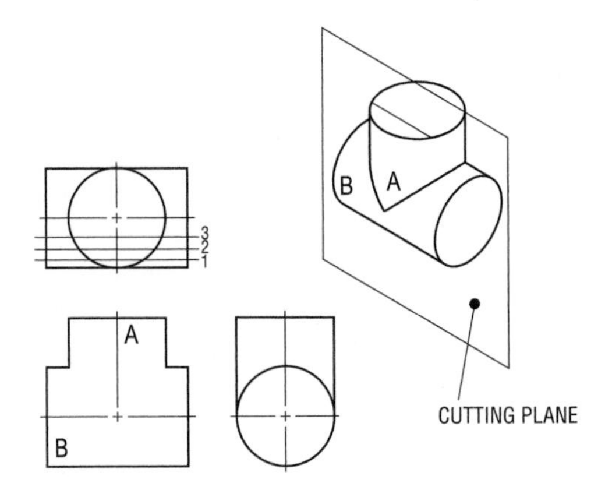

Figure 7.35 Vertical cutting planes will create corresponding lines on both cylinders

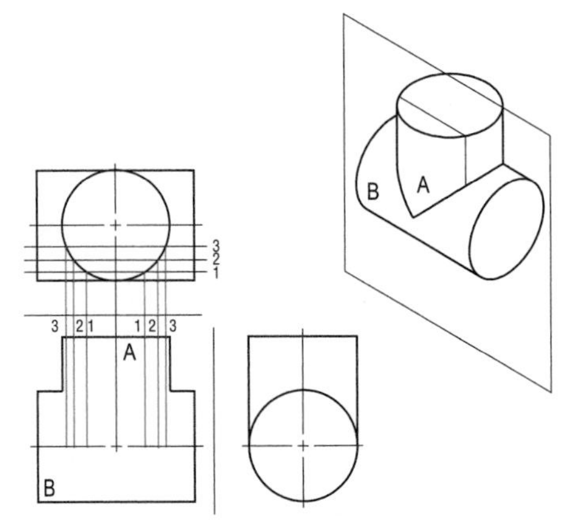

Figure 7.36 Surface lines on vertical cylinder projected from the plan view

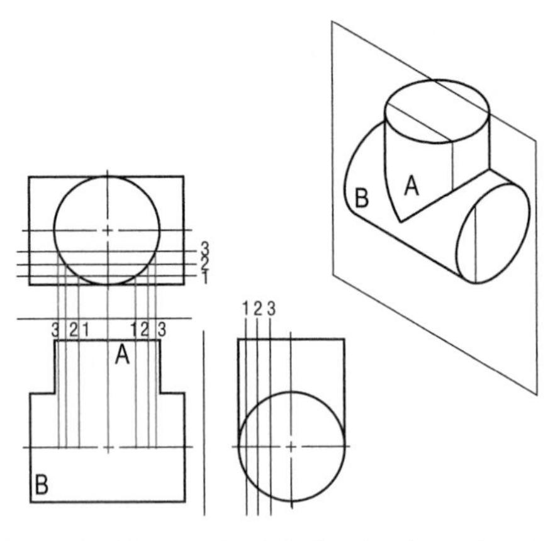

Figure 7.37 Locating surface lines on pipe B by locating the cutting planes on pipe B

Example 7.5

4. Project the surface lines from the right-side view to the front view (**Figure 7.38**).
5. Identify the intersection points of corresponding lines in the front view. **Figure 7.39** shows the intersection points.
6. Join the intersection points with a smooth curve. **Figure 7.40** shows the line of intersection.

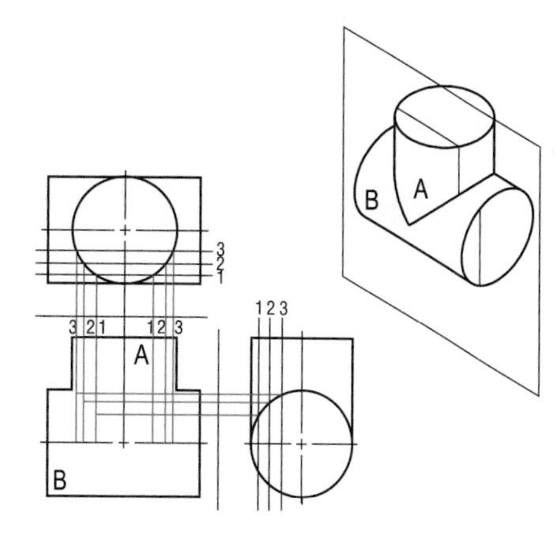

Figure 7.38 Locating surface lines in the front view by projecting from the side view

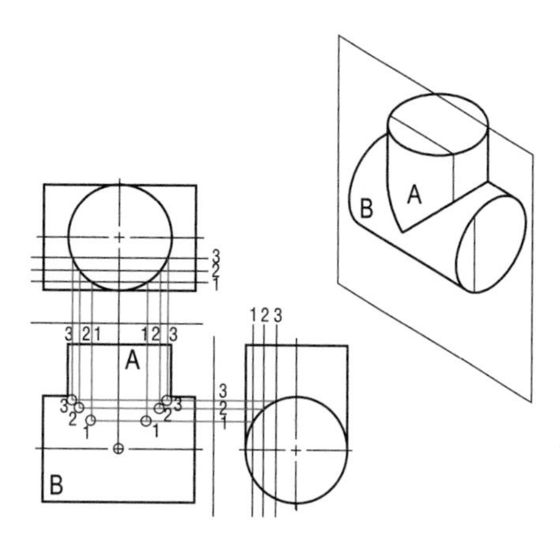

Figure 7.39 Intersection points of corresponding surface lines

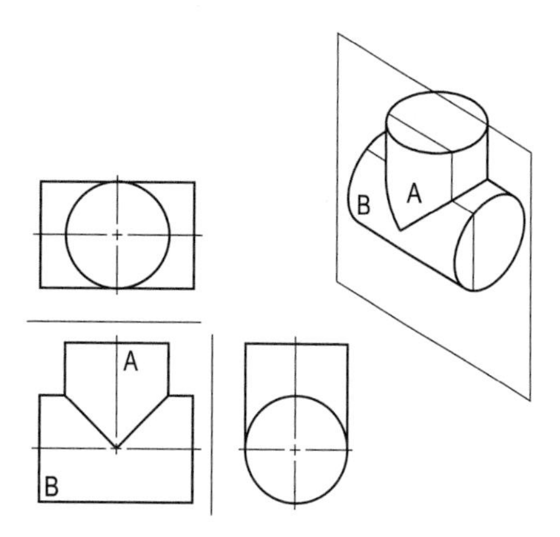

Figure 7.40 Complete line of intersection

Intersection of a Cylinder and a Cone

Example 7.6

A solid can be separated from a liquid with a simple device called a cyclone separator. This is an inverted cone with an inlet line on the side and outlet lines at top and bottom. Fluid (gas or liquid) and solid enter the side of the separator in a tangential direction. Solid particles are thrown to the side and fall to the bottom, and the fluid flows out through an outlet on the top of the cone. Separation efficiencies of over 90 percent can be attained. **Figure 7.41** shows the arrangement of inlet and outlet lines. Find the line of intersection so that a hole can be cut in the body and the inlet line connected. Draw the line of intersection between the inlet line and the body.

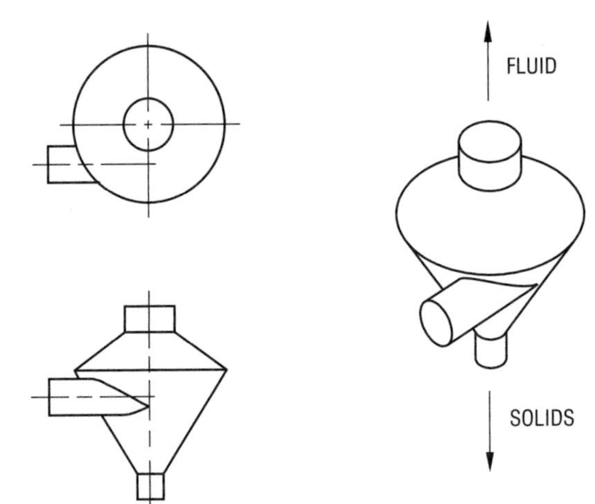

Figure 7.41 Cyclone separator

1. Select cutting planes that will give lines on the surface of the cylinder and on the surface of the cone. Horizontal cutting planes will create straight lines on the cylinder and circles on the surface of the cone. **Figure 7.42** shows edge views of the cutting planes in the front view. The top portion of the separator has been removed for clarity.

   Horizontal cutting planes are used instead of vertical ones because the resulting shapes on the conical body (circles) and the round inlet line (straight lines) are easy to draw. Vertical cutting planes would result in straight lines on the inlet line and parabolas on the body. Circles are easier to draw than parabolas.

2. In the plan view, draw circles corresponding to each cutting plane. The radius of each circle is obtained from the front view. **Figure 7.43** shows the circles on the body corresponding to each cutting plane. You are looking down on the inside of the conical body.

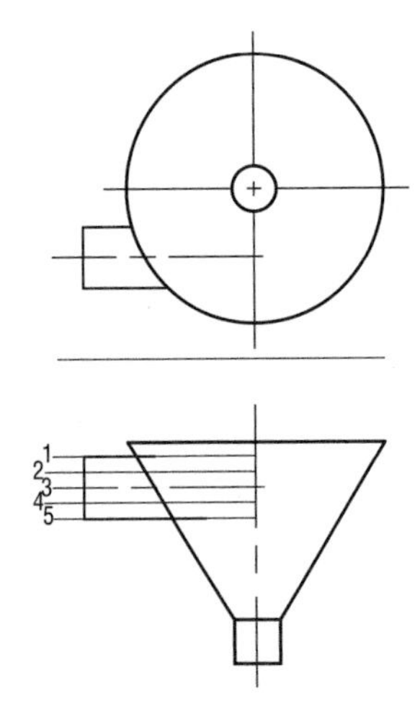

Figure 7.42 Horizontal cutting planes used to create surface lines on the cone and cylinder

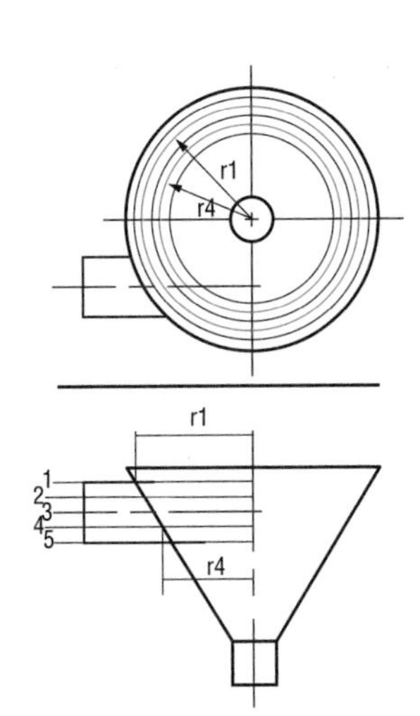

Figure 7.43 Lines on the conical body are seen as circles in the plan view

3. Draw an auxiliary view showing the axis of the inlet line as a point view. The centre of the inlet line is located using the distance from the folding line seen in the front view. This view enables surface lines to be located on the inlet line. **Figure 7.44** shows this step.
4. Locate the cutting planes in the auxiliary view using the distance from the folding line seen in the front view. Lines on the cylinder are seen in point view on the auxiliary view. **Figure 7.45** shows these point views on the inlet line. Cutting planes 1 and 5 are at the top and bottom of the cylinder and are not shown in the auxiliary view.
5. Project the surface lines on the inlet from the auxiliary view to the plan view. They will intersect corresponding circles on the body and define points on the line of intersection in the plan view.
6. Project intersection points from the plan view to corresponding planes in the front view. These points define the line of intersection in the front view. (**Figure 7.46**)

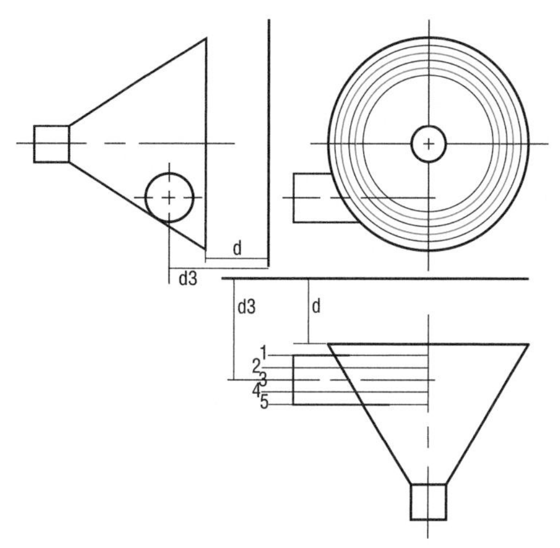

Figure 7.44 Auxiliary view showing the inlet line axis as a point view

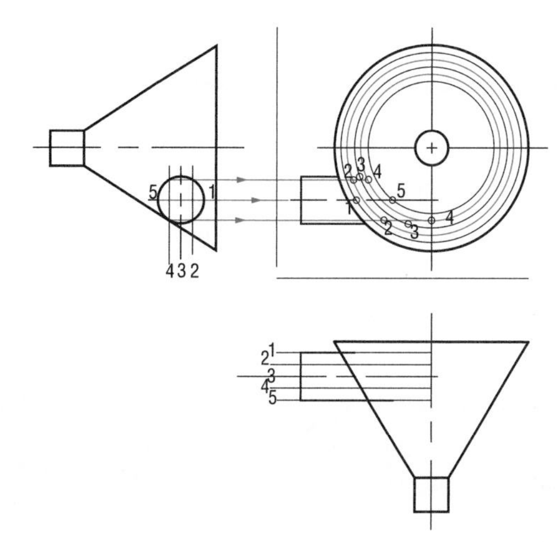

Figure 7.45 Surface lines on the inlet and body

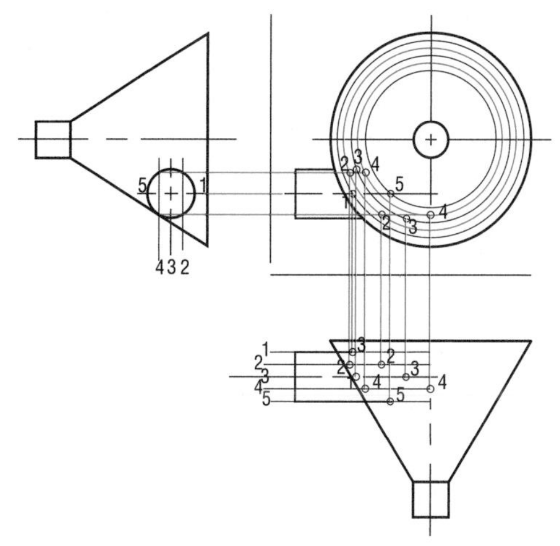

Figure 7.46 Points on the line of intersection in the front view

Example 7.6

7. Join the intersection points to get the line of intersection in the plan and front views as shown in **Figure 7.47**.

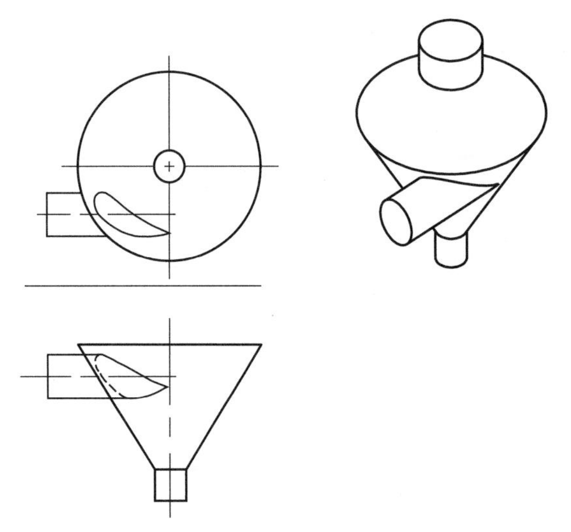

Figure 7.47 Complete line of intersection in plan and front views

Intersection of a Plane and a Sphere

The procedure for finding a line of intersection is the same for all cases: Find a point, or points, where a line on one surface intersects a corresponding line on the other surface. Lines on each surface are created by passing a plane through each surface. This ensures that the lines actually intersect. A series of intersecting lines gives points on the line of intersection.

Example 7.7

Figure 7.48 shows the plan view of a sphere cut by a vertical plane, XX. Draw the line of intersection in the front view.

Any plane passing through a sphere will create a closed curve on the surface. The curve is a circle; however, it could appear as an ellipse in some views. (In this example, the line of intersection would be seen as an ellipse in the front view.)

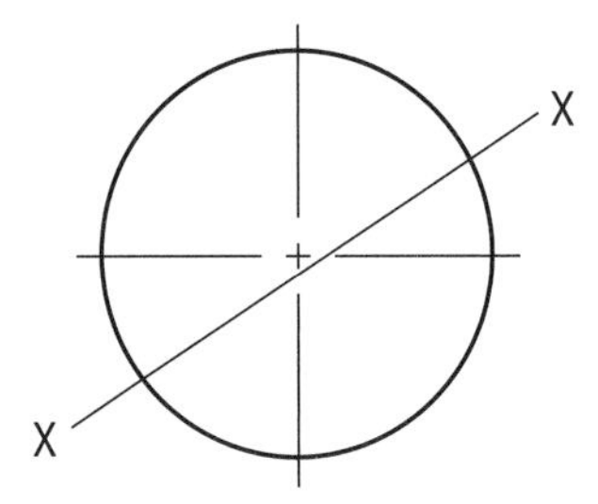

Figure 7.48 Plan view of sphere cut by a vertical plane

You will find the line of intersection in the same manner as in the previous problems, by determining the intersection of lines on the surface of the sphere with the plane.

1. Pass vertical cutting planes through the sphere to create lines on the surface. These surface lines will appear as circles in the front view. Edge views of these cutting planes are seen in **Figure 7.49**. The intersection of the surface lines with the plane are numbered (only three of the planes are numbered).

   You could use horizontal planes in the front view, but you would first have to draw the front view of the sphere; however, you can place vertical planes in the plan view so less work is involved.

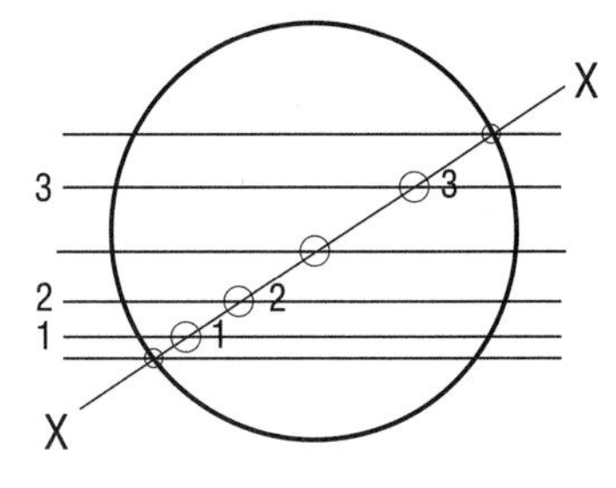

Figure 7.49 Vertical cutting planes create surface lines that will appear as circles in the front view

2. Draw circles corresponding to the cutting planes in the front view. The radius of each circle is found from the plan view, as shown in **Figure 7.50**. The centre is located by projecting from the front view.
3. Project the intersection points from the plan view to the corresponding circle in the front view. There are intersection points on the top and bottom halves of the sphere. **Figure 7.51** shows the points defining the line of intersection.
4. Join the intersection points to give the line of intersection. **Figure 7.52** shows the line of intersection. The shape formed by the plane XX has been shaded. This is not seen in true shape in the front view.

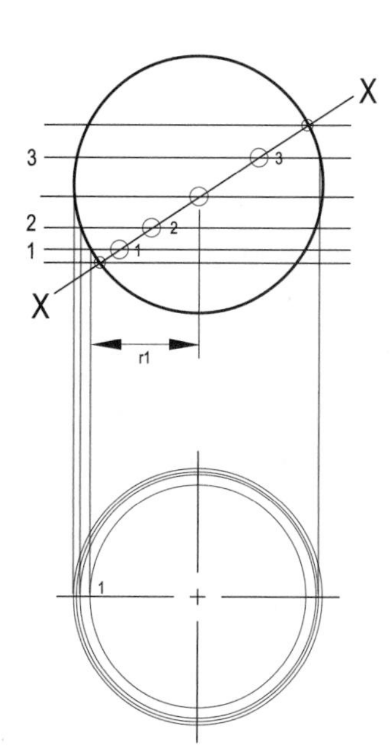

Figure 7.50 Circles corresponding to the cutting planes are lines on the surface of the sphere

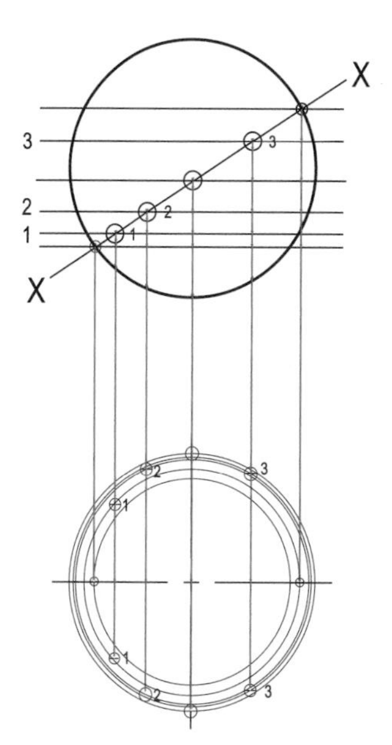

Figure 7.51 Intersection points are projected from the plan view to the corresponding circle in front view

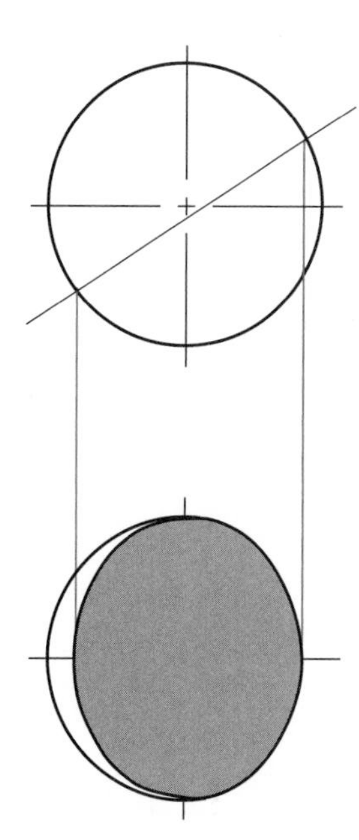

Figure 7.52 The line of intersection drawn in the front view

Flat Surfaces

Rectangular ducts are common in heating, ventilating, and air-conditioning applications. Many structural steel shapes are rectangular. If a rectangular section (or other shape made up of plane surfaces) meets a curved surface, the technique for finding the line of intersection is the same as that for curved surfaces but is usually simpler.

Example 7.8

Figure 7.53 shows a round column supported by a rectangular support. Determine the line of intersection between the top and bottom surfaces of the support and the column. If the top and bottom surfaces are cut on the line of intersection, they will fit tightly against the column and can be welded.

The line of intersection can be seen in the plan and front views, but it is not a true shape in either. We will draw the true shape of the line of intersection so it can be cut.

1. Choose points on the line of intersection in the plan view. Intersection points are circled in **Figure 7.54**. These points correspond to vertical cutting planes labelled 1 to 7. The points (or the cutting planes) can be located anywhere on the line of intersection.
2. Project these points on the top (and bottom) surfaces of the support to the corresponding surface in the front view. Only the points on the top surface are shown in **Figure 7.55**.

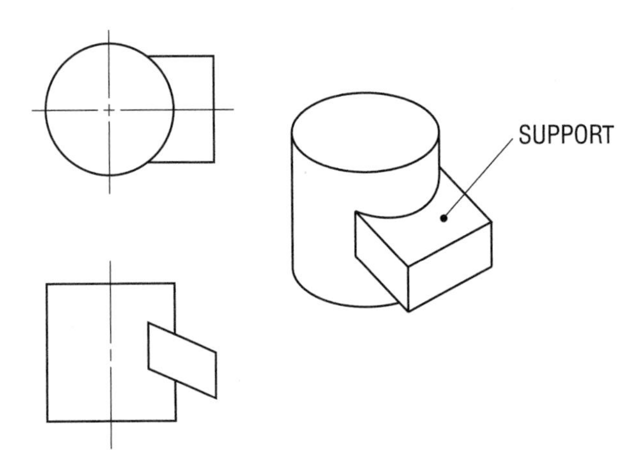

Figure 7.53 Cylindrical column supported by a rectangular support

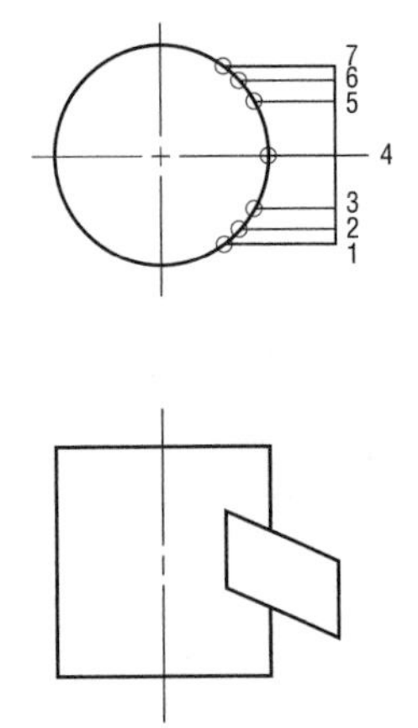

Figure 7.54 Points can be chosen anywhere on the line of intersection

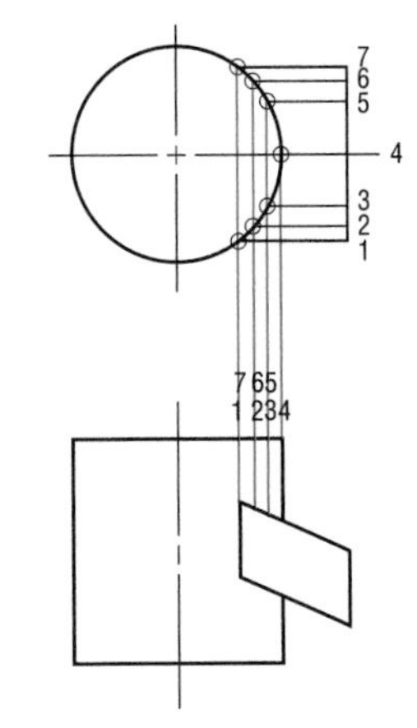

Figure 7.55 Points on line of intersection projected to the front view (only the points on the top surface are shown for clarity)

3. We will now draw the true shape of the top surface of the support. The true shape of the bottom surface can also be seen in this view but is not shown. Since both top and bottom surfaces are seen as edge views in the front view, the true shape can be projected from them. Points on the line of intersection can then be projected to the true shape. Lines locating intersection points are located from the side of the support in **Figure 7.56**. The lengths of these lines are not known.
4. Project intersection points from the front view to the corresponding line in the auxiliary view showing the true shape of the top surface. The line of intersection on the bottom surface is not shown in **Figure 7.57**, but it has the same shape as the line of intersection on the top surface.

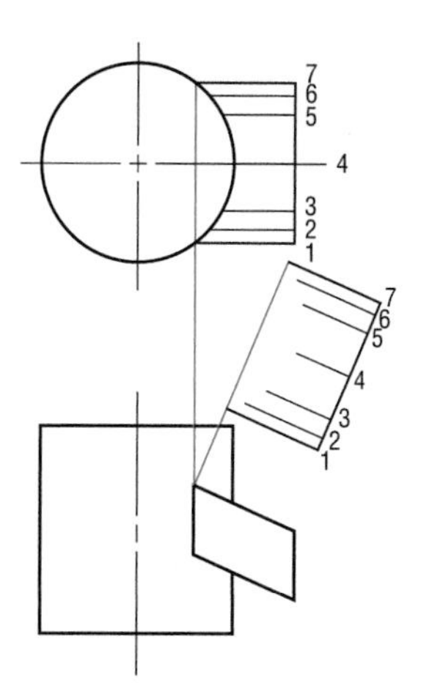

Figure 7.56 True shape of top surface with lines to locate intersection points

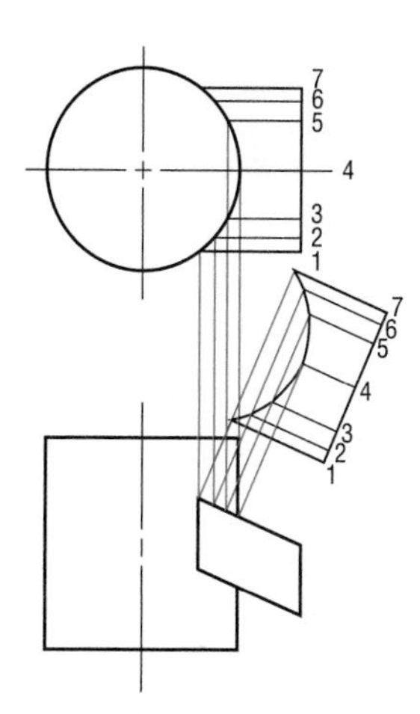

Figure 7.57 True shape of the line of intersection on the top surface of the support

Here are some things you need to know when finding the intersection between two solids.

- The line of intersection is defined by the intersection of lines on the surface of one solid with corresponding lines on the surface of the other solid. The corresponding lines must be in the same plane. See Figure 7.33 (page 255).
- Corresponding lines on each surface are found by passing cutting planes through both solids and drawing the lines on the surfaces created by these cutting planes. See Figures 7.35 to 7.40 (pages 256 and 257).

Now that you know a little more about the intersection of curved surfaces, try the following problems.

Problems

17. **Figure 7.58** shows a 50 mm diameter cylinder intersecting a 100 mm diameter cylinder at 90°. Draw the line of intersection between these cylinders.

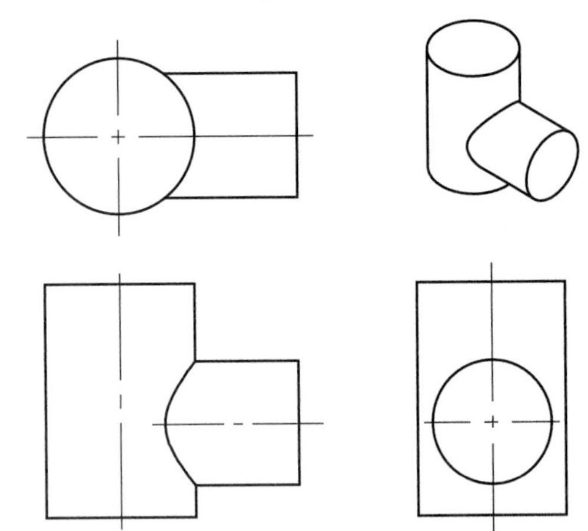

Figure 7.58

18. **Figure 7.59** shows a portion of a support for a pier. The frame is to be made from a 300 mm diameter tube. One member slopes at 30°. Draw the line of intersection between the tubes. Scale is 1:10.

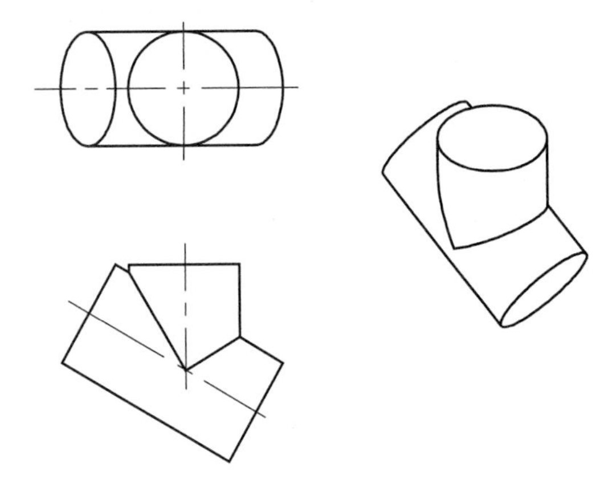

Figure 7.59

19. A horizontal cylinder (50 mm diameter) meets a cone (height 100 mm, base diameter, 100 mm) such that the sides of the cylinder are tangent to the cone. Draw the line of intersection. Scale is 1:2.

20. **Figure 7.60** shows plan and front views of the end of a heat exchanger. A 300 mm diameter discharge line leaves the end vertically in the location shown. The end of the heat exchanger is a 700 mm diameter hemisphere. Draw the line of intersection between the discharge line and the end of the heat exchanger. Scale is 1:10.

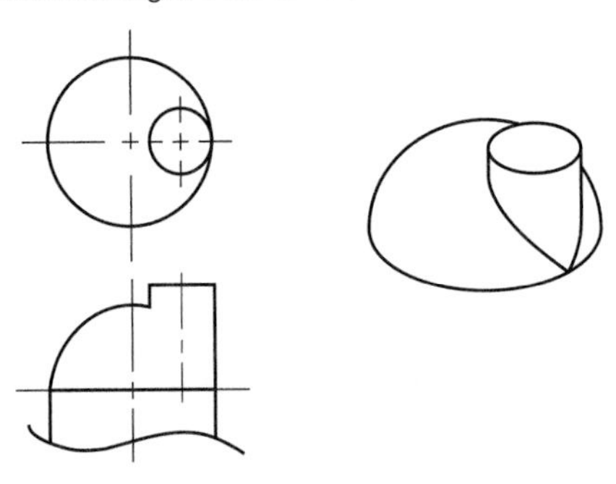

Figure 7.60

21. **Figure 7.61** shows a spherical tank supported by sloping legs. There are four legs spaced at 90° (only one is shown). The diameter of the sphere is 2000 mm, and the support is 500 mm in diameter. The centre of the support is offset 500 mm from the centreline of the sphere, and the support has a slope of 75°. Draw the line of intersection between the support and the tank. Scale is 1:20.

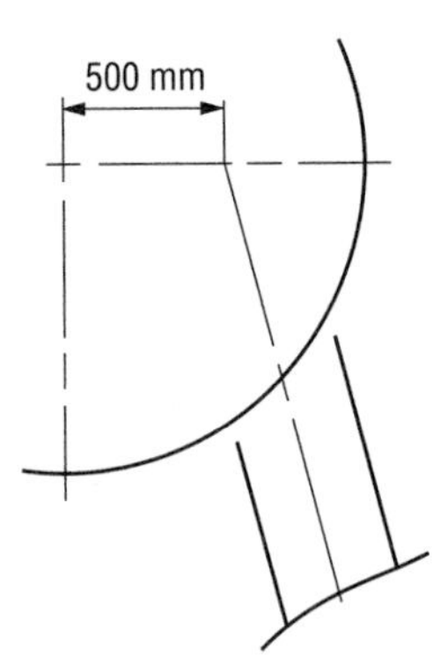

Figure 7.61

DEVELOPMENT

If two pipes are to be joined, the end of one, and sometimes of both, must be shaped so they meet. Some procedure must be used to mark the line of intersection (or what will be the line of intersection). A 2-D pattern is made and wrapped around the end of the pipe to mark the cut. The process of making a 2-D pattern that can be rolled or folded into a 3-D shape is called **development.**

You have seen many applications of developed objects, although you probably did not think of them this way. A cereal box is made by folding a flat piece of cardboard. The cardboard is cut so that the final shape is formed after folding (Figure 7.62).

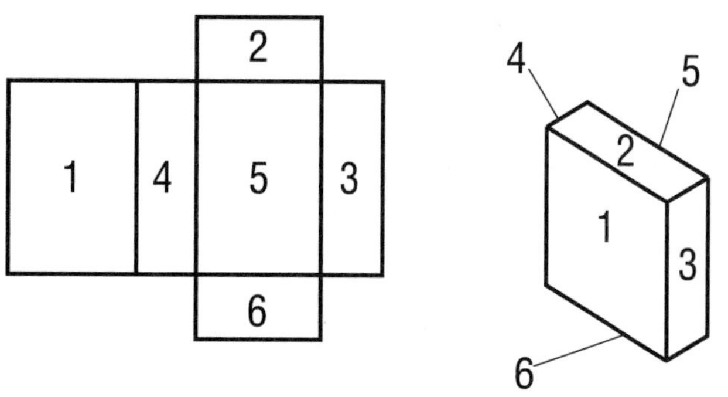

Figure 7.62 A flat pattern that can be folded to form a box

Large diameter pipe is made from a flat piece of steel by rolling and welding the seam. Mailing tubes are made by rolling a flat piece of cardboard.

A line of intersection must be found before a surface can be developed. This process has been discussed; however, when a development is required, surface lines must be equally spaced. Equal spacing is important only if development is required.

Development of a Cylinder

A cylinder is made by joining the edges of a rectangle, as shown in Figure 7.63. The lines drawn parallel to the axis at the seam are called **generator lines**. Two other generator lines are shown on the flat pattern.

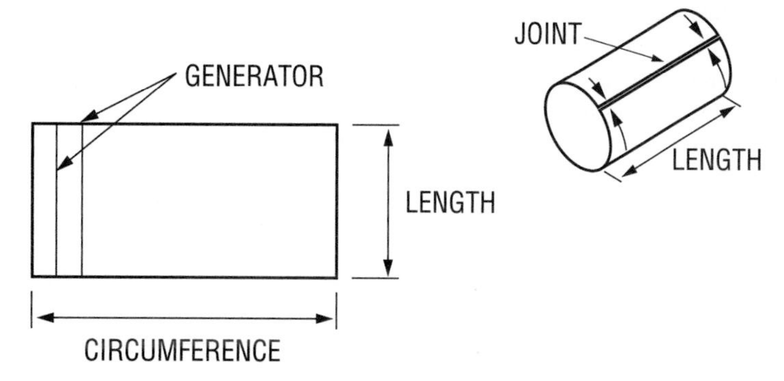

Figure 7.63 A rectangle can be rolled to form a cylinder

If a specific length and diameter are required, as is usually the case, the material must be the right size. One side must equal the length and the other the circumference.

The development of a cylinder with square ends (ends that are perpendicular to the axis) as shown in Figure 7.63 is simple. All generator lines are the same length. If one end is at an angle that is not 90°, all generator lines will not be the same length, and the pattern is not so simple.

Example 7.9

Develop a pattern for the end of the cylinder shown in **Figure 7.64** so that it can be cut to meet the wall. An auxiliary view showing the axis of the cylinder as a point view is shown in Figure 7.64.

A pattern for the end of this cylinder can be developed by starting with a rectangle, as in Figure 7.63, and marking the length of each generator line so that when rolled it will produce the cylinder in Figure 7.64. The problem is reduced to finding the length of a line (actually several lines). The first three steps are similar to those used previously to find the line of intersection, except that generator lines must be equally spaced around the circumference of the cylinder.

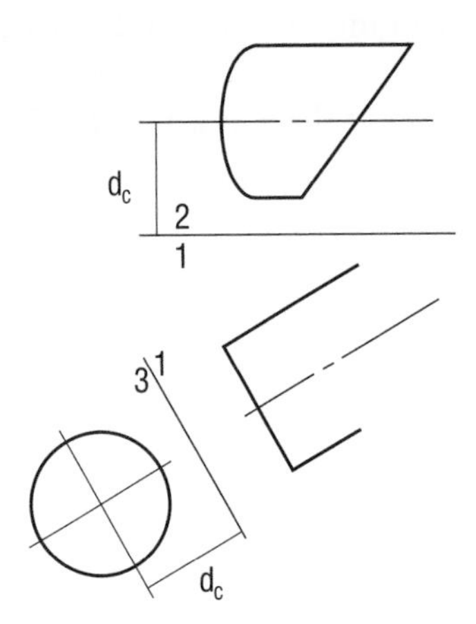

Figure 7.64 Plan and front views of a cylinder with end cut at an angle, plus an auxiliary view showing the point view of the cylinder axis

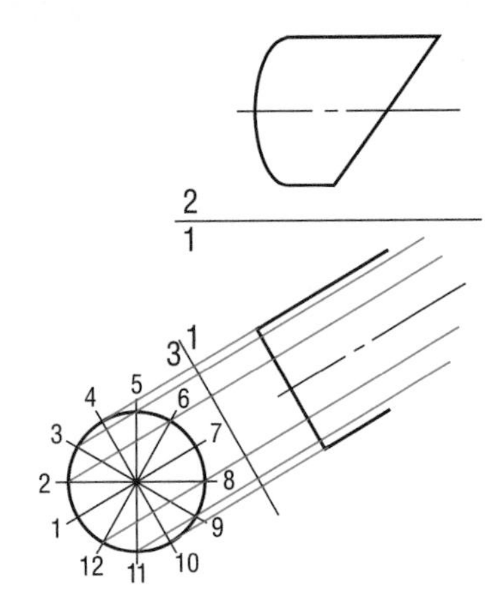

Figure 7.65 Line of intersection as seen in the front view

1. Locate equally spaced generator lines on the surface of the cylinder. This cannot be done in the plan view. Generator lines can only be positioned in a view showing the true shape of the cylinder (i.e., a circle). They will appear as points in this view. Generator lines have been located around the circumference by spacing them at equal angles (**Figure 7.65**). A protractor is used to do this. Thirty degrees is a good spacing if the cylinder is small (or has been drawn to scale). Generator lines are numbered for reference. The numbers usually start at the seam (where the two edges are joined). Generator lines have been projected to the front view in Figure 7.65. Since they are seen in point view in the auxiliary view, they will be true length in the front view.

2. Locate the generator lines in the plan view to find where each line intersects the wall. They are located in the plan view using the distance from the folding line in the auxiliary view. **Figure 7.66** shows how this distance is transferred to the plan view.

3. Project all intersection points from the plan view to the corresponding line in the front view, and join them with a smooth curve to get the line of intersection. The line of intersection is shown in **Figure 7.67**. An arbitrary datum has been added. The length of each generator line, which is true length in the front view, will be measured from this datum. Datum line can be in any convenient location. It helps to number generator lines.

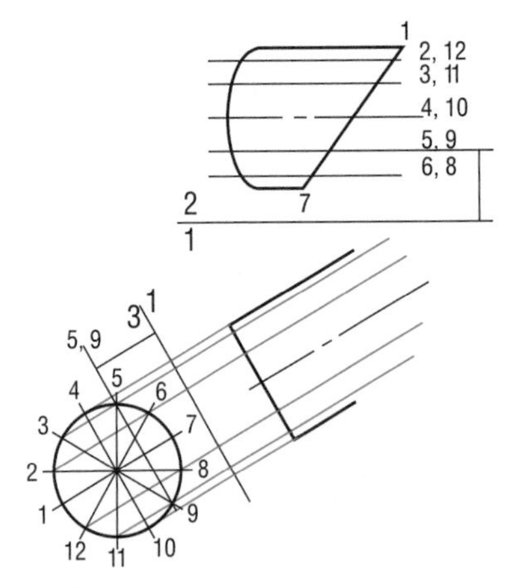

Figure 7.66 Distance transferred to plan view using the distance from the folding line in the auxiliary view

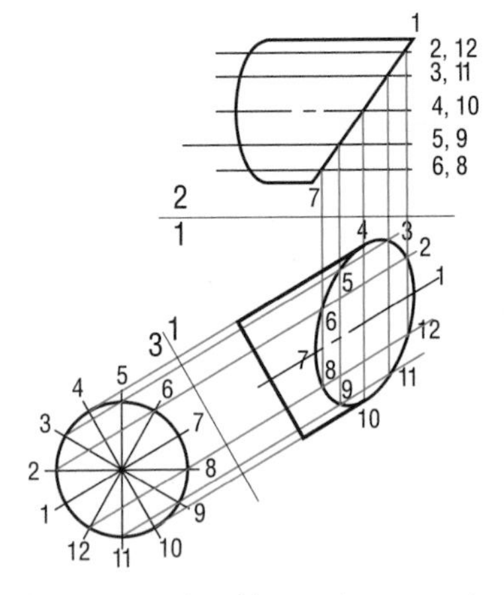

Figure 7.67 Line of intersection as seen in the front view

4. Lay out the circumference (πd). This is shown as line AB in **Figure 7.68.** Generator lines must now be located along the circumference.

5. Divide the circumference into 12 equal parts so that generator lines can be located at equal intervals. You can calculate the circumference and divide it by 12, but it will probably not divide evenly, making it difficult to divide the line accurately. There is an easy way to divide the line into 12 equal parts. The process is shown in **Figure 7.69**.

   a. Draw a line, at any angle and any length (make it as long as possible), from one end of AB (see **Figure 7.69a**). Small angles do not work well for this. Use an angle of about 30°. There is no need to measure it.
   b. Mark off 12 equal lengths along this line, using a scale (see **Figure 7.69b**). Use any convenient interval, but it should not be too small. Ten millimetres is a good interval.
   c. Draw a line from the 12th division to the end of AB (see **Figure 7.69c**).
   d. Draw lines parallel to the line drawn in Figure 7.69c from each of the measured divisions. The circumference is now divided into 12 equal divisions (see **Figure 7.69d**).

   Figure 7.70 shows the circumference divided into 12 parts using this method. Generator lines have been drawn at each division. The length of each generator line must now be laid out.

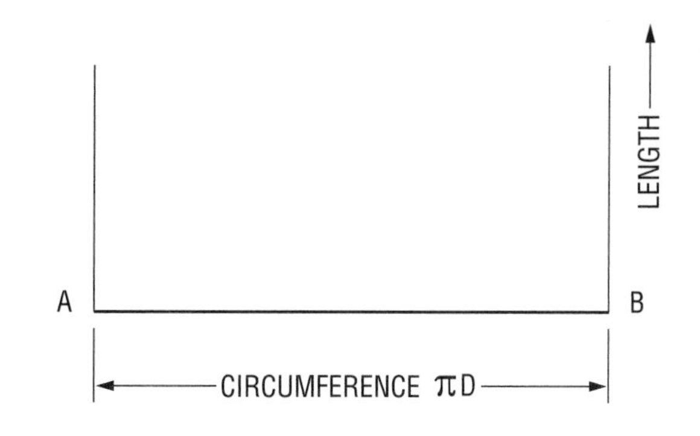

Figure 7.68 Circumference laid out

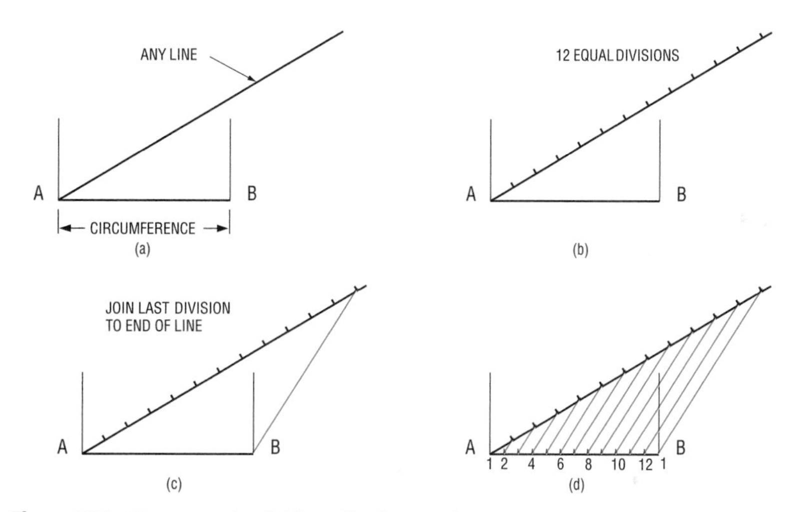

Figure 7.69 The process for dividing a line into equal parts

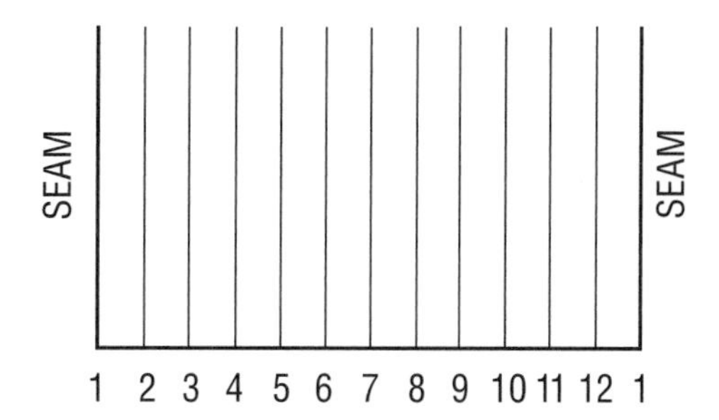

Figure 7.70 Circumference, showing generator lines at equal divisions

Example 7.9

6. Mark off the length of each generator line on the pattern. Length, measured from the arbitrary datum, is taken from the front view. The front view must be used because generator lines are true length only in the front view. The lengths of generator lines 5 and 12 are shown in the front view. **Figure 7.71** shows the length of all generator lines on the pattern.

7. Draw a smooth curve through the end of each generator line to finish the pattern (**Figure 7.72**). When the seam (generator line 1) is joined, the end of the cylinder will be the correct shape to meet the wall.

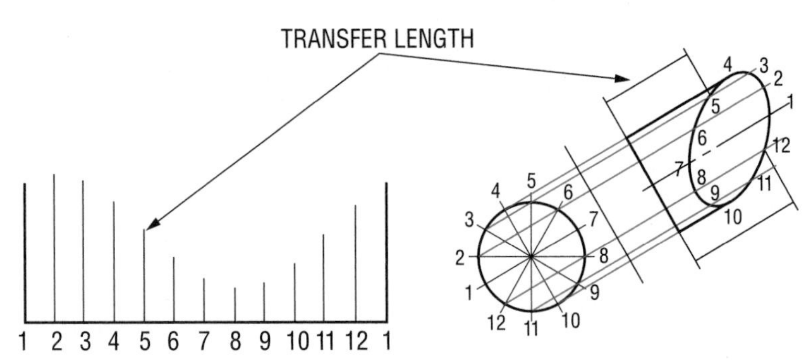

Figure 7.71 Generator line length laid off on pattern using lengths taken from the front view

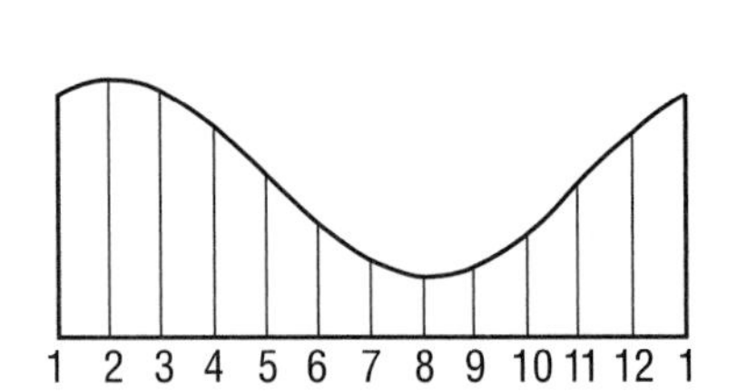

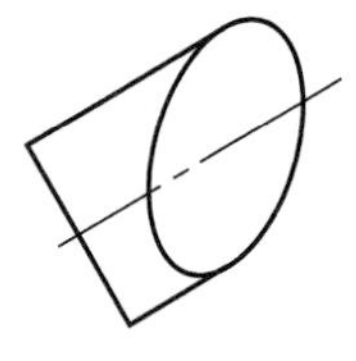

Figure 7.72 Pattern for the end of cylinder

Development of Two Curved Surfaces

When two cylinders (or other curved surfaces) meet, development is done in the same way as for the intersection with a plane surface. The circumference is first divided into equal divisions and the line of intersection is found. The circumference is laid out and the length of each generator line determined.

Example 7.10

Figure 7.73 shows a joint in a frame made from large diameter pipe. A pattern must be made so that the vertical pipe can be cut to fit the sloping pipe.

You will create intersecting lines on the surface of both cylinders. The points where the lines intersect define the line of intersection. The pattern can then be made. **Figure 7.74** shows how corresponding lines will be seen on both surfaces.

1. Draw a plan view, and divide the circumference of the vertical pipe into equal parts to give equally spaced generator lines. Point views of these generator lines are numbered in **Figure 7.75**.

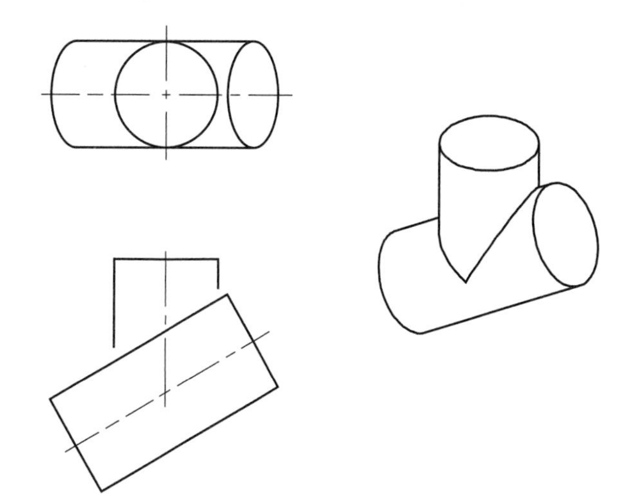

Figure 7.73 Intersection of two cylinders at an angle which is not 90°

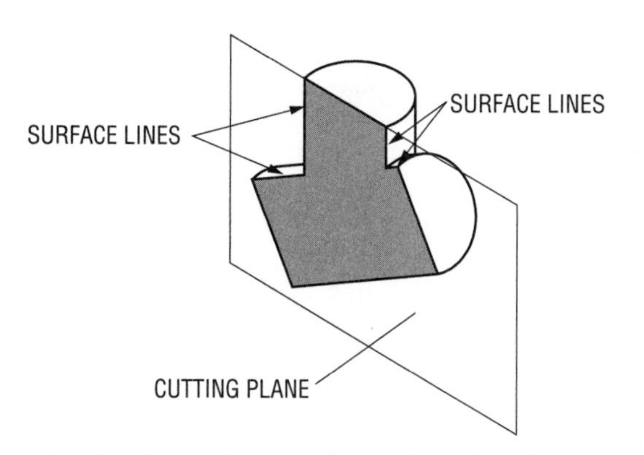

Figure 7.74 A cutting plane used to create lines on the surface of intersecting cylinders

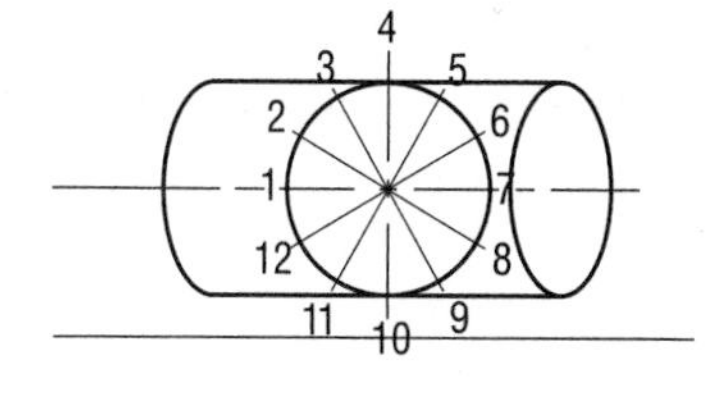

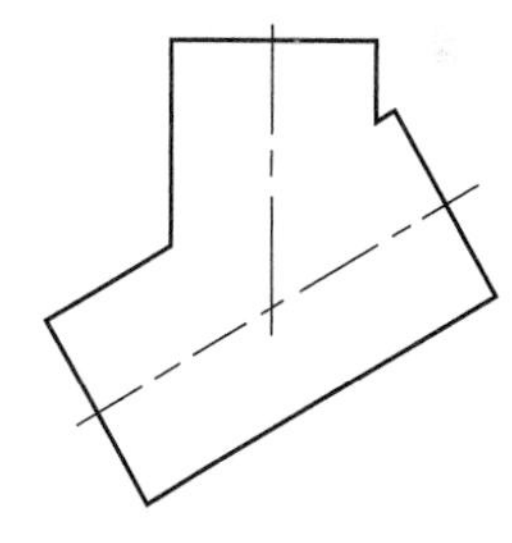

Figure 7.75 Circumference of vertical cylinder divided into equal segments to give equally spaced generator lines

2. Put cutting planes through the generator line locations. Cutting planes are seen in edge view in the plan view and are numbered to correspond to the generator lines (**Figure 7.76**).

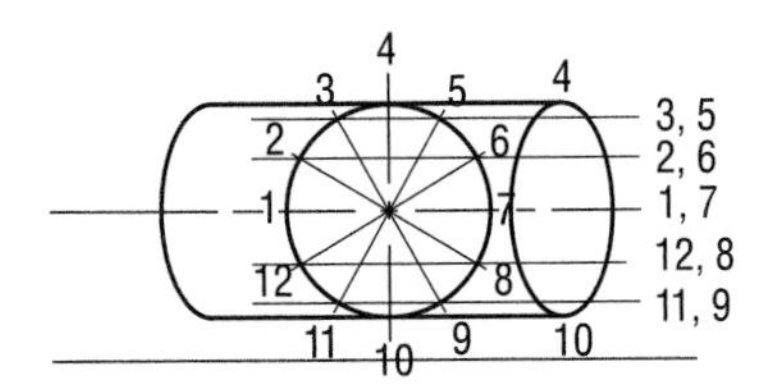

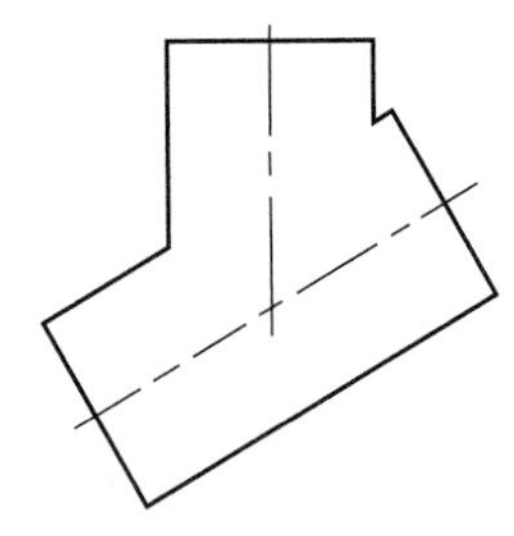

Figure 7.76 Cutting planes, parallel to the folding line, are located at generator line locations

3. Draw an auxiliary view showing the centreline of the sloping pipe as a point. Locate the cutting planes in this view using the distance from the folding line in the plan view. This creates lines on the sloping pipe that correspond to those on the vertical pipe. They are seen as points in this auxiliary view (**Figure 7.77**).

4. Project the surface lines on the sloping pipe to the front view and the generator lines on the vertical cylinder from the plan, and identify the intersection points of corresponding lines (**Figure 7.78**).

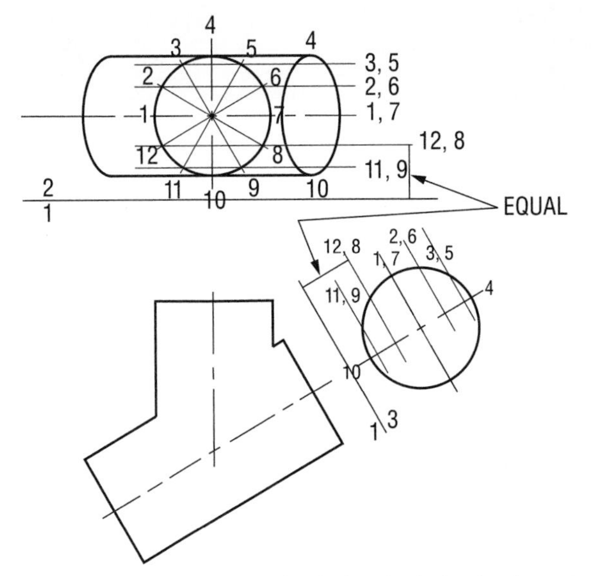

Figure 7.77 Creating corresponding lines on the sloping pipe by locating the cutting planes in the auxiliary view

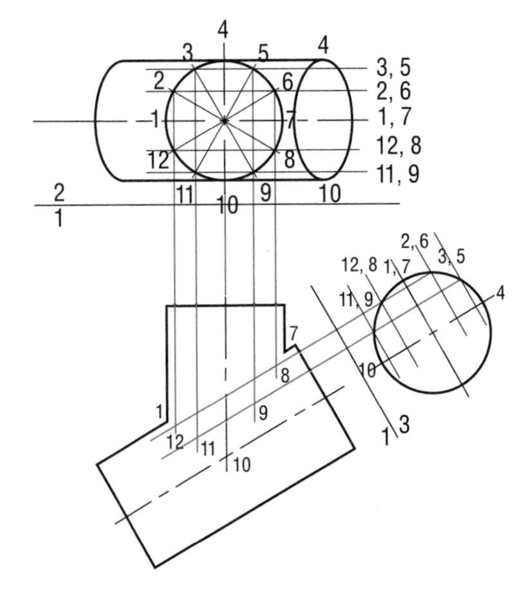

Figure 7.78 Intersection points defined by the intersection of corresponding lines in the front view

Example 7.10

5. Draw a smooth curve through these points to show the line of intersection (**Figure 7.79**).
6. Lay out the circumference of the vertical pipe, and divide it into 12 parts, corresponding to the generator lines (**Figure 7.80**). The method of doing this is shown in Figure 7.69 (page 267).
7. Mark off the length of each generator line on the pattern. The length of these lines is seen in the front view and is measured from an arbitrary datum. The length of each generator line is projected from the front view (**Figure 7.81**).
8. Join the ends of the generator lines with a smooth curve (**Figure 7.82**).

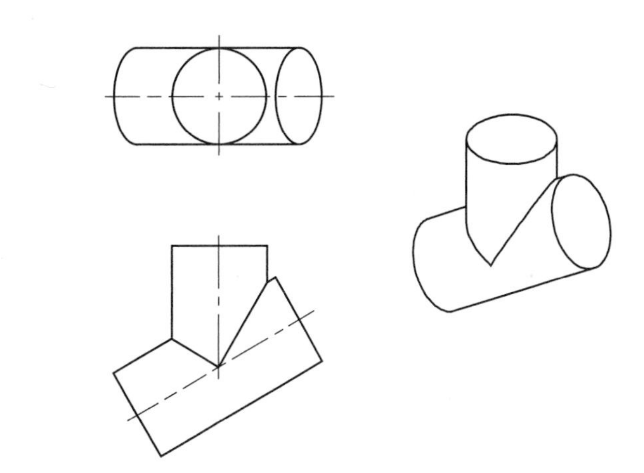

Figure 7.79 Line of intersection

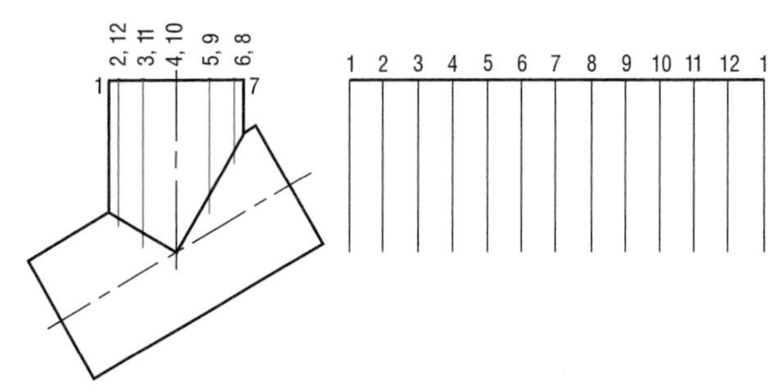

Figure 7.80 Generator lines located on the pattern using the method shown in Figure 7.69

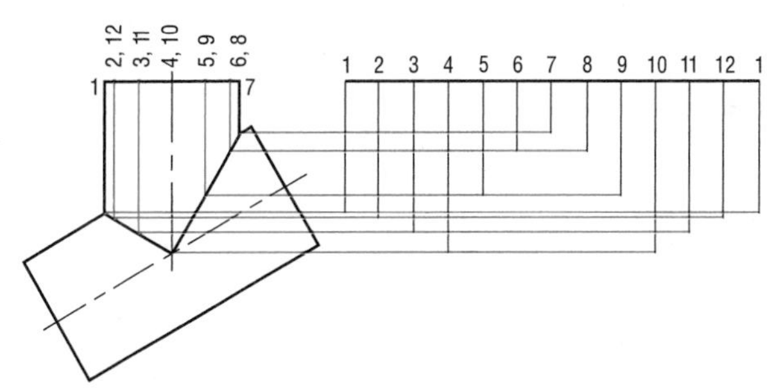

Figure 7.81 Generator line lengths laid out on pattern by projecting the true length from the front view

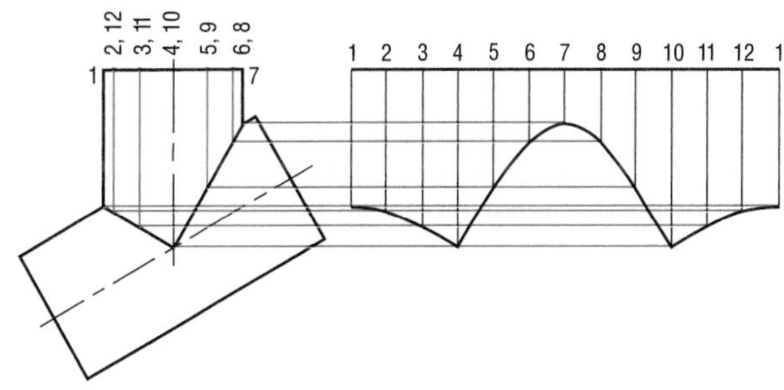

Figure 7.82 Completed pattern for the end of the vertical pipe

Example 7.10

Here are some things you need to know about development.

- The development (pattern) for the end of a cylinder is found by finding the length of lines (generator lines) equally spaced around the circumference of the cylinder. See Figures 7.75 to 7.82 (pages 269 to 271).
- Equally spaced generator lines are created by dividing the circumference of the cylinder into equal segments, in the view that shows the axis as a point view. See Figure 7.75 (page 269).

Now that you know a little more about development, try the following problems.

Problems

22. **Figure 7.83** shows a tank with hemispherical ends filled with a horizontal line 600 mm in diameter. The tank diameter is 1500 mm. The edge of the inlet line is on the centreline of the tank. Develop a pattern for the end of the inlet line so that it can be welded to the tank. Scale is 1:20.

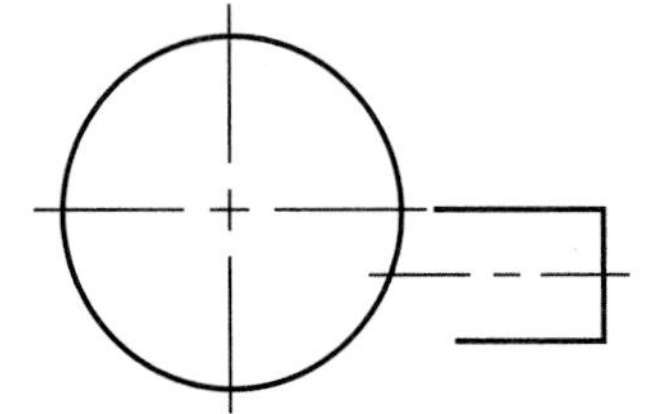

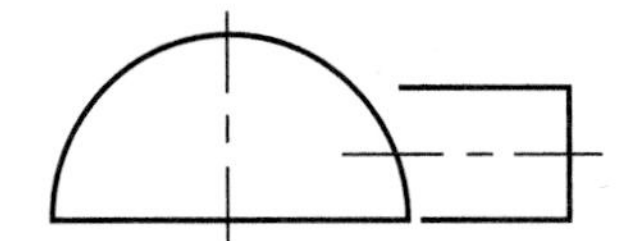

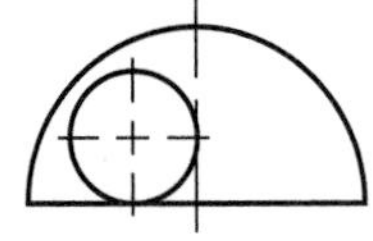

Figure 7.83

23. A front view of a 90° elbow supported by a cylindrical support is shown in **Figure 7.84**. The centrelines of the elbow and the support are in the same vertical plane. The cross-section of the elbow is circular, with a diameter of 400 mm. The diameter of the support is 300 mm. Develop a pattern for the support so that it can be welded to the elbow. Scale is 1:10.

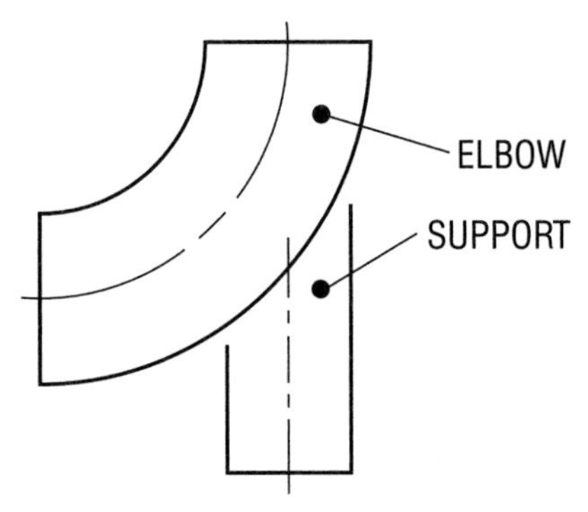

Figure 7.84

24. **Figure 7.85** shows the end of a cylindrical tank, diameter 600 mm, with a hemispherical end. There is a 360 mm discharge line at the top. The discharge line is in the plane of the centreline of the tank and slopes at 40°. Develop a pattern for the discharge line. Scale is 1:10.

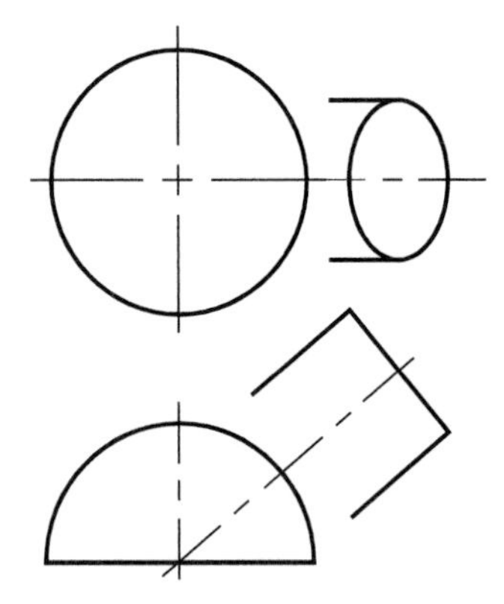

Figure 7.85

25. **Figure 7.86** shows a 500 mm diameter cylinder supporting a 1000 mm diameter column. The centrelines of the two cylinders are in the same plane. The support makes an angle of 45° with the ground. Develop a pattern for the support. Both ends must be developed, and the pattern must be the correct length. Locate the seam on the underside of the support. Choose a datum in the middle of the support, and measure the length to the line of intersection to the top (at the column) and the bottom (at the ground). Scale is 1:20.

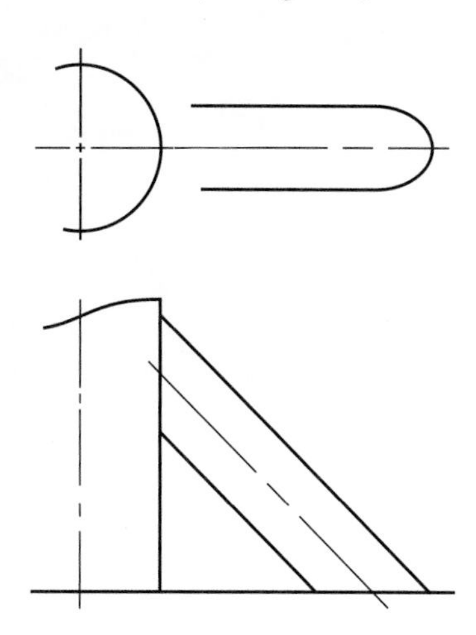

Figure 7.86

26. Plan and elevation views of a portion of a spherical tank supported by vertical legs are shown in **Figure 7.87**. Supports are spaced around the tank at 45° (only one is shown). The tank diameter is 8 m, and the support diameter is 2 m. Develop a pattern for the top of the support so that it can be welded to the tank. Scale is 1:50.

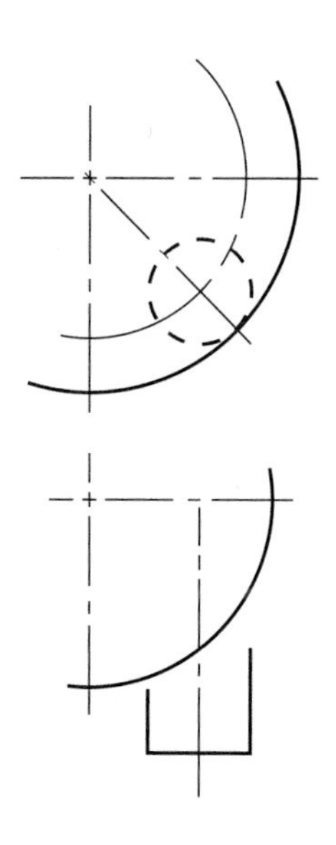

Figure 7.87

27. **Figure 7.88** shows a 400 mm diameter feed line entering a 1000 mm diameter tank. The line enters the tank at an angle of 30° to the horizontal and is tangent to the side of the tank. Develop a pattern for the end of the feed line. Scale is 1:10.

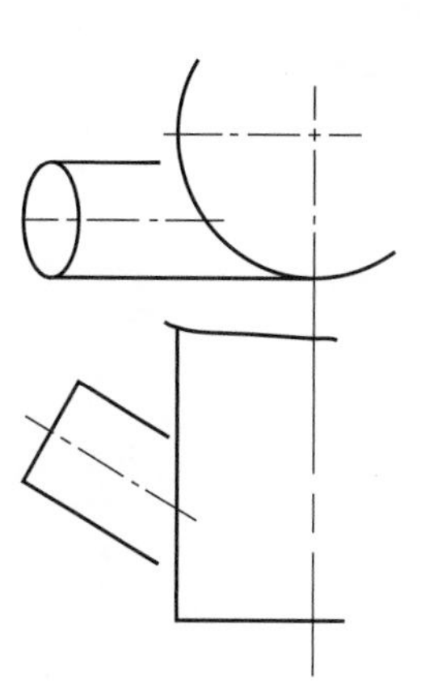

Figure 7.88

> *If you have two equally likely solutions to a problem, pick the simplest.*
>
> *Occam's razor*

8 Engineering Design

WHAT IS ENGINEERING DESIGN?

Engineering design is about solving problems and creating new and better things. These can be mechanical devices, chemical systems, computer chips, or clean water. Problem solving requires design, and design solves problems.

If science is to benefit society, it must be applied. Engineering design fills that role by applying science for the benefit of society. Engineers make up about 1 percent of the population, but the effect of their work is enormous. Cars, planes, computers, lights, and practically everything we take for granted in a modern society are the results of the engineering application of scientific principles. Science can provide explanations as to why things work. Engineering makes them work to our benefit.

The Canadian Council of Professional Engineers gives a broad definition of engineering design:

> *Engineering design integrates mathematics, basic sciences, engineering sciences and complementary studies in developing elements, systems and processes to meet specific needs. It is a creative, iterative and often open-ended process subject to constraints which may be governed by standards or legislation to varying degrees depending on the discipline. These constraints may relate to economic, health, safety, environmental, social or other pertinent interdisciplinary factors.*
>
> *Canadian Engineering Accreditation Board,* Accreditation Criteria and Procedures, *2003*

Engineering design is the integration of physics, mathematics, engineering, and natural sciences to create goods and services and solve problems for the benefit of society. It is a creative process that requires divergent thinking and innovation. The solution to an engineering design problem is a compromise, requiring the consideration of economics, ethics, safety, and community values, to name only a few. Intuition, creativity, and judgment are important aspects of engineering design. You learn these by practising them, not by reading about them.

Engineering design is a process but is not a linear process. There is no formula for a successful design, and one cannot follow a set of rules to arrive at a suitable design.

There are many possible solutions to an engineering design problem. Individual designers will have different approaches and methods for solving the same problem or creating a design to suit a specific task. They are all "right," but some may be better than others. The solution must consider the technical as well as the nontechnical aspects of any problem.

A real engineering problem is not like the textbook problems you worked on in your mathematics, physics, chemistry, and other courses, where a problem is carefully formulated and all relevant information is given. This is necessary for learning. A real problem, however, is often poorly defined. A problem may not even be correctly stated. Your first challenge may be to determine what exactly the problem is and define it in such a way that it can be solved. There could be little or no information given, or there could be so much information that you have to sort out what is useful and what is not. Some of the information may be irrelevant, and some may be wrong. Safety is always a major concern and is rarely considered in a textbook problem. These are some of the challenges that make engineering interesting. You apply the material you learned in solving problems using new and innovative ways.

DIFFERENT TYPES OF ENGINEERING DESIGN

When we think of engineering design, we often think of mechanical design—cars, airplanes, machinery—but there are many types of design, involving all fields of engineering. Civil, chemical, metallurgical, and electrical engineers, to name a few, are all involved in the creation of new and different things. Most designs are based on existing designs and are not original. Some designs use existing technology in new ways.

Original Design

Though the design of a completely new product or system is relatively rare, some engineers are asked to, or have the vision to, develop something original. One example is the transistor, which had a profound effect on society by creating a new means of mass communication. The transistor radio enabled simple, inexpensive mass communication all over the world. Radios became cheap enough to be purchased by almost everyone and small enough to be taken anywhere. Before transistors were used, portable radios were significantly larger, more expensive, and heavier. The transistor was a revolutionary design.

Modification and Redesign

A large number of engineering designs are redesigns or modifications of an existing item or process system to meet a different need or extend the use of something. This is probably the most common type of engineering design. For example, it was required to increase the power of a small hydro station without adding another penstock. The existing penstock was modified to supply two turbines. A mechanism to raise the seat on a wheelchair may require redesign to adapt it to a new wheelchair.

Adding new features, or adapting new features to an existing system, can be as challenging as creating a new piece of equipment. There can be increased constraints, since the new features must not interfere with the existing system and may have to fit into the existing space. Modifications and additions may have to be done without shutting down. In the case of the wheelchair seat lift, nothing that would void the guarantee can be done.

THE ENGINEERING DESIGN PROCESS

The design process has a number of steps, but not all steps are followed for every problem. Usually, you will not work your way continuously through the steps but may have to come back to a previous step a number of times. Following the steps in the design process is a good way to keep on track. When the process gets "bogged down," and it will, a good way to get it going again is to go back to your problem definition in order to refocus on what you want to achieve.

The steps in the design process are:

- recognize a need,
- define the problem,
- decide on the design criteria,
- gather information,
- search for possible solutions,
- evaluate possible solutions,
- recommend a solution, and
- detail design and implementation.

None of these steps is done in isolation, and all overlap. You may find that a technological solution is not required or that you have to redefine the problem after reviewing the design criteria. You may even find that the particular problem is impossible to solve. You will gather information throughout the project. Testing and analysis may show faults in your proposed solution, and this may require more information or going back to a previous step. You may have to return to a previous step a number of times.

Recognize a Need

If you perceive a need for a new product or service, the first step would be to ensure that there actually is a demand for it. Developing a new product can be a long and expensive process, and if there is insufficient demand, you are not likely to profit from all your efforts. There may or may not be a market study (or the client has already determined a need), and you may not be involved in this phase of the project. There may be no need for a market study if a client has identified a problem and hires you to fix it.

Define the Problem

The first step is to determine what the actual problem is. It is often the case that a client will come to you with an idea for solving a problem and hire you to design and develop it. The client may be asking for the impossible—for example, a perpetual motion machine. The client may describe what he or she perceives as the problem, but it may be something else. Thus, your first task is to determine what the real problem is in specific terms that both you and your client agree on before you can formulate any plan for solving it. For example, a client may have a problem with solids depositing in a piping system and ask you to design a system for heating the pipe in the region where material is being deposited. In this case, the problem was in the construction of the pipeline. Heaters would not solve the problem and would result only in warm deposits. The pipeline was modified and the problem was solved.

You will have to do some information gathering and investigation before you can define the problem clearly and attempt to solve it. Time spent in carefully defining the problem is time well spent and pays off in the long run. The desire to "get going" is strong but should be resisted. Good problem solvers spend more time planning what they are going to do than in actually doing it. John Roebling, a famous bridge designer, said, "When the first sod is turned, the project is half over."

Time is required to gather information, and there are costs associated with this. You may never be able get all the information you want, and the project must still be completed. You must have some information to proceed, and some may not affect your course of action even if you have it. You have to stop somewhere. These are some of the nontechnical decisions you will have to make.

One obvious source of information is the client. Your client may not have all the information you want, but this is a good place to start. If the problem has been presented as a situation, you may be able to find out details of what caused it. This may give a clue to the identification and redefinition of the problem.

If you are to design, or redesign, equipment to do a particular task, talk to the people who will use the equipment. What is wrong with existing equipment, or what do they like about it? If new equipment has to be constructed, what features would the prospective user like? Listen carefully, but do not believe all you hear. The people who use the machine may not know how or why it operates the way it does. They probably just know how to use it. The people you talk to may not be well informed, and their ideas may be based on what someone else has told them. They could be biased, may not tell you the whole story, or they simply may not know. They may tell you what they think you want to hear. You could, however, get a lot of background from people who use, or will use, the equipment, and this may result in a redefinition or expansion of the problem. Get the facts yourself.

Decide on the Design Criteria

Design criteria and specifications on which the success of your design depends must be determined before you start, but they must always be examined, and changed if required. You must decide on what you want and what you do not want. The customer may want every imaginable feature, some impractical and some impossible. (You cannot get more energy out than you put in, although many people try.) Examine the customer's wishes carefully. Design criteria must be specific and described with numbers that can be measured. "It should not be too heavy," is not a satisfactory design criterion. How heavy is "too heavy"—too heavy to lift with a crane, or too heavy to lift by hand? By whom—a weight lifter or a senior citizen? A subjective criterion (too heavy, quiet, small) cannot be evaluated. Someone will always find something wrong with the criterion, but if, for example, you specify a maximum mass of 20 kg, it is easy to determine if that requirement has been met. If something must fit into a specific space, the exact size of the space must be known before you start. Remember that everything has a tolerance. If your client tells you that the machinery must handle steel as it comes from the mill, and the steel will be straight, you must know how straight is straight. Check applicable standards to determine the allowable tolerances. The steel may look straight, but it can be curved and still be within the allowable tolerance.

Gather Information

We have already discussed gathering information from the client and the people who may use your design. This information is often subjective and must therefore be confirmed before you accept it. You may need factual information or more background information. Where do you get it? One obvious way is go into the field to observe and take measurements. When you do this, take a camera. Photographs make remembering easier.

Standards

Standards define what is acceptable. There are hundreds of standards written by various organizations. Appendix D provides a list of standards as well as of organizations that write standards. Some standards become law and are referred to as codes—for example, the ASME Boiler and Pressure Vessel Code, which specifies how pressure vessels must be designed. Equipment that does not meet the standard will not be approved for use, nor will it be purchased. Codes and standards that apply to your design will aid in defining the design requirements. In the case of codes, it is mandatory that they are followed. Your design may have to be approved by appropriate governing bodies before it can be used; it will not be approved if it does not meet applicable codes. A search for applicable standards before you start will give you information on what requirements must be satisfied. Your client will undoubtedly be concerned if applicable standards are not met.

Patents

Patents are an excellent source of information and ideas. Patents issued by different countries can be found on the World Wide Web. A search may show that there is a patent on the item you want. You may be able to purchase or license the appropriate equipment or system. This could save considerable design and development time. Find out how others have solved similar problems and how previously designed equipment works. Remember, a patent does not guarantee that the item will work or is practical.

Books and Journals

Books and engineering journals are good sources of information. There are hundreds of technical journals. Engineering organizations, such as the ASME, SAE, and ASHRAE (American Society of Heating, Ventilating and Refrigeration Engineers), to name only a few, publish papers and journals on a range of topics. These can usually be found on the organizations' Web sites, and there are many online journals.

Search for Possible Solutions

"The way to get good ideas is to get lots of ideas and throw the bad ones away."

Linus Pauling

There is no unique solution to an engineering design problem. There is no "best solution" that works every time. Revisions and changes are part of engineering design. Engineering design requires creativity, but what if you think you are not creative enough? Some people are more creative than others, but all of us can be creative in different degrees. Later, we will present some techniques to aid creativity, and with a little practice, you will be surprised what you can do. Often, creativity can mean just another way of looking at things.

The more solutions you can come up with, the greater the chances are that some of them will be good (or better, anyway). Brainstorming is one technique you can use to generate possible solutions. A group of those familiar with the problem will suggest possible solutions, which are recorded. There should be no discussion, evaluation, or criticism of any suggestion. Most of the solutions suggested will not be suitable, but you will determine that later. The task in a brainstorming session is to generate lots of ideas. When one member of the team suggests a possible solution, it often triggers an idea from another team member, who perhaps suggests a modification or something completely new. Often, one idea of a possible solution leads to another, and you continually improve your design. Do not get fixated on a design and close your mind to other possibilities. Keep an open mind, and do not be afraid to present your idea even if you think it is not a good one. It may trigger another, perhaps better, idea from someone else. When the session is over, all that has been recorded can be evaluated and possible solutions investigated further.

In your investigation into alternative solutions, you should look frequently at your original problem statement and goals and ask if they are still applicable. They should be revised if they no longer apply. Your client may also request changes that will require you to rework the problem definition. This happens regularly.

Time is a major constraint in any engineering project. There is never enough time. One can look at alternative solutions forever, but you have to stop somewhere and investigate what you think are the best. Some suggested solutions are easy to throw out, perhaps because of cost or availability of material.

Evaluate Possible Solutions

When you have decided on a number of possible designs, you will analyze them in more detail and get more information. Presumably each design meets the design criteria, some perhaps better than others; then you will have to decide what features are most important to you. Your analysis will include technical and nontechnical items. Nontechnical factors, such as cost and delivery, may have more impact. As a designer, you strive for a design that meets requirements, and it would be nice if details like cost could be ignored. However, cost is something that must be considered. Lengthy design and implementation time increases cost. A project requiring a long development time may not proceed if the uncertainty does not justify the return. Environmental effects and sustainability may also have to be considered.

Your analysis may show that your design requires material that is not easily available. This may not be a problem if the increased cost of getting the material custom made is justified. This consideration may result in changing the design so that more easily available materials are used. For example, if you want a special section extruded from aluminum, you can find someone who will make it for you. That will require a new extrusion die, and the supplier will require you to buy a minimum length to cover the cost of making the die. There could be a long wait before you get what you want. If the minimum order is 500 m, and you need only 5 m, you should revise your design.

Recommend a Course of Action

This follows from the conclusions reached in your evaluation of possible solutions and is supported by your investigations. Recommend a course of action in clear concise statements.

Detail Design and Implementation

Your recommendations may be accepted, but there will no doubt be some changes to them. The budget may be lower than expected, and the scope of the project may therefore have to be changed. This will result in revisiting the problem definition to ensure that the same goals apply. Perhaps these goals cannot be met with the new budget constraint. So, even at this stage, when you think you have a final design, there can be changes that may require looking back. Changes occur at all stages of an engineering project, even during construction.

DESIGN CONSIDERATIONS

The following could be called types of design, but all of them must be considered to some degree in every design.

Design for Safety

If you read the Code of Ethics of your local Association of Professional Engineers, you will find that safety is one of the first things mentioned. Safety is paramount, and it must therefore be an integral part of any design. You cannot rely on a sign or written instructions to indicate that something is not safe. If there is a possibility that something could be used in an unsafe way, someone will use it that way. Design the object such that it cannot be used in an unsafe way. If something should only be used in one way, design it so that it cannot be used in any other way.

Assume that something will fail (something always does eventually), and consider the consequences of the failure. For example, should a backup pump be included in case the first one fails? Some situations require triple redundancy so that even if two pumps failed, the system would still operate. The pumps could have three separate power sources—regular electrical power, diesel-powered generators, and battery power—for a critical application.

Design for Manufacture and Assembly

Design your system or equipment such that it can be manufactured in the most efficient and cost-effective way. Manufacturing aspects must be considered at the design stage. Designers must know something about manufacturing methods, but they cannot be expected to know everything. This is why a manufacturing specialist should be on the design team or consulted early in the design process. **Concurrent design**, where all phases—from design, to manufacture, to distribution, to recycling—are considered by specialists in each area working together, helps ensure that problems are considered at the design stage. It is much easier and cheaper to fix problems at the design stage. Consider the costs of recalling thousands of items because of a design defect.

Manufacturing methods will depend on how many units are to be manufactured. A one-off prototype will be manufactured much differently from the final version. If you are designing something that will be made once for a specific purpose, it will be made in a different way than something that is intended for mass production. It is cost effective to add design features to assist manufacture if the production run is large. For example, include holding or locating features, used during manufacture, to reduce set up time. If you are only going to make one or two units, less time and money can be spent considering the manufacturing method. This does not mean that manufacturing is not important. It just means that possible cost savings must be balanced against increased design effort.

Assembly (and disassembly) must be considered at the design stage, whether the product is assembled automatically by robots in the factory or manually in the field.

Machine assembly can be complex, and the product must be designed with this in mind. It is easier and cheaper to include features for machine assembly at the design stage than to change the design later in the process. Special jigs and fixtures used in the assembly process should be considered at the design stage. These make assembly easier, faster, and repeatable. Instead of designing one large assembly, small modules can be used to simplify the process. Modular assembly can also make shipping easier. Large items and systems are assembled in the field by technically competent persons, but they may not be familiar with all the details of the product. Clear, explicit drawings and instructions are therefore required to facilitate problem-free assembly.

Design for Maintenance

If maintenance is not considered at the design stage, it can be very difficult and time consuming when it is required. The fact that parts wear out and have to be replaced must be considered at the design stage so that provision can be made for efficient replacement and downtime is minimized. The cost of lost production can be huge.

You know which parts wear out and will need replacing and suppliers will usually tell you the life expectancy for the use you have in mind. Design the equipment such that these parts can be inspected and replaced easily. Parts should be replaced before they wear out so that when they fail, it will not cause damage and lost production.

Design for Humans

It may seem obvious that people have to use the machines that are designed, and yet it is surprising that many machines are not designed with this in mind. Adapt the design to the human operator, and do not expect the operator to adapt to the machine. Repeated motion, even without much force, leads to injury—muscular skeletal injury (MSI), soft tissue injury, and repetitive strain injury (RSI). Examples include carpal tunnel syndrome, tendonitis, and tennis elbow (epicondylitis).

These types of injuries result from awkward postures, material handling and lifting, repetitive motion, and poorly designed work stations. If you design something that is going to be used by a human operator, you have to ensure that your product or system is suited to human use. The study of human capabilities and limitations to the design of products and systems is called **ergonomics**, or human factors engineering. Information on capabilities, limitations, and sizes of humans is available.

TEAMWORK

Teamwork is superior to individual work. I had good ideas, but could only think about one of them at a time. By discussing our ideas we came up with the best idea between us.

Teamwork requires participation. I learned to be a leader and a speaker.

Student comment

The Engineering Team

A discussion of the design team is included because most engineers work in teams and employers want people who can work with a team. Only a small engineering project would be handled by one person, and even then, one must be able to communicate one's ideas. A large project will have a large team working on different parts of the design. The organization and scheduling of a large engineering project can be a challenging task, as it involves a large number of people. Obviously, teamwork and the ability to communicate with all members of the team are important.

A project team is not usually self-selected unless there are only a few people available in a small firm and there is not much choice who one works with. Team size will vary depending on the project, and all members may not be engineers. There could be architects, a financial person dealing with budget and costs, and a representative from the client, to name only a few. The team must work together to achieve a common goal—the satisfactory completion of the project.

No doubt you have been on teams that worked well together and teams that did not. What about a soccer team where one player never passed the ball? If the team is not working well, sometimes you can simply leave it, but with an engineering project team, this is often not a possibility. You have to make it work. Understanding team dynamics will aid you in handling problems which inevitably arise.

Group dynamics and behaviour have been studied extensively and show that there are a few things that must be in place for a group to function effectively. Some of these may seem like common sense, but even on a small project, and particularly on a large one, they must be done. These tasks are the responsibility of all team members, and at some stage, all members of a student team will take on these roles.

> *In a group project you need one member to oversee the whole project. We did not do this, and I now see that this is time consuming but essential. Someone must set schedules, divide work fairly and maintain communication between group members. This person must be responsible for bringing the project together.*
>
> *Student comment*

Initiating Activity

This helps the group get started. There will be times when the group is at a standstill, and the "initiator" can suggest new solutions, other ways to look at the problem, or a redefinition of tasks.

Getting Information

Information is required at all stages of the work. Someone must get it or assign some one else to get it and then ensure that it is received on schedule. All group members will be required to find information at one time or another and present it to the group. If someone fails to get what is needed, the process will be disrupted. If there are reasons that someone cannot find what is required, they must be made known as soon as possible. Perhaps other team members can help.

Seeking Opinion

Generally, for a group to be effective, everyone must work together to reach a consensus. One member should not dominate and force his or her ideas and opinions on other members. Opinions of all members are important. It is the duty of the group to hear all opinions. Some members may be reluctant to speak out. The group atmosphere must be such that they will feel comfortable to speak out, confident that they will be heard. Someone must facilitate this, for example, by asking "What do you think we should do?"

Giving Opinion

Giving your opinion is often difficult, particularly if it differs from that of the rest of the group. Remember, your opinion is what you think, based on how you see the facts. Differing opinions can give the team other ways of looking at a problem and can thus make a valuable contribution. Others may not agree with your opinion, and this should not be taken as a personal attack on you. There is usually disagreement among team members because of their backgrounds and how they see things; this diversity means a better solution because the problem is seen in different ways.

Elaborating

Because we all perceive a problem differently, our views may not be clear to everyone. Team members have the responsibility to ensure that everyone understands what is presented. Perhaps you can elaborate on what someone else presents so that others will understand it. Often, rephrasing something will help understanding.

Coordinating

The tasks required of team members must be coordinated to ensure that they contribute to reaching the common goal. Some things must be done before others in order to keep the process on track. New tasks must be assigned, and human resources diverted to particular tasks if any team member needs help. Some tasks turn out to be larger than expected.

Summarizing

At some stage in the process, the team will have so much information coming in that they will not know what to do with all of it. This is the time to summarize what has been done and what has to be done and sort out how to get back on track. It could be the task of the group leader or another member to step back from the immediate details and bring cohesion to the process.

Record Keeping

Someone must be responsible for keeping written records of everything the group does. Write down whatever happens or is discussed immediately, and record the date. Decisions and actions, the person responsible, and the timeline must be recorded. Sketches and drawings must be saved. Every project involves periodic reports and a final report—to the client, your supervisor, and others—and your day-to-day records will make it much easier to prepare these reports. Every member should have his or her own log book and should record everything that is done and what is not done and the reasons—calculations, sketches, phone calls, and anything else that pertains to the project.

Things That Maintain the Group Spirit

Encouragement

The group atmosphere must encourage all members to contribute. Listen to other members' ideas and opinions without judgment (that can come later) and without interrupting. You must be a speaker as well as a listener. Encourage a diffident or shy team member to express his or her opinion. Some are reluctant to speak for fear of being criticized or making a mistake. Recall that since there are no right answers, it is hard to make a mistake; besides, in the creative process, "mistakes" often lead to the best solution.

Group Standards

The group should set standards for evaluating decisions and for the expectations of the team (e.g., do not interrupt when someone is speaking). Read the Code of Ethics of your local Professional Engineering Association to find out what standards are expected of a professional engineer.

Things That Can Sabotage the Group

Here are some of the characters that can sabotage the group effort. No doubt you have encountered some of these. Some of these types seem to be in every group, and knowing how to recognize these characters is the first step in dealing with them.

The Aggressor

The aggressor disrupts the group by attacking other members and being angry, sarcastic, and demeaning. Other members will not speak out because they do not want to be criticized and attacked by this person. Point out the inappropriate behaviour. Establish group standards at the start of the project that rule out this sort of hurtful and disruptive behaviour, and remind the aggressor of them.

The Blocker

Blockers are negative and stubborn. They will unreasonably oppose suggestions and attempt to go back to previous issues that have already been dealt with. You may not be able to convince a blocker to change his or her mind so that work can move forward. Remind this person of the goals of the group and what needs to be done to reach them. Remind him or her of the needs of others in the group. If there is one area that is continuously disrupted, try to find out why. Suggest that this person walk away from the discussion of the offending issue, since it arouses such negative feelings.

The Dominator

This character tries to assert authority over all or part of the group by assuming a superior status and directing actions, while ignoring the wishes and opinions of others. Get the dominator to focus on the group standards and on the importance of all opinions being heard.

The Pessimist

The statement most commonly heard from the pessimist is "We can't do that." This can prevent the group from moving forward. Problems should be investigated before deciding they cannot be solved. Break the problem into smaller parts, and try to solve the small parts. Point out that the group goal must be achieved, and ask your pessimist to define what has to be done to solve the problem. For example, ask the pessimist to complete the following sentence: "The problem cannot be solved, unless..." Thus, while you acknowledge that it "cannot be done," you leave a way open to find the solution. The pessimist may be right (some things just cannot be done), and the group can save time by rejecting impossible or impractical solutions. Your pessimist can be useful in pointing out what can go wrong so that you can work that possibility into the design.

The Noninvolved

This character contributes nothing to the group effort, being "too busy" doing other things. Other group members have to do more work to make up for this person's lack of effort. Your aim is to get some contribution out of this character. Point out the facts—things not done, meetings and deadlines missed, and so on. It is hard for the noninvolved to dispute facts. For this person, try to find tasks that have to be done at a specific time. If you still encounter noncooperation, you could suggest that he or she leave the group, although this may not always be possible. Suggest that since there has been no contribution to the group goals, the person's name will not be on the final report and that no credit for the group's achievement will be given to that person. You may need outside help with this character, so talk to your instructor or supervisor.

The Irrelevant Person

This person loves to talk, but not about the project. The irrelevant person will direct discussion toward something he or she is interested in but completely extraneous to the project. This is easy to do, and we all tend to do it in varying degrees. The leader should point out that this is getting off the topic and steer the discussion in the right direction. The leader (or another group member) should point out that time is limited and so it is important to keep to the topic. You can put a time limit on your meetings—say, one hour.

CREATIVITY

> *Creativity does not necessarily mean completely original. Creativity is the ability to see connections and relationships, and bring together existing concepts in a new way.*
>
> *Student comment*

There is no satisfactory definition of "creativity" and no formula for being creative. The process is not logical or rational. It requires hard work, conflict, timing, luck, and an open mind. Just looking at a problem from a different angle can lead to a creative solution. Creativity requires thinking versus memory, divergent thinking versus convergent thinking. Creative people have many ideas, but only a few good ones. Thomas Edison had over 1300 patents, but he is remembered for only two. There is no tried and true formula for creativity, but there are things you can do to stimulate creativity.

The heuristic methods given here will help guide your thinking. "Heuristic" comes from a Greek word meaning to invent or to discover. These are techniques you can use to discover, solve, and find out things for yourself. These methods are an aid to creativity in that they give a plan for looking at the problem in different ways. Students have commented that these ideas are common sense but are surprised to see how many they end up using in their design projects.

The Repeated Element Heuristic

Try repeating an element that works in another place and repeating it as many times as necessary. It could be a suitable building block for other applications. Is there something you have seen somewhere else that could be adapted for use in your application? Or is there a feature in your design that can be used more than once?

The Fine-Tuning Heuristic

When all the parts are in place, try rearranging them for better performance or ease of assembly or disassembly. Could features be changed to make it easier and faster to make your product? Could standard parts be used instead of manufactured parts? Standard parts are less expensive and can be purchased off the shelf, and there are thousands of them available.

The Deleted Feature Heuristic

We often start or end up with unnecessarily complex items. Can a component be eliminated or simplified? Albert Einstein said, "Things should be as simple as possible, but not simpler." If there are two possible solutions to a problem, choose the simplest.

> *A simple design with large tolerances is much easier and cheaper to build, and is often more reliable, than a complicated and difficult one.*
>
> *Student comment*

The Added Feature Heuristic

Can something be added that will enable the design to meet the objectives? But be careful not to add too much and increase complexity.

The Scale Heuristic

Try changing the size of one or more features. This could open new possibilities. Increasing or decreasing the size could expand the uses.

The Matching Heuristic

The way features are linked is fundamental, and changing the links could open new possibilities. For example, would a chain drive be more suitable than a gear drive in a particular application?

The Interpolation Heuristic

Look for intermediate states in a line of development of a product. Can these provide inspiration for something new, building on what has been developed? Is there something you have observed that could be adapted to your situation?

The Extrapolation Heuristic

Look for trends in the development of a product. What is the next step in the development? Could the product be applied to other uses with easy modifications? Use some of the other heuristics to look at possible changes. What is the natural progression of your design? Could it be improved with only a small change?

The Shared Property Heuristic

If your product shares any attributes with existing products, can they be connected or combined to enhance value, eliminate redundancy, or minimize space? Minimizing space can save packaging and shipping costs.

The Simplification Heuristic

Look for ways to make your product simpler. This does not necessarily mean deleting a feature, though that could be one way. Adding a "more complex" feature can result in simplification of the whole item. For example, if you replaced a mechanical timing device with a "more complex" electronic one, you could reduce cost, size, and maintenance. Eliminate unnecessary complication, the key word being "unnecessary." Keep it simple. However, "simple" is a relative term, and what would be called a simple device in one application could appear very complex in another.

Design requires focused, disciplined thinking and hard work. The heuristics above will help. Studies of creative people have shown that their insight, or breakthrough, did not come when they were thinking about it but when they were doing something else (driving a car, taking a shower, reading a book). Learn from these creative people. Learn all you can about the situation, know your problem and your goal, and then forget about it. Do something else. The problem, however, will remain planted firmly in your unconscious mind. Something will trigger an idea for a solution, and it will come suddenly when you least expect it. You may never know what triggered the solution. It may sound strange that solutions to problems come when you are not thinking consciously about them, but they do.

ENGINEERING DESIGN PROJECTS

Requirements for these problems are stated in general terms and are subject to interpretation. You will have to decide on some of the criteria, and do some research to determine some of the variables (standard sizes, material weights) in order to fill in the missing information. Design questions are posed for some of the problems. Prepare drawings to aid you in your design.

1. Design a device to open and close a gate so that a farmer can drive from one field to another without getting out of his or her vehicle to open and close the gate. There is no electricity available.

2. Design a device that can be used by a wheelchair-bound person to get into and out of a bathtub.

3. Design a folding picnic table that can seat a minimum of four adults. Consider size, weight, ease of setup, and stability against overturning and collapse.

4. Design a portable holder for a flagpole. The holder will rest on the ground and hold a flagpole 3 m long and 50 mm in diameter at the base. It should be easy for one person to remove the flagpole from the holder and to move the holder. The flagpole must be held in a vertical position and must not fall over under normal conditions.

   Since the base must be moved by one person, how much weight can an average person carry? The angle at which the flagstaff can be tipped before falling depends on the location of the centre of gravity and the size of the base. How can the base be made stable?

5. Design a device to pour liquid from a 200 L steel drum (a 45 gallon drum.) The drum must be lifted and tilted so that all liquid in the drum can be poured into a 1500 mm high tank. The liquid in the drum has a density 10 percent greater than water. The lifting/tilting device must be easily attached and removed from the drum by one person. A crane is available for lifting the drum. The unit must be easily and safely operated by one person. What are the dimensions of a 200 L drum, and how much will it weigh when full?

6. Design a device to pick up a 500 mm long by 150 mm diameter cylinder from the bottom of a 5 m deep concrete tank filled with clear water. The cylinder contains uranium and is radioactive. It must be kept at least 3 m below the surface. The lifter must be operated by one person. How much does the cylinder weigh? What will the lifting device weigh?

7. Design a device to hold a bicycle when doing maintenance and repairs at home. The device would be held in a vise or rest on the floor and should be easy to operate by one person. It should be capable of firmly holding a bicycle in different positions.

8. Design a simple easel that could be used by children three to five years old for painting or colouring at home. Incorporate some method, which can be used by the child, for holding a paper supply. The easel should fold easily for storage.

9. Design a play table for children to use when playing with small toys. The age range of the children is four to seven years, and the children may sit or stand at the table. Bear in mind that children may climb on the table, and it must be strong enough to allow this.

10. Design a device that will lift a 50 kg crate from the floor to a height of 1 m, where it is slid onto a roller conveyer and taken away. The device must be easy to operate, and the crate must be easily slid onto the roller conveyer. The crate is 300 mm wide, 450 mm long, and 230 mm high.

11. It is often useful to have extra light in a localized area—for example, when painting. Design a fixture that will hold a standard two-tube fluorescent light. The fixture should be able to hold the light so that it illuminates a horizontal surface, such as a table, or a vertical surface, such as an easel. The fixture should be easily movable and take up as little space as possible.

12. Design a support that will hold the central processing unit (CPU) of a personal computer under a table. The holder should provide security against theft and protection in the event of an earthquake. All controls and cable connections must be easily accessible.

13. Design a method of measuring the level of water in a rectangular tank. The tank is made of wood, and no holes of any sort can be put in the walls. A wooden cover is used, and since it is always above the water level, small holes can be made in the cover. Anyone wanting to see what the water level is must be able to do so 6 m from the tank. The tank is in a remote area without electricity, and no battery-powered devices can be used. It should be possible to do assembly and maintenance with simple tools by unskilled workers.

ENGINEERING DESIGN CHALLENGES

The following challenges are to be done by student teams, which may be self-selected or specified. Teams of four are suggested. A small and relatively simple machine will be designed, built, and tested. There are six challenges, and each team of students will be randomly assigned to one of them. A preliminary design should be submitted for review before detailed design is started.

Materials and Restrictions

Your mechanism may take any form you wish, with the following restrictions:

1. Your machine must fit within and perform its task within a (300 + 25) mm cube.
2. The net cost of your machine must be no more than $(40.00 + 2.00)
3. You must design and build your machine yourself. Purchasing a machine capable of performing the required task is not acceptable.

Project Report

This report should detail the progress and process of your design. Be sure to include the following:

- Summary
- Introduction
- At least three of your preliminary designs, evaluated and briefly discussed (about one page each)
- Your design selection criteria, and how each of your design alternatives rated
- Design calculations and assumptions
- Performance expectations
- Final production drawings
- Conclusions
- Detailed budget
- Timesheets for all group members

Challenges

1. *Potential Energy*
   Move the maximum possible mass from A to B using only the stored potential energy of a 100 g mass. Point B is 355 mm away from A and 200 mm above A.

2. *Materials Handling*
   Sort nickels from an equal mix of quarters, dimes, nickels, and pennies.

3. *Sensor/Control Development*
   Devise a rotation counter capable of performing a control action. Your rotation counter must accept input from a rotating shaft, count rotations, and perform a control action after a specified number of rotations.
   - **a.** Discrete rotation counter—your device must count only complete rotations, ignoring incomplete rotations. Perform a control action after six complete rotations.
   - **b.** Continuous rotation counter—your device must count rotations on a continuous basis. Perform a control action after 5.25 rotations.

4. *Pre-existing Element Integration*
   Move a 100 g mass 250 mm horizontally with external power being applied by a vertical, down-facing linear actuator with a maximum stroke of 50 mm. The mass must remain at its final position.

5. *Alternative Energy*
   Lift a 750 g mass 200 mm vertically using only wind power derived from an ordinary hair dryer. Once your mass reaches its 200 mm elevation, it must remain there without applied outside forces.

6. *Repetitive Motion*
   Create a device capable of diverting every single, second, or third equally spaced nickel from a moving belt.

> *…as is made clear and obvious in the illustration because I cannot so well set it forth in words as I see it in my mind's eye. But the picture will show it.*
>
> *Guido da Vigevano, "Guido's Texaurus," 1335*

CHAPTER

9 Solid Modelling

We saw in Chapter 1 how objects can be created by combining simple shapes to form complex objects, and we have used this concept in reverse in Chapter 2 to analyze objects by breaking them into simple shapes. The process of creating complicated objects by building them up from simple shapes is called solid modelling. The most common shapes (called primitives) are shown in Figure 9.1.

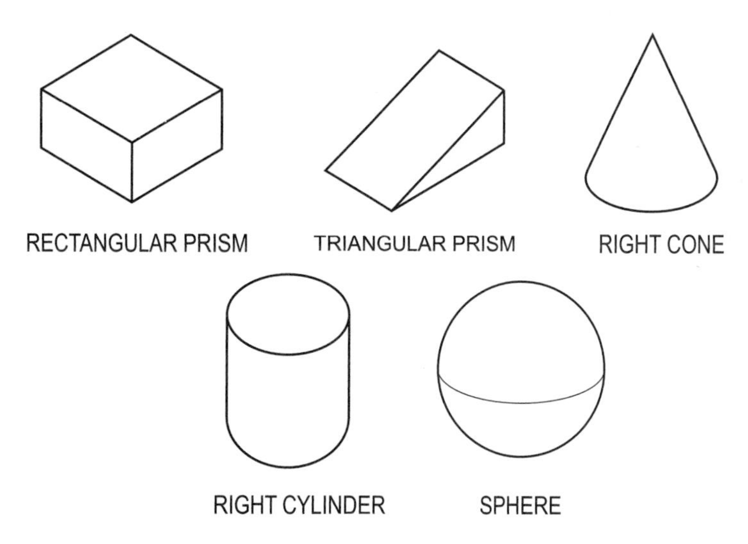

Figure 9.1 Common shapes that are combined to create complex objects

A solid model is made from three-dimensional (3-D) shapes and shows a 3-D representation that is easy to visualize. You can also see it from any viewpoint of your choice by rotating the object.

TYPES OF SOLID MODELS

There are several ways to represent 3-D shapes—wire frame, surface, and true solid models.

Wire Frame Model

The wire frame model shows only the edges. There is no surface joining these edges and thus no distinction between the inside and the outside of the object. You can see through the object. A simple block is shown in Figure 9.2.

There are no surfaces on a wire frame model, so all the edges can be seen; there is no clear distinction which edge is at the front and which is behind. Even a simple object like that shown in Figure 9.2 can be confusing. The wire frame model of a cube in Figure 9.3a shows how a simple model can be confusing, even when you know what it looks like. The top and bottom surfaces appear to change positions as you continue to look at the model, and sometimes you see the top surface and other times the bottom surface.

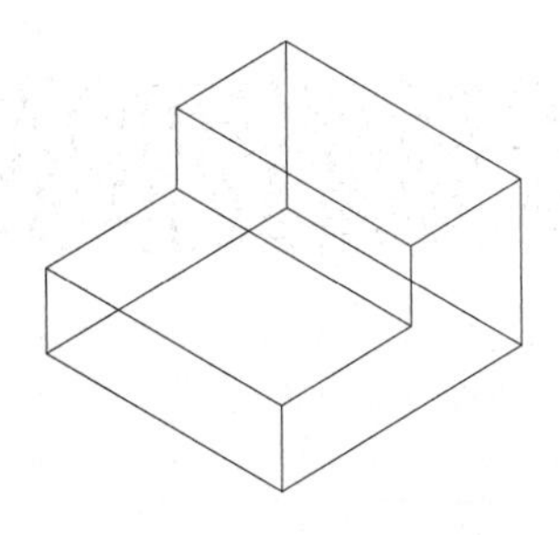

Figure 9.2 A wire frame model of a simple block shows all the edges

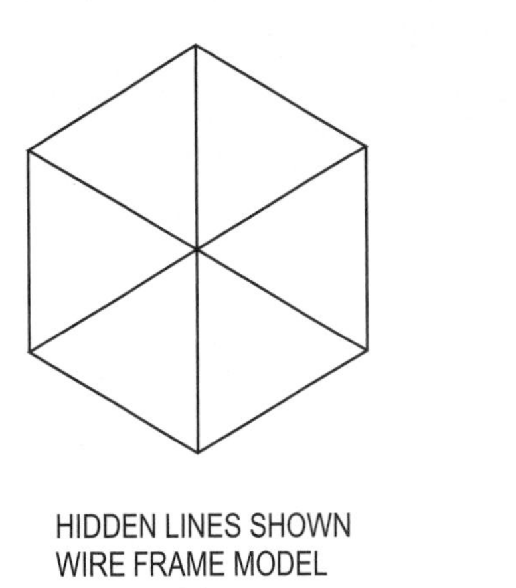

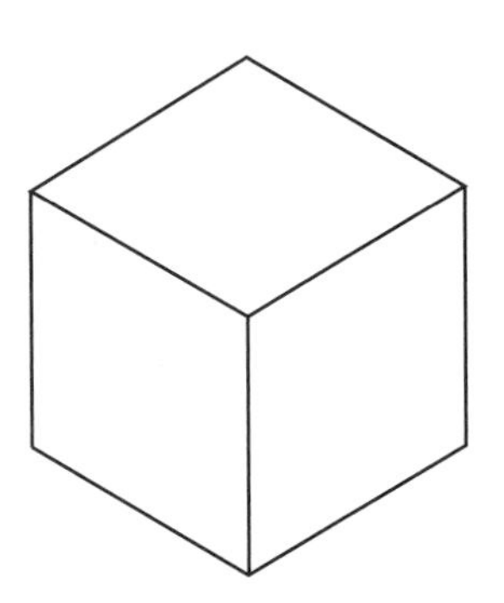

Figure 9.3 Wire frame and surface models of a cube

Wire frame models, however, have the advantage that they can be done quickly. They are often used during the early stages of a design so you can quickly check what you are doing. It helps to know what you are looking at.

Surface Model

A surface model shows only the surface and edges of the object. If you could make a physical model, the inside would be empty. Figure 9.4 shows the surfaces of the object seen in Figure 9.2, and Figure 9.3b shows how adding surfaces to a cube eliminates the confusion associated with a wire frame model.

There are no hidden lines, so there is no confusion as to which are visible and which are not. A surface model shows the shape of the object, but there is no other information about the object, since the model is hollow.

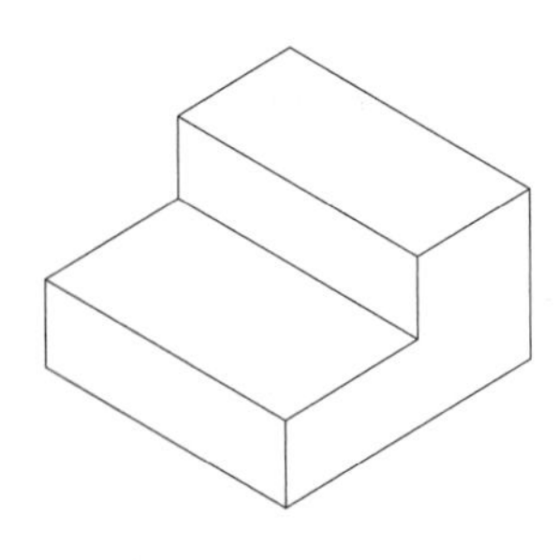

Figure 9.4 A surface model shows surfaces and edges

Solid Model

A solid model also shows surfaces, but there is information about the inside. For example, if the density of the material from which the part is made is specified, the mass could be found. The mass of parts made from different materials could be compared easily. Other properties, such as centre of gravity and moment of inertia, can be determined from the database. Interior details can be seen if sections are taken. This cannot be done with a surface or wire frame model.

How Solid Models Are Made

A solid model can be created either by combining simple solid shapes (called constructive solid geometry, CSG) or by boundary representation (B-rep), which describes the surface of a solid. An orientation convention is used to show which side of the surface is the outside and which the inside. A CSG model has a closed surface and is always valid. A B-rep model is easily rendered on a graphic display. Most modelling programs combine these methods.

When you create a solid model, you create a "part." You can also make an assembly of several parts and a three-view orthographic working drawing giving dimensions and other information needed for manufacturing the part. All of these are created using the database built up when you make the model.

Commands for creating a solid model vary with different programs, but the goal is the same for all modelling programs. You use them to create solid shapes (parts) and combine them to create assemblies. A new design can be created by combining simple solids using some "recipe" that is specified by the user. This creates a design history (often called a design tree) that specifies the shapes used and the processes used to combine them.

Primitives can be selected from a catalogue of available shapes (blocks, cylinders, cones). Parameters are then assigned by the user to each shape. A cylinder, for example, would be assigned a diameter and a length.

It is more common to create shapes yourself, and these do not have to be a simple shapes. They are created by defining an area and then moving this area along a defined path to create a volume. There are several ways to do this, and the method used will depend on the shape that you want to create. A common method of creating shapes is extrusion or sweeping. These are similar, but extrusion usually refers to a straight path and sweeping to a curved path. Different programs may use different names for these processes.

Extrusion

Figure 9.5 shows how a block (a parallelepiped) and a cylinder are created by extruding a rectangle to create the block and a circle to create the cylinder. (This process is also called protrusion.) You must specify the area by defining the cross-section and a path through which it is moved and sometimes the direction in which the area is to be extruded. Extrusion refers to a forming process in which a metal or plastic shape is formed by forcing material through a hole in a piece of metal (called a die) that has the shape you want to make.

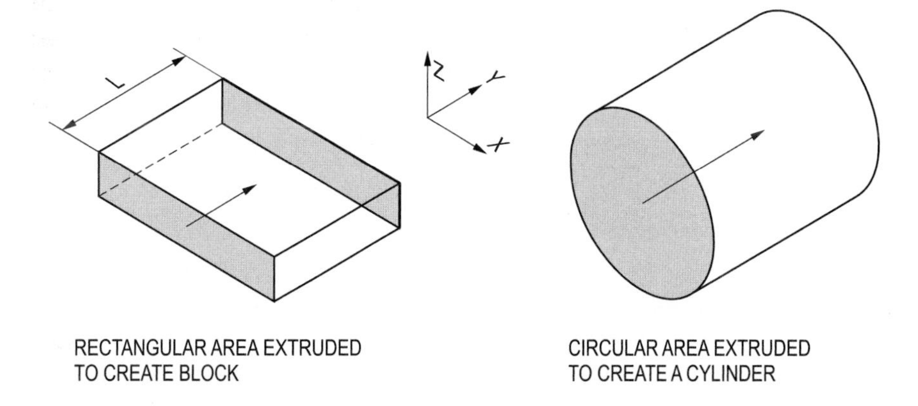

Figure 9.5 Extrusion used to create a block and a cylinder

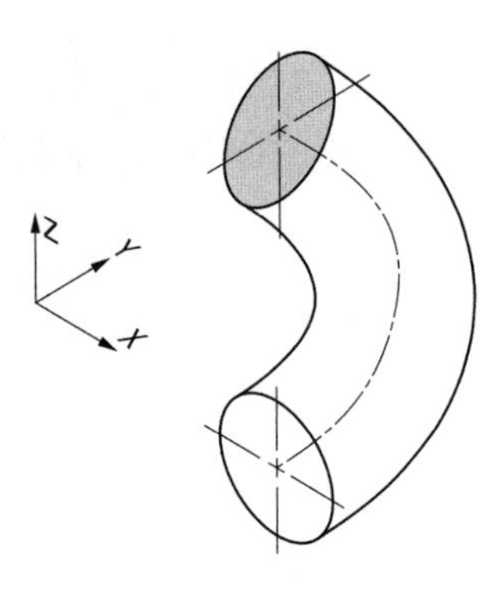

Figure 9.6 Sweeping a circular area to create a curved part

The area to be extruded is drawn in the *x-z* plane and extruded in the *y*-direction. You must choose the plane in which the area is drawn and the extrusion direction to suit the object you want to create.

Sweeping

Sweeping refers to an extrusion along a curved path. Figure 9.6 shows a curved part created by sweeping a circular area along a curved path. The path does not have to be a circular arc.

A circle in a vertical plane (*y-z* plane) is swept along a curved path in a horizontal plane (*x-y* plane) to create a curved part. The area and the sweep path must be is different planes.

Example 9.1

Create a block 100 mm × 50 mm × 15 mm thick.

You will draw a rectangle in the horizontal plane and then extrude it to the required thickness.

1. Select a plane to draw on. The horizontal plane (*x-y*) is chosen for this object.
2. Sketch a rectangle about 50 mm × 100 mm in the horizontal plane. The dimensions do not have to be exact.
3. Specify the length of each side of the rectangle. These dimensions will appear on your sketch, and the length will adjust to the value you specify (**Figure 9.7a**). You may also align the area with some datum, say, at a corner or the centre. You do this by specifying the distance from two edges to the datum. If you dimension some feature twice, your program will tell you that the drawing is overdefined. Your drawing must generally be completely defined (but not overdefined) before you can work with it.
4. Select the rectangle as the area to extrude.
5. Use the extrude command to extrude this area to a height of 15 mm (**Figure 9.7b**). The block has now been created, and other features can be added to it. Dimensioning in this context is not the same as dimensioning an engineering drawing as discussed in Chapter 4. The purpose is to specify the parameters that define some feature of the part. You will specify the length of a line, the diameter of a circle, and the distance from some datum, and any other parameters needed to define the profile.

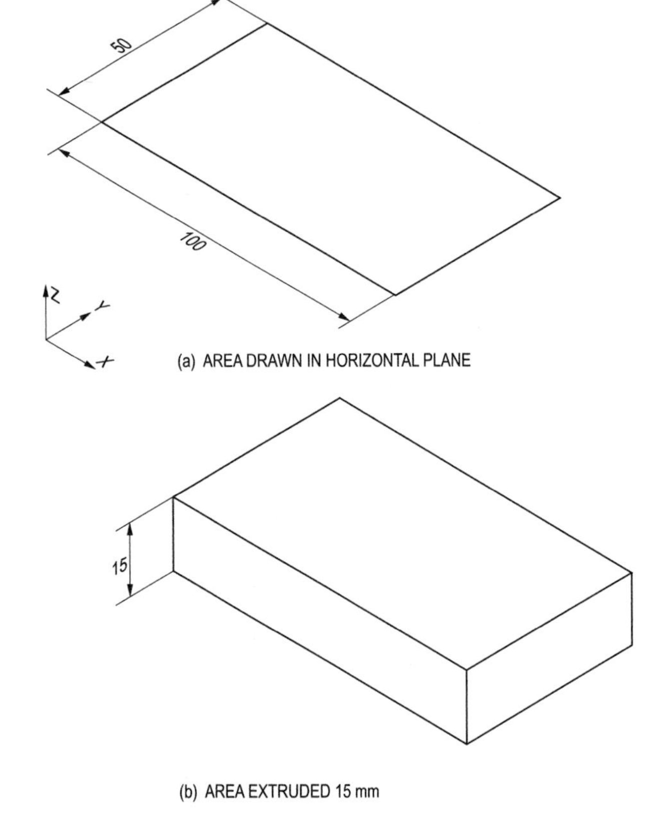

Figure 9.7 Creating a volume by extrusion

Example 9.2

Create a 90° bend, for a 50 mm diameter bar.

A 50 mm diameter circle will be swept along an arc with an included angle of 90°. The circle must be drawn in one plane and then swept through 90° in another plane that makes a right angle with the plane of the circle. **Figure 9.8** shows the bend and the planes used to create it.

1. Choose a vertical plane (right plane is used), and sketch a circle about 50 mm in diameter. Specify the circle diameter (see **Figure 9.8a**).
2. Choose a plane at 90° to the plane in which the circle is drawn (the horizontal plane [*x-y*]), and draw an arc starting at the centre of the circle. The included angle is 90°, and the radius must be large enough so that the area does not intersect with itself when it is swept along the path. Specify the radius of the arc (see **Figure 9.8b**).
3. You will sweep the circle along the arc to create the elbow. Select the circle and then the arc. The sweep command is used to move the circle along the arc.

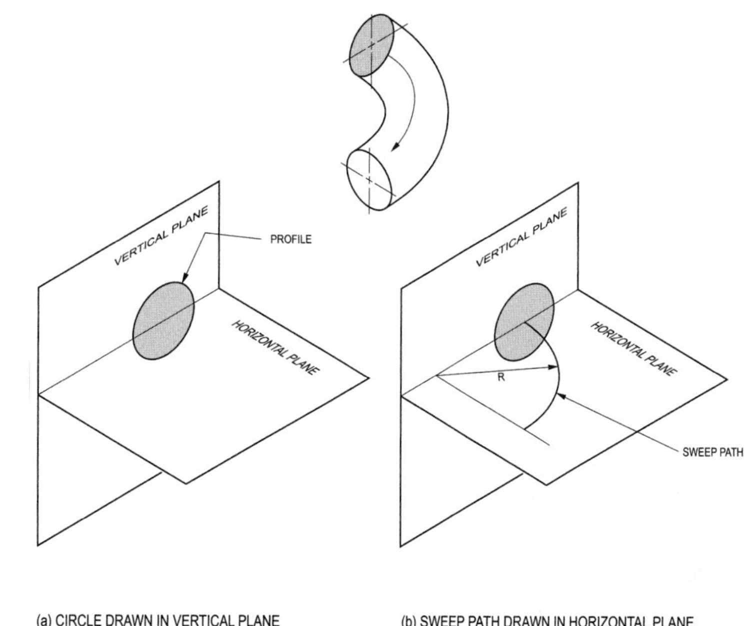

Figure 9.8 Creating a bend with a sweep

The sweep or extrusion path must be in a different plane than the area being swept or extruded. The path does not have to be perpendicular to the area, but it cannot be in the same plane. It could not be in the vertical plane in this example. That would create an area, not a volume. If the path is a circular arc, it is simpler to create it by revolution.

Revolving

Another way to create a volume is by revolving an area around an axis. An area, which must be closed, depending on the modelling program, is revolved around an axis to create a 3-D shape. (Some modelling programs can use an open area.) Figure 9.9 shows an area (triangle) revolved around an axis that coincides with one side.

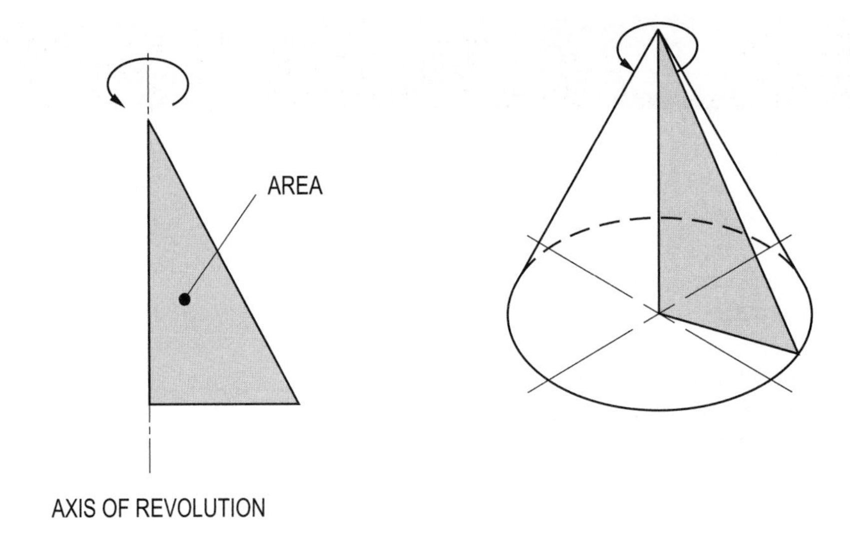

Figure 9.9 Revolving an area to create a cone

The angle through which the area is revolved is specified by the user and need not be 360°, nor does the axis have to pass through the area. Figure 9.10 shows a torus (doughnut shape) created by revolving a circle around a vertical axis.

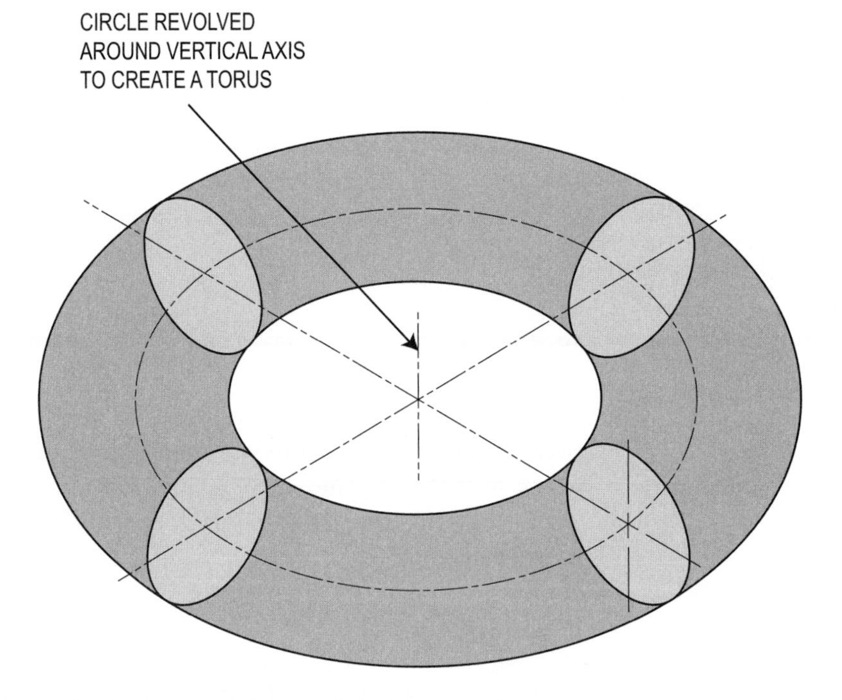

Figure 9.10 A circle revolved about a vertical axis to create a torus

Example 9.3

A simple 2-D shape will be sketched and revolved around a vertical axis. The starting area is determined by the user. **Figure 9.11** shows the shape to be created. A profile will be revolved around a vertical axis to do this.

Sketch the outline of the area you want in the front plane. There are some restrictions on the profile. It must be an area that can be revolved without interfering. That is, it cannot intersect itself. If the area you want to revolve is not suitable, your program will indicate this. **Figure 9.12** shows how the shape is created.

1. Draw a vertical line about 50 mm long, and add straight sections at the top and bottom. Dimension all of these lines (see **Figure 9.12a**).
2. Add a three-point arc starting at the end point of the top horizontal line. Dimension this arc as 10 mm radius. The included angle should be 180° (see **Figure 9.12b**).
3. Add the "triangular" centre section. The vertex is 25 mm from the base and 25 mm from the axis of rotation. The dimensions are shown in **Figure 9.12c.** Join the triangular section to the base, as shown in **Figure 9.12d**, and add a centreline. Use the revolved boss/base command to revolve the area 360° to create the shape.

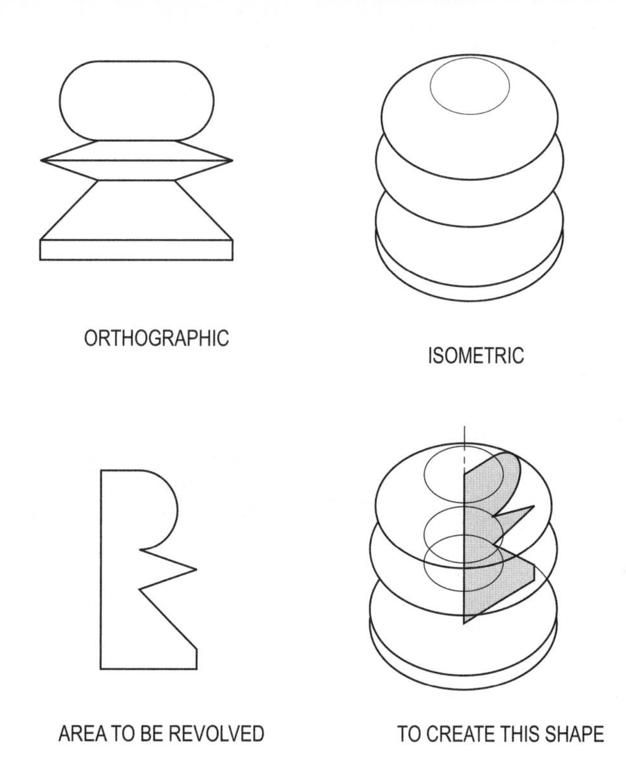

Figure 9.11 Shape created by revolving an area

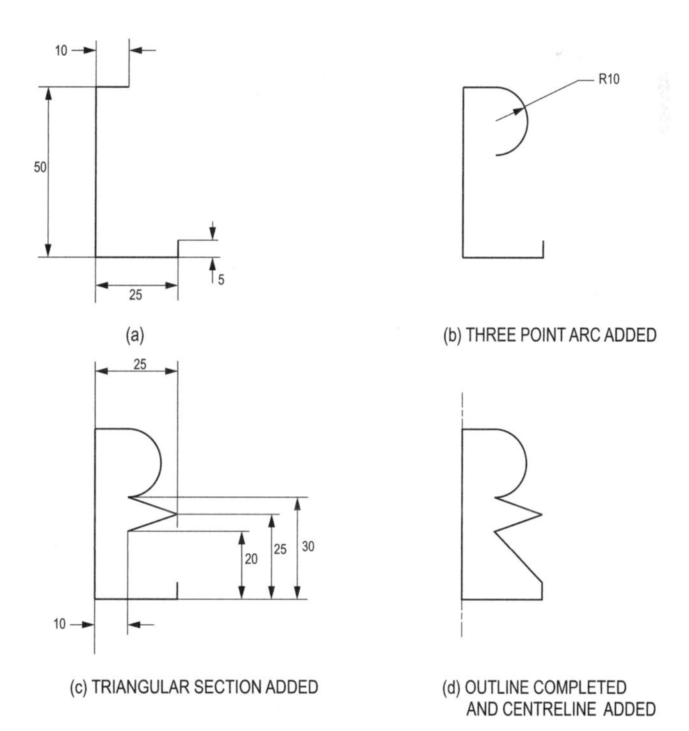

Figure 9.12 Steps in creating a solid by revolving an area

Example 9.3

The axis used for a "revolve" does not have to pass through the area being revolved.

Figure 9.13 shows how a wheel can be created by revolving an area around an axis.

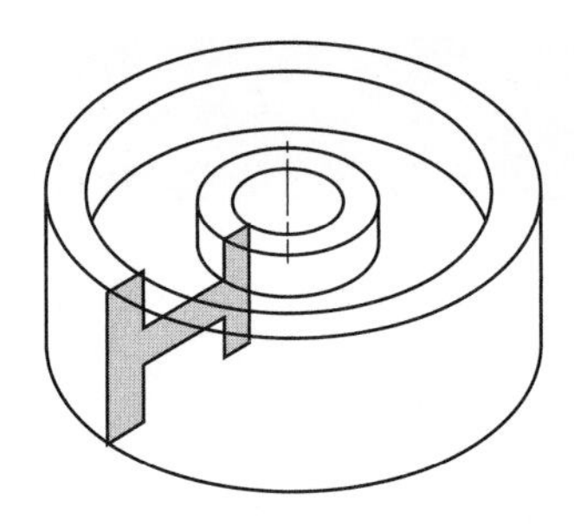

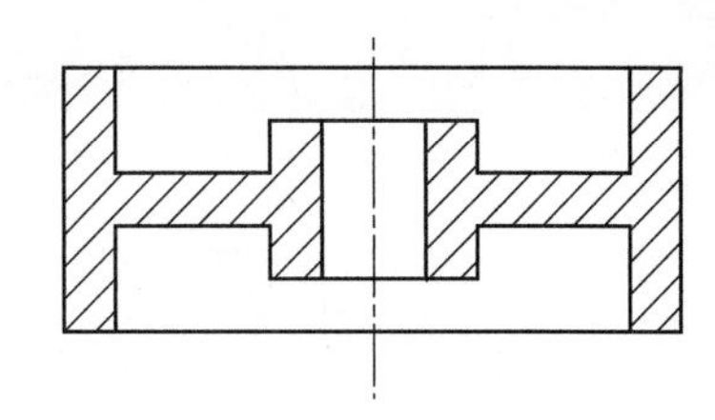

WHEEL CREATED BY REVOLVING

SECTION OF WHEEL

Figure 9.13 Revolving an area to create a wheel

Example 9.4

Create a model of a wheel with the cross-section shown in Figure 9.13. The outside diameter is 100 mm and the inside diameter 20 mm.

The wheel is created by drawing the cross-section in a vertical plane and revolving this area about a vertical axis that does not intersect the area.

1. Sketch a cross-section of the wheel with the dimensions given. The sketch does not have to be exact, since dimensions will be specified (**Figure 9.14a**).
2. Specify the dimensions (see **Figure 9.14b**).
3. Draw a centreline on the vertical axis about which the area is to be revolved (see **Figure 9.14c**).
4. Use the revolved command to revolve the area through 360°.

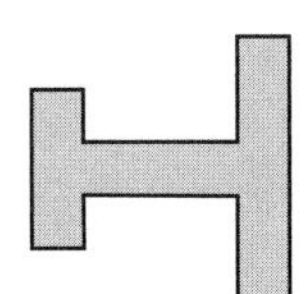

(a) PROFILE DRAWN IN FRONT PLANE

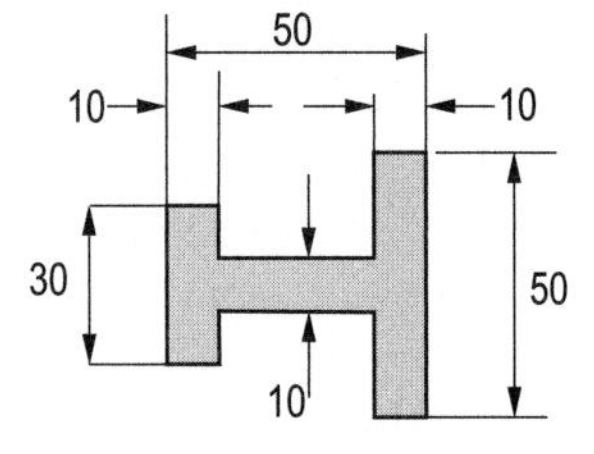

(b) PROFILE DIMENSIONED

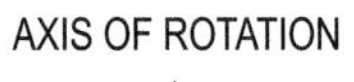

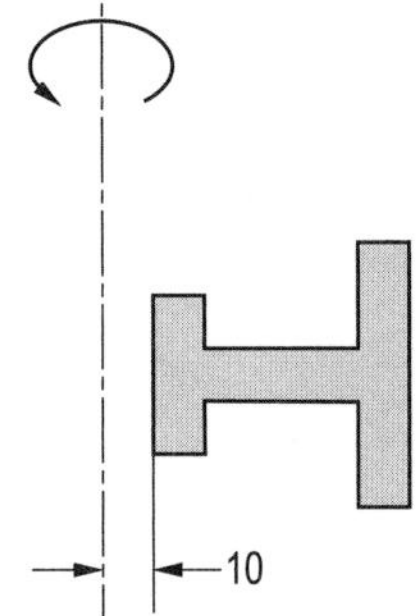

(c) AXIS OF REVOLUTION ADDED

Figure 9.14 Steps in creating a wheel

Combining Shapes

Shapes must be combined to form more complicated shapes, and this is done using Boolean operations of union, difference, and intersection.

The union operation joins two primitives. The difference operation subtracts one from the other, and intersection defines a volume common to both. These are illustrated in Figure 9.15.

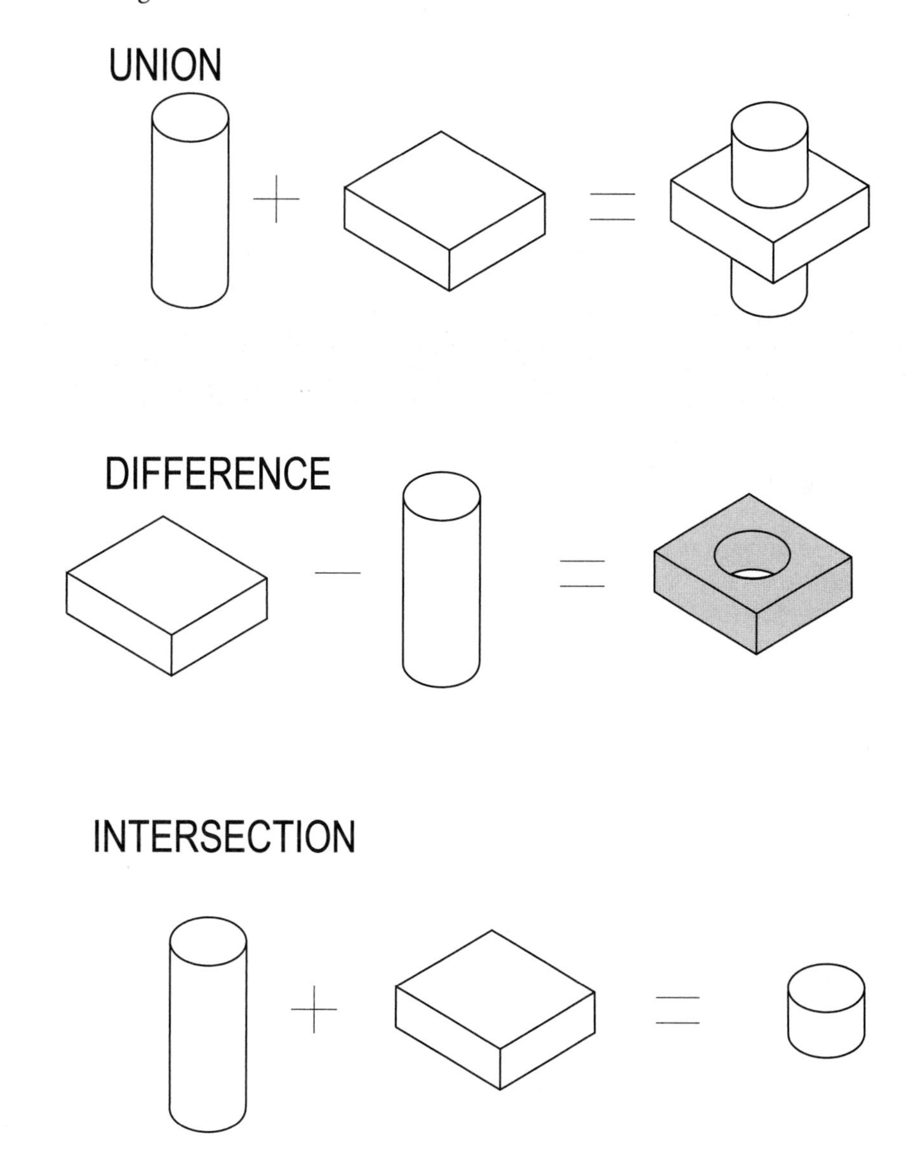

Figure 9.15 Boolean operations: union, difference, and intersection

The three operations are shown with a cylinder and a block. With a union operation, the cylinder and the block are joined. This is like welding the cylinder to the block. The difference operation removes a cylinder from the block. Think of the cylinder as a negative volume or a round hole in the block. The intersection operation results in a cylinder whose length is equal to the thickness of the block. It is like cutting a short section out of the cylinder.

The modelling program you use will make these Boolean operations invisible to the user; however, you must know what you are doing because the order in which operations are done will affect the outcome. You will be concerned with defining the shapes you want to combine and specifying their locations. There are two steps in creating a shape. You must draw the area you want to use to create a specific volume and specify how you want to use it, usually with the appropriate command. The area is drawn with what is usually referred to as a sketch, meaning that you draw the general shape, but not necessarily the correct size. You then dimension your sketch to define the parameters that define the area.

It is fairly easy to specify shapes, but it requires planning to create the most appropriate shapes and combine them in an efficient way. The way in which a model is created is important, or you may not get what you want. The process used to create the model can make it difficult to change later.

Once you have defined the shape, you combine it with another part of your drawing by specifying where it is to be located. The command you use will specify how it is to be combined—whether an addition or a subtraction—in the case of a hole.

Example 9.5

A simple object is created by combining a base, a boss, and two holes. A **boss** is a cylindrical shape attached to a surface. It is a common term used in solid modelling programs. **Figure 9.16** shows the part to be created. The boss is solid.

The order in which parts are combined can determine the resulting solid, and you should plan what you want to do. You must always start with a base feature and add or remove features from it. In this simple example, it is obvious what the base feature must be. We will create the base, add a boss, and cut two round holes. The base will be created by extruding a rectangle. The boss is created by extruding a circle, and the holes will be cut from the base.

The base is 100 mm × 50 mm and 15 mm thick. The boss is 15 mm diameter and 15 mm high. It is 20 mm from the left edge and 20 mm from the near edge. Both holes go through the base. They are 10 mm diameter, 15 mm from the right edge, and equally spaced about the centreline of the base.

Tree → Base + boss – holes

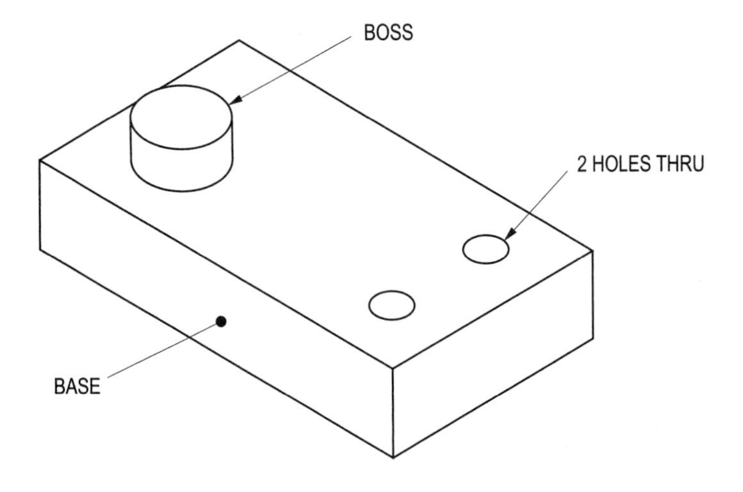

Figure 9.16 Part to be created as a solid model

The base is created by drawing a rectangle in the vertical plane (*y-z* plane) and extruding it in the *x*-direction. You should carefully choose the plane in which you will draw the base feature, since it determines the orientation of the part. You could draw a rectangle in the *x-y* plane and extrude it vertically.

1. Draw a rectangle and dimension the sides (50 × 15). The rectangle does not have to be drawn to the dimensions given, but it is good practice to draw to the approximate dimensions (these are shown as you move the cursor). This gives you a better picture of what the part will look like. The exact dimensions are specified, and the profile adjusts to the correct size.
2. Extrude this area through 100 mm (**Figure 9.17a**).
3. Add the boss. Specify face on which you want to put the boss (the top face), and draw a circle in the approximate location of the boss. It is not necessary to draw the exact diameter or the exact location. These parameters will be specified.
4. Specify the location and diameter of the boss (15 mm) (see **Figure 9.17b**). Dimension the distance from the centre of the circle to the left edge of the base (20 mm). Dimension the distance from the centre of the circle to the near edge of the base (20 mm).
5. Indicate what feature you want to extrude (the circle representing the boss). Specify the length of the extrusion (15 mm) (see **Figure 9.17c**).
6. Holes are created by cutting them from the base. You must specify the size, location, and depth of the hole. In this case, the holes go through the base. This process can be thought of as extruding a negative area through the base.

   Draw circles at the approximate locations of the holes.

   Dimension the diameter of the circles (10 mm).

   Dimension the location of the centre from the right edge (15 mm).

   Dimension the location of the centre from the near edge (15 mm) (see **Figure 9.17d**).

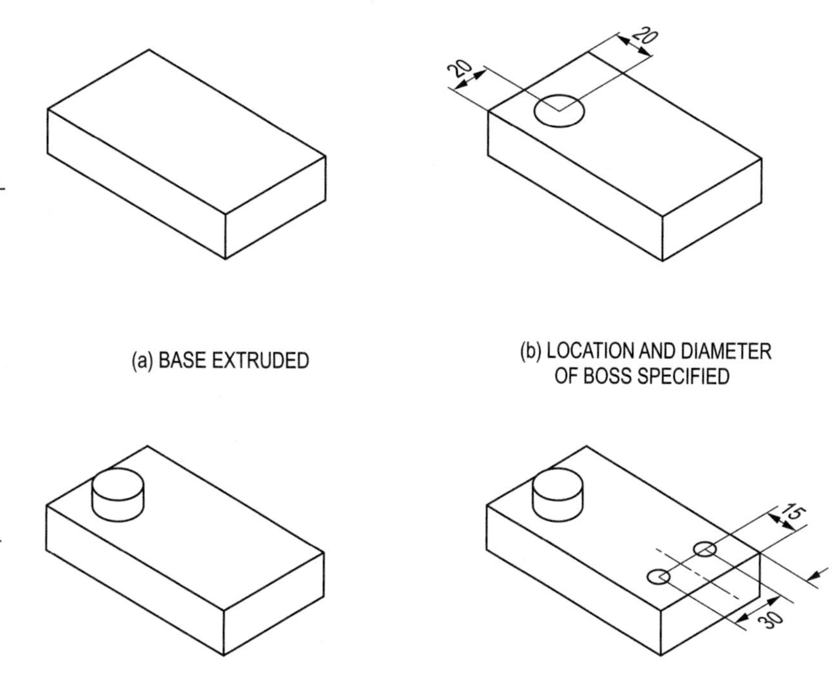

Figure 9.17 Steps in creating a model by combining and removing

Example 9.5

Since these are to be holes, you will use a "cut" through the base to make them. Since the holes go all the way through the base, they can be specified as **through holes** (see Figure 9.18).

Both holes should be added as one feature instead of individually. This way, if the size is changed, it will be changed on both holes. If they are specified individually, any subsequent changes must be made to each hole.

FEATURES

A feature could be anything you create, but there are features that are commonly used and are catalogued in the modelling program. You need to only specify the feature, size, and location. These are sometimes called "intelligent shapes." Dimensions for features are commonly added when the feature is specified.

One of the simplest common features is a round hole. Figure 9.18 shows a through hole and a blind hole. All that is required is to specify the hole location, the type (THRU or blind), diameter, and depth for a blind hole. There is no need to specify the depth of a through hole, since it is determined by the part that the hole is in. Other examples of commonly used features are hole shapes (e.g., counterbored, counterdrilled) used for different fasteners. (These were shown in Chapter 4.)

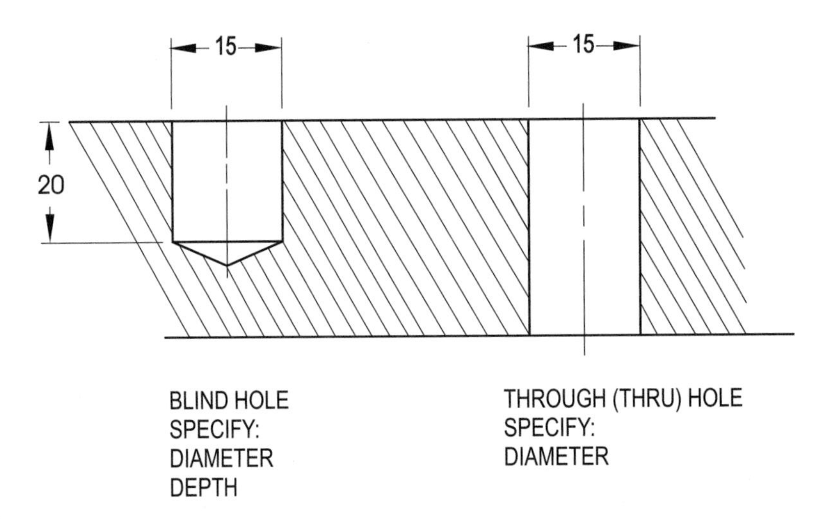

Figure 9.18 A simple feature is a blind hole or a through hole

Other commonly used features are fillets, chamfers, and rounds. Figure 9.19 shows these features added to the edges of a part. These shapes are catalogued in the modelling program, and the user has only to call them up and specify the location and size. The location of these features is specified by selecting the desired edge or edges. It is generally better to add fillets after the part has been created. This speeds up redrawing the part while you are designing it.

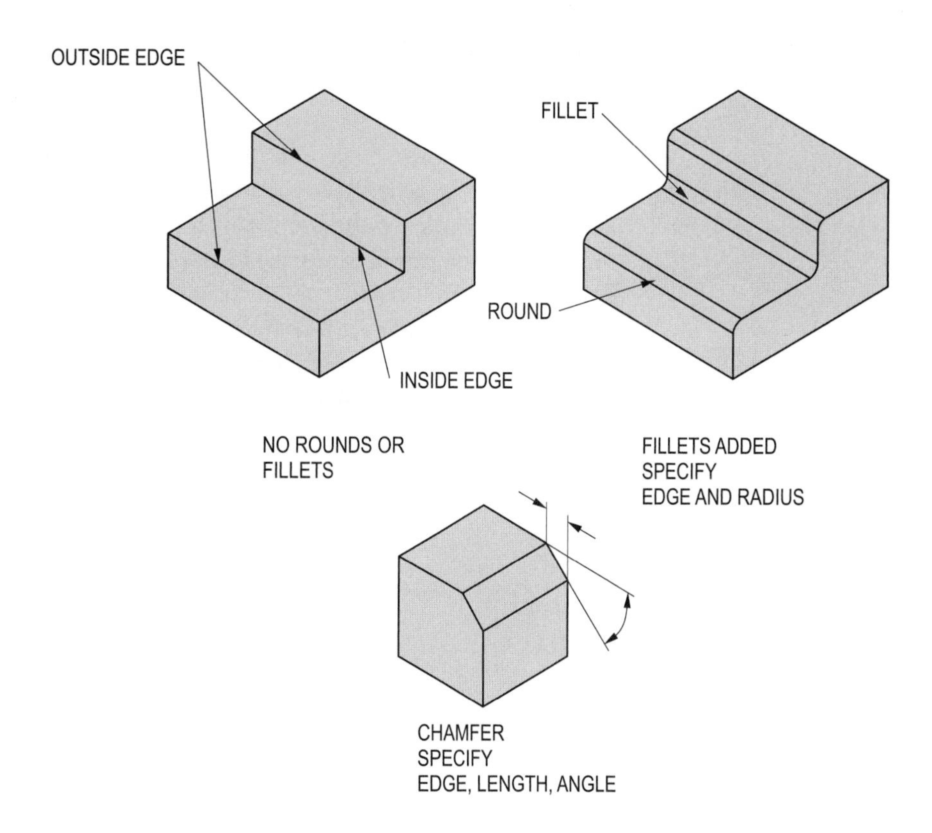

Figure 9.19 Fillets and chamfers are added by specifying an edge and parameters

Thin-Walled Objects

Shell

A thin-walled part—a box, for example—is created by removing the interior from a solid. The object is created as a solid and then "hollowed out" using a shell command. The face to be "removed" and the wall thickness are specified to create the hollow object. Figure 9.20 shows a box created with a shell command. The bottom of the box has the same thickness as the sides.

A 50 × 50 mm square was drawn on the top plane and extruded 40 mm vertically to create the solid. A shell was made by identifying the top surface as the face to be removed, and a wall thickness of 2 mm was specified.

You can shell inward or outward. Shelling inward means that the surface you drew on the original shape becomes the outside surface. If you shell outward, the outside dimension increases, and the original surface becomes the inside surface. The box in Figure 9.20 was shelled inward, and so the outside dimensions are 50 × 50 mm. The inside dimensions would be 50 × 50 mm, and the outside dimensions would be 54 × 54 mm, if shelled outward.

All sides of a thin object do not have to be the same thickness. Different thicknesses can be specified for different walls with some programs.

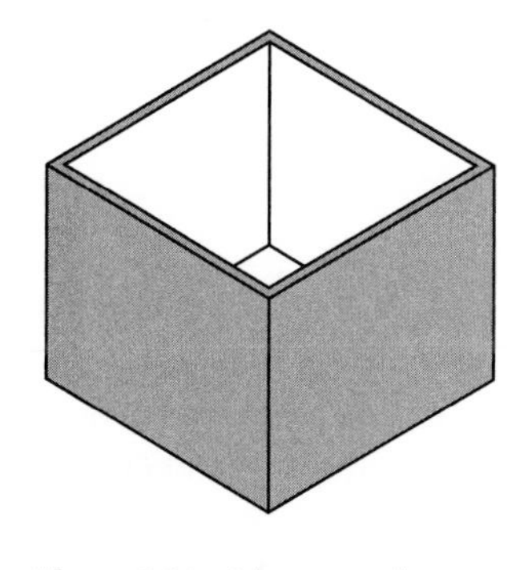

Figure 9.20 A box created as a shell

Extruding a Thin Object

A thin-walled object, such as a tube, is not a shell. The best way to create a part like this is to extrude a thin feature. A profile is created and then extruded as a thin feature

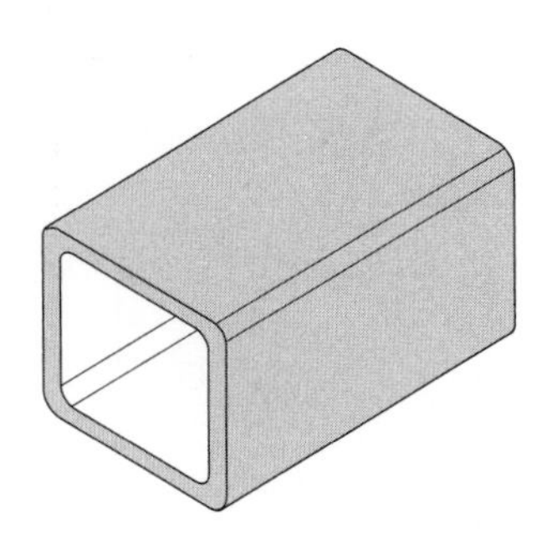

Figure 9.21 A hollow section created by extruding as a thin feature

by specifying a wall thickness. Figure 9.21 shows an example of a hollow structural section created by extruding as a thin section. Fillets and rounds were added after extruding.

The shape could have been drawn with rounded corners and then extruded, but this is a more complex shape than a square. It takes longer to draw and dimension and contains more information, which must be processed as you build the part. You should keep the profile you plan to extrude as simple as possible and add features to it. A simple sketch is easier to create and to edit, and it is easier to change individual features than to change a complex sketch.

A hollow part, such as a tube or pipe, can be created by sweeping a thin feature along a curved path. The 90° tube bend was created by sweeping a circle drawn in a vertical plane along a curved sweep path in a horizontal plane as a thin feature (Figure 9.22).

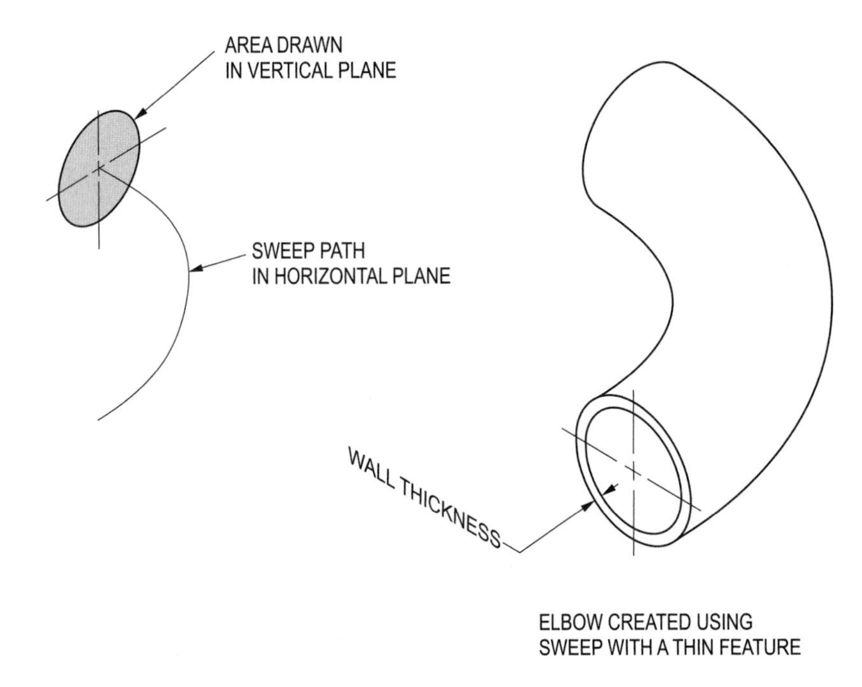

Figure 9.22 A hollow fitting created by sweeping a thin feature

There are other ways to create hollow features. For example, a system of pipes and tanks can be converted from solid to thin-walled using a modified shell command. This depends on the specific program being used

Creating a Drawing

Solid models are shown as pictorial views because they are easy to visualize. Isometric views are commonly used, but there are other types of pictorials. This is useful for visualization, and the model can be viewed from any position, but for construction, orthographic working drawings with materials and dimension are required. These can be created from the database and presented as a multiview drawing.

How this is done depends on the program you are using, but it usually means creating a working drawing that can be dimensioned. Modelling programs can do this after you have finished the solid model of the part. Generally, you create a drawing and specify the views you want. You can also add a pictorial to the drawing to aid visualization. Figure 9.23 shows a drawing created from the model created in Figure 9.16 (page 300).

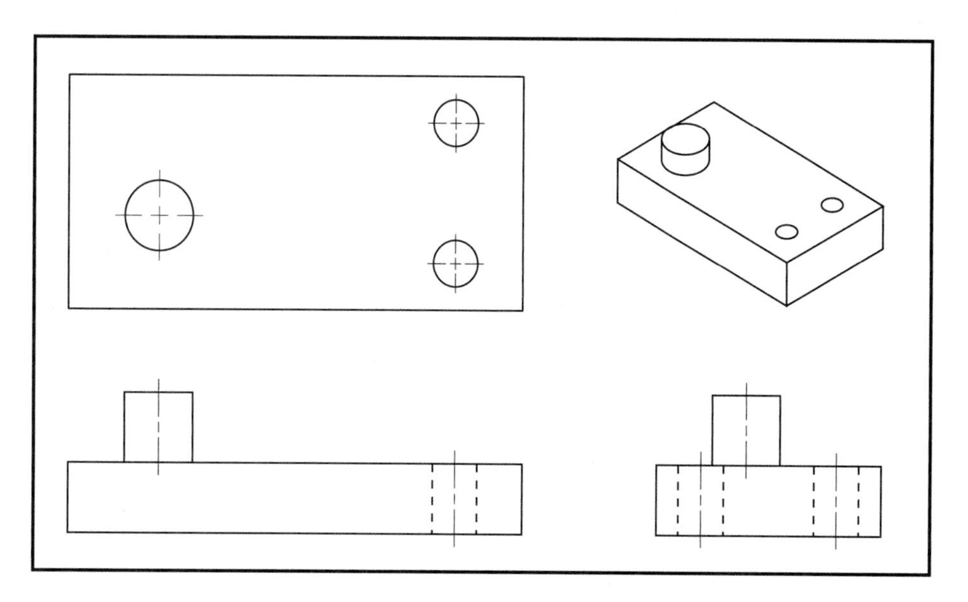

Figure 9.23 A multiview drawing created from a solid model

Dimensions and title block have not been added to the drawing. Dimensions can be added, but they may require editing so that they meet the standards discussed in Chapter 4. A solid model can be created in many ways, and the location of dimensions can be affected by the way the model is created.

Section Views

You can make section views using another Boolean operation. This is a "cut" operation. A cutting plane (vertical or horizontal as required) is created where you want to see a section, and the portion of the object on one side of the cutting plane is removed. The vertical plane can be a frontal plane or a right-side plane. The user specifies which way the section is to be viewed, and the portion of the object in front of the cutting plane is removed. Figure 9.24 shows how this is done for the object shown in Figure 9.23. A section is to be taken through the two holes to show that they go through. Only the plan view and the resulting section are shown.

A right-side cutting plane is required to see the desired section, and the location is specified in the plan view (see Figure 9.24a). The section is to be viewed from the right, and this direction is specified. Arrows on the cutting plane show the direction from which the section is viewed. The resulting section is shown in Figure 9.24b.

Assembly Drawing

Modelling programs can also be used to create an assembly of parts. An assembly shows the relation between parts and how they go together. An exploded assembly shows the relative positions of the parts, but they are separated so that more details can be seen. Assembly drawings were discussed in Chapter 5, with several examples. Assemblies and exploded assemblies can be created with a modelling program. The details of doing this vary depending on the program and will not be discussed in detail here.

The parts to be assembled relate to each other in some way, and this is specified when you assemble them. For example, a bearing and a shaft would be mating parts that share the same centreline (concentric), and this relationship would be specified. The assembly is built up by adding parts and specifying their relation to other parts.

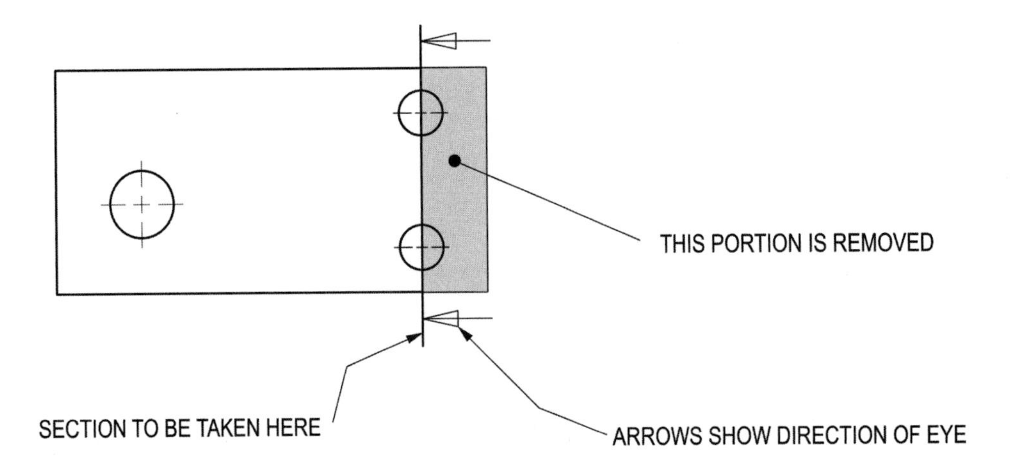

(a) SPECIFYING WHERE SECTION IS TAKEN AND VIEW DIRECTION

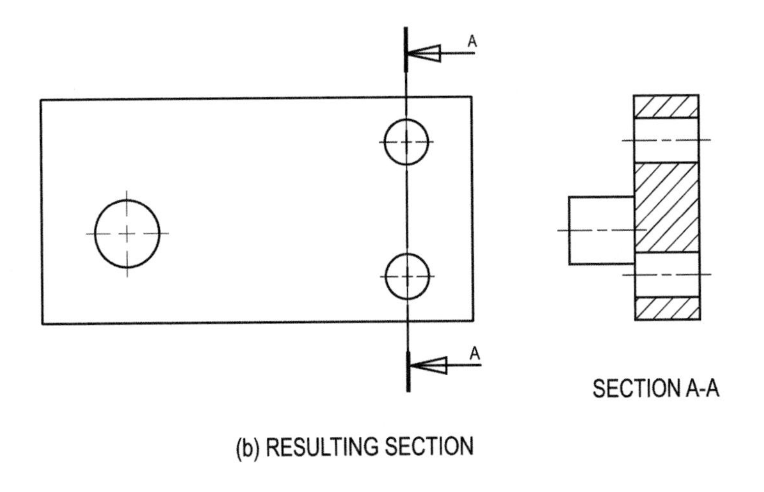

(b) RESULTING SECTION

Figure 9.24 A section can be created anywhere in the object

If a plane (or edge) on one part must be parallel (or perpendicular) to a plane (or edge) on a mating part, this relationship is specified. If parts must be a specific distance apart (the distance between the centrelines of two shafts, for example), this distance is specified. If planes must be parallel, this relation can be specified. Some programs have the facility to check for interference between parts. The relation between features and parts should be determined in the planning stage.

You can assemble parts into complex assemblies or you can start with an assembly and work back to design the parts to suit. You can see how parts fit together and, with some programs, how they move relative to other parts. This gives you a very powerful design tool.

Solid modelling programs are powerful tools for engineering design and drawing, and because they are powerful, they are complex. Only the basic features have been described here. You can easily create simple models as we have done here, but a complex part or assembly requires planning to do it well. Changes are inevitable, and it can be difficult to make changes. Planning your work before you start increases efficiency and reduces wasted time.

Problems

The only way to learn how to do solid modelling well is to do it, starting with simple shapes and trying different techniques. It is suggested that you use the objects shown in the problems in Chapters 1 and 2 as a place to start. You can then work up to building solid models of the more complex objects in Chapters 3 and 4.

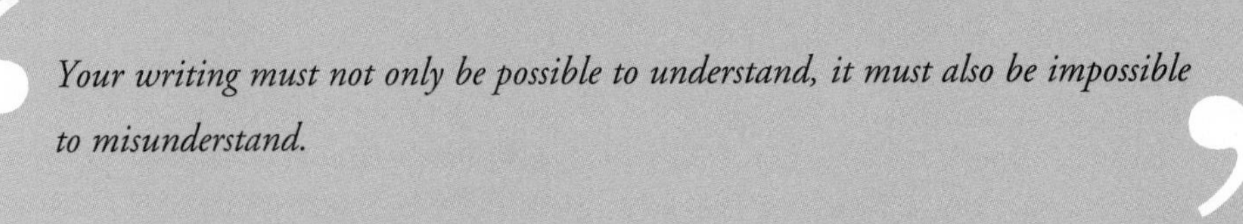

Your writing must not only be possible to understand, it must also be impossible to misunderstand.

CHAPTER

10 Presenting Technical Information

THE IMPORTANCE OF BEING UNDERSTOOD

If you ask engineering managers what they want in a new graduate, they will tell you that they want engineers who can communicate effectively with other engineers and clients.

Engineers do not spend all of their time in front of computers creating drawings and doing calculations. In fact, they spend only a small portion of their time drawing and calculating. Engineers spend 60 to 70 percent of their time communicating: talking, writing, and listening. Ideas and problem solutions are of little value unless they can be communicated effectively to others. Records must be kept, and work must be documented. Information must be collected and questions asked. Documentation, even for a small project, can reach hundreds of pages. Whether information is stored electronically or on paper, someone must write it, and the engineer is often the only person who understands a project well enough to document it.

Information is of no use unless it can be understood. If a prospective client does not understand what you are trying to say, you will probably not get the job. A request that is not understood is unlikely to get the results you want. Records are of no use if they cannot be understood. Time and money are wasted if readers cannot understand what you are trying to tell them. How many times have you read something twice because you did not understand it the first time?

Assembling the Right Information

It is standard practice to document and file all information pertaining to an engineering project. To ensure that you gather all of the information you need, you may want to use the *Who? What? When? Why? How? and Where? process.* With this process, you ask yourself such questions as:

Process Question	Description
Who?	• Who was involved in the project? • Who made the decisions?
What?	• What did the project entail? • What was said? • What decisions were made? • What deadlines (timelines) were established? • What estimates were made? • What assumptions were made? • What action was taken?
When?	• When were the decisions made? • When was the project due to be completed? • When did the meeting take place?
Why?	• Why were the decisions made? • Why was the project delayed? • Why did the rafters not meet properly?
How?	• How were the decisions arrived at? • How was the problem resolved?
Where?	• Where is the project located?

By asking such questions, you can start to build a complete picture of a project. Always keep in mind that you may not be around to answer questions, so your documentation must reflect the five Cs of writing.

Using the Five Cs of Writing

1. **Clear** Ensure that what you are writing is going to be interpreted correctly by the reader; for example, place the word "only" in the correct position; keep the subject and verb together, etc.
2. **Concise** Ensure that your communication is not wordy.
3. **Correct** Ensure that your communication is accurate as to factual content (names, locations, sums of money, etc.), spelling, grammar, etc.
4. **Concrete** Ensure that your statements are specific: for example, "The P95 valve was not working on April 9" as opposed to "The valve was not working."
5. **Courteous** Ensure that all communications are polite, whatever you may feel. Choose words that will not offend the reader.

Keep in mind that records are valuable in preventing and resolving misunderstandings. Documents are often used years after a project has been completed.

It is common practice for engineers to keep their own files of documents and correspondence relating to each project they work on. If well-written records are kept, there is no need remember everything that has been said or done (an impossible task).

WHAT DO ENGINEERS WRITE?

Engineers write everything from letters to instruction manuals. The most common written communications are memorandums, letters, reports, specifications, standards, and codes.

Memorandums

A **memorandum**, often called a **memo**, is an informal, internal communication that you can send to one or more persons or simply file as a record. It is one of the most used methods of written communication. Figure 10.1a illustrates a standard memorandum format.

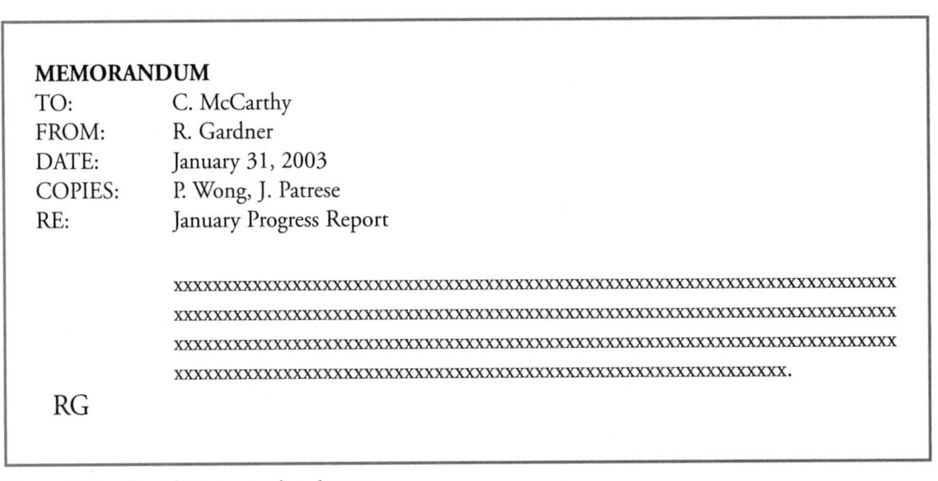

MEMORANDUM
TO: C. McCarthy
FROM: R. Gardner
DATE: January 31, 2003
COPIES: P. Wong, J. Patrese
RE: January Progress Report

xxxxxxxxxxxxxxxxxxxxxxxxxxxxxxxxxxxxxxxxxxxxxxxxxxxxxxxxxxxxxxxxxxxxxxxxxxxxxxxx
xxxxxxxxxxxxxxxxxxxxxxxxxxxxxxxxxxxxxxxxxxxxxxxxxxxxxxxxxxxxxxxxxxxxxxxxxxxxxxxx
xxxxxxxxxxxxxxxxxxxxxxxxxxxxxxxxxxxxxxxxxxxxxxxxxxxxxxxxxxxxxxxxxxxxxxxxxxxxxxxx
xxxxxxxxxxxxxxxxxxxxxxxxxxxxxxxxxxxxxxxxxxxxxxxxxxxxxxxxxxxxxxxxxx.

RG

Figure 10.1a Sample memorandum format

The format of a memorandum is relatively standard. The guide words at the top of the memorandum are usually TO:, FROM:, DATE:, COPIES:, and SUBJECT: or RE: (meaning "regarding"). They may be organized in any order, though the subject or re: line should be the last item.

Many organizations have pre-printed forms. If your firm does not, ask one of the clerical staff in your office whether there is a memorandum template on the computer system. To save time, many organizations use initials in memorandums (instead of full names) (Figure 10.1b).

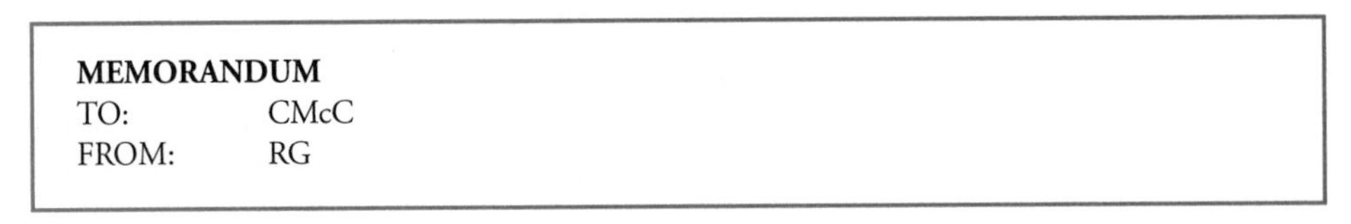

MEMORANDUM
TO: CMcC
FROM: RG

Figure 10.1b Sample memorandum format using initials

There is no restriction on the length of memos, but normally they are short. While there is no need to sign a memorandum, some writers like to initial them.

Letters

Letters are formal communications between two or more people and are signed by the persons that wrote them. The format of a business letter is standardized, although there are variations. Business letters will be dealt with in more detail later in this chapter.

Reports

Reports are written for a variety of reasons, but the main ones are:

- to present the results of an investigation,
- to present information on an event,
- to describe a piece of equipment, and
- to make recommendations to a client on the solution to a problem.

All reports have a similar format. They may vary in length from a few pages to several thousand pages.

Engineering reports will be looked at in more detail later in this chapter.

Manuals

Manuals are written to instruct someone how to do something. An assembly manual, for example, tells a person how to put something together. If steps are missing, the item will either be assembled incorrectly or not at all.

Maintenance or instruction manuals that cannot be understood or have missing instructions are not only frustrating but can also have serious legal and safety consequences.

Proposals

Engineering firms submit **proposals** to prospective clients to indicate how they would do the work, how long the work would take, and how much it would cost.

An unclear or ambiguous proposal is not likely to be accepted. If your proposal is not accepted, it means loss of work. An engineering company that does not get jobs will not survive.

A proposal that has spelling mistakes, poor grammar, and missing words will not inspire confidence in the firm that submitted it.

Specifications

Specifications are written to define how something must be done or how it must perform. If you want bids on a heat exchanger, for example, you would write a specification giving the quantity of heat to be exchanged, the fluids (water, engine oil, steam, etc.), temperatures, and other information and ask heat exchanger manufacturers to quote on the project.

A specification on a building would outline the details of the building and what the contractors must do.

Standards

There are **standards** for almost everything in engineering; for example, there are standards that define the content and strength of metals and alloys. The American Society for Testing and Materials (ASTM) publishes standards for testing materials and for the materials themselves. This ensures that a designer can specify a material and knows how much load it can carry. Appendix D lists some organizations that write standards.

Codes

Codes are similar to standards. They define construction and performance criteria for the protection of the general public. For example, the American Society of Mechanical Engineers (ASME) Boiler and Pressure Vessel Code specifies how vessels operating at high pressures and temperatures must be built and tested. Pressure vessels that do not meet this code will not be licensed by local authorities.

National and local building codes define how public buildings must be built.

THE WRITING PROCESS

Writing well is not an easy task. It requires the same thought and preparation as other engineering tasks. It also takes time. Do not start the night before a written communication is due and expect to do a good job. Even experienced writers revise their work several times.

Writing starts with you, the writer, knowing what you want to say and why you want to say it. This seems obvious, but beginning writers often start with no clear idea of what they want to say, and this is reflected in what they write. The result is a communication that is confusing and difficult to read. This is not surprising: Good writing cannot come from a confused mind. You must be clear and specific about what you want to say. (Remember those five Cs of writing mentioned earlier.)

You may find the following six-step process useful:

Step 1: Identify Your Objective
Step 2: Prepare an Outline
Step 3: Identify Your Audience
Step 4: Prepare a First Draft
Step 5: Edit Your Communication
Step 6: Review the Final Product

Step 1: Identify Your Objective

Start by defining the objective of your communication. What is your reason for writing?

- Do you want to explain something?
- Do you want to recommend something?
- Do you want to describe something?
- Do you want to ask for information?
- Do you want to give information?

Then, ask yourself what you must tell or ask the reader to do in order to meet this objective. For instance, if you need to explain something, what does the reader need to know? If you are making recommendations, what information must you provide to support your recommendations? What would you like the reader to do after reading your work? If the reader must make a decision based on your work, what information must the reader have to make the decision? Write down the answers to these and other questions like them in as much detail as possible. More items can be added later, if required.

If you can come up with a clear, precise written statement of your reasons for writing, you will make your work easier. Read this statement from time to time to stay focused on your objective.

Step 2: Prepare an Outline

When you have a clear idea of your objective and what you want to say, you can plan how you want to say it. To do this, create an **outline**. Figure 10.2 shows a sample outline.

Introduction
Project Specification
Project Objectives
Preliminary Design Solutions
    Gravity Launcher
    Pulley Launcher
Project Design I
    Launcher Description
    Testing and Adjustments
    Calculations and Assumptions
    Performance Evaluation
    Cost
Project Design II
    Launcher Description
    Testing and Adjustments
    Calculations and Assumptions
    Performance Evaluation
    Cost
Conclusion

Figure 10.2 Sample outline

The following explains the process of preparing a draft outline:

- The first thing on your outline will normally be an **introduction** or **background**. How detailed this introduction or background is depends on what the reader knows about the subject, but a short introduction is a good way to lead into your topic.
- Next, list the main topics you want to cover. Do not worry, if you leave something out, you can always add it later.
- Then, look at each of the main topics, and make notes under each heading. These notes can be subtopics, words, or phrases. It does not matter. Note in Figure 10.2, that the subtopics are indented. This would help you to keep yourself organized.

The first outline will be a "skeleton," or table of contents, of what you want to write. It will give you an idea of how you can organize the material for maximum impact. You can move items around on your outline until you are happy with the "flow."

We will look at various organizational methods a little later in this chapter.

Step 3: Identify Your Audience

Engineering communications are intended to be read. There is not much point in writing something that is never going to be read or understood. Even records, written and filed, are intended to be read at some time or another.

When you are writing, the reader and the needs of the reader must be kept in mind at all times. If the reader (or readers—there is often more than one) does not understand what you have written, you have wasted both the reader's time and your own.

You will usually know who the readers will be (or at least their background), and you must write with this in mind. A reply to a question from a Grade 9 student would not be written in the same way as it would be if it were to another engineer.

It is reasonable to assume that most of the readers of an engineering report have some technical background, but it may be in another field of engineering. Your readers should not have to guess at the meaning of any terms. A footnote at the bottom of the page, an endnote at the end of a section, or a glossary of terms can be used to help readers. Readers will try to follow your meaning, but eventually, even the most persistent give up if your meaning is not clear. Make it easy for the readers. Make sure that they remember you for clarity, not confusion.

Step 4: Prepare a First Draft

You are now ready to write the first draft.

The objective of the first draft is to get your ideas on paper, not to create the finished work. Do not worry too much about how you express yourself. Use your outline. Write short, concise paragraphs under each heading. This will help you stay focused. Prepare the first draft as quickly as you can (following your outline). There is always time to work on the sentence structure and flow in the second and subsequent drafts.

When you have finished your first draft, use the grammar- and spell-checker on your word processor to clean up your work. Print a copy of your first draft, and put it aside until the next day.

Step 5: Edit Your Communication

The next day, read your first draft. It helps to read your work out loud. You will write several drafts of an important communication before it is sent out. Even experienced writers do not consider the first draft a finished product.

As you read, edit your work for flow, extra words, slang and idioms, gender-inclusive language, and spelling and grammar.

Flow

When a document flows, it is easy to read. Edit your work so that one idea flows naturally into another. Make sure that each paragraph in your communication expresses one main idea.

Extra Words

Edit your work to remove redundant words. Quantity does not equal quality.

The average first draft can be shortened by one-third by removing excess words. The easiest way to identify excess words is to remove words and see what happens. Here is an example:

The following sentence contains frequently used words:

> *However, it should be noted that extra words are a waste of time to read and write.*

Removing redundant words reduces this to:

Extra words are a waste of time to read and write.

The number of words has been reduced by 35 percent, but the meaning has not changed.

Make every word count. If you can say it in five words, do not use ten. Your readers are busy. Do not waste their time by making them read unnecessary words.

Slang and Idioms

Avoid slang, such as "There's no way that this project will fly," and idioms, such as "I think we need to emphasize that they must keep their eye on the ball." While you may understand what you are saying, remember that engineers work in a global environment. What may seem everyday, understandable language to you may be totally "foreign" to the overseas reader.

Gender-Inclusive Language

As you are editing, watch for any stereotypical references—for instance, referring to an engineer as "he." If you reword the sentence into the plural, you can eliminate this problem—for example, "Engineers...they..."

Spelling and Grammar

A letter or report full of spelling or grammatical errors does not inspire confidence in the writer. If you are sloppy with your spelling and grammar, what about your engineering work? Poor quality work reflects badly on both the writer and the firm. You cannot use such excuses as "the typist made those spelling errors." The person responsible for finding and correcting the errors is the person who sends out the communication.

A spell-checker may find errors if the words are misspelled but not if the wrong word is used. What would a spell-checker do with the following sentence?

Hoe wood ewe two this problem?

There are no spelling mistakes, but the sentence makes no sense. These are all common words that are in the "dictionary" of your word processor. Errors like this can be found by proofreading your work. If you received a letter with errors like the above from someone doing engineering work for you, what would your impression be?

Use the spell-checker, but do not rely on it completely. The same thing applies to the grammar-checker. Your best solution is to proofread your work *several times.* Buy a good dictionary and style manual, and refer to them whenever you are in doubt.

Proofreading

Ensure that your message is clear. The active voice should be used instead of the passive voice, wherever possible. The active voice is more concise and "punchy." An example is:

Passive voice: The man was bitten by the dog.
Active voice: The dog bit the man.

When your word processor identifies passive voice sentences, look at the suggestions provided in the grammar-checker. You do not have to follow the suggestions, but they are useful.

The following proofreading hints may be of assistance to you:

Proofread for:	Hints
Spelling	• Check for spelling errors, even if a spell-checker has been used. • Check for out-of-context words. A spell-checker will not find words used out of context, e.g., "there" instead of "their." • Check for repeated words. Some spell-checkers will not find repeated words, e.g., "This is is the radius." • Check names of people and places carefully.
Grammatical Errors	• Look for incorrect changes in verb tenses, e.g., changing from the present tense to the past tense. • Ensure that subjects and verbs match, e.g., check that singular nouns have singular verbs. • Check that every sentence is, in fact, a sentence. Each sentence must contain a verb. • Check parallelism, e.g., bulleted items (such as in this list) start with the same part of speech (verb); that words in a series match.
Accuracy of Facts	• Check numbers. • Check names and dates. • Check page and section references; e.g., if the report text states that something is on page 14, check that it is really there.
Hyphenation	• Ensure that the last word of a paragraph or page is not hyphenated. • Check that words are correctly hyphenated. Use a dictionary or word division reference book if you are uncertain. • Ensure that there are no more than three consecutive lines ending with a hyphen. If there are, edit the text.
Punctuation	• Check the spacing after commas, periods, colons, semicolons, etc. • Check that all question marks are inserted. • Check that all commas, semicolons, parentheses, brackets, apostrophes, and quotation marks are correctly placed. • Ensure that underscores do not extend beyond the words to be underlined.

	• Ensure that parentheses or brackets around bold or italicized text are not bold or italicized.
Page Layout and Alignment	• Ensure that paragraphs are indented the same distance (if an indented style is used). • Check that all columns align at the bottom of the page if you are using a multicolumn format. • Check that all figures in a column are aligned. Dollar amounts should align on the decimal point. • Check that all margin settings are the same for each page. • Ensure that all pages have the same format. • Check all pages for balance. • Ensure that there are no "widows" or "orphans": one line of a paragraph left at the bottom of a page or the top of a page.
Consistency	• Check that headings of the same level use the same font and format (use styles). • Check that all captions are labelled in the same manner. • Check that all sections and units are referenced in the same manner, e.g., Section 4 in all places, not Section IV in one place and Section 4 in another. • Check that all tables are formatted the same, e.g., that all have centred column headings or all have left-aligned column headings. • Check numbered items or lists for missing numbers. • Check headers and footers to ensure that they relate to the material on the page, section of the report, or the report itself. • Ensure that the page numbers are in order and that none are missing.
Graphics	• Ensure that any graphics are referenced in the text. • Ensure that the captions on graphics match the graphics themselves. • Check that the graphics are there.

(Adapted from Porozny, G. H. J. *Desktop Publishing Design Basics and Applications.* Toronto: Copp Clark Pitman Ltd., 1993)

Keep editing your communication until it is clear, concise, correct, concrete, and courteous; however, avoid the perfectionist trap, where you keep editing without any improvement to quality.

Step 6: Review the Final Product

When you have thoroughly proofread your final draft, print it on good-quality white bond paper (or letterhead, if the communication is a letter).

Review the final product:

- Is there plenty of white space on the page, or is the text cramped?
- Are the top, bottom, and side margins too wide or too narrow?
- Are there any messy spots or fingerprints on the printout?
- Is the printout clear, or do you need to add toner or ink to your printer?
- Does this communication look as if it has been prepared by a professional organization?
- Would you like to receive this communication?

If you are satisfied with the quality of your communication, sign it (if appropriate).

BUSINESS LETTERS

The **business letter** is a formal communication between two people. It is the most common form of communication in business.

Business letters all share a similar format. A standard format enables the reader to find out who the letter is from, who it is intended for, and what it is about without reading the entire letter. If the subject of the letter is such that it should be passed to others for action, this can be done without taking the time to read the letter.

In addition to a common format, most business letters are short.

The most common business letter format is the block style illustrated in Figure 10.3.

Advanced Analysis Group Inc. 1421 Albert Street Vancouver, British Columbia, Canada V7R 3E7 Phone: (604) 425-9871 Fax: (604) 425-1327 e-mail: advanced@sympatico.bc.ca	**Letterhead**
Our File 04-A56	
July 20, 2003	
Ms. Jane P. Yee, President International Equipment Supply Limited 1587 Vernon Street Vancouver, BC V6T 1Z9	**Inside Address**
Dear Ms. Yee:	**Salutation**
Re: X-14 Diesel Engine Oil Cooler Investigation	**Subject Line**
We enclose our report, "Failure of the X-14 Diesel Engine Oil Cooler," which you requested on June 15, 2006. Our investigation showed that the cooler failed because of poor installation. The events leading to this failure are detailed, and suggestions are made to prevent it happening again.	**Body**
We will be pleased to discuss this with you at your convenience.	
Yours truly,	**Complimentary Closing**
ADVANCED ANALYSIS GROUP INC.	
W. A. Baxter Wm. A. Baxter, Ph.D., P.Eng. President	**Signature Block**

Figure 10.3 Block style letter

Letterhead

The letterhead, which is normally pre-printed, gives the name of the company sending the letter, the address, the telephone and fax numbers, and often the e-mail or Web site address. Some companies print the company name in the letterhead and the remaining information at the bottom of the page.

If a letter is more than one page long, the letterhead is used only for the first page. Continuation paper is used for the second and subsequent pages. Figure 10.4 shows a sample continuation page.

Advanced Analysis Group Inc.

Ms. Jane P. Yee
Page 2
July 20, 2003

events leading to this failure are detailed ...

Figure 10.4 Continuation page

Inside Address

The inside address gives the name and title of the person receiving the letter and the complete address.

Salutation

If you are writing to a specific person, as in Figure 10.3, then the name of the person is used in the salutation.

If you are not writing to a specific person, you can use a general salutation, such as:

Dear Sir:
Dear Sirs:
Dear Madam:

Subject Line

The subject line tells the reader what the letter is about. It should be specific. If the reader has no interest in the subject, there is no need to read further. The letter can be forwarded to the appropriate person.

Body

The body of a letter contains what you want to say. It should be short and precise. Ensure that it has a logical flow: introduction, middle, and ending.

Complimentary Closing

Keep the complimentary closing simple. "Yours truly" and "Sincerely yours" are commonly used.

Signature Block

The signature block contains your name and title. You sign in the space between the company name and your own.

There are variations on this format and some companies have their own formats.

ENGINEERING REPORTS

Reports are required at all stages of a project. They may be formal or informal, but at some stage, a formal report must be prepared.

Format

An engineering report can vary from a one-page letter to thousands of pages. The longer the report, the more important it is that the reader be able to find information quickly. A report need not be read from start to finish and usually is not. For example, one reader may read only the conclusions, while another may be interested only in costs.

A well-formatted report is divided into short sections and subsections to make it easy to refer to and easy to read.

The following are the main sections of a report:

- Cover
- Title page
- Abstract
- Table of contents
- List of figures (if necessary)
- List of tables (if necessary)
- Introduction
- Discussion
- Conclusions (if necessary)
- Recommendations (if necessary)
- Appendix(ces) (if necessary)
- References (if necessary)
- Bibliography (if necessary).

The purpose of your report will determine which sections you include; however, all reports have a cover page and a title page.

The following table briefly outlines the various parts of a report.

Report Section	Description
Cover	• The cover of a report usually has nothing on it other than the report title. The title must be specific and give the reader as much information as possible in about eight words or less. • Many companies have standard report covers with a simple design and company logo on them. The title is added to the cover. • Many report covers are printed on coloured lightweight cardboard or heavyweight paper.
Title Page	• The title page format varies from company to company. However, it will have on it: – the report title (same as on the cover) – who wrote the report – who the report was written for – when the report was written. • The title page is page "i" of the report, but the number is not printed on the page.
Abstract	• Sometimes called a **summary**, the abstract is a "report-in-miniature." It contains an accurate summary of the topic, important results, and conclusions. • The abstract must be brief and concise. The reader must be able to understand the key elements of the report by reading the abstract. • The abstract should contain no reference to the report itself such as "this report contains…," nor should it contain any reference to sections, figures, or tables. • The abstract is normally placed immediately after the title page (as page ii), unless there is a preface.

	• Many people consider the abstract to be the most important part of a report. Consider writing it last.
Table of Contents	• The table of contents functions as an outline of the report. It lists all headings and major topics, and indicates the pages on which they can be located. • The table of contents usually starts at page iii. • If you are using word processing software, you can generate a table of contents automatically, provided that you have assigned styles to the headings in your report.
List of Figures	• The list of figures gives the figure number, its caption, and the page number where it can be found. • A short report does not require a list of figures.
List of Tables	• The list of tables gives the table number, its caption, and the page number where it can be found. • A short report does not require a list of tables.
Introduction	• The introduction states why the report is being written, what it is intended to achieve, the scope of the report, the plan of development, and the general conclusion or recommendations. • Some background on the project may be included, depending upon the intended reader; for instance, if the report is intended for someone who is familiar with the project, less background material is required. • The introduction should start on a right-hand page and be numbered (page 1).
Discussion	• Sometimes called the **body** of the report, the discussion is normally the longest section of the report. • It presents facts, data, and arguments to support the conclusions. • All information is relevant. • All information is presented in an organized, logical manner. The following are some sample organizational methods: – **Problem-analysis-solution** (presents the problem, analyzes it, and then proposes a solution) – **Cause-and-effect** (presents the cause of the problem and what effects the problem will

have; you may or may not want to include recommendations)

- **General-to-specific** (starts with a general statement and provides specific examples)
- **Simple-to-complex** (moves the reader from a simple concept through to a more complex one)
- **Question-and-answer** (starts with a question and then proceeds to answer it)
- **Chronological** (presents information in date order or as a sequence of events/steps)
- **Geographical** (divides the material accordingly to physical location).

• Supporting information such as calculations (unless essential for reader understanding) is put into an appendix or series of appendices and referenced in the discussion; for example, "...see Appendix B-1."

Conclusions

• Conclusions must be based entirely on the material contained in the report.

• New information, or information you knew before you wrote the report, is not a conclusion.

• Put each conclusion into a separate paragraph. Consider numbering each conclusion for easy reference.

• Some reports do not have conclusions; for example, a report that describes something does not have conclusions.

Recommendations

• Recommendations must follow logically from the conclusions and be supported by the material in the discussion.

• Recommendations must be clearly stated so that there are no misunderstandings.

Appendices

• An appendix or series of appendices contain supporting information that is not essential to the understanding of the report.

• Each appendix deals with just one subject, and is assigned a letter; for example:
  - Appendix A: Calculations
  - Appendix B: Equipment
  - Appendix C: Physical Properties.

• If you have appendices with related information, you may assign a letter and number; for example:
  - Appendix A-1: Materials List – Warehouse
  - Appendix A-2: Materials List – Production Facility I

	– Appendix A-3: Materials List – Production Facility II.
References	• The references section lists, in alphabetical order, all sources you have *referred to in your report.* Each entry includes the author(s), the title of the material, the publisher's city and province/state and publisher's name, and the date and year.
Bibliography	• The bibliography lists, in alphabetical order, all sources you have *used in writing your report* (which you may or may not have referred to in your report). Each entry includes the author(s), the title of the material, the publisher's city and province/state and publisher's name, and the date and year.

Presenting Your Information

How you present your report has an effect upon how the reader receives it. A poorly formatted report, crammed onto "dog-eared" or finger-marked pages, or a computer presentation (such as PowerPoint®) with difficult-to-read visuals will not be well received. While you may have the best proposal or solution to a problem, a poor presentation will not make a good impression.

You need to combine the elements of your communication: text, drawings, graphs, photographs, to give the best possible visual effect. You do not need to be a professional desktop publisher to create a visually pleasing document. The following are some tips from "the professionals" that you may find useful:

Text

Here are some hints for formatting your text:

- Use double spacing between lines of type. Use at least quadruple spacing if you are preparing a PowerPoint® presentation.
- Use easy-to-read fonts (Times New Roman, Arial, etc.).
- Consider using serif fonts (ones with little arms and legs on the letters), such as Times New Roman, for body text because they are easier to read.
- Consider using sans-serif fonts (ones without little arms and legs on the letters), such as Arial, for captions and labels on figures.
- Avoid using more than three font styles in your communication.
- Take careful advantage of font attributes, such as size, bold, italics.
- Use a readable size of font (normally somewhere between 10 and 13). Use a larger font if you are preparing a PowerPoint® presentation because it normally has to be viewed from a distance.
- Use uppercase (capital letters) sparingly: only for labels, single words, or very short lines.
- Use a consistent paragraph format style (aligned left or justified).
- Use coloured text and backgrounds with care when preparing a PowerPoint® presentation; for example, blue text on a blue background will not work.

Compare the single-spaced text below the graph in Example 10.1 with the double-spaced text in Example 10.2. Note how much easier Example 10.2 is to read. The additional **white space** between the lines of text helps readability and makes the page look "lighter."

Example 10.1

$K := \text{slope}(x, F)$ $\quad K = 297.009 \ \frac{N}{m}$

$b := \text{intercept}(x, F)$ $\quad b = -92.39 \ N$

$y := K .X + b$ $\quad$ (best fit line calculated form linear regression)

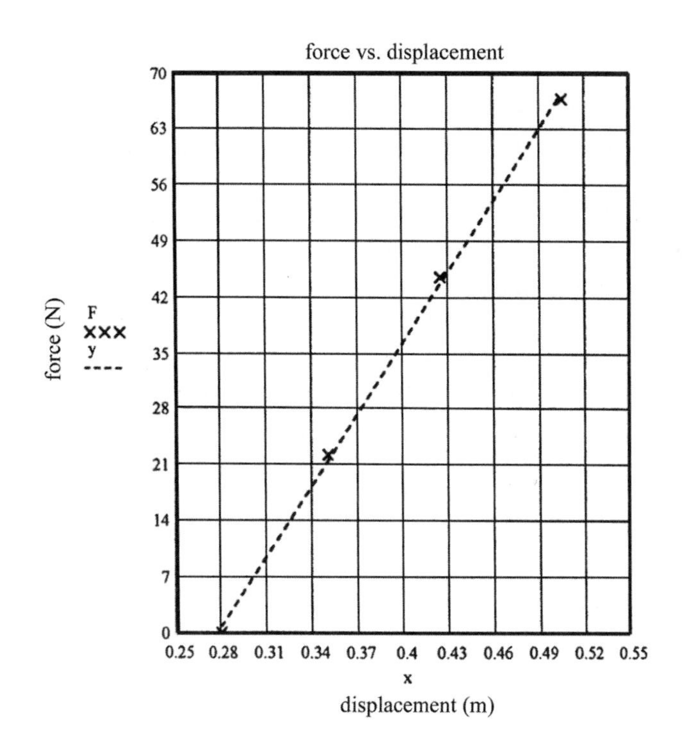

From the best fit line of the data point we obtain a spring constant of 297.009 N/m. To calculate the energy stored in the spring we take the integral of the force over the displacement of the spring. In our launching device, the spring is stretched somewhat at the point of release, so our bounds of integration will be from the displacement of the spring at ball release, to the total displacement of the fully stretched spring.

Displacement of spring at ball release $\quad x1 := \frac{(39.5 - 28)}{100}$

Total displacement of spring $\quad x2 := \frac{(71 - 28)}{100}$

Mechanical Engineering 251 - Project II: Projectile Launcher

(Reprinted, with minor modifications, with the kind permission of Patty Lee, Cari Orbeck, Darren Rafferty, and Carola Veloso, Projectile Launcher Project Report, November 1998)

Example 10.2

Projectile Launcher 2

calculated that the piston could not generate enough velocity to accelerate the combined weight of the piston, cup, and squash ball. Even with more pressure, the piston would not be able to generate enough velocity to launch the ball ten meters. The cylinders are also only rated to 150 psi, so we do not want to be working in the upper ranges of the allowable pressures of the cylinder. A new idea incorporating some mechanical advantage would need to be considered for the launcher to be successful.

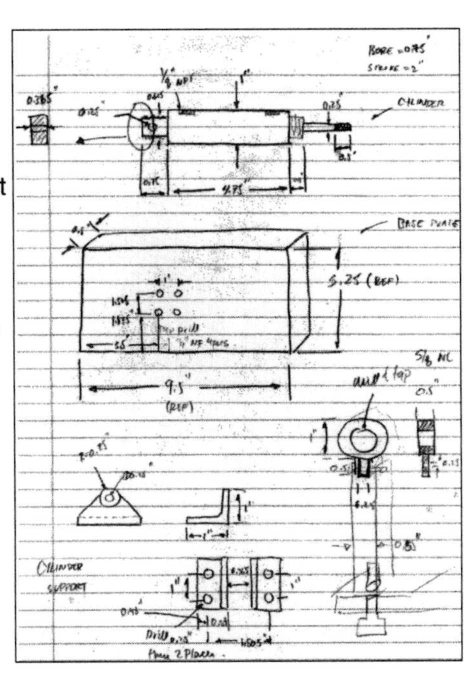

Figure 2: Initial experiment sketches

1.2 REVISED DESIGN

We still wanted to use the air cylinder and solenoid idea, since we saw how accurate it was, but needed some kind of leverage to increase the ball's initial velocity. One idea was to put a spring in the launcher cup so that it would compress as the piston shot out and then the spring would extend as the piston stopped. This would give the ball an extra push as it was released. This idea might have worked, but it would have been too hard to time the release of the spring to the stopping of the piston. Another idea we had was to have the piston shoot backwards and hit a lever that would catapult the ball towards the target. Figure 3 is a copy of our preliminary sketch. The cup used in this design turned out to have too high edges, resulting in a vacuum within the cup as the ball left it. To eliminate this, we cut away the back edge, doubling the distance of the launch.

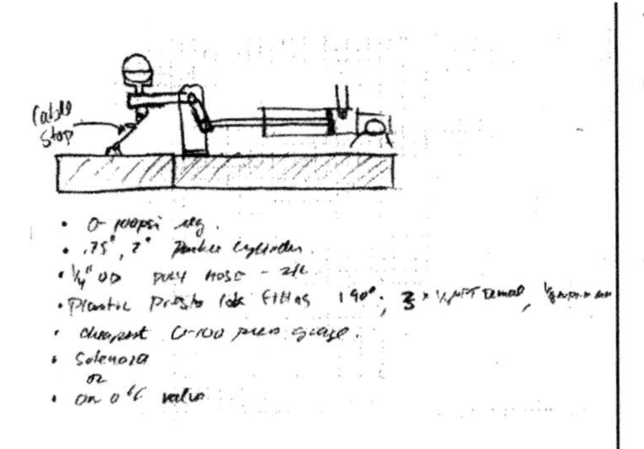

Figure 3: Preliminary Sketch

(Reprinted, with minor modifications, with the kind permission of Neil Allyn, Kurtis Guggenheimer, Alex Tielker, and Matei Ghelesel, Projectile Launcher Technical Report, 1998)

When preparing materials for a PowerPoint® presentation, make sure that you have plenty of white space or background colour and that your text is not crammed on the visual. Text will not be crammed on the visuals if you do not put too much information on each one. Generally, there should be no more than 3 to 5 points per visual, and if this is done, you will have ample white space.

It does not matter what paragraph style you use. Both Examples 10.1 and 10.2 use an aligned left style (the left side of the text is blocked, and the right side is ragged). This book uses a justified style (both left and right sides of the text are blocked). If you select a justified style but your word processing software leaves too much space between words, use a left-aligned style. Most word processing programs have various methods to rectify the spacing problem, but if you are not a word processing expert and do not have time to become one, then use the simplest style you know that looks good.

It does not matter whether you indent the first line of your paragraphs or not. If you do, make sure that your indentations are consistent (1/2 inch or 12 mm is normal).

Look again at Examples 10.1 and 10.2. Do you notice that the paragraph of text below the graph in Example 10.1 looks "heavy"? Do you realize that your eyes are drawn to this "heaviness" as opposed to the graph that is "light and airy"? You usually do not want this.

Spend a few minutes looking at the text in this book. Ask yourself the following questions:

- Is the text easy to read?
- Are the fonts simple?
- Is a serif font used for body text and a sans-serif font used for captions and labels?
- Are the font sizes and attributes used sparingly?
- Are capital letters (uppercase) used sparingly?
- Is there plenty of white space?

Single-Page versus Double-Page Spread

Before you start preparing a report, decide whether you are going to print on one side (single-page spread) or both sides of the publication (double-page spread). It will make a difference as to the page setup you select on your word processing program.

This book uses a double-page spread. There are two pages in front of you when it is open.

Here are some hints for setting up a double-page spread report:

Hints

- Page numbers must alternate (odd page numbers–1, 3, 5, etc.–on the right-hand pages, even page numbers–2, 4, 6, etc.–on the left-hand pages).
- Page numbers must be on the outside of the alternating header or footer, unless you are using a centred page number style.
- Use an alternating page setup so that the left margin on the left-hand page and the right margin on the right-hand page are the same width (often called *mirror margins*).
- Ensure that you allow a sufficient "gutter" (right margin on the left-hand page and left margin on the right-hand page) so that the report can be three-hole punched without damaging the text.
- Ensure that the gutter is wide enough so that the text does not disappear down the centre of the two-page spread.

If you are using a single-page spread, you do not need to worry about alternating headers and footers (including page numbers) or gutters.

Example 10.3

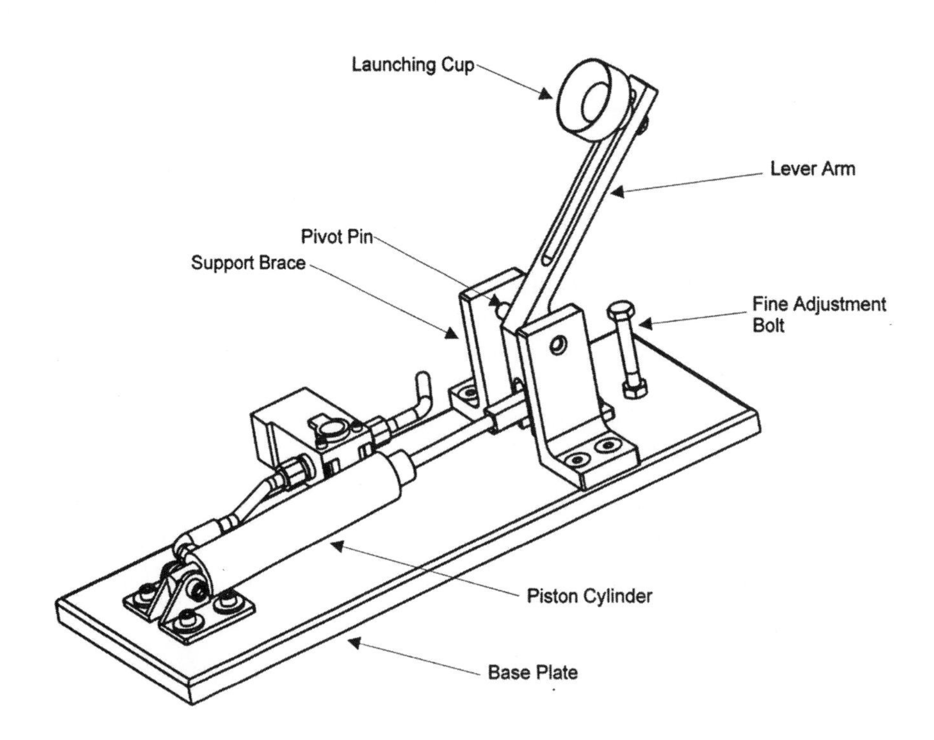

Figure 4: 3D representation

2.0 FINAL DESIGN

In the following section we will talk about our final design, detailing how and why it was made in certain ways. Figure 4 shows a 3D representation of our final design.

2.1 MANUFACTURING

We roughly calculated how much mechanical advantage we would need to increase the ball's velocity to the minimum necessary, and then decided on a length for this lever arm.

The first lever arm that we built was made out of thin-walled, square-sectioned aluminum bar-stock. We mitered two pieces of bar-stock at 22.5°, and then jigged and welded them together to make a piece with a 45° angle. We mounted the piston and the lever so that the distance ratio of the cup and the piston was about 3.5:1. We made our launcher

(Reprinted, with minor modifications, with the kind permission of Neil Allyn, Kurtis Guggenheimer, Alex Tielker, and Matei Ghelesel, Projectile Launcher Technical Report, 1998)

Example 10.4

Appendix B: Drawings

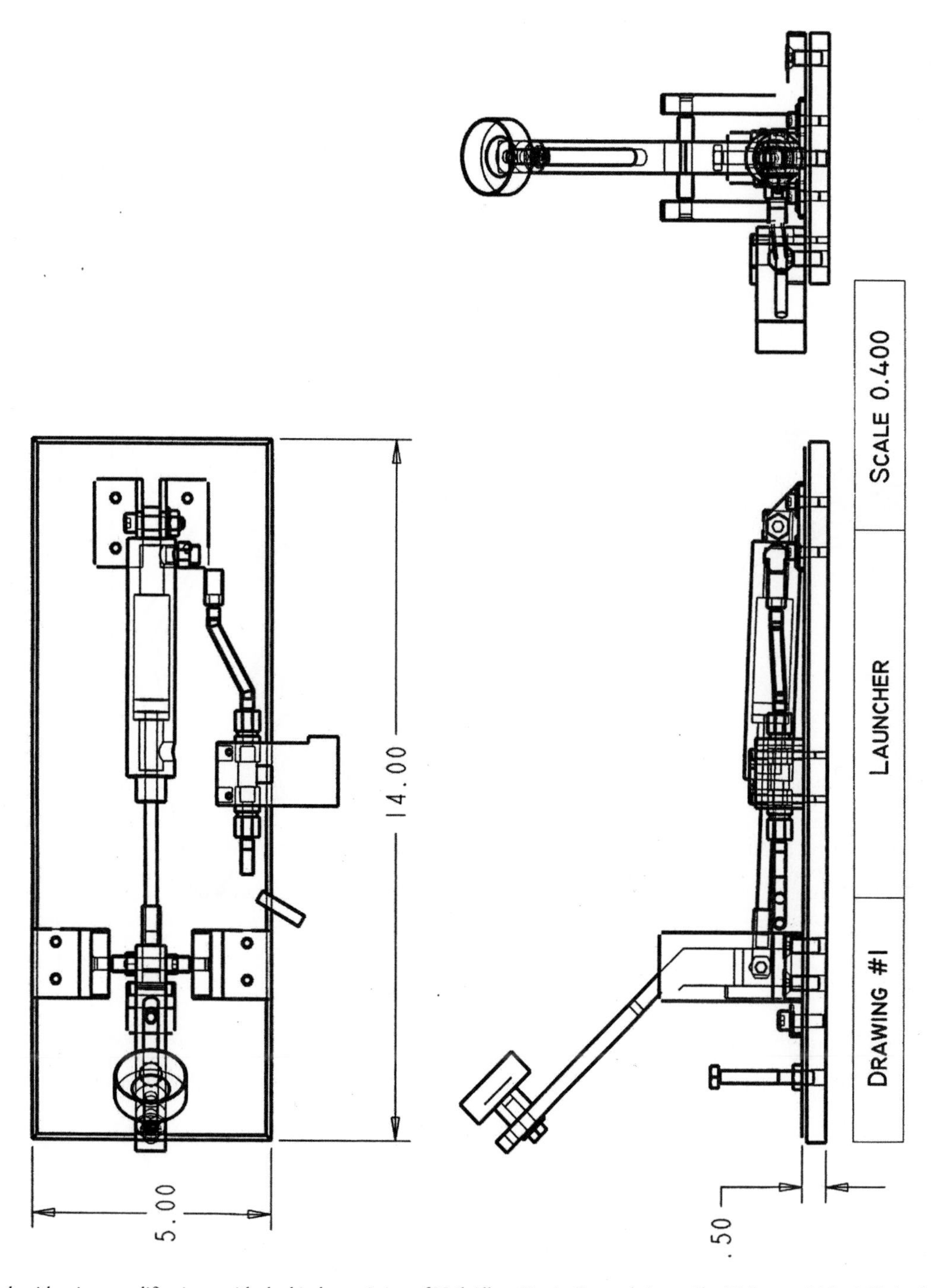

(Reprinted, with minor modifications, with the kind permission of Neil Allyn, Kurtis Guggenheimer, Alex Tielker, and Matei Ghelesel, Projectile Launcher Technical Report, 1998)

Example 10.5

Projectile Launcher 11

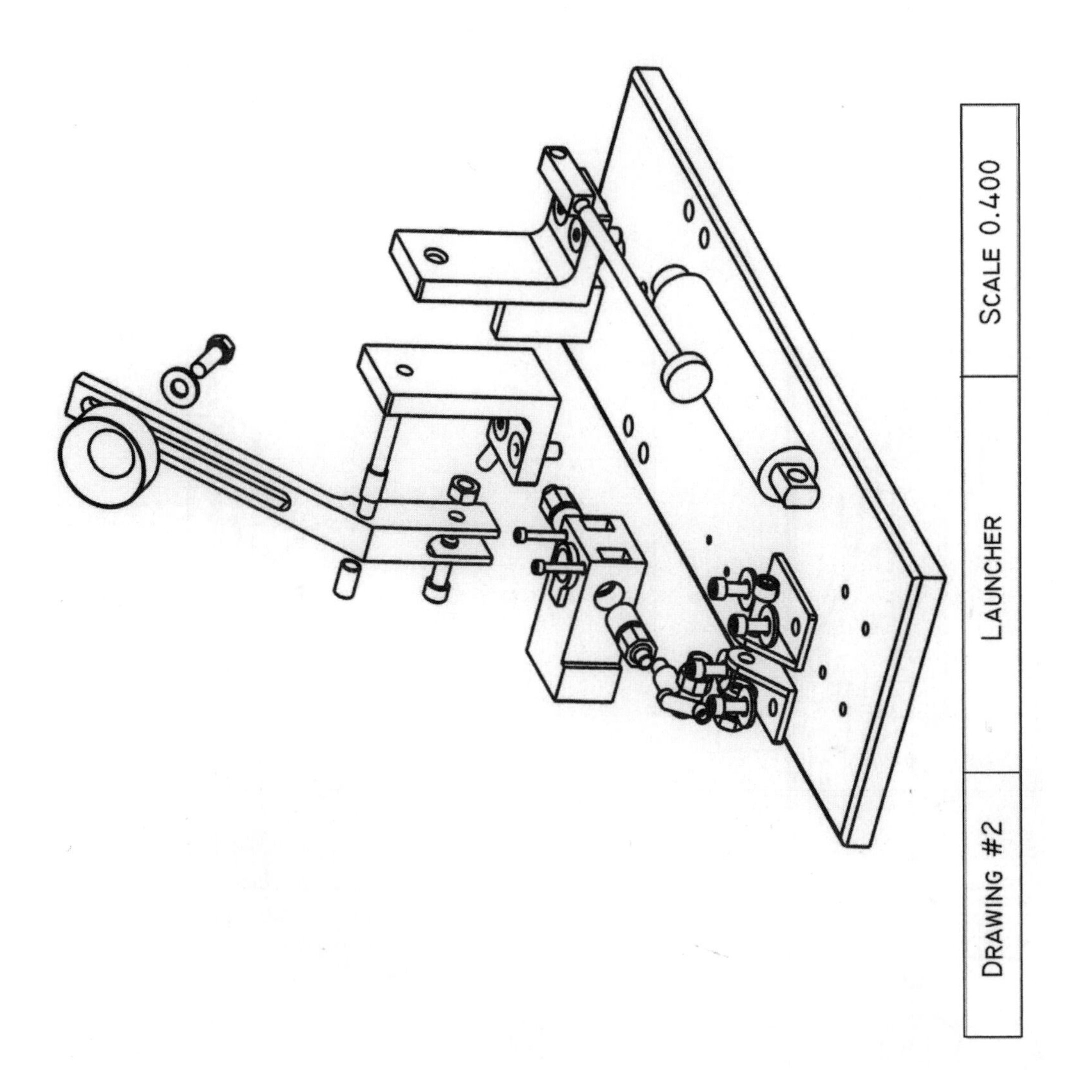

(Reprinted, with minor modifications, with the kind permission of Neil Allyn, Kurtis Guggenheimer, Alex Tielker, and Matei Ghelesel, Projectile Launcher Technical Report, 1998)

Graphics

Graphics comprise anything that is not text, such as photographs, diagrams, drawings, sketches, maps, and graphs. They can be displayed in a number of different ways. Look back at the graphics in Examples 10.1 and 10.2, as well as the ones in Examples 10.3, 10.4, and 10.5. Examples 10.4 and 10.5 illustrate engineering drawings and exploded assembly drawings included in an appendix to a report.

The following are some hints for graphics presentation:

Hints

- Reference the graphic in the text ("Figure 3 illustrates . . . .")
- Place the graphic as close to the text reference as possible (always after or beside the text, never before it).
- Put the graphic number in the caption (Figure 3).
- Put a brief description in the caption (Figure 3: Profile of the mountain).
- Be consistent with capitalization and punctuation in the caption:
  - Figure 3: Profile of the mountain
  - Figure 4: Topographical Map. [Incorrect]
  - Figure 4: Topographical map [Correct]
- Centre graphics horizontally on the page (see Example 10.1).
- Balance graphics diagonally on the page for an attractive page layout (see Example 10.2).
- Wrap text around graphics, where appropriate.
- Use appropriately sized graphics to balance with the text on the page (see Example 10.3).
- Place a graphic less than a third of a page deep at the top of a page and one more than a third of a page deep at the bottom of a page: This will ensure good page balance.
- Use landscape mode (sideways display) as opposed to portrait mode (vertical display) to fit detailed drawings onto the page (see Examples 10.4 and 10.5).
- Ensure all labels on graphics are easy to read and consistent as to size and capitalization (see Example 10.3).
- Save time by creating graphics in the simplest computer program. For instance, the graph in Example 10.1 has been prepared in a spreadsheet and inserted into the report; the freehand sketches in Example 10.2 have been scanned and inserted into the report.
- Consider linking (as opposed to embedding) spreadsheets and graphs into your report; then, if you have to make changes to your spreadsheet or graph, the changes will automatically be updated in your report.
- Ensure that the values on the axes are appropriate so that the graph is not distorted, either horizontally or vertically. Keep in mind that the width of your report page may be less than your spreadsheet printout page. Graphs should be "open" and easy to read.
- Ensure that your pie charts contain no more than seven segments.
- Avoid cluttering graphs with unnecessary grid lines or labels.
- Avoid using a variety of bold patterns as fills on graphs or charts.
- Avoid cluttered graphics when preparing a PowerPoint® presentation. Labels may have to be read from a distance, and so keep your reader in mind when selecting a label or diagram type size.

While it may seem that there are many things to think about when preparing communications, remember that the time will be well spent. Professional-looking communications project a professional image.

Problems

1. Write a proposal describing your proposed design for one of the projects in Chapter 8 (pages 278–289) for a prospective client who is not an engineer. Illustrate your proposal with clear sketches or drawings.
2. Choose a common object, and write a report on how it works. Select something that can be obtained easily at little or no cost (e.g., an old toaster, an old power tool), dismantle it, and with the aid of clear sketches and drawings, describe the various components and how they work together. Find out the correct names for the various parts. Photographs are a good method of showing what the components look like and how they are assembled.
3. Write a set of instructions to explain how to program a DVD or VCR to perform a function, such as how to record at a specific time. Write your instructions, illustrated with sketches or drawings, so that someone who has little or no experience operating a VCR can perform the function.
4. Write a description of how some piece of equipment operates: for example, a gasoline engine, an air compressor, a refrigerator, or a gas turbine. Ensure that your description can be understood by a Grade 9 student.
5. Write a set of instructions for conducting a detailed engineering search on the Internet. Write your instructions so that someone who has only a little knowledge of the Internet or someone who is searching for something specific can use them.

APPENDIX

A Geometric Construction

Many graphical problems require solutions that use geometric construction techniques. This appendix describes the most common techniques that can be used in both manual and CAD construction.

LINES

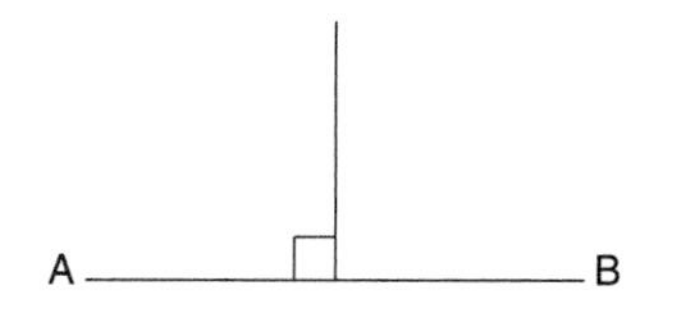

Figure A.1

A **line** is the shortest distance between two points. A line constructed equidistant from the line is said to be parallel or offset. A line at right angles to it is perpendicular (Figure A.1). For accuracy, object snap points should always be used in CAD.

ARCS

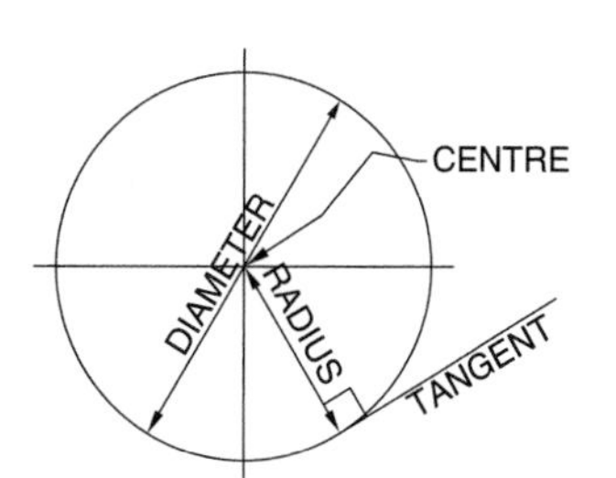

Figure A.2

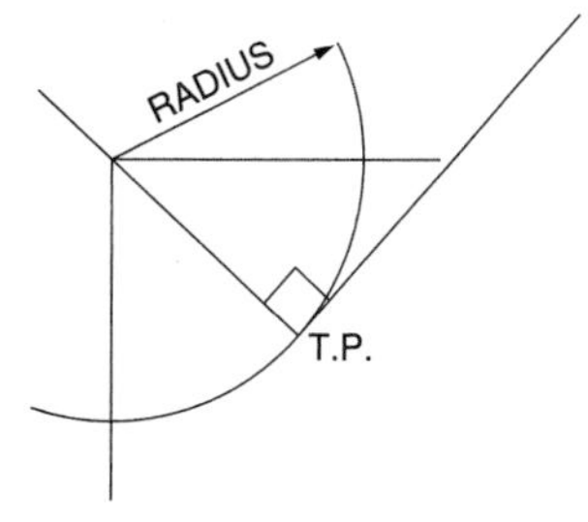

Figure A.3

The terms associated with a circle are given in Figure A.2. An **arc** is an incomplete circle of given radius. Both circles and arcs are constructed from a given centrepoint. Often, in CAD, it is easier to construct a circle and trim away to form an arc.

A **point of tangency** is a point where a line joins an arc or where two arcs join without crossing. A construction line drawn perpendicular to the line and passing through the centre of the circle will define the tangent point (TP) (Figure A.3).

To construct an arc tangent to two nonparallel lines, follow the example in Figure A.4. Draw construction lines parallel to the subject lines at a distance equal to the radius of the arc. The centre of the arc is the intersection of the construction lines. Lines drawn from the intersection, perpendicular to the subject lines, will identify the tangent points.

To create an arc tangent to two intersecting lines, see Figure A.5. Short arcs can easily be constructed in CAD using a fillet.

To create an arc (radius R) tangent to a circle (or arc) and a line, see Figure A.6. Draw a construction line parallel to the tangent line at a distance equal to the arc's radius (R). Add radius (R) to radius (r), and swing new radius (R + r) from the centre (A) to cross the construction line. The intersection is the centre for the required arc.

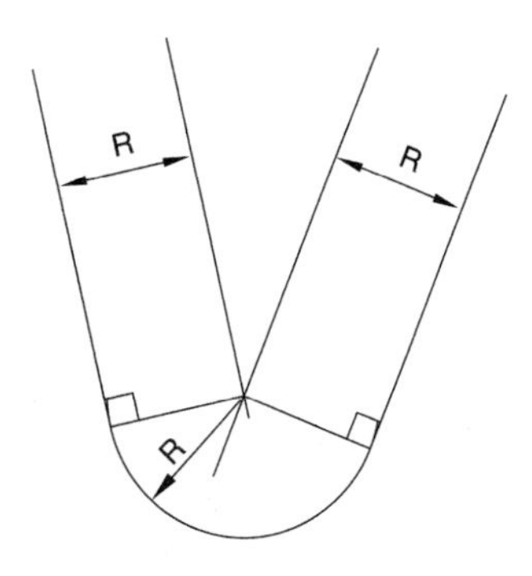

Figure A.4

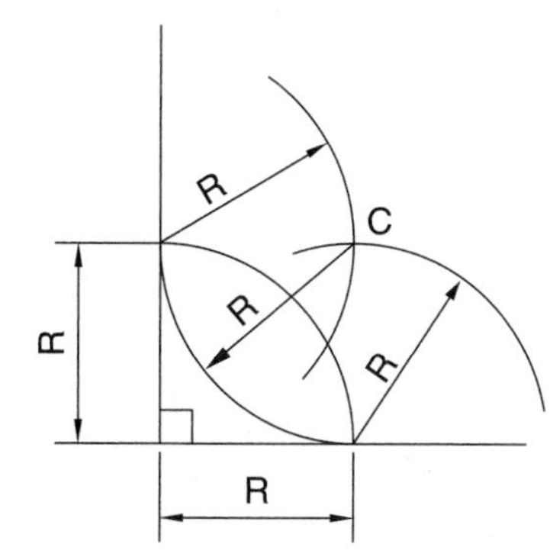

Figure A.5

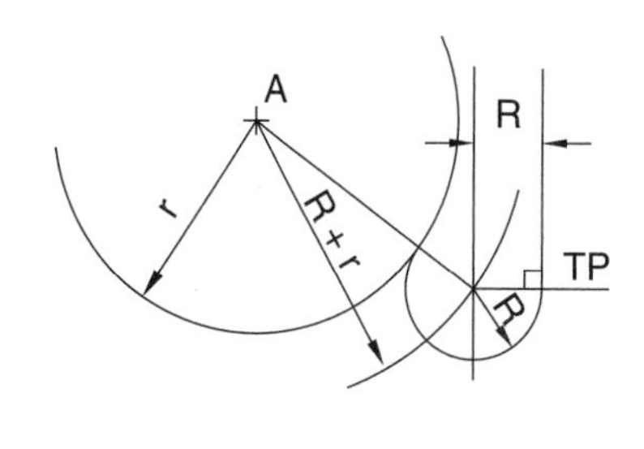

Figure A.6

To create an arc (radius R) tangent to two arcs (r1 and r2), see Figure A.7. Add radius R to r1 and swing from the centre (A). Subtract radius R from radius r2 and swing from centre B. The intersection (C) is the centre for radius R.

To draw a *convex arc* (radius R) tangent to two arcs (r1 and r2), see Figure A.8. Add radius R to r1 and swing from the centre (A). Repeat for r2. The intersection (C) is the centre for arc R.

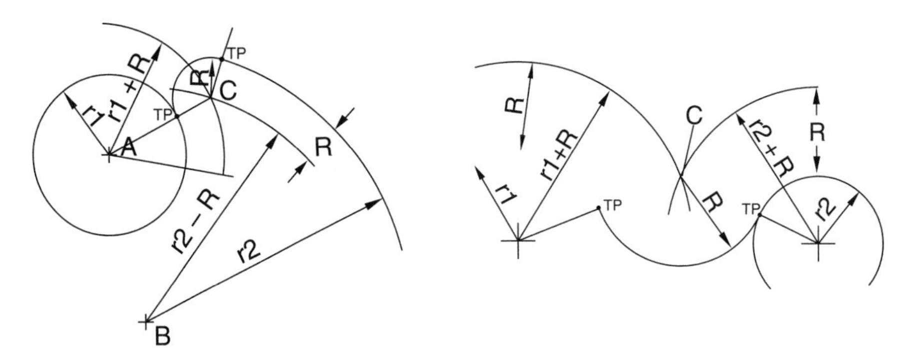

Figure A.7

Figure A.8

To draw a *concave arc* (radius R) tangent to two arcs (r1 and r2), see Figure A.9. Subtract r1 from R and swing from centre A. Subtract r2 from R and swing from centre B. The intersection of the arcs (C) is the centre of arc R.

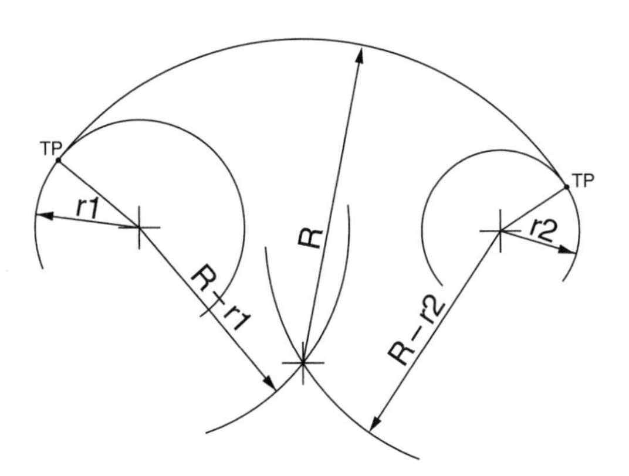

Figure A.9

When using CAD to create arcs, you may create a circle using Tangent, Tangent, Radius options, and trim.

ELLIPSE

An **ellipse** is formed by the line of intersection between a plane and a right circular cone at an angle to the axis that is greater than the angle between the axis and the sides. The major diameter is the largest diameter and is the true length. The shortest diameter is called the minor diameter and is perpendicular to the major diameter (Figure A.10).

Usually, an ellipse would be hand drawn using templates or using an ellipse command in CAD. Templates are plastic and have various diameter ellipses drawn at intervals of 5 degrees.

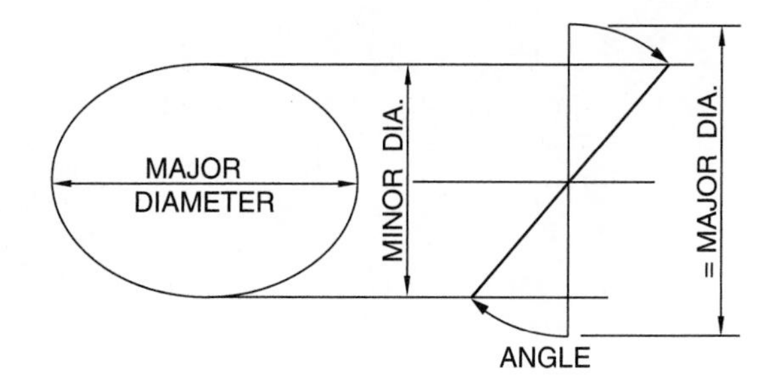

Figure A.10

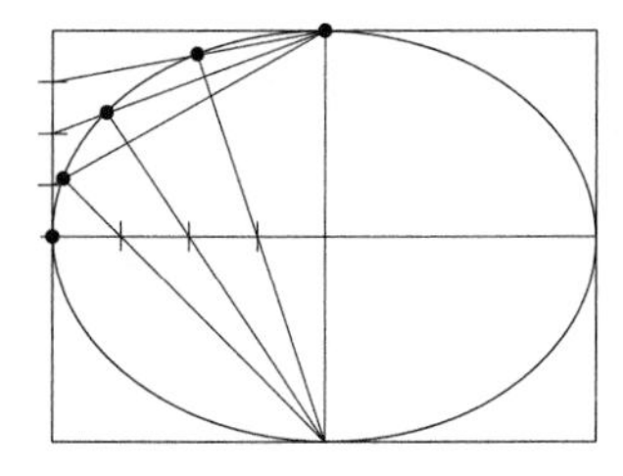

Figure A.11

There are several methods for constructing an ellipse. One method is given in Figure A.11. Draw a rectangle equal to the major and minor diameters. Divide the vertical line in one quadrant into four equal spaces. Divide the horizontal line in one quadrant into four equal spaces. Construct the curve in the quadrant as shown and repeat for the other quadrants.

POLYGON

Regular **polygons** (equal sides) can be constructed inside or outside the circumference of a circle. Each vertex intersects with a line passing through the centre at a prescribed (equal) angle. CAD programs have a polygon command capable of generating over 1000-side polygons. Figure A.12 shows a hexagon construction.

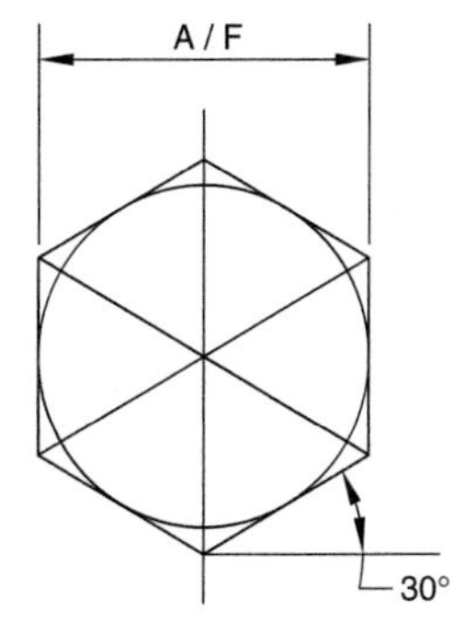

Figure A.12

HELIX

A helix is a three-dimensional curve formed by a point moving uniformly around a cylinder or cone. Examples of use are a screw thread, drill bit, spring, and corkscrew. Figure A.13 shows the construction of a helix. In CAD, a similar construction method can be used along with a 3-D polyline connecting through the intersecting points.

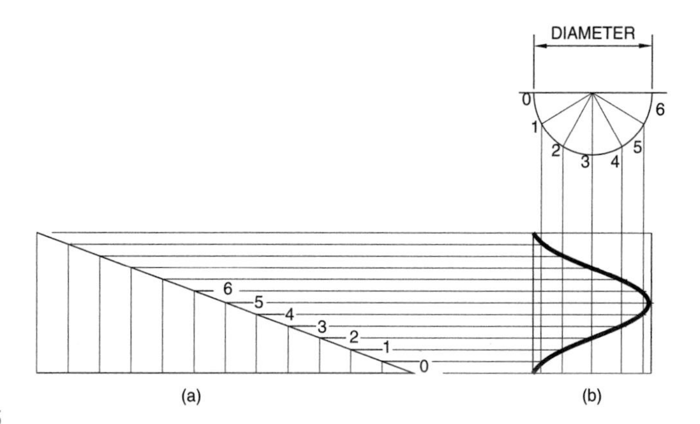

Figure A.13

APPENDIX

B Vector Analysis

VECTORS

A **vector** is a quantity that is defined by two things: magnitude and direction. **Force** is a vector quantity; it has magnitude and direction. It is important to know in which direction a force acts; for instance, if you want to lift something off the ground, you do not push down on it. **Velocity** is another vector quantity; it has magnitude, speed (kilometres per hour), and direction. It can make a big difference if you are moving north instead of south.

We will deal with the graphical representation of vectors; however, vectors can be represented mathematically. The graphical representation of a vector quantity must indicate both the magnitude and the direction of the vector quantity, say, a force. Figure B.1 shows horizontal force acting on a body. The magnitude of the force is 50 N.

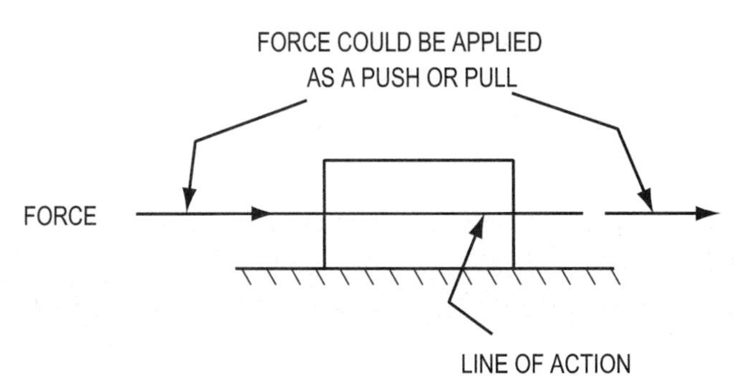

Figure B.1 Horizontal force acting on a body at rest

The force is shown on the line on which it acts (the line of action), and the arrowhead shows the direction. The effect on the object is the same whether it is pushed or pulled, and it does not matter where the force is shown on the line of action as long as the direction is indicated.

The force is represented graphically by an arrow pointing in the direction of the force. The length of the arrow represents the magnitude of the force. The vector is drawn to some scale (e.g., 1 cm = 10 N; 1 cm on the paper represents a force of 10 N). The magnitude is read directly from the scale, and there is no arithmetic to do. More on this later. That is all there is to it. Draw a line of specific length in the direction of the force.

Now let's look at adding vectors.

ADDING VECTORS

We usually deal with more than one vector, and we must know how to combine them. When two forces act on an object, we may want to replace these forces with one equivalent force by adding them. Vectors are added by joining them "nose-to-tail." Figure B.2 shows two forces, A and B, acting at a point. The forces are on their lines of action but are not drawn to scale.

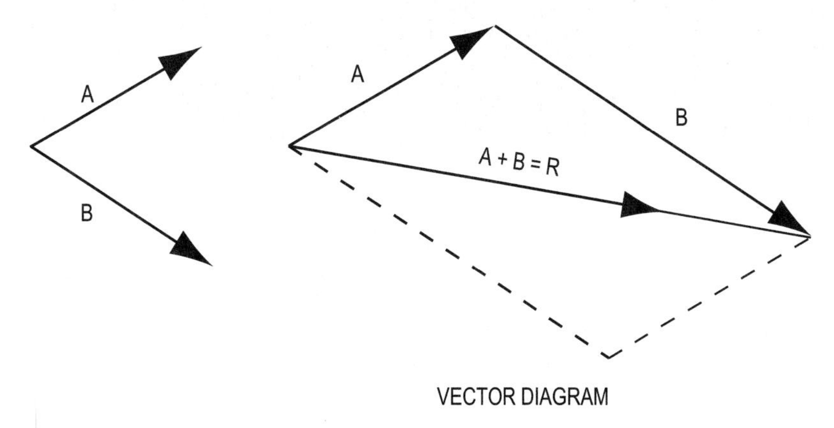

Figure B.2 Adding vectors

Vector A representing force A is drawn to scale, at the same angle as the force. The length represents the magnitude of the force. The vector B is drawn at the same angle as force B, starting from the end of A. The resultant is found by joining the starting point to the end point. Forces A and B could be replaced by one force of magnitude R, acting in the direction of the resultant (the arrow is shown in midline for clarity). Vectors can be added in any order. The result is the same if force B is drawn first and A is added at the end, as shown in Figure B.2. This is called the **parallelogram law**.

Now let's look at a practical application of this information.

Example B.1

A boat leaves a pier on one side of a river flowing at 0.5 m/s in the direction shown. The operator heads straight for the opposite shore. If the river is 500 m wide, where will the boat hit the opposite shore if it moves at 1 m/s? **Figure B.3** shows the layout.

The boat heads toward the opposite shore at a velocity of 1 m/s (the direction and magnitude are known), but the river will carry it downstream. The resultant velocity is the sum of boat velocity and river velocity. Figure B.3 shows the addition of these velocities and the resulting boat velocity.

The boat moves downstream as it crosses the river and hits the opposite bank at a point 246 m (500 tan 26.2°) downstream of the starting point.

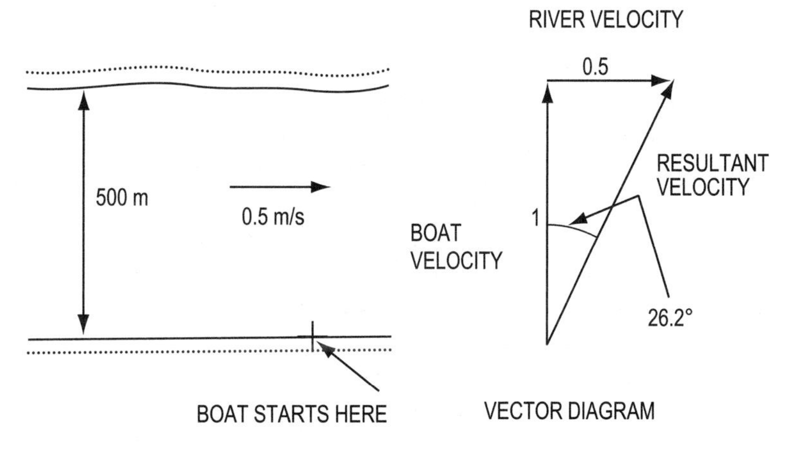

Figure B.3

The process of adding vectors is the same for any number of vectors. Figure B.4 shows a frame loaded with four forces. This is a **space diagram** that shows the frame is drawn to scale (e.g., 1:100) and the forces are shown on their lines of action. The forces are not drawn to scale. The space diagram shows where they act and the direction. The space scale must be shown.

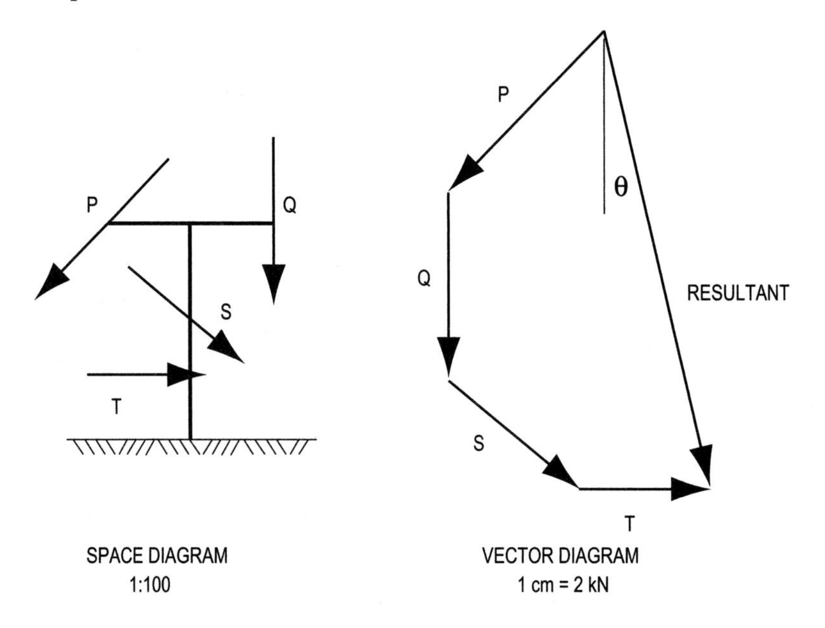

Figure B.4

The resultant of these forces is found by adding the force vectors nose-to-tail in a vector diagram. Each force is drawn to scale (e.g., 1 cm = 2 kN) in the same direction as the line of action. The forces can be added in any order.

Measure the resultant using the appropriate scale to find the magnitude. The direction, in this case, is downward at angle θ from the vertical.

Now let's move on to look at freebody diagrams.

FREEBODY DIAGRAMS

The first step in any analysis involving forces is to identify the forces and their lines of action. A freebody diagram is used to do this. A **freebody diagram** shows an object as if it were floating in air, supported only by the forces acting on it. Figure B.5 shows a beam supported at each end, and carrying a load, W. The left end is supported by a pin joint, and the right end by a roller support. A pin joint can take a force in any direction; a roller support can carry a load only in a normal direction (perpendicular to the surface). The weight of the beam is small relative to the applied force and can be neglected.

A freebody diagram is drawn by removing the supports and replacing them with forces. The other force acting is the load, W. As a check, and to plan the next step, sketch a vector diagram. Since the object is in equilibrium (not moving), the forces acting on it must balance and the vector diagram must close. (It must start and finish at the same point, and there is no resultant force.) If it does not close, you know there is an error somewhere. The three forces are all vertical, and they have been displaced for clarity in Figure B.5. At this stage, a sketch is sufficient to determine whether the diagram closes.

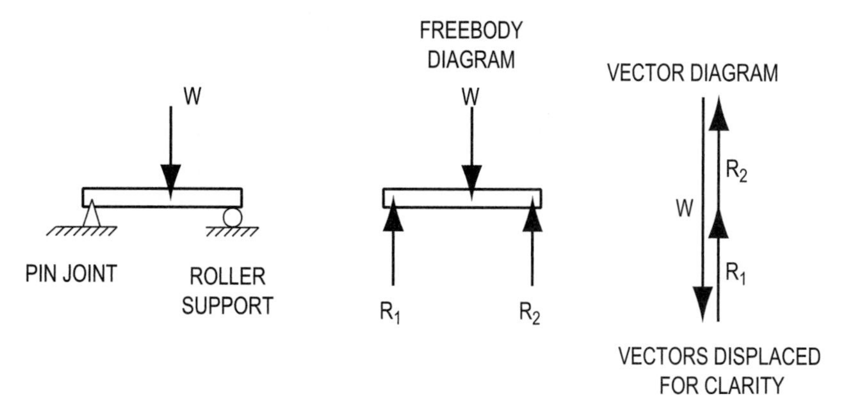

Figure B.5

Now change the loading and see how the fact that the system is in equilibrium can be used. Figure B.6 shows the load and freebody diagrams for a beam with a nonvertical load, L, added.

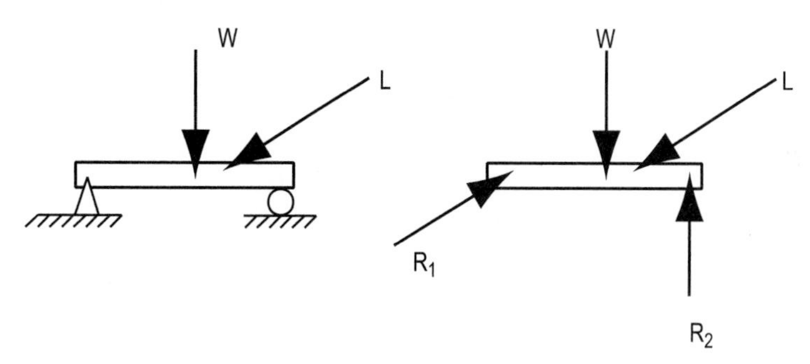

Figure B.6

The direction of the force at the left end is not known, but we know it is not vertical, since there must be some force to balance the nonvertical load, L. If this force is not balanced, the system will move to the left, and since it is not moving, there must be some force acting to the right.

The direction of force R_1 can be found by drawing the vector diagram. The directions of the other three forces are known. We can sketch the vector diagram to see how R_1 can be found. Since we have no values, all we can do is a sketch, but this will be sufficient to determine what to do. Figure B.7 shows how the vector diagram is constructed.

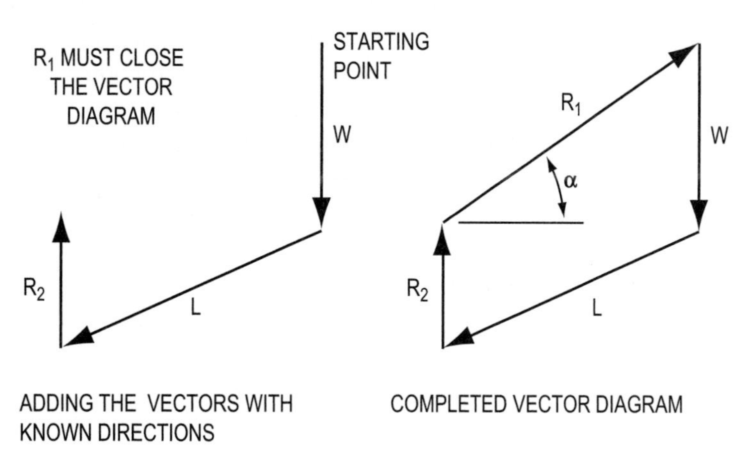

Figure B.7

Forces W, L, and R_2 can be added because their directions are known. Since the system is in equilibrium, the vector diagram must close. Therefore, the line representing force R_1 must go from the nose of R_2 back to the starting point. This would establish the magnitude and direction of R_1, if the magnitudes of the other forces were known. Sketching the vector diagram shows how to approach the problem. The vector diagram, drawn to scale, is used to find the magnitude and direction of R_1.

Now let's look at friction.

FRICTION

If a body is pushed on a surface, there is a friction force opposing the motion. If the force pushing the object is removed, the friction force will bring the object to a stop. Friction always opposes motion. Friction force depends only on the **coefficient of static friction, μ,** and the weight of the object. The areas of the two surfaces have no effect on friction force. The coefficient of friction depends on the two surfaces in contact. We will deal with the coefficient of static friction that applies when a body is just about to move. A **coefficient of sliding friction** applies when an object is sliding on a surface, and a **coefficient of rolling friction** applies to a wheel rolling on a surface. Some typical coefficients of static friction are:

Hard steel on hard steel	0.78
Mild steel on mild steel	0.74
Aluminum on mild steel	0.61
Brass on mild steel	0.51
Teflon on steel	0.04

(Source: Baumeister, T., and L.S. Marks (eds.). *Standard Handbook for Mechanical Engineers,* Seventh Edition. New York: McGraw-Hill Book Company, 1967.)

Consider a block on a flat surface. The coefficient of friction for the two surfaces is μ. If a small horizontal force is applied to the block, it will not move because the force is less than the friction force opposing motion. If the force is gradually increased, the block will move when the horizontal force is large enough to just overcome the friction force. The limiting friction force depends on the coefficient of friction and the normal force of the block on the surface. (The normal force may, or may not, be equal to the weight of the block.) When motion is impending (the block is just about to move) friction force, f, is equal to:

$$f = \mu N,$$

where N is the normal force. Figure B.8 shows the block, freebody, and vector diagrams. The line of action of the friction force is on the surface.

Normal and friction force can be added to give a resultant and if the coefficient of friction is known, the direction of the resultant can easily be found. Figure B.9 shows how this is done.

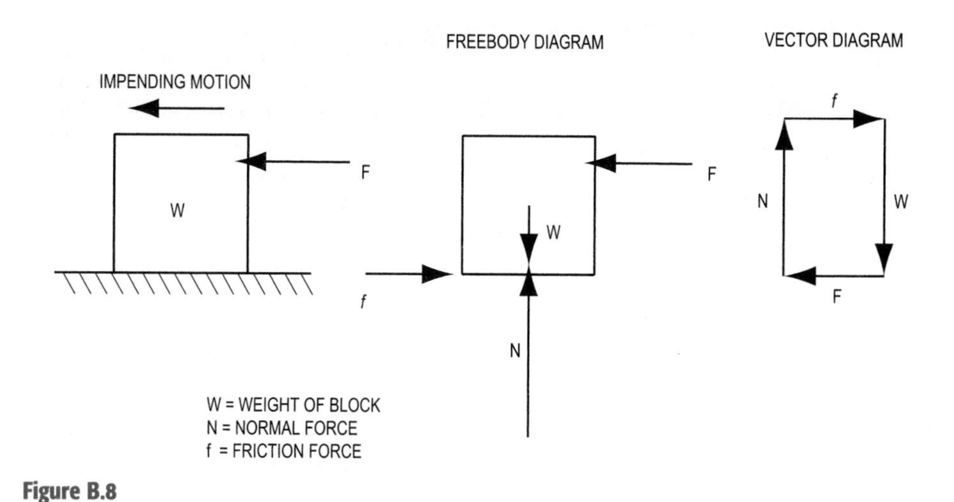

Figure B.8

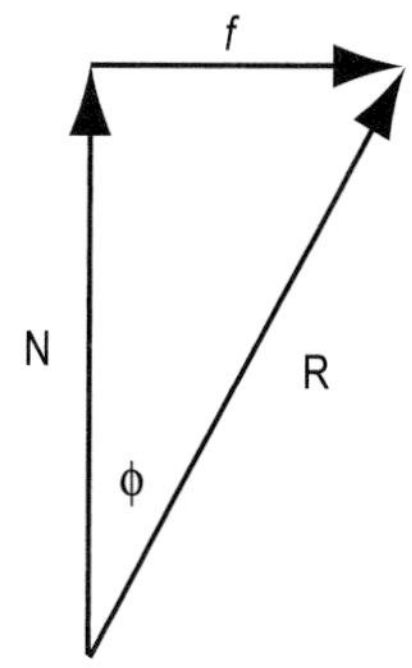

f

N

R

ϕ

$f = \mu N$

$\tan\phi = \frac{f}{N} = \mu$

Figure B.9

The fact that the direction can be found, even though the magnitude, ϕ, and the magnitude of the resultant are not usually important and are seldom required. Since the tangent is known ($\tan \phi = \mu$), the angle can be drawn, giving the direction of the resultant. Figure B.10 shows how to construct an angle when you know its tangent.

$$\tan \phi = \frac{f}{N} = \mu$$

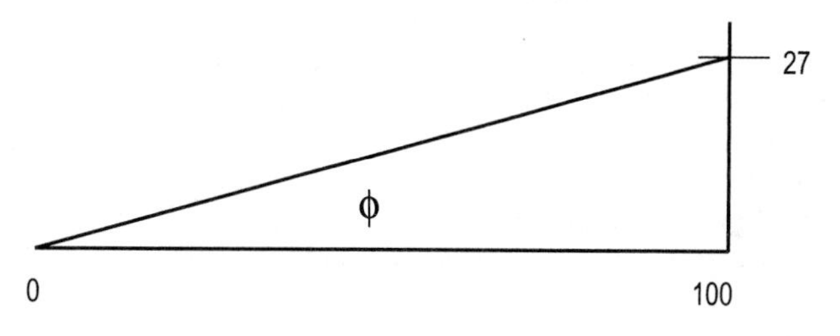

Figure B.10 Constructing an angle knowing the tangent

1. Draw a line 100 units long. Any scale can be used, but a long line (about 100 mm) gives better accuracy.
2. Erect a perpendicular at the end of the line.
3. Measure 27 units (at the same scale) on the perpendicular and join this point with the starting point.

It is more accurate to construct the angle this way than calculate and measure it with a protractor. The angle corresponding to $\mu = 0.27$ is 15.11°, which is difficult to read accurately on a protractor. The figures 100 and 27 are easy to read on a scale.

Let's see how we can use all this to draw a vector diagram knowing the magnitude of one force and all of the directions.

Example B.2

Figure B.11 shows two blocks connected by a rigid link. Determine the weight of block A that will just prevent slipping in the position shown. Block B weighs 220 N and the coefficient of friction between all surfaces is 0.22.

We want to find the weight of block A that will just hold the system in the position shown. If the weight of block A is reduced a small amount, the system will move to the left. The following are the steps to find the weight.

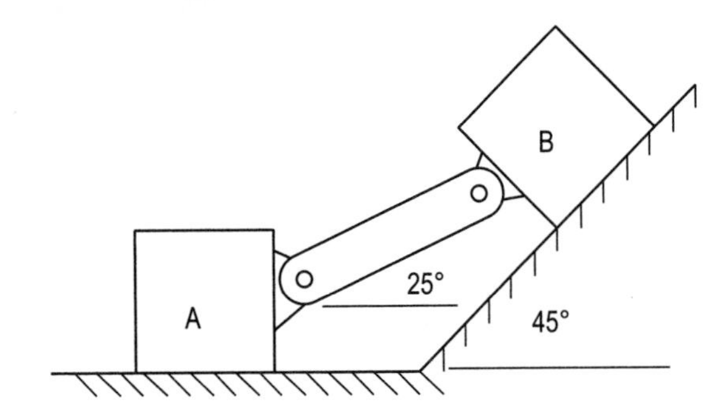

Figure B.11

1. Sketch the freebody diagram for all components, blocks A and B, and the link, to identify all forces acting and their directions for the condition of motion impending to the left (**Figure B.12**).

2. Sketch vector diagrams for all components to see how scale drawings should be constructed to find the required information. Since these are sketches, there is no scale and no values are shown. Show the angular relation (Ø) between the friction force and the associated normal force (**Figure B.12**).

   While vectors can be added in any order, some combinations do not help in solving the problem. Always add the friction force to the end of the normal force so the angle Ø can be shown.

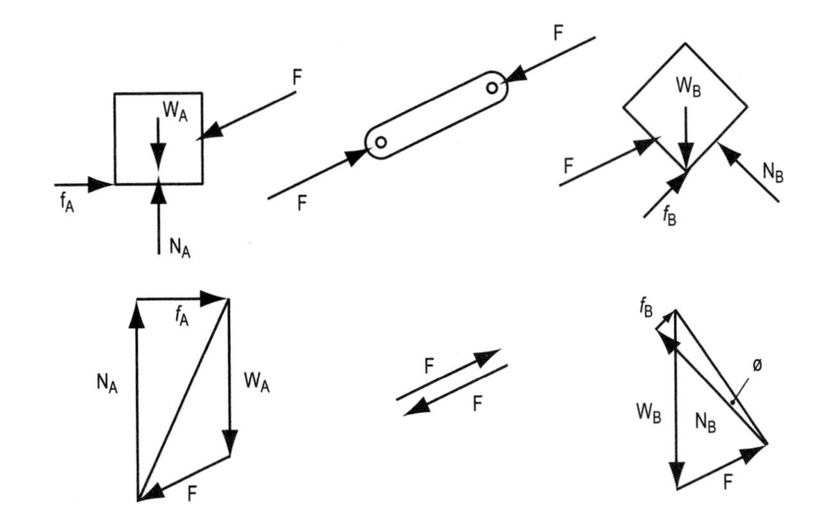

Figure B.12 Freebody and vector diagrams

3. If the vector diagram for A is drawn to scale, W_A can be found, but there is not enough information to do this. The directions of all forces are known, but no magnitudes. If the magnitude of one force were known, the vector diagram could be drawn, and W_A found by measuring the length of the line representing W_A.

   The force on the link, F, is common to vector diagrams for both A and B, and can be found using the vector diagram for block B.

4. Draw the vector diagram for B. Since only W_B is known, it must be the starting point (**Figure B.13**). A scale of 1 cm = 50 N is used. Values can be read directly off the 1:5 metric scale.
   a. Draw W_B to scale.
   b. Add force F at the end of W_B. The starting point for F is known, but not the length. Draw it any length in the correct direction.
   c. The vector diagram must close, since the system is in equilibrium. We must get from some point on F back to the starting point by adding friction force, f_B, and normal force, N_B. The directions of these are known, but where they start on F is not. Work back from the starting point of the vector diagram (the tail of W_B). Look at the sketch to see how this is done. The sketch is repeated in **Figure B.14**.
   d. Construct the angle Ø starting at the end of W_B by drawing a line, 100 units long, parallel to the normal. At the end of this line, draw a perpendicular 22 units long (μ = 0.22). Complete the triangle to find where the resultant intersects F. The magnitude of F is now known. There is no need to find the magnitude of N_B or f_B. There is no need to find the magnitude of F either, since all we need is the scale length.
   e. Draw the vector diagram for A. **Figure B.15** shows the process. Draw a vertical line representing the direction of W_A. The starting point is not known. Add F at the end of W_A (**Figure B.15a**). At the nose of F, draw a line in the direction of the resultant of f_A and N_A using the method described above. The intersection of the resultant and W_A gives the length of W_A. Measure the length to scale. W_A = 440 N (**Figure B.15b**).

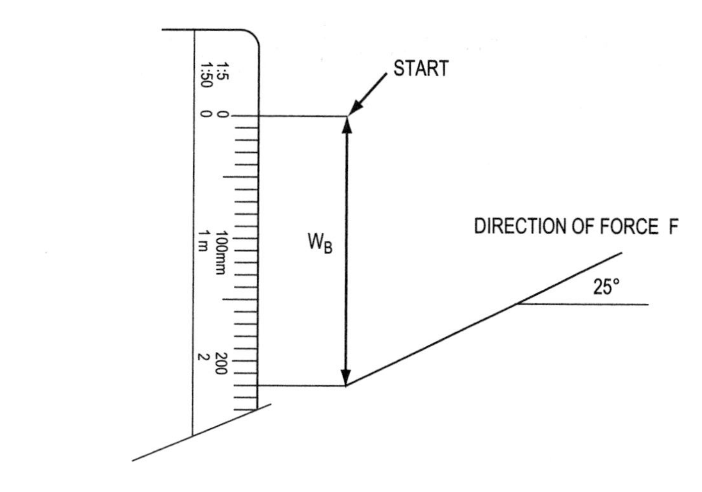

Figure B.13

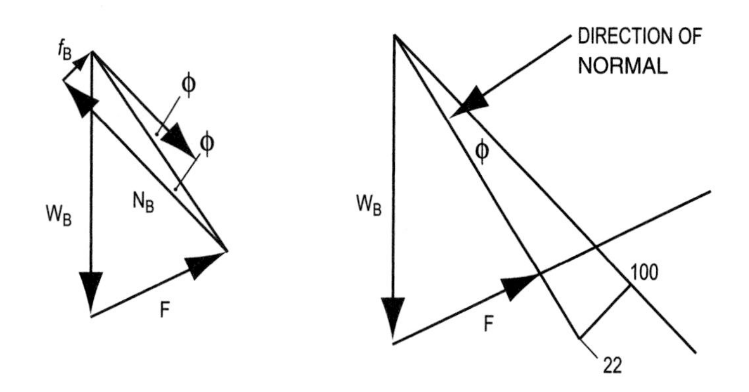

Figure B.14

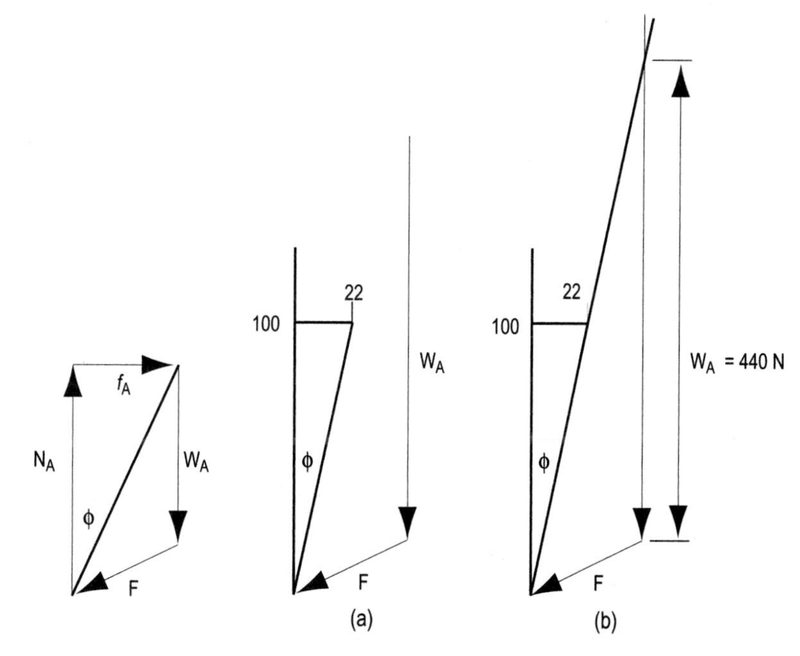

Figure B.15

Example B.2

Now let's look at finding the line of action of the resultant.

FINDING THE LINE OF ACTION OF THE RESULTANT

It is usually necessary to determine what effect forces have on a system. To do this, all lines of action must be known. For example, the net force acting on a system may result in its moving in an undesirable direction or in failure. It is often necessary to find where the line of action of the resultant is or where some force acts.

We saw how two vectors can be added to find a resultant. The resultant acts through the intersection of the lines of action of the vectors. Similarly, a vector quantity can also be represented by two components. You may have used this technique to replace a vector quantity with Cartesian components (components aligned with x and y axes), as shown in Figure B.16.

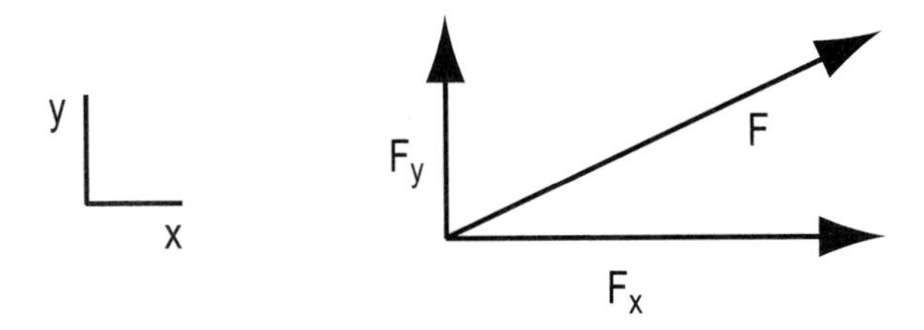

Figure B.16 Cartesian components

Cartesian components are a special case of components at 90°, but components can be at any angle. Figure B.17 shows a force, F, replaced by arbitrary components. The line on top of the letters in the equation indicates that this is a vector addition. The arrowhead on the resultant has been moved away from the end of the line for clarity.

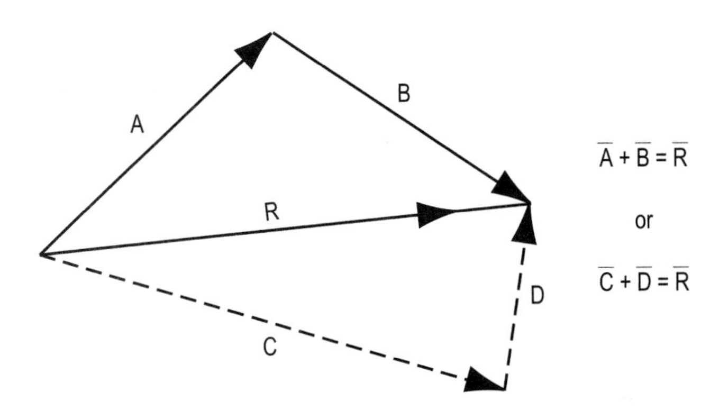

Figure B.17

The use of vector components to locate the line of action of a resultant can be shown best by an example.

Example B.3

A barge, shown in **Figure B.18**, is acted on by three forces. These forces move the barge. How it moves depends on the location of the line of action of the resultant. The craft could move in the direction of the resultant or it could rotate depending on the location of the line of action.

1. Find the magnitude and the direction of the resultant by adding the forces. **Figure B.19** shows the vector diagram used to find this information.
2. Find two components of resultant, R. The line of action of R passes through the intersection of these components. **Figure B.20** shows the components of R. One component is F_3. The other component, R_1, is the resultant of F_1 and F_2.
3. Replace F_1 and F_2 by resultant, R_1, on the space diagram. The resultant acts through the intersection of the lines of action of F_1 and F_2 (**Figure B.21**).
4. Find the intersection of R_1 and F_3 by projecting R_1 and F_3. Resultant R passes through this point (**Figure B.21**).
5. Draw R on the space diagram.

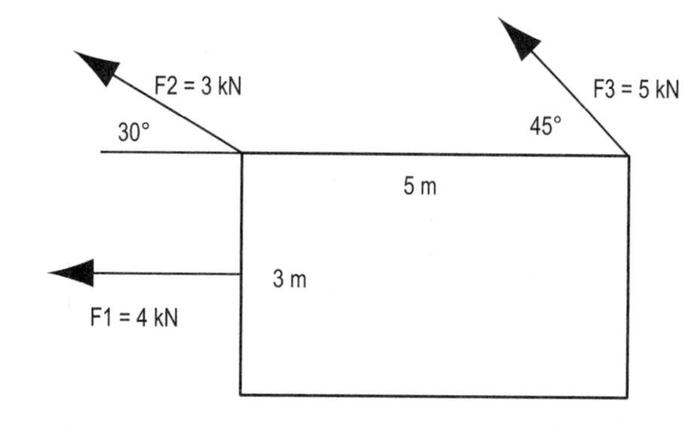

Figure B.18

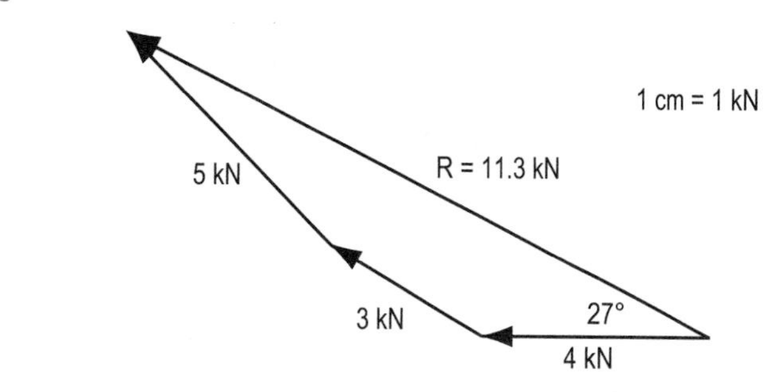

Figure B.19

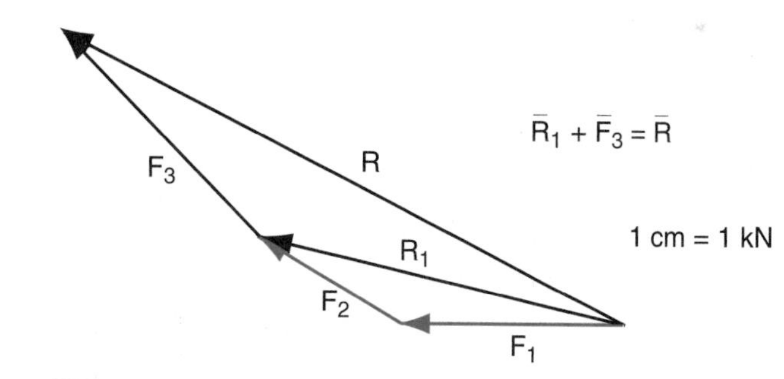

Figure B.20

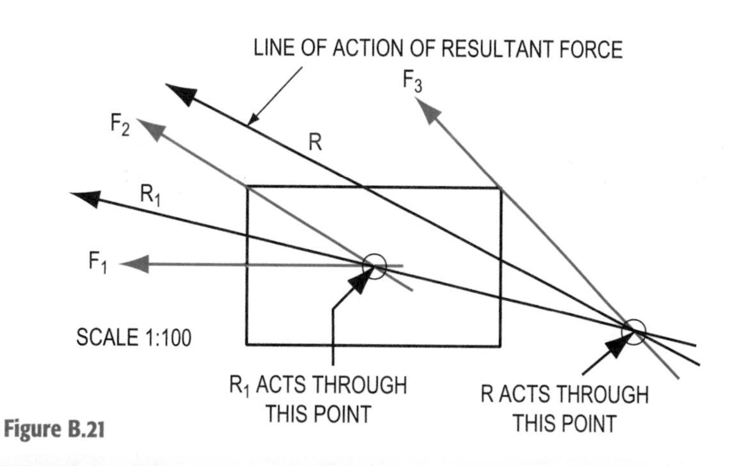

Figure B.21

Example B.4

Sketch freebody and vector diagrams for the three components of the cargo sling.

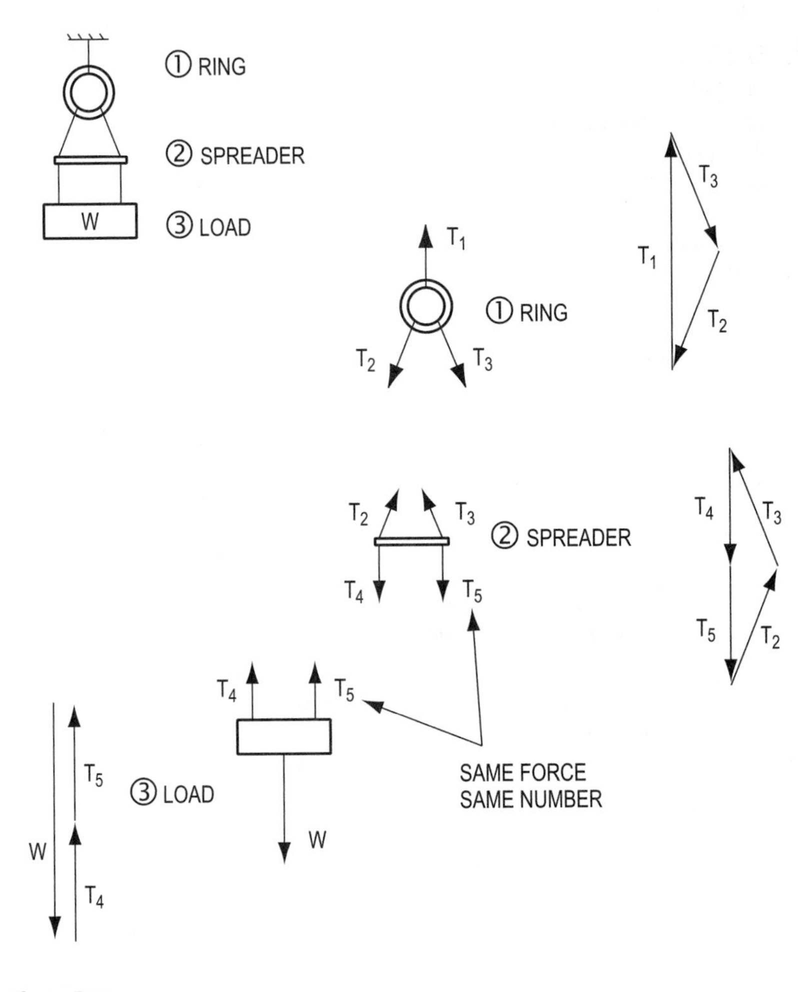

Figure B.22

Problems

For all problems, sketch the freebody and vector diagrams for all components. Then draw vector diagrams, to scale, to find the required information.

1. Sketch freebody and vector diagrams for the objects shown in Figure B.23. Sketches should be done for all numbered components and not for the assembly, as shown in Example B.4. Ignore the weight of the part unless weight is indicated by W.

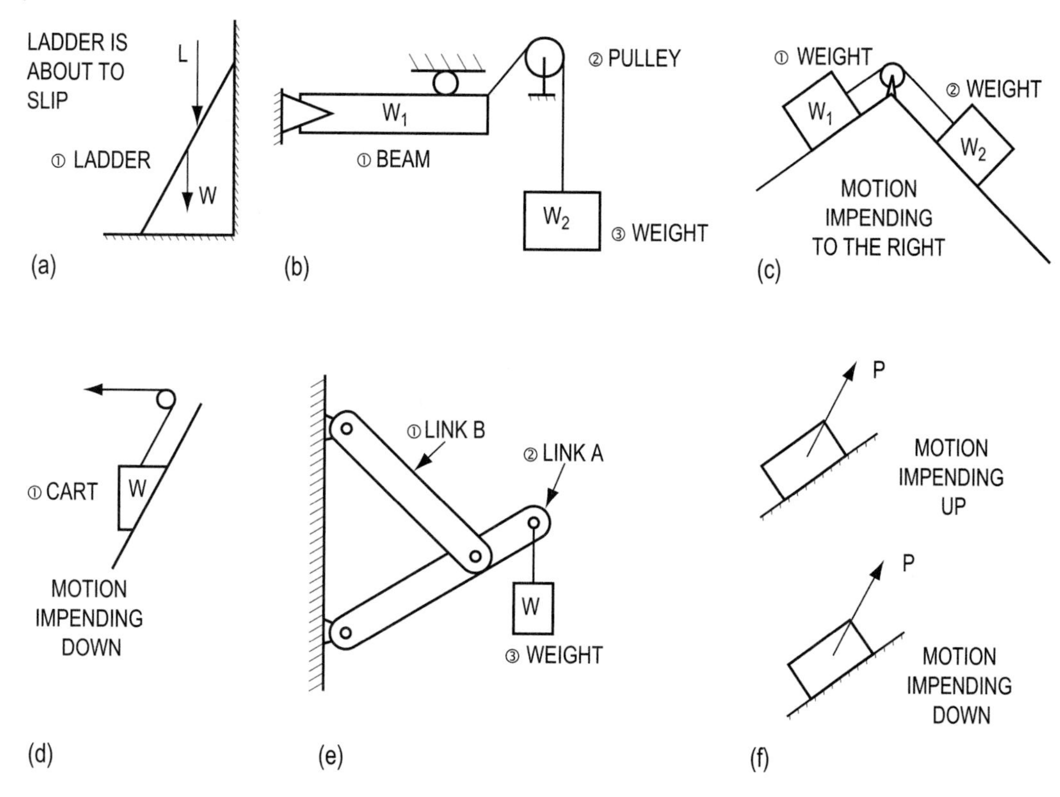

Figure B.23

2. Figure B.24 shows a system of wedges, A and B, used for positioning heavy parts. The part weighs 8.5 kN, and the coefficient of friction between the part and the floor is 0.18. The weight of A and B is negligible, and the coefficient of friction between A and B and any surface over which they slide is 0.12. Find the force, P, that will just move the part to the right. The scale is 1 cm = 0.5 kN.

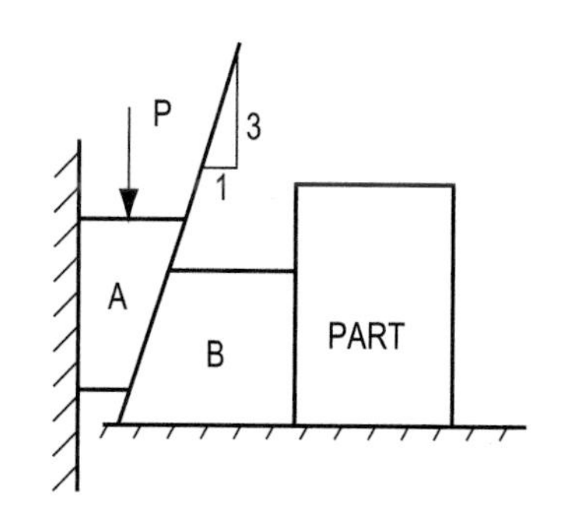

Figure B.24

3. **Figure B.25** shows a system of wedges intended to move C to the right against a force of 4 kN. The weight of parts A, B, and C is negligible. The coefficient of friction for all surfaces is 0.18. Find the force, P, that will just cause the system to move. The scale is 1 cm = 200 N.

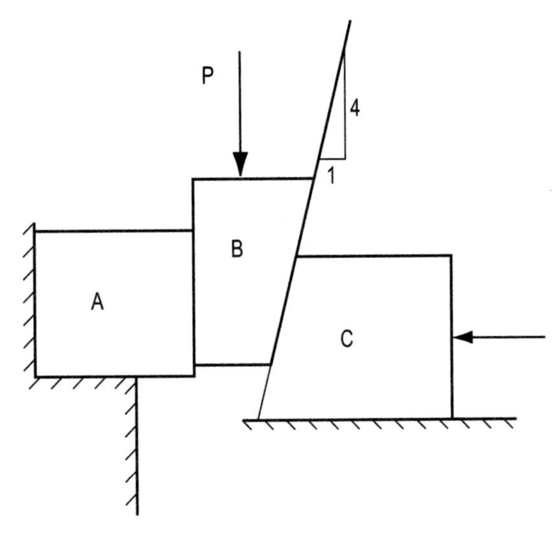

Figure B.25

4. What is the force, P, that will just move the system in **Figure B.26** to the left? Blocks A and B weigh 1000 N and 1500 N, respectively. The coefficient of friction for all surfaces is 0.23. Ignore friction in the pulley. The scale is 1 cm = 200 N.

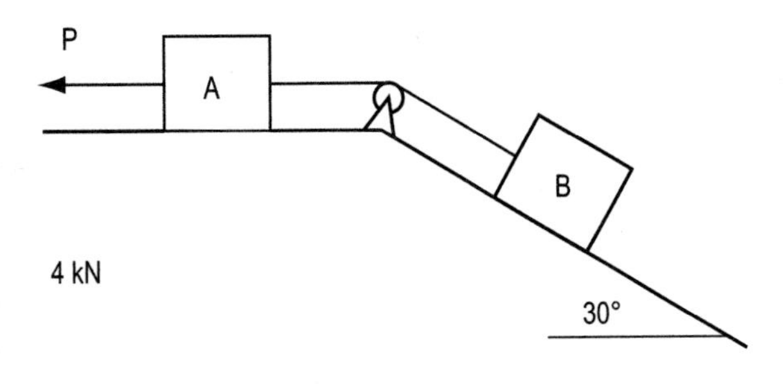

Figure B.26

5. A mechanical system can be represented as shown in **Figure B.27**. A force, P, that will just move the system to the left must be applied to A. Blocks A and B weigh 700 N and 950 N, respectively. The coefficient of friction is 0.2 for all surfaces. The pulley is frictionless and the rope connecting the blocks is parallel to the surface. Specify the magnitude and direction of the minimum force, P, that will just move the system to the left. The broken line indicates that the line of action is not known and must be specified. The scale is 1 cm = 100 N.

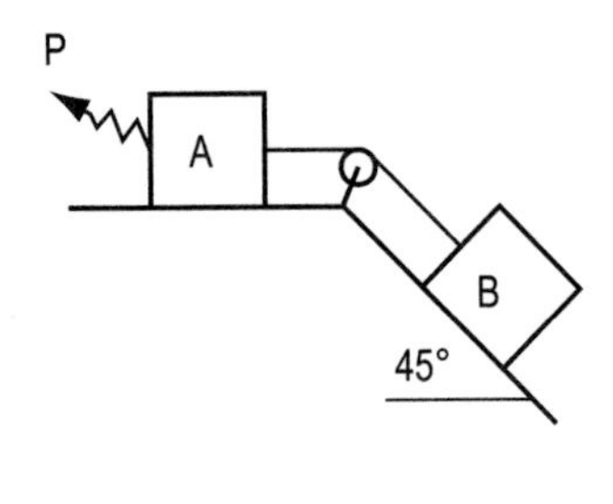

Figure B.27

6. Find the minimum force, P, that will just move the system shown in **Figure B.28** to the left. Block A weighs 450 N and block B weighs 1400 N. The coefficient of friction for all surfaces is 0.18. The broken line showing where force P is applied indicates that the line of action is not known, and must be determined. The scale is 1 cm = 100 N.

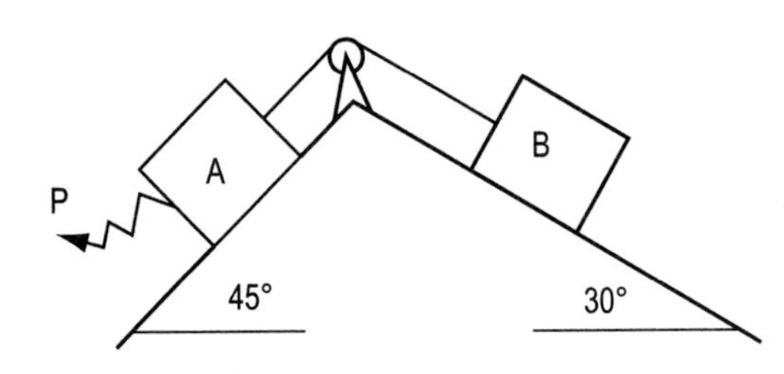

Figure B.28

Problems

7. What is the weight of A that will just cause the system described in problem 6 to slide to the left with no applied force?
8. For the system described in problem 6, find the minimum force, P, that will just prevent the system slipping to the right. The scale is 1 cm = 100 N.
9. **Figure B.29** shows a proposed mechanism for raising a table and a part to be machined. The force, F, will be applied by a hydraulic cylinder. Force, F, must be known so that the cylinder can be specified. The table and the part to be machined weigh 500 N. The coefficient of friction for all surfaces is 0.25. The weights of parts A and B are negligible. The scale is 1 cm = 50 N.
   a. Find the force that will just raise the table and part.
   b. Is this proposed mechanism a good one for this application?

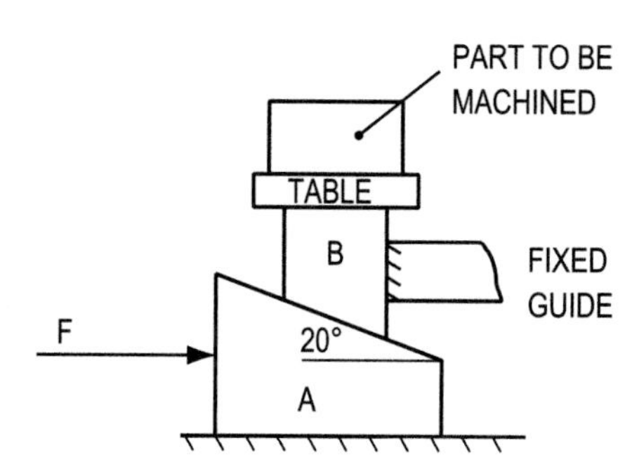

Figure B.29

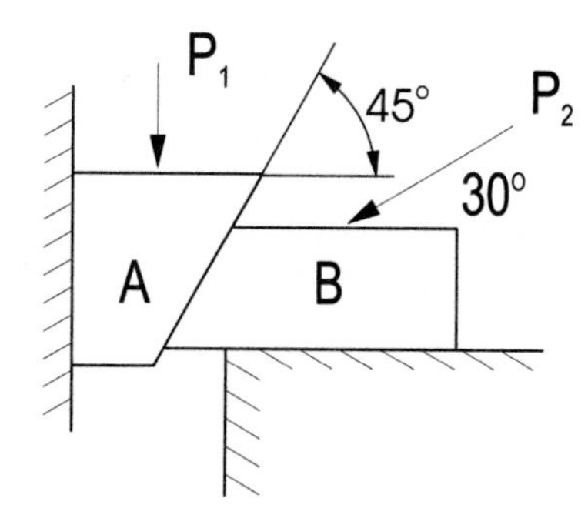

Figure B.30

10. Block A weighs 200 N and B weighs 300 N (**Figure B.30**). The coefficient of friction for all surfaces is 0.23. Force P_2 is 550 N, and P_1 is just sufficient to move A down. Find the force, P1, that will just cause A to move. The scale is 1 cm = 100 N.
11. A worker claims he was injured when climbing up the ladder shown in **Figure B.31** because it slipped. Investigate this case, and determine whether the worker has a valid claim. The worker weighs 800 N. The coefficient of friction between the ladder and the ground is 0.62 and between the ladder and the wall 0.38. The ladder weighs 100 N. The space scale is 1:50. The force scale is 1 cm = 100 N.

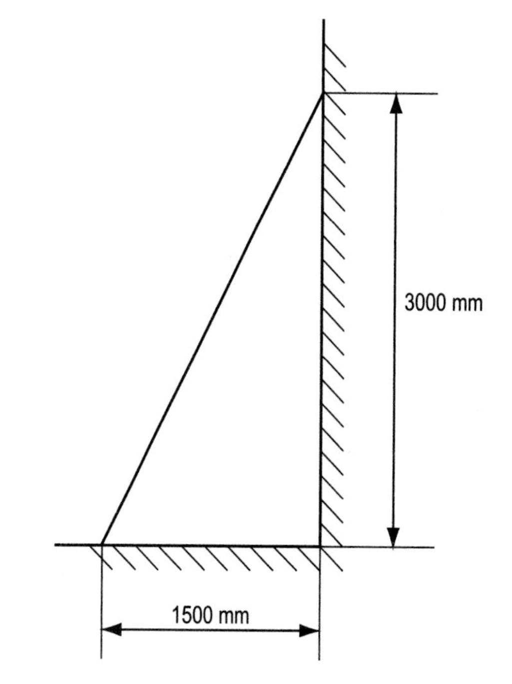

Figure B.31

12. A building mounted on skids is to be moved across a smooth, level parking lot (**Figure B.32**). Three tractors will be used to move the building. One tractor applies a force of 9 kN, due west at corner A. Another applies 20 kN due west at corner B. A third tractor applies a force of 13 kN northwest at C. The combined effort is sufficient to move the building slowly across the parking lot.

    Determine the magnitude and direction of the resultant of the three forces. Dimension the distance from C where the line of action of the resultant crosses CD. The space scale is 1:200. The force scale is 1 cm = 2 kN.
13. A raft is acted on by the forces shown in **Figure B.33**. The raft must be moved alongside a pier by the minimum possible force applied at A. The resultant of all forces acting on the raft must be to the right and downward at 30°.
    a. Find the magnitude and direction of the force applied at A.
    b. Determine the magnitude of the resultant force.
    c. Show where the line of action acts on the space diagram.
14. **Figure B.34** shows a 3 kN cylinder being pulled up a ramp. The coefficient of friction at A and B is 0.4. The ramp weighs 1.2 kN. At some point in the travel, the ramp will slip. How far up the ramp can the cylinder be pulled before the ramp slips? Ignore friction between the cylinder and the ramp.
    a. Sketch the freebody diagrams for the cylinder and the ramp.
    b. Sketch the vector diagrams for the cylinder and the ramp.
    c. Draw the vector diagram for the cylinder and determine the normal force between the cylinder and ramp. The scale is 1 cm = 500 N.
    d. Draw the vector diagram for the ramp, and determine RA and RB. The scale is 1 cm = 500 N.
    e. How far from the bottom of the ramp is the centre of the cylinder when the ramp slips? Show this distance along the ramp.

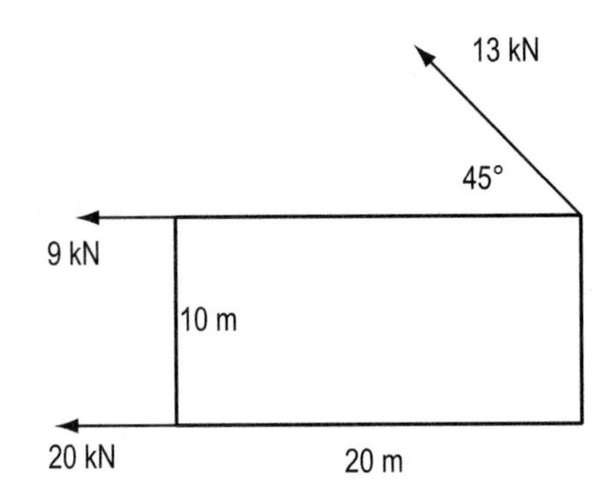

Figure B.32

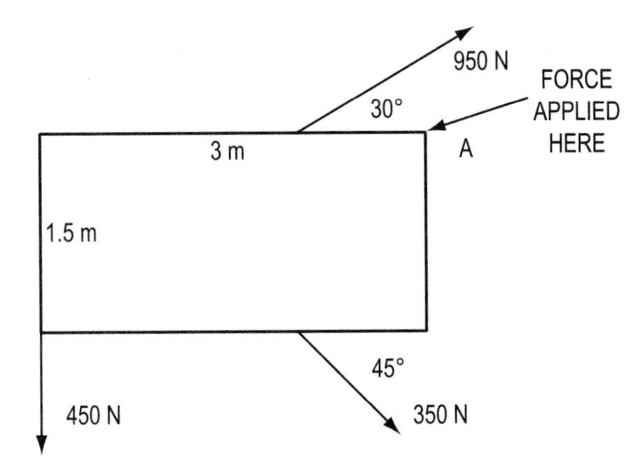

Figure B.33

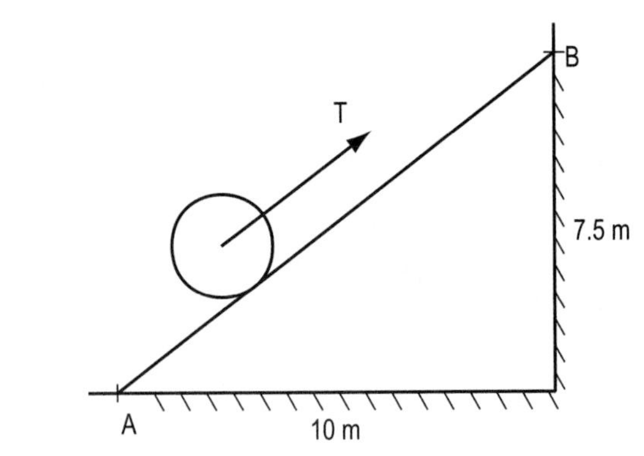

Figure B.34

15. A load-limiting device is shown in Figure B.35. It is designed to prevent the load, F, from exceeding 50 kN. If this load is exceeded, the system will slip. Specify the coefficient of friction required to meet these conditions. The coefficient of friction is the same for all sliding surfaces. The scale is 1 cm = 10 kN.

$$W_A = 70 \text{ kN}$$

$$W_B = 60 \text{ kN}$$

$$\alpha = 30°$$

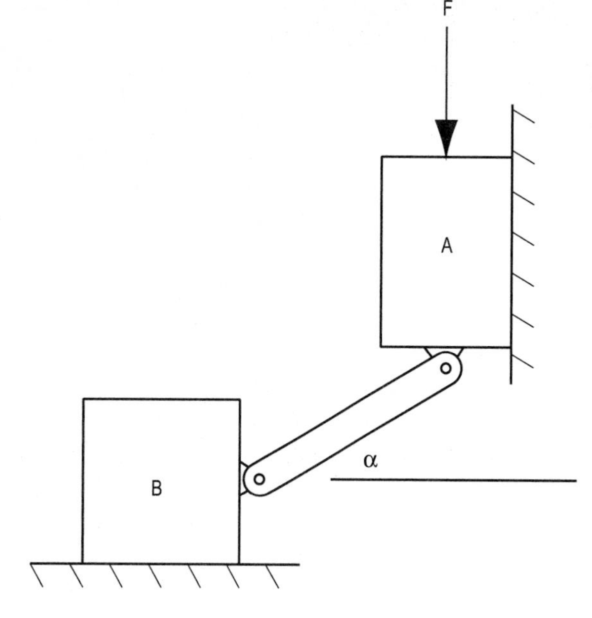

Figure B.35

APPENDIX

C Common Sectioning Symbols and Designations

There are sectioning symbols for many materials. Following are some of the common ones.

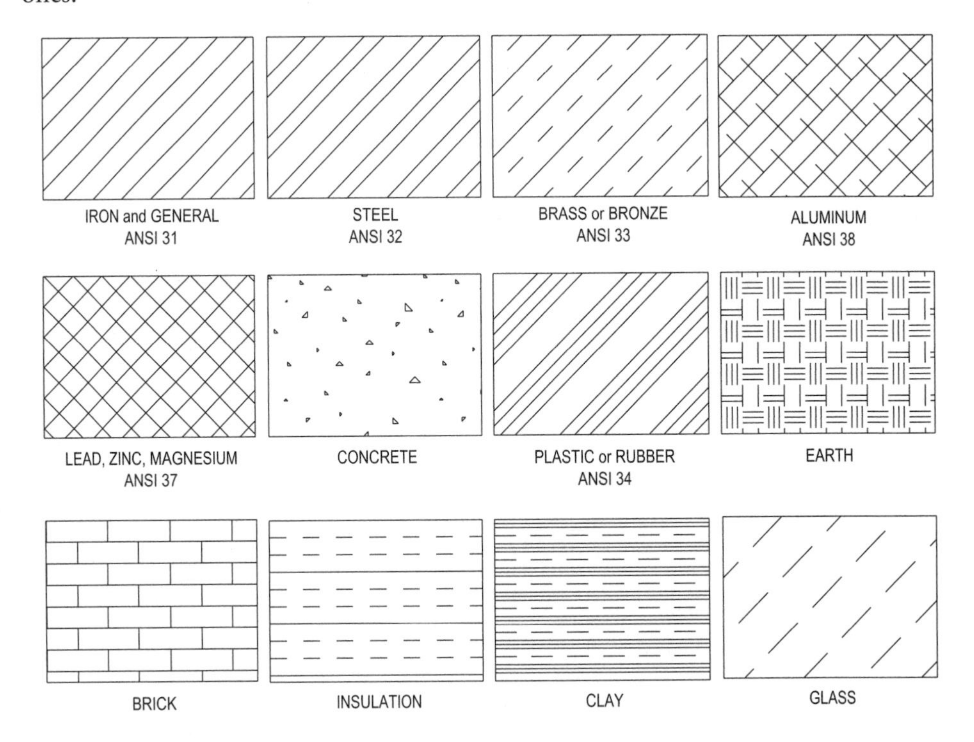

APPENDIX

D Standards List

There are hundreds of standards in engineering. This is a very small list, pertaining mainly to drawings and fasteners. More information on standards can be found from the following Web sites. Standards can be ordered over the Internet from most of these organizations. There are many organizations that issue standards, but only a few are listed here.

American National Standards Institute	www.ansi.org
American Society for Testing and Materials	www.astm.org
American Society of Mechanical Engineers	www.asme.org
Canadian Standards Association	www.csa-international.org
Institute of Electrical and Electronics Engineers	www.standards.ieee.org
International Organization for Standardization	www.iso.ch
National Standards System Network	www.nssn.org
Society of Automotive Engineers	www.sae.org

ASME Y1.1-1989

Abbreviations for Use on Drawings and in Text
Standard abbreviations for use on engineering drawings

ASME Y14.1-1995

Decimal Inch Sheet Size and Format
Sheet size and format for engineering drawings

ASME Y14.1M-1995

Metric Drawing Sheet Size and Format
Metric sheet size and format for engineering drawings

ASME Y14.2M-1992(R1998)

Line Conventions and Lettering
Line and lettering practices used on engineering drawings

ASME Y14.3M-1994

Multiview and Sectional View Drawings
Requirements for selection and arrangement of orthographic views, section views, auxiliary views, and conventional drawing practices

ASME Y14.4M-1989(R1994)
Pictorial Drawing
Defines and illustrates the use of various kinds of pictorial drawings used in mechanical drawings

ASME Y14.5M-1994
Dimensioning and Tolerancing
Methods of dimensioning and specifying tolerances on engineering drawings

CSA B78.1
Technical Drawing, General Principles

CSA B78.2
Dimensioning and Tolerancing of Technical Drawings

ASME Y14.6-1978(R1993)
Screw Thread Representation
Standards for representing, dimensioning, and specifying screw threads

ASME Y14.6M-1981(R1998)
Screw Thread Representation (Metric Supplement)
Standards for representing, dimensioning, and specifying metric threads

ASME Y32.2.3-1949(R1994)
Graphic Symbols for Pipe Fittings, Valves, and Piping

ASME Y32.2.4-1949(R1998)
Graphic Symbols for Heating, Ventilating, and Air Conditioning

ASME Y32.2.6-1950(R1998)
Graphic Symbols for Heat-Power Apparatus

ASME Y32.4-1977(R1994)
Graphic Symbols for Plumbing Fixtures for Diagrams Used in Architecture and Building Construction

ASME Y32.7-1972(R1994)
Graphic Symbols for Railway Maps and Profiles

ASME Y32.10-1967(R1994)
Graphic Symbols for Fluid Power Diagrams

ASME Y32.11-1961(R1998)
Graphic Symbols for Process Diagrams in Petroleum and Chemical Industries

IEEE315A-86
Graphic Symbols for Electrical and Electronic Diagrams

IEEE91-84
Graphic Symbols for Logic Diagrams

AISC S340 (American Institute of Steel Construction)
Metric Properties of Structural Steel Shapes with Dimensions According to ASTM A6M

AISC P648
Fundamentals of (Metric) Structural Shop Drafting—CISC (Revised December 1998)

ASME B1.1-1998
Unified Inch Screw Threads (UN and UNR Thread Form)
Designations and other information for unified threads

ASME B1.13-1995
Metric Screw Threads—M Profile
General standards for metric threads

ISO 262:1973
ISO General Purpose Metric Screw Threads—Selected Sizes for Screws, Bolts, and Nuts

ISO 68:1973
ISO General Purpose Screw Threads—Basic Profile

ISO 724:1993
ISO General Purpose Screw Threads—Basic Dimensions

ANSI/ASME B1.13M
Metric Screw Threads for Commercial Mechanical Fasteners

ASME B18.2.1-1996
Square and Hex Bolts and Screws (Inch Series)
General information and dimensions

ASME B18.2.2-1987(R1993)
Square and Hex Nuts
General information and dimensions

ASME B18.3-1998
Socket, Cap, Shoulder, and Set Screws, Hex, and Spline Keys
General information and dimensions

ASME B18.5-1990(R1998)
Round Head Bolts (Inch Series)
General information and dimensions

ASME B18.6.2-1972(R1993)
Slotted Head Cap Screws, Square Head Set Screws, and Slotted Headless Set Screws
General information and dimensions

ASME B18.6.3-1972(R1997)
Machine Screws and Machine Screw Nuts
General information and dimensions

ASME B1.20.1-1983(R1992)
Pipe Threads, General Purpose (Inch)
Dimensions and gauging of pipe threads

ASME B5.10-1984
Machine Tapers-Self Holding and Steep Taper Series

CSA B232-75
Keys, Keyseats, and Keyways

ASME A13.1-81R85
Scheme for the Identification of Piping Systems

APPENDIX E Standard Dimensions (Cap and Machine Screws)

ISO Metric Screw Thread Standard Series

Normal Size Dia. (mm)			Pitches (mm)															Nominal Size Dia. (mm)
Column[a]			Series with Graded Pitches		Series with Constant Pitches													
1	2	3	Course	Fine	6	4	3	2	1.5	1.25	1	0.75	0.5	0.35	0.25	0.2	
0.2			0.075														0.25
0.3			0.08														0.3
	0.35		0.09														0.36
0.4			0.1														0.4
	0.45		0.1														0.45
0.5			0.125														0.5
	0.55		0.125														0.55
0.6			0.15														0.6
	0.7		0.175														0.7
0.8			0.2														0.8
	0.9		0.225														0.9
			0.25													0.2	1
	1.1		0.25													0.2	1.1
1.2			0.25													0.2	1.2
	1.4		0.3													0.2	1.4
1.6			0.35													0.2	1.6
	1.8		0.35													0.2	1.8
2			0.4												0.25		2
	2.2		0.45												0.25		2.2
2.5			0.45											0.35			2.5
3			0.5											0.35			3
	3.5		0.6											0.35			3.5
4			0.7										0.5				4
	4.5		0.75										0.5				4.5
5			0.8										0.5				5
		5.5											0.5				5.5
6			1									0.75					6
		7	1									0.75					7
8			1.25	1							1	0.75					8
		9	1.25								1	0.75					9
10			1.5	1.25						1.25	1	0.75					10
		11	1.5								1	0.75					11
12			1.75	1.25					1.5	1.25	1						12
	14		2	1.5					1.5	1.25[b]	1						14
		15							1.5		1						15
16			2	1.5					1.5		1						16
		17							1.5		1						17
	18		2.5	1.5				2	1.5		1						18
20			2.5	1.5				2	1.5		1						20
	22		2.5	1.5				2	1.5		1						22
24			3	2				2	1.5		1						24
		25						2	1.5		1						25
		26							1.5		1						26
	27		3	2				2	1.5		1						27
		28						2	1.5		1						28

ISO Metric Screw Thread Standard Series

Normal Size Dia. (mm)			Pitches (mm)														Nominal Size Dia. (mm)
Column[a]			Series with Graded Pitches		Series with Constant Pitches												
1	2	3	Course	Fine	6	4	3	2	1.5	1.25	1	0.75	0.5	0.35	0.25	0.2	
30			3.5	2			(3)	2	1.5		1						30
		32						2	1.5								32
	33		3.5	2			(3)	2	1.5								33
		35[c]							1.5								35[c]
36			4	3				2	1.5								36
		38							1.5								38
	39		4	3				2	1.5								39
		40					3	2	1.5								40
42			4.5	3		4	3	2	1.5								42
	45		4.5	3		4	3	2	1.5								45
48			5	3		4	3	2	1.5								48
		50					3	2	1.5								50
	52		5	3		4	3	2	1.5								52
		55				4	3	2	1.5								55
56			5.5	4		4	3	2	1.5								56
		58				4	3	2	1.5								58
	60		5.5	4		4	3	2	1.5								60
		62				4	3	2	1.5								62
64			6	4		4	3	2	1.5								64
		65				4	3	2	1.5								65
	68		6	4		4	3	2	1.5								68
		70			6	4	3	2	1.5								70
72					6	4	3	2	1.5								72
		75				4	3	2	1.5								75
	76				6	4	3	2	1.5								76
		78						2									78
80					6	4	3	2	1.5								80
		82						2									82
	85				6	4	3	2									85
90					6	4	3	2									90
	95				6	4	3	2									95
100					6	4	3	2									100
	105				6	5	4	3	2								105
110					6	5	4	3	2								110
	115				6	5	4	3	2								115

a Thread diameter should be selected from columns 1, 2, or 3, with preference being in that order.
b Pitch 1.25 mm in combination with diameter 14 mm has been included for sparkplug applications.
c Diameter 35 mm has been included for bearing locknut applications.

The use of pitches shown in parentheses should be avoided wherever possible.
The pitches enclosed in the bold frame, together with the corresponding nominal diameters in columns 1 and 2, are those combinations which have been established by ISO Recommendations as a selected "coarse" and "fine" series for commercial fasteners.

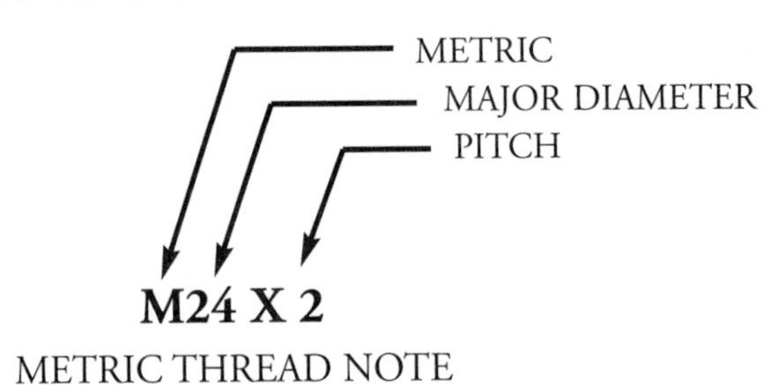

METRIC THREAD NOTE

AMERICAN STANDARD HEX HEAD BOLTS

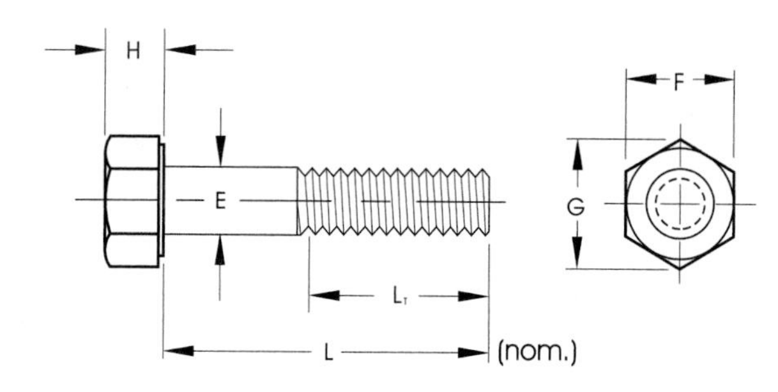

DIMENSIONS OF HEX CAP SCREWS

Nominal Size Or Basic Product Diamenter		E Body Diameter		F Width Across Flats			G Width Across Corners		H Head Height			L_T Thread Length For Screw Lengths	
		Max.	Min.	Basic	Max.	Min.	Max.	Min.	Basic	Max.	Min.	6 in. and Shorter Nom.	Over 6 in. Nom.
1/4	0.2500	0.2500	0.2450	7/16	0.438	0.428	0.505	0.488	5/32	0.163	0.150	0.750	1.000
5/16	0.3125	0.3125	0.3065	1/2	0.500	0.489	0.577	0.557	13/64	0.211	0.195	0.875	1.125
3/8	0.3750	0.3750	0.3690	9/16	0.562	0.551	0.628	0.628	15/64	0.243	0.226	1.000	1.250
7/16	0.4375	0.4375	0.4305	5/8	0.625	0.612	0.698	0.698	9/32	0.291	0.272	1.125	1.375
1/2	0.5000	0.5000	0.4930	3/4	0.750	0.736	0.840	0.840	5/16	0.323	0.302	1.250	1.500
9/16	0.5625	0.5625	0.5545	13/16	0.812	0.798	0.910	0.910	23/64	0.371	0.348	1.375	1.625
5/8	0.6250	0.6250	0.6170	15/16	0.938	0.922	1.083	1.051	25/64	0.403	0.378	1.500	1.750
3/4	0.7500	0.7500	0.7410	1 1/8	1.125	1.100	1.299	1.254	15/32	0.483	0.455	1.750	2.000
7/8	0.8750	0.8750	0.8660	1 5/16	1.312	1.285	1.516	1.465	35/64	0.563	0.531	2.000	2.250
1	1.0000	1.0000	0.9900	1 1/2	1.500	1.469	1.732	1.675	39/64	0.627	0.591	1.250	2.500
11/8	1.1250	1.1250	1.1140	1 11/16	1.688	1.631	1.949	1.859	11/16	0.718	0.658	2.500	2.750
1 1/4	1.2500	1.2500	1.2390	1 7/8	1.875	1.812	2.165	2.006	25/32	0.813	0.749	2.750	3.000
1 3/8	1.3750	1.3750	1.3630	2 1/16	2.062	1.994	2.382	2.273	27/32	0.878	0.810	3.000	3.250
1 1/2	1.5000	1.5000	1.4880	2 1/4	2.250	2.175	2.598	2.480	1 5/16	0.974	0.902	3.250	3.500
1 3/4	1.7500	1.7500	1.7380	2 5/8	2.625	2.538	3.031	2.893	1 3/32	1.134	1.054	3.750	4.000
2	2.0000	2.0000	1.9880	3	3.000	2.900	3.464	3.306	1 7/32	1.263	1.175	4.250	4.500
2 1/4	2.2500	2.2500	2.2380	3 3/8	3.375	3.262	3.897	3.719	1 3/8	1.423	1.327	...	5.000
2 1/2	2.5000	2.5000	2.4880	3 3/4	3.750	3.625	4.330	4.133	1 17/32	1.583	1.479	...	5.500
2 3/4	2.7500	2.7500	2.7380	4 1/8	4.125	3.988	4.763	4.546	1 11/16	1.744	1.632	...	6.000
3	3.0000	3.0000	2.9880	4 1/2	4.500	4.350	5.196	4.959	1 7/8	1.935	1.815	...	6.500

AMERICAN STANDARD HEX HEAD NUTS

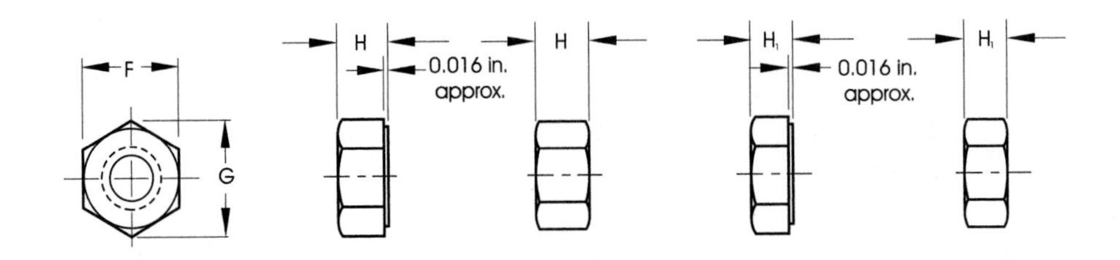

DIMENSIONS OF HEX NUTS AND HEX JAM NUTS

Nominal Size Or Basic Major Diam. of Thread		F Width Across Flats			G Width Across Corners		H Thickness Hex Nuts			H_1 Thickness Hex Jam Nuts		
		Basic	Max.	Min.	Max.	Min.	Basic	Max.	Min.	Basic	Max.	Min.
1/4	0.2500	7/16	0.438	0.428	0.505	0.488	7/32	0.226	0.212	5/32	0.163	0.150
5/16	0.3125	1/2	0.500	0.489	0.577	0.557	17/64	0.273	0.258	3/16	0.195	0.180
3/8	0.3750	9/16	0.562	0.551	0.650	0.628	21/64	0.337	0.320	7/32	0.227	0.210
7/16	0.4375	11/16	0.688	0.675	0.794	0.768	3/8	0.385	0.365	1/4	0.260	0.240
1/2	0.5000	3/4	0.750	0.736	0.866	0.840	7/16	0.448	0.427	5/16	0.323	0.302
9/16	0.5625	7/8	0.875	0.861	1.010	0.982	31/64	0.496	0.473	5/16	0.324	0.301
5/8	0.6250	15/16	0.938	0.922	1.083	1.051	35/64	0.559	0.535	3/8	0.387	0.363
3/4	0.7500	1 1/8	1.125	1.088	1.299	1.240	41/64	0.665	0.617	27/64	0.446	0.398
7/8	0.8750	1 5/16	1.312	1.269	1.516	1.447	3/4	0.776	0.724	31/64	0.510	0.458
1	1.0000	1 1/2	1.500	1.45	1.732	1.653	55/64	0.887	0.831	35/64	0.575	0.519
11/8	1.1250	1 11/16	1.688	1.631	1.949	1.859	31/32	0.999	0.939	39/64	0.639	0.579
1 1/4	1.2500	1 7/8	1.875	1.812	2.165	2.066	1 1/16	1.094	1.030	23/32	0.751	0.687
1 3/8	1.3750	2 1/16	2.062	1.994	2.382	2.273	1 11/64	1.206	1.138	25/32	0.815	0.747
1 1/2	1.5000	2 1/4	2.250	2.175	2.598	2.480	1 9/32	1.317	1.245	27/32	0.880	0.808

AMERICAN STANDARD SQUARE BOLTS

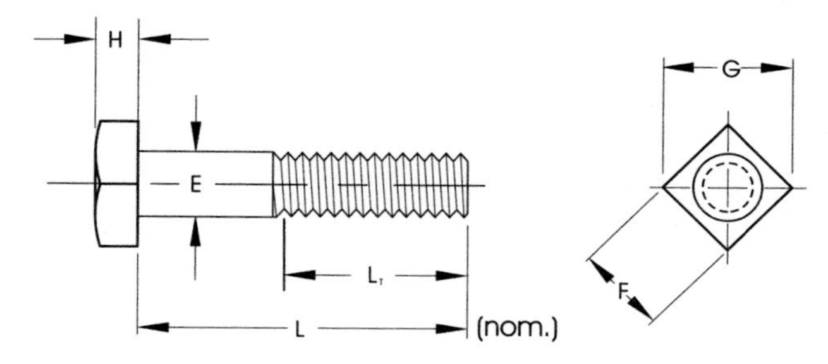

DIMENSIONS OF SQUARE BOLTS

Nominal Size Or Base Product Diameter		E Full Size Body Diameter		F Width Across Flats			G Width Across Corners			H Head Height		L_T Thread Length For Screw Lengths	
												6 in. and Shorter	Over 6 in
		Max.	Min.	Basic	Max.	Min.	Max.	Min.	Basic	Max.	Min.	Nom.	Nom.
1/4	0.2500	0.260	0.237	3/8	0.375	0.362	0.530	0.498	11/64	0.188	0.156	0.750	1.000
5/16	0.3125	0.324	0.298	1/2	0.500	0.484	0.707	0.665	13/64	0.220	0.186	0.875	1.125
3/8	0.3750	0.388	0.360	9/16	0.562	0.544	0.795	0.747	1/4	0.268	0.232	1.000	1.250
7/16	0.4375	0.452	0.421	5/8	0.625	0.603	0.884	0.828	19/64	0.316	0.278	1.125	1.375
1/2	0.5000	0.515	0.482	3/4	0.750	0.725	1.061	0.995	21/64	0.348	0.308	1.250	1.500
5/8	0.6250	0.642	0.605	15/16	0.938	0.906	1.326	1.244	27/64	0.444	0.400	1.500	1.750
3/4	0.7500	0.768	0.729	1 1/8	1.125	1.088	1.591	1.494	1/2	0.524	0.476	1.750	2.000
7/8	0.8750	0.895	0.852	1 5/16	1.312	1.269	1.856	1.742	19/32	0.620	0.568	2.000	2.250
1	1.0000	1.022	0.976	1 1/2	1.500	1.450	2.121	1.991	21/32	0.684	0.628	2.250	2.500
1 1/8	1.1250	1.149	1.098	1 11/16	1.688	1.631	2.386	2.239	3/4	0.780	0.720	2.500	2.750
1 1/4	1.2500	1.277	1.223	1 7/8	1.875	1.812	2.652	2.489	27/32	0.876	0.812	2.750	3.000
1 3/8	1.3750	1.404	1.345	2 1/16	2.062	1.994	2.917	2.738	29/32	0.940	0.872	3.000	3.250
1 1/2	1.5000	1.531	1.470	2 1/4	2.250	2.175	3.182	2.986	1	1.036	0.964	3.250	3.500

AMERICAN STANDARD SQUARE NUTS

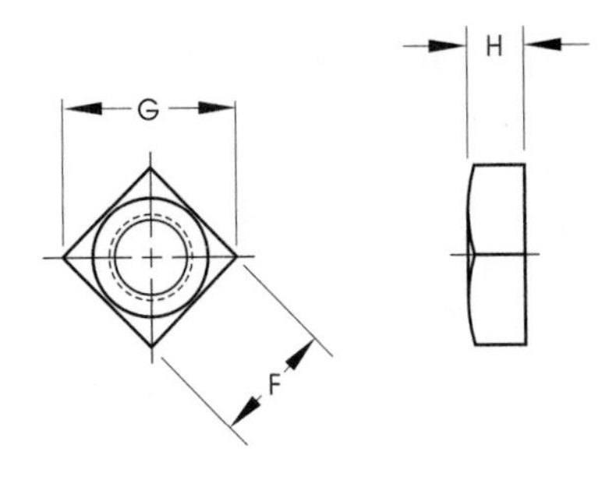

DIMENSIONS OF SQUARE NUTS

Nominal Size or Basic Major Diam. of Thread		F Width Across Flats			G Width Across Corners			H Thickness	
		Basic	Max.	Min.	Max.	Min.	Basic	Max.	Min.
1/4	0.2500	7/16	0.438	0.425	0.619	0.554	7/32	0.235	0.203
5/16	0.3125	9/16	0.562	0.547	0.795	0.721	17/64	0.283	0.249
3/8	0.3750	5/8	0.625	0.606	0.884	0.802	21/64	0.346	0.310
7/16	0.4375	3/4	0.750	0.728	1.061	0.970	3/8	0.394	0.356
1/2	0.5000	13/16	0.812	0.788	1.149	1.052	7/16	0.458	0.418
5/8	0.6250	1	1.000	0.969	1.414	1.300	35/64	0.569	0.525
3/4	0.7500	1 1/8	1.125	1.088	1.591	1.464	21/32	0.680	0.632
7/8	0.8750	1 5/16	1.312	1.269	1.856	1.712	49/64	0.792	0.740
1	1.0000	1 1/2	1.500	1.450	2.121	1.961	7/8	0.903	0.847
1 1/8	1.1250	1 11/16	1.688	1.631	2.386	2.209	1	1.030	0.970
1 1/4	1.2500	1 7/8	1.875	1.812	2.652	2.458	1 3/32	1.126	1.062
1 3/8	1.3750	2 1/16	2.062	1.994	2.917	2.708	1 13/64	1.237	1.169
1 1/2	1.5000	2 1/4	2.250	2.175	3.182	2.956	1 5/16	1.348	1.278

SLOTTED FILLISTER HEAD CAP SCREWS

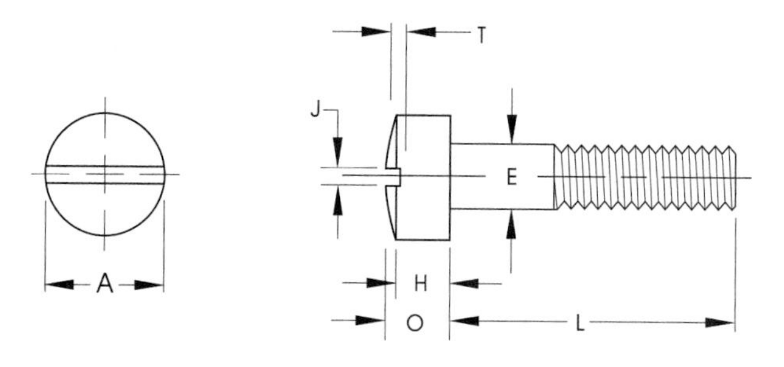

DIMENSIONS OF SLOTTED FILLISTER HEAD CAP SCREWS

Nominal Size or Basic Screw Diameter		E Body Diameter		A Head Diameter		H Head Side Height		O Total Head Height		J Slot Width		T Slot Depth	
		Max.	Min.	Max.	Min.	Max.	Min.	Max.	Min.	Max.	Min.	Max.	Min.
1/4	0.2500	0.2500	0.2450	0.375	0.363	0.172	0.157	0.216	0.194	0.075	0.064	0.097	0.077
5/16	0.3125	0.3125	0.3070	0.437	0.424	0.203	0.186	0.253	0.230	0.084	0.072	0.115	0.090
3/8	0.3750	0.3750	0.3690	0.562	0.547	0.250	0.229	0.314	0.284	0.094	0.081	0.142	0.112
7/16	0.4375	0.4375	0.4310	0.625	0.608	0.297	0.274	0.368	0.336	0.094	0.081	0.168	0.133
1/2	0.5000	0.5000	0.4930	0.750	0.731	0.328	0.301	0.413	0.376	0.106	0.091	0.193	0.153
9/16	0.5625	0.5625	0.5550	0.812	0.792	0.375	0.346	0.467	0.427	0.118	0.102	0.213	0.168
5/8	0.6250	0.6250	0.6170	0.875	0.853	0.422	0.391	0.521	0.478	0.133	0.116	0.239	0.189
3/4	0.7500	0.7500	0.7420	1.000	0.976	0.500	0.466	0.612	0.566	0.149	0.131	0.283	0.223
7/8	0.8750	0.8750	0.8660	1.125	1.098	0.594	0.556	0.712	0.668	0.167	0.147	0.334	0.264
1	1.0000	1.0000	0.9900	1.312	1.282	0.656	0.612	0.803	0.743	0.188	0.166	0.371	0.291

SLOTTED ROUND HEAD CAP SCREWS

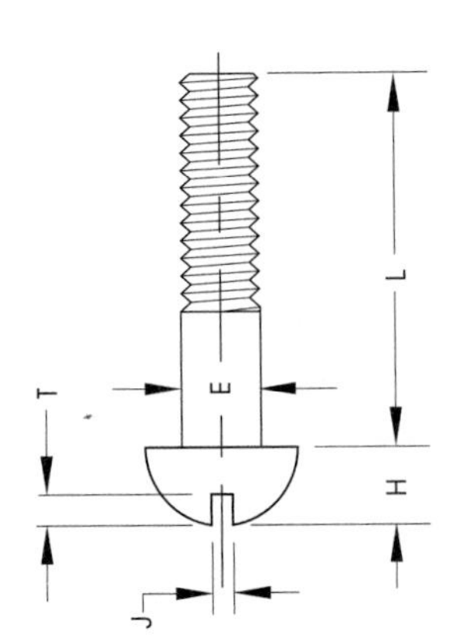

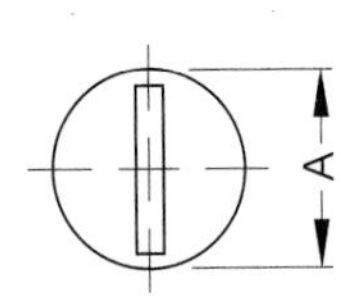

DIMENSIONS OF SLOTTED ROUND HEAD CAP SCREWS

Nominal Size or Basic Screw Diameter		E Body Diameter		A Head Diameter		H Head Height		J Slot Width		T Slot Depth	
		Max.	Min.	Max.	Min.	Max.	Min.	Max.	Min.	Max.	Min.
1/4	0.2500	0.2500	0.2450	0.437	0.418	0.191	0.175	0.075	0.064	0.117	0.097
5/16	0.3125	0.3125	0.3070	0.562	0.540	0.245	0.226	0.084	0.072	0.151	0.126
3/8	0.3750	0.3750	0.3690	0.625	0.603	0.273	0.252	0.094	0.081	0.168	0.138
7/16	0.4375	0.4375	0.4310	0.750	0.725	0.328	0.302	0.094	0.081	0.202	0.167
1/2	0.5000	0.5000	0.4930	0.812	0.786	0.354	0.321	0.106	0.091	0.218	0.178
9/16	0.5625	0.5625	0.5550	0.937	0.909	0.409	0.378	0.118	0.102	0.252	0.207
5/8	0.6250	0.6250	0.6170	1.000	0.970	0.437	0.405	0.133	0.116	0.270	0.220
3/4	0.7500	0.7500	0.7420	1.250	1.215	0.546	0.507	0.149	0.131	0.338	0.278

SLOTTED FLAT COUNTERSUNK HEAD CAP SCREWS

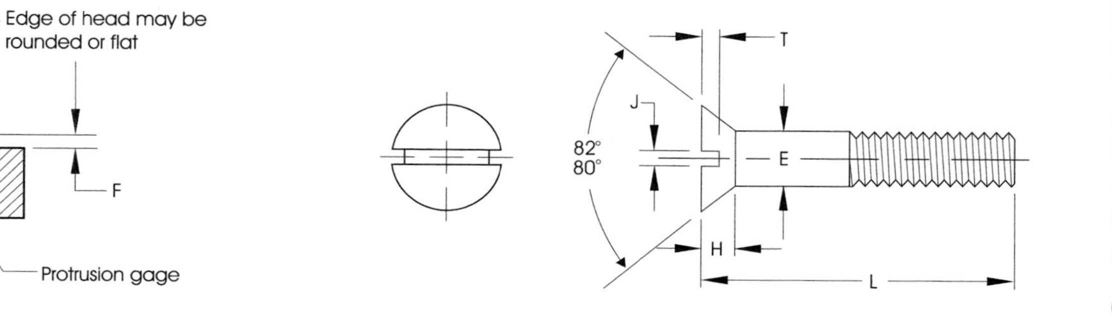

DIMENSIONS OF FLAT COUNTERSUNK HED CAP SCREWS

Nominal Size or Basic Screw Diameter		E Body Diameter		A Head Diameter		H Head Height	J Slot Width		T Slot Depth		F Protrusion Above Gaging Diameter		G Gaging Diameter
		Max.	Min.	Max., Edge Sharp	Min., Edge Rounded or Flat	Ref.	Max.	Min.	Max.	Min.	Max.	Min.	
1/4	0.2500	0.2500	0.2450	0.500	0.452	0.140	0.075	0.064	0.068	0.045	0.046	0.030	0.424
5/16	0.3125	0.3125	0.3070	0.625	0.567	0.177	0.084	0.072	0.086	0.057	0.053	0.035	0.538
3/8	0.3750	0.3750	0.3690	0.750	0.682	0.210	0.094	0.081	0.103	0.068	0.060	0.040	0.651
7/16	0.4375	0.4375	0.4310	0.812	0.736	0.210	0.094	0.081	0.103	0.068	0.065	0.044	0.703
1/2	0.5000	0.5000	0.4930	0.875	0.791	0.210	0.106	0.091	0.103	0.068	0.071	0.049	0.756
9/16	0.5625	0.5625	0.5550	1.000	0.906	0.244	0.118	0.102	0.120	0.080	0.078	0.054	0.869
5/8	0.6250	0.6250	0.6170	1.125	1.020	0.281	0.133	0.116	0.137	0.091	0.085	0.058	0.982
3/4	0.7500	0.7500	0.7420	1.374	1.251	0.352	0.149	0.131	0.171	0.115	0.099	0.068	1.208
7/8	0.8750	0.8750	0.8660	1.625	1.480	0.423	0.167	0.147	0.206	0.138	0.113	0.077	1.435
1	1.0000	1.0000	0.9900	1.875	1.711	0.494	0.188	0.166	0.240	0.162	0.127	0.087	1.661
1 1/8	1.1250	1.1250	1.1140	2.062	1.880	0.529	0.196	0.178	0.257	0.173	0.141	0.096	1.826
1 1/4	1.2500	1.2500	1.2390	2.312	2.110	0.600	0.211	0.193	0.291	0.197	0.155	0.105	2.052
1 3/8	1.3750	1.3750	1.3630	2.562	2.340	0.665	0.226	0.208	0.326	0.220	0.169	0.115	2.279
1 1/2	1.5000	1.5000	1.4880	2.812	2.570	0.742	0.258	0.240	0.360	0.224	0.183	0.124	2.505

Reprinted from ASME B18.6.2-1998 by permission of the American Society of Mechanical Engineers.

SLOTTED FLAT COUNTERSUNK HEAD MACHINE SCREWS

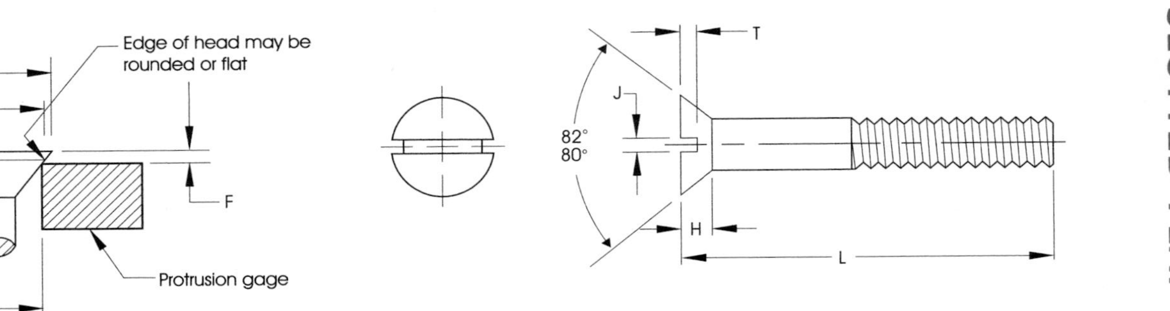

DIMENSIONS OF SLOTTED FLAT COUNTERSUNK HEAD MACHINE SCREWS

Nominal Size[1] or Basic Screw Diameter		L[2] These Lengths or Shorter are Undercut	A Head Diameter Max., Edge Sharp	A Head Diameter Min., Edge Rounded or Flat	H[3] Head Height Ref.	J Slot Width Max.	J Slot Width Min.	T Slot Depth Max.	T Slot Depth Min.	F Protrusion Above Gaging Diameter Max.	F Protrusion Above Gaging Diameter Min.	G Gaging Diameter
0000	0.0210	....	0.043	0.037	0.011	0.008	0.004	0.007	0.003	*	*	*
000	0.0340	....	0.064	0.058	0.016	0.011	0.007	0.009	0.005	*	*	*
00	0.0470	....	0.093	0.085	0.028	0.017	0.010	0.014	0.009	*	*	*
0	0.0600	1/8	0.119	0.099	0.035	0.023	0.016	0.015	0.010	0.026	0.016	0.078
1	0.0730	1/8	0.146	0.123	0.043	0.026	0.019	0.019	0.012	0.028	0.016	0.101
2	0.0860	1/8	0.172	0.147	0.051	0.031	0.023	0.023	0.015	0.029	0.017	0.124
3	0.0990	1/8	0.199	0.171	0.059	0.035	0.027	0.027	0.017	0.031	0.018	0.148
4	0.1120	3/16	0.225	0.195	0.067	0.039	0.031	0.030	0.020	0.032	0.019	0.172
5	0.1250	3/16	0.252	0.220	0.075	0.043	0.035	0.034	0.022	0.034	0.020	0.196
6	0.1380	3/16	0.279	0.244	0.083	0.048	0.039	0.038	0.024	0.036	0.021	0.220
8	0.1640	1/4	0.332	0.292	0.100	0.054	0.045	0.045	0.029	0.039	0.023	0.267
10	0.1900	5/16	0.385	0.340	0.166	0.060	0.050	0.053	0.034	0.042	0.025	0.313
12	0.2160	3/8	0.438	0.389	0.132	0.067	0.056	0.060	0.039	0.045	0.027	0.362
1/4	0.2500	7/16	0.507	0.452	0.153	0.075	0.064	0.070	0.046	0.050	0.029	0.424
5/16	0.3125	1/2	0.635	0.568	0.191	0.084	0.072	0.088	0.058	0.057	0.034	0.539
3/8	0.3750	9/16	0.762	0.685	0.230	0.094	0.081	0.106	0.070	0.065	0.039	0.653
7/16	0.4375	5/8	0.812	0.723	0.223	0.094	0.081	0.103	0.066	0.073	0.044	0.690
1/2	0.5000	3/4	0.875	0.775	0.223	0.106	0.091	0.103	0.065	0.081	0.049	0.739
9/16	0.5625	....	1.000	0.889	0.260	0.118	0.102	0.120	0.077	0.089	0.053	0.851
5/8	0.6250	....	1.125	1.002	0.298	0.133	0.116	0.137	0.088	0.097	0.058	0.962
3/4	0.7500	....	1.375	1.230	0.372	0.149	0.131	0.171	0.111	0.112	0.067	1.186

1 Where specifying nominal size in decimals, zeros preceding decimal and in the fourth decimal place shall be omitted.
2 Screws of these lengths and shorter shall have undercut heads.
3 Tabulated values determined from formula for maximum H.
* Not practical to gage.

SLOTTED FILLISTER HEAD MACHINE SCREWS

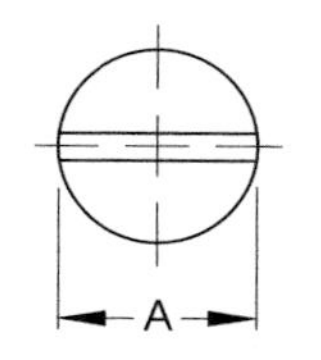

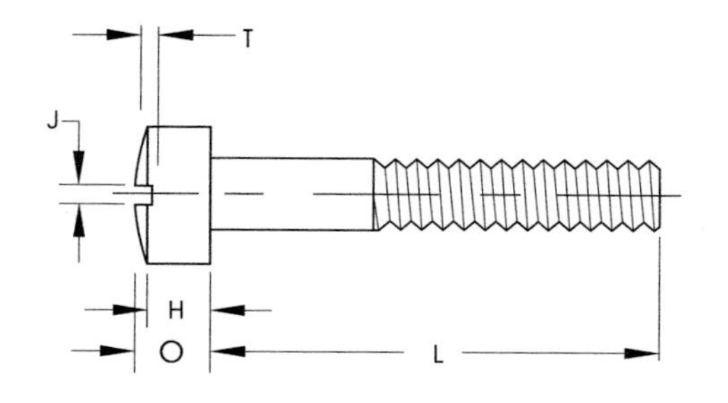

DIMENSIONS OF SLOTTED FILLISTER HEAD MACHINE SCREWS

Nominal Size[1] or Basic Screw Diameter		A Head Diameter		H Head Side Height		O Total Head Height		J Slot Width		T Slot Depth	
		Max.	Min.	Max.	Min.	Max.	Min.	Max.	Min.	Max.	Min.
0000	0.0210	0.038	0.032	0.019	0.011	0.025	0.015	0.008	0.004	0.012	0.006
000	0.0340	0.059	0.053	0.029	0.021	0.035	0.027	0.012	0.006	0.017	0.011
00	0.0470	0.082	0.072	0.037	0.028	0.047	0.039	0.017	0.010	0.022	0.015
0	0.0600	0.096	0.083	0.043	0.038	0.055	0.047	0.023	0.016	0.025	0.015
1	0.0730	0.118	0.104	0.053	0.045	0.066	0.058	0.026	0.019	0.031	0.020
2	0.0860	0.140	0.124	0.062	0.053	0.083	0.066	0.031	0.023	0.037	0.025
3	0.0990	0.161	0.145	0.070	0.061	0.095	0.077	0.035	0.027	0.043	0.030
4	0.1120	0.183	0.166	0.079	0.069	0.107	0.088	0.039	0.031	0.048	0.035
5	0.1250	0.205	0.187	0.088	0.078	0.120	0.100	0.043	0.035	0.054	0.040
6	0.1380	0.226	0.208	0.096	0.086	0.132	0.111	0.048	0.039	0.060	0.045
8	0.1640	0.270	0.250	0.113	0.102	0.156	0.133	0.054	0.045	0.071	0.054
10	0.1900	0.313	0.292	0.130	0.118	0.180	0.156	0.060	0.050	0.083	0.064
12	0.2160	0.357	0.354	0.148	0.134	0.205	0.178	0.067	0.056	0.094	0.074
1/4	0.2500	0.414	0.389	0.170	0.155	0.237	0.207	0.075	0.064	0.109	0.087
5/16	0.3125	0.518	0.490	0.211	0.194	0.295	0.262	0.084	0.072	0.137	0.110
3/8	0.3750	0.622	0.590	0.253	0.233	0.355	0.315	0.094	0.081	0.164	0.133
7/16	0.4375	0.625	0.589	0.265	0.242	0.368	0.321	0.094	0.081	0.170	0.135
1/2	0.5000	0.750	0.710	0.297	0.273	0.412	0.362	0.106	0.091	0.190	0.151
9/16	0.5625	0.812	0.768	0.336	0.308	0.466	0.410	0.118	0.102	0.214	0.172
5/8	0.6250	0.875	0.872	0.375	0.345	0.521	0.461	0.133	0.116	0.240	0.193
3/4	0.7500	1.000	0.945	0.441	0.406	0.612	0.542	0.149	0.131	0.281	0.226

1 Where specifying nominal size in decimals, zeros preceding decimal and in the fourth decimal place shall be omitted.

Reprinted from ASME B18.6.3-1972 by permission of the American Society of Mechanical Engineers.

SLOTTED PAN HEAD MACHINE SCREWS

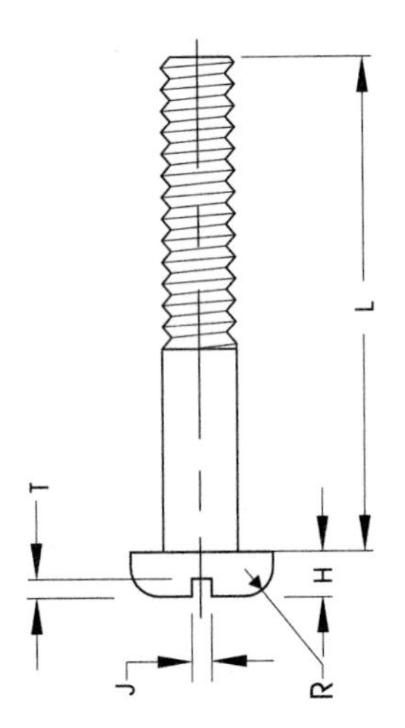

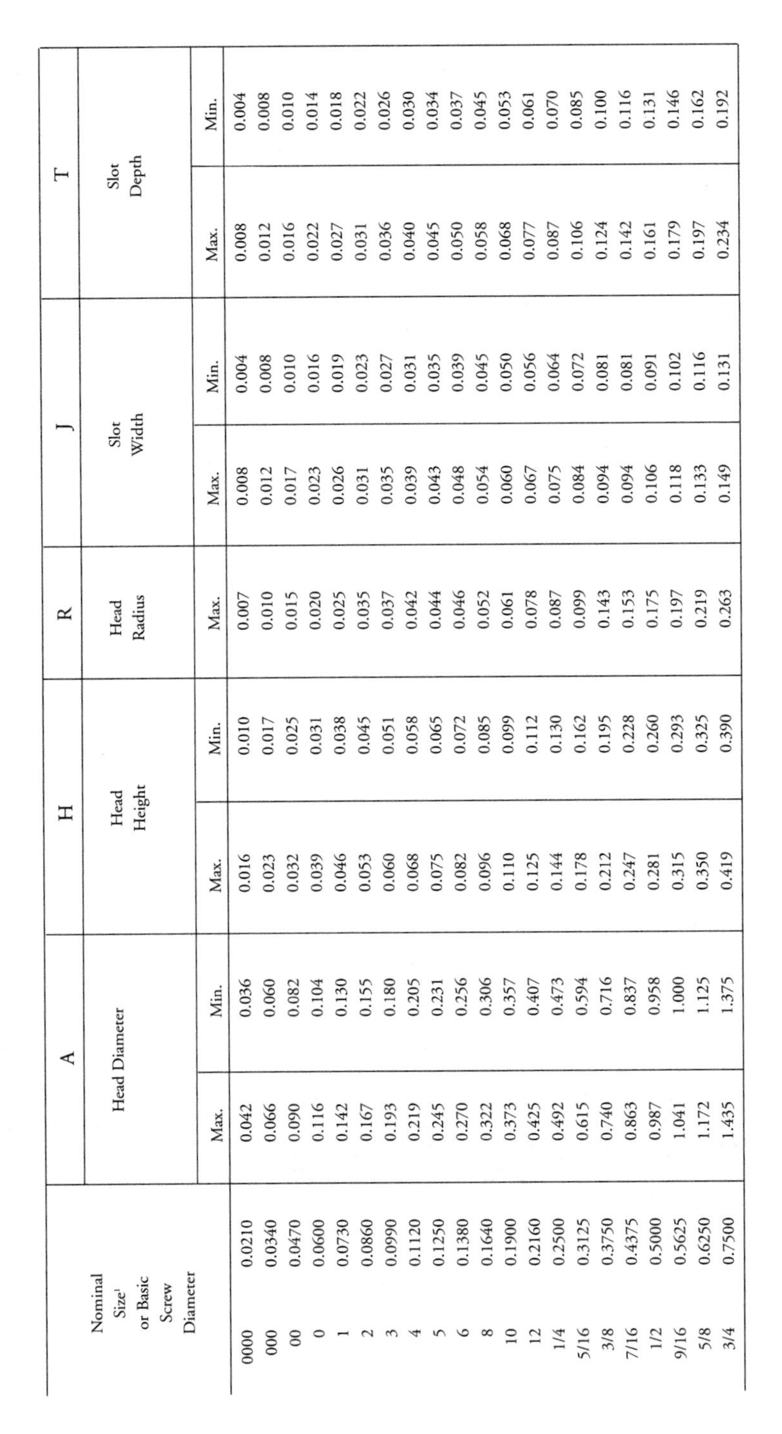

Nominal Size[1] or Basic Screw Diameter		A Head Diameter		H Head Height		R Head Radius	J Slot Width		T Slot Depth	
		Max.	Min.	Max.	Min.	Max.	Max.	Min.	Max.	Min.
0000	0.0210	0.042	0.036	0.016	0.010	0.007	0.008	0.004	0.008	0.004
000	0.0340	0.066	0.060	0.023	0.017	0.010	0.012	0.008	0.012	0.008
00	0.0470	0.090	0.082	0.032	0.025	0.015	0.017	0.010	0.016	0.010
0	0.0600	0.116	0.104	0.039	0.031	0.020	0.023	0.016	0.022	0.014
1	0.0730	0.142	0.130	0.046	0.038	0.025	0.026	0.019	0.027	0.018
2	0.0860	0.167	0.155	0.053	0.045	0.035	0.031	0.023	0.031	0.022
3	0.0990	0.193	0.180	0.060	0.051	0.037	0.035	0.027	0.036	0.026
4	0.1120	0.219	0.205	0.068	0.058	0.042	0.039	0.031	0.040	0.030
5	0.1250	0.245	0.231	0.075	0.065	0.044	0.043	0.035	0.045	0.034
6	0.1380	0.270	0.256	0.082	0.072	0.046	0.048	0.039	0.050	0.037
8	0.1640	0.322	0.306	0.096	0.085	0.052	0.054	0.045	0.058	0.045
10	0.1900	0.373	0.357	0.110	0.099	0.061	0.060	0.050	0.068	0.053
12	0.2160	0.425	0.407	0.125	0.112	0.078	0.067	0.056	0.077	0.061
1/4	0.2500	0.492	0.473	0.144	0.130	0.087	0.075	0.064	0.087	0.070
5/16	0.3125	0.615	0.594	0.178	0.162	0.099	0.084	0.072	0.106	0.085
3/8	0.3750	0.740	0.716	0.212	0.195	0.143	0.094	0.081	0.124	0.100
7/16	0.4375	0.863	0.837	0.247	0.228	0.153	0.094	0.081	0.142	0.116
1/2	0.5000	0.987	0.958	0.281	0.260	0.175	0.106	0.091	0.161	0.131
9/16	0.5625	1.041	1.000	0.315	0.293	0.197	0.118	0.102	0.179	0.146
5/8	0.6250	1.172	1.125	0.350	0.325	0.219	0.133	0.116	0.197	0.162
3/4	0.7500	1.435	1.375	0.419	0.390	0.263	0.149	0.131	0.234	0.192

1 Where specifying nominal size in decimals, zeros preceding decimal and in the fourth decimal place shall be omitted.

PLAIN WASHERS

ASME B18.22.1–1965(R1998)

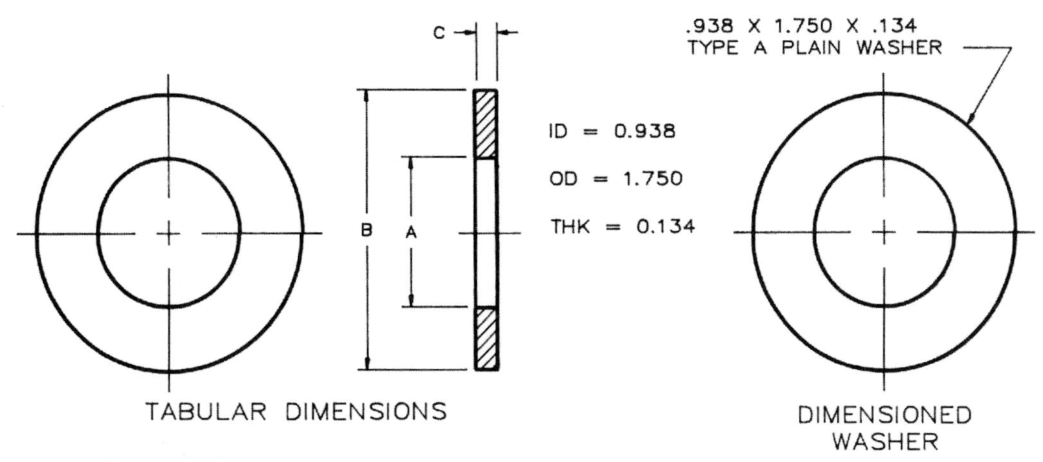

Dimensions of Preferred Sizes of Type A Plain Washers[a]

When specifying washers on drawings or in notes, give the inside diameter, outside diameter, and the thickness. *Example:* 0.938 × 1.750 × 0.134 TYPE A PLAIN WASHER.

Nominal Washer Size[b]			Inside Diameter A			Outside Diameter B			Thickness C		
			Basic	Tolerance Plus	Tolerance Minus	Basic	Tolerance Plus	Tolerance Minus	Basic	Max	Min
—	—		0.078	0.000	0.005	0.188	0.000	0.005	0.020	0.025	0.016
—	—		0.094	0.000	0.005	0.250	0.000	0.005	0.020	0.025	0.016
—	—		0.125	0.008	0.005	0.312	0.008	0.005	0.032	0.040	0.025
No. 6	0.138		0.156	0.008	0.005	0.375	0.015	0.005	0.049	0.065	0.036
No. 8	0.164		0.188	0.008	0.005	0.438	0.015	0.005	0.049	0.065	0.036
No. 10	0.190		0.219	0.008	0.005	0.500	0.015	0.005	0.049	0.065	0.036
$\frac{3}{16}$	0.188		0.250	0.015	0.005	0.562	0.015	0.005	0.049	0.065	0.036
No. 12	0.216		0.250	0.015	0.005	0.562	0.015	0.005	0.065	0.080	0.051
$\frac{1}{4}$	0.250	N	0.281	0.015	0.005	0.625	0.015	0.005	0.065	0.080	0.051
$\frac{1}{4}$	0.250	W	0.312	0.015	0.005	0.734[c]	0.015	0.007	0.065	0.080	0.051
$\frac{5}{16}$	0.312	N	0.344	0.015	0.005	0.688	0.015	0.007	0.065	0.080	0.051
$\frac{5}{16}$	0.312	W	0.375	0.015	0.005	0.875	0.030	0.007	0.083	0.104	0.064
$\frac{3}{8}$	0.375	N	0.406	0.015	0.005	0.812	0.015	0.007	0.065	0.080	0.051
$\frac{3}{8}$	0.375	W	0.438	0.015	0.005	1.000	0.030	0.007	0.083	0.104	0.064
$\frac{7}{16}$	0.438	N	0.469	0.015	0.005	0.922	0.015	0.007	0.065	0.080	0.051
$\frac{7}{16}$	0.438	W	0.500	0.015	0.005	1.250	0.030	0.007	0.083	0.104	0.064
$\frac{1}{2}$	0.500	N	0.531	0.015	0.005	1.062	0.030	0.007	0.095	0.121	0.074
$\frac{1}{2}$	0.500	W	0.562	0.015	0.005	1.375	0.030	0.007	0.109	0.132	0.086

Preferred sizes are for the most part from series previously designated "Standard Plate" and "SAE." Where common sizes existed in the two series, the SAE size is designated "N" (narrow) and the Standard Plate "W" (wide). These sizes as well as all other sizes of Type A Plain Washers are to be ordered by ID, OD, and thickness dimensions.

[b] Nominal washer sizes are intended for use with comparable nominal screw or bolt sizes.

[c] The 0.734 in., 1.156 in., and 1.469 in. outside diameters avoid washers which could be used in coin-operated devices.

Cont.

PLAIN WASHERS (CONT.)

Nominal Washer Size[b]			Inside Diameter A			Outside Diameter B			Thickness C		
			Basic	Tolerance Plus	Tolerance Minus	Basic	Tolerance Plus	Tolerance Minus	Basic	Max	Min
$\frac{9}{16}$	0.562	N	0.594	0.015	0.005	1.156[c]	0.030	0.007	0.095	0.121	0.074
$\frac{9}{16}$	0.562	W	0.625	0.015	0.005	1.469[c]	0.030	0.007	0.109	0.132	0.086
$\frac{5}{8}$	0.625	N	0.656	0.030	0.007	1.312	0.030	0.007	0.095	0.121	0.074
$\frac{5}{8}$	0.625	W	0.688	0.030	0.007	1.750	0.030	0.007	0.134	0.160	0.108
$\frac{3}{4}$	0.750	N	0.812	0.030	0.007	1.469	0.030	0.007	0.134	0.160	0.108
$\frac{3}{4}$	0.750	W	0.812	0.030	0.007	2.000	0.030	0.007	0.148	0.177	0.122
$\frac{7}{8}$	0.875	N	0.938	0.030	0.007	1.750	0.030	0.007	0.134	0.160	0.108
$\frac{7}{8}$	0.875	W	0.938	0.030	0.007	2.250	0.030	0.007	0.165	0.192	0.136
1	1.000	N	1.062	0.030	0.007	2.000	0.030	0.007	0.134	0.160	0.108
1	1.000	W	1.062	0.030	0.007	2.500	0.030	0.007	0.165	0.192	0.136
$1\frac{1}{8}$	1.125	N	1.250	0.030	0.007	2.250	0.030	0.007	0.134	0.160	0.108
$1\frac{1}{8}$	1.125	W	1.250	0.030	0.007	2.750	0.030	0.007	0.165	0.192	0.136
$1\frac{1}{4}$	1.250	N	1.375	0.030	0.007	2.500	0.030	0.007	0.165	0.192	0.136
$1\frac{1}{4}$	1.250	W	1.375	0.030	0.007	3.000	0.030	0.007	0.165	0.192	0.136
$1\frac{3}{8}$	1.375	N	1.500	0.030	0.007	2.750	0.030	0.007	0.165	0.192	0.136
$1\frac{3}{8}$	1.375	W	1.500	0.045	0.010	3.250	0.045	0.010	0.180	0.213	0.153
$1\frac{1}{2}$	1.500	N	1.625	0.030	0.007	3.000	0.030	0.007	0.165	0.192	0.136
$1\frac{1}{2}$	1.500	W	1.625	0.045	0.010	3.500	0.045	0.010	0.180	0.213	0.153
$1\frac{5}{8}$	1.625		1.750	0.045	0.010	3.750	0.045	0.010	0.180	0.213	0.153
$1\frac{3}{4}$	1.750		1.875	0.045	0.010	4.000	0.045	0.010	0.180	0.213	0.153
$1\frac{7}{8}$	1.875		2.000	0.045	0.010	4.250	0.045	0.010	0.180	0.213	0.153
2	2.000		2.125	0.045	0.010	4.500	0.045	0.010	0.180	0.213	0.153
$2\frac{1}{4}$	2.250		2.375	0.045	0.010	4.750	0.045	0.010	0.220	0.248	0.193
$2\frac{1}{2}$	2.500		2.625	0.045	0.010	5.000	0.045	0.010	0.238	0.280	0.210
$2\frac{3}{4}$	2.750		2.875	0.065	0.010	5.250	0.065	0.010	0.259	0.310	0.228
3	3.000		3.125	0.065	0.010	5.500	0.065	0.010	0.284	0.327	0.249

APPENDIX

F Table of Notations for Common Structural Steel Shapes

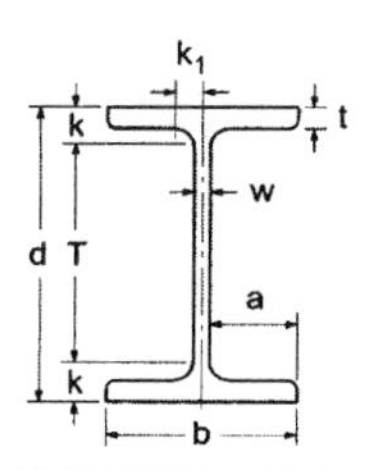

W SHAPES
W690 - W610

DIMENSIONS AND SURFACE AREAS

Nominal Mass	Theo-retical Mass	Depth d	Flange Width b	Flange Thick-ness t	Web Thick-ness w	Distances a	T	k	k_1	d-2t	Surface Area (m^2) per metre of length Total	Minus Top of Top Flange	Imperial Designation
kg/m	kg/m	mm	mm	mm	mm	mm	mm	mm	mm	mm			
802	802.6	826	387	89.9	50.0	169	603	112	45	646	3.10	2.71	W27x539
548	548.6	772	372	63.0	35.1	168	602	85	38	646	2.96	2.59	W27x368
500	500.5	762	369	57.9	32.0	169	603	80	36	646	2.94	2.57	W27x336
457	458.2	752	367	53.1	29.5	169	602	75	35	646	2.91	2.55	W27x307
419	419.1	744	364	49.0	26.9	169	602	71	34	646	2.89	2.53	W27x281
384	384.7	736	362	45.0	24.9	169	602	67	33	646	2.87	2.51	W27x258
350	351.0	728	360	40.9	23.1	168	603	63	32	646	2.85	2.49	W27x235
323	324.4	722	359	38.1	21.1	169	602	60	31	646	2.84	2.48	W27x217
289	289.1	714	356	34.0	19.0	169	602	56	30	646	2.81	2.46	W27x194
265	265.7	706	358	30.2	18.4	170	602	52	30	646	2.81	2.45	W27x178
240	241.1	701	356	27.4	16.8	170	603	49	29	646	2.79	2.44	W27x161
217	218.9	695	355	24.8	15.4	170	602	47	28	645	2.78	2.42	W27x146
192	191.4	702	254	27.9	15.5	119	603	50	28	646	2.39	2.14	W27x129
170	169.9	693	256	23.6	14.5	121	602	45	28	646	2.38	2.13	W27x114
152	152.1	688	254	21.1	13.1	120	602	43	27	646	2.37	2.11	W27x102
140	139.8	684	254	18.9	12.4	121	603	41	27	646	2.36	2.11	W27x94
125	125.6	678	253	16.3	11.7	121	602	38	26	645	2.34	2.09	W27x84
551	551.1	711	347	69.1	38.6	154	529	91	40	573	2.73	2.39	W24x370
498	498.2	699	343	63.0	35.1	154	529	85	38	573	2.70	2.36	W24x335
455	454.1	689	340	57.9	32.0	154	530	80	36	573	2.67	2.33	W24x306
415	415.5	679	338	53.1	29.5	154	529	75	35	573	2.65	2.31	W24x279
372	372.3	669	335	48.0	26.4	154	529	70	34	573	2.63	2.29	W24x250
341	340.4	661	333	43.9	24.4	154	530	66	33	573	2.61	2.27	W24x229
307	307.3	653	330	39.9	22.1	154	530	62	31	573	2.58	2.25	W24x207
285	285.3	647	329	37.1	20.6	154	529	59	31	573	2.57	2.24	W24x192
262	261.1	641	327	34.0	19.0	154	529	56	30	573	2.55	2.23	W24x176
241	241.7	635	329	31.0	17.9	156	529	53	29	573	2.55	2.22	W24x162
217	217.9	628	328	27.7	16.5	156	529	50	29	573	2.54	2.21	W24x146
195	195.6	622	327	24.4	15.4	156	530	46	28	573	2.52	2.19	W24x131
174	174.3	616	325	21.6	14.0	156	529	43	27	573	2.50	2.18	W24x117
155	154.9	611	324	19.0	12.7	156	529	41	27	573	2.49	2.17	W24x104
153	153.6	623	229	24.9	14.0	108	530	47	27	573	2.13	1.91	W24x103
140	140.1	617	230	22.2	13.1	108	529	44	27	573	2.13	1.90	W24x94
125	125.1	612	229	19.6	11.9	109	529	41	26	573	2.12	1.89	W24x84
113	113.4	608	228	17.3	11.2	108	530	39	26	573	2.11	1.88	W24x76
101	101.7	603	228	14.9	10.5	109	530	37	26	573	2.10	1.87	W24x68
91	89.9	598	227	12.7	9.7	109	529	35	25	573	2.08	1.86	W24x61
84	83.1	596	226	11.7	9.0	109	529	34	25	573	2.08	1.85	W24x56

Note: Items in grey indicate restricted availability.

Adapted from the *Handbook of Steel Construction*, 7th edition. Canadian Institute of Steel Construction, 2000, p. 6-25. For further information go to www.cisc.ca.

APPENDIX

G Common Pipe Sizes and Symbols

DIMENSIONS AND WEIGHTS OF WELDED AND SEAMLESS WROUGHT STEEL PIPES

Customary Units				Identification		SI Units		
Nominal pipe size	Outside Diameter in.	Wall Thickness in.	Plain End Weight lb/ft	Standard(STD) ExtraStrong(XS) Double Extra-Strong (XXS)	Schedule No.	Outside Diameter mm	Wall Thickness mm	Plain End Mass kg/m
1/8	0.405	0.068	0.24	STD	40	10.3	1.73	0.37
1/8	0.405	0.095	0.31	XS	80	10.3	2.41	0.47
1/4	0.540	0.088	0.42	STD	40	13.7	2.24	0.63
1/4	0.540	0.119	0.54	XS	80	13.7	3.02	0.80
3/8	0.675	0.091	0.57	STD	40	17.1	2.31	0.84
3/8	0.675	0.126	0.74	XS	80	17.1	3.20	1.10
1/2	0.840	0.109	0.85	STD	40	21.3	2.77	1.27
1/2	0.840	0.147	1.09	XS	80	21.3	3.73	1.62
1/2	0.840	0.188	1.31	---	160	21.3	4.78	1.95
1/2	0.840	0.294	1.71	XXS	---	21.3	7.47	2.55
3/4	1.050	0.113	1.13	STD	40	26.7	2.87	1.69
3/4	1.050	0.154	1.47	XS	80	26.7	3.91	2.20
3/4	1.050	0.219	1.94	---	160	26.7	5.56	2.90
3/4	1.050	0.308	2.44	XXS	---	26.7	7.82	3.64
1	1.315	0.133	1.68	STD	40	33.4	3.38	2.50
1	1.315	0.179	2.17	XS	80	33.4	4.55	3.24
1	1.315	0.250	2.84	---	160	33.4	6.35	4.24
1	1.315	0.358	3.66	XXS	---	33.4	9.09	5.45
1 1/4	1.660	0.140	2.27	STD	40	42.2	3.56	3.39
1 1/4	1.660	0.191	3.00	XS	80	42.2	4.85	4.47
1 1/4	1.660	0.250	3.76	---	160	42.2	6.35	5.61
1 1/4	1.660	0.382	5.21	XXS	---	42.2	9.70	7.77
1 1/2	1.900	0.145	2.72	STD	40	48.3	3.68	4.05
1 1/2	1.900	0.200	3.63	XS	80	48.3	5.08	5.41
1 1/2	1.900	0.281	4.86	---	160	48.3	7.14	7.25
1 1/2	1.900	0.400	6.41	XXS	---	48.3	10.15	9.56
2	2.375	0.154	3.65	STD	40	60.3	3.91	5.44
2	2.375	0.218	5.02	XS	80	60.3	5.54	7.48
2	2.375	0.344	7.46	---	160	60.3	8.74	11.11
2	2.375	0.436	9.03	XXS	---	60.3	11.07	13.44

2 1/2	2.875	0.141	4.12	STD	40	73.0	5.16	8.63
2 1/2	2.875	0.276	7.66	XS	80	73.0	7.01	11.41
2 1/2	2.875	0.375	10.01	---	160	73.0	6.35	10.44
2 1/2	2.875	0.552	13.69	XXS	---	73.0	14.02	20.39
3	3.500	0.216	7.58	STD	40	88.9	5.49	11.29
3	3.500	0.300	10.25	XS	80	88.9	7.62	15.27
3	3.500	0.438	14.32	---	160	88.9	11.13	21.35
3	3.500	0.600	18.58	XXS	---	88.9	15.24	27.68
3 1/2	4.000	0.226	9.11	STD	40	101.6	5.74	13.57
3 1/2	4.000	0.318	12.50	XS	80	101.6	8.08	18.63
4	4.500	0.237	10.79	STD	40	114.3	4.78	12.91
4	4.500	0.337	14.98	XS	80	114.3	8.56	22.32
4	4.500	0.531	22.51	---	160	114.3	13.49	33.54
4	4.500	0.674	27.54	XXS	---	114.3	17.12	41.03
5	5.563	0.258	14.62	STD	40	141.3	6.55	21.77
5	5.563	0.375	20.78	XS	80	141.3	9.53	30.97
5	5.563	0.625	32.96	---	160	141.3	12.70	40.28
5	5.563	0.750	38.55	XSS	---	141.3	19.05	57.43
6	6.625	0.250	17.02	STD	40	168.3	6.35	25.36
6	6.625	0.432	28.57	XS	80	168.3	10.97	42.56
6	6.625	0.719	45.35	---	160	168.3	18.26	67.56
6	6.625	0.864	53.16	XXS	---	168.3	21.95	79.22
8	8.625	0.322	28.55	STD	40	219.1	8.18	42.55
8	8.625	0.500	43.39	XS	80	219.1	12.70	64.64
8	8.625	0.875	72.42	XSS	---	219.1	22.23	107.92
8	8.625	0.906	74.69	---	160	219.1	23.01	111.27
10	10.750	0.365	40.48	STD	40	273.0	9.27	60.31
10	10.750	0.500	54.74	XS	60	273.0	12.70	81.55
10	10.750	0.594	64.43	---	80	273.0	15.09	96.01
10	10.750	1.000	104.13	XXS	140	273.0	25.40	155.15
10	10.750	1.125	115.64	---	160	273.0	28.58	172.23
12	12.750	0.375	49.56	STD	---	323.8	9.53	73.88
12	12.750	0.406	53.52	---	40	332.8	10.31	79.73
12	12.750	0.500	65.42	XS	---	332.8	12.70	97.46
12	12.750	0.688	88.63	---	80	332.8	17.48	132.08
12	12.750	1.000	125.49	XXS	120	332.8	25.40	186.97
12	12.750	1.312	160.27	---	160	332.8	33.32	238.76

H Graphic Symbols for Electrical and Electronic Diagrams

ANSI STANDARD GRAPHIC SYMBOLS FOR ELECTRICAL DIAGRAMS

Single-line symbols are shown at the left, complete symbols at the right, and symbols for both purposes are centred in each column. Listing is alphabetical.

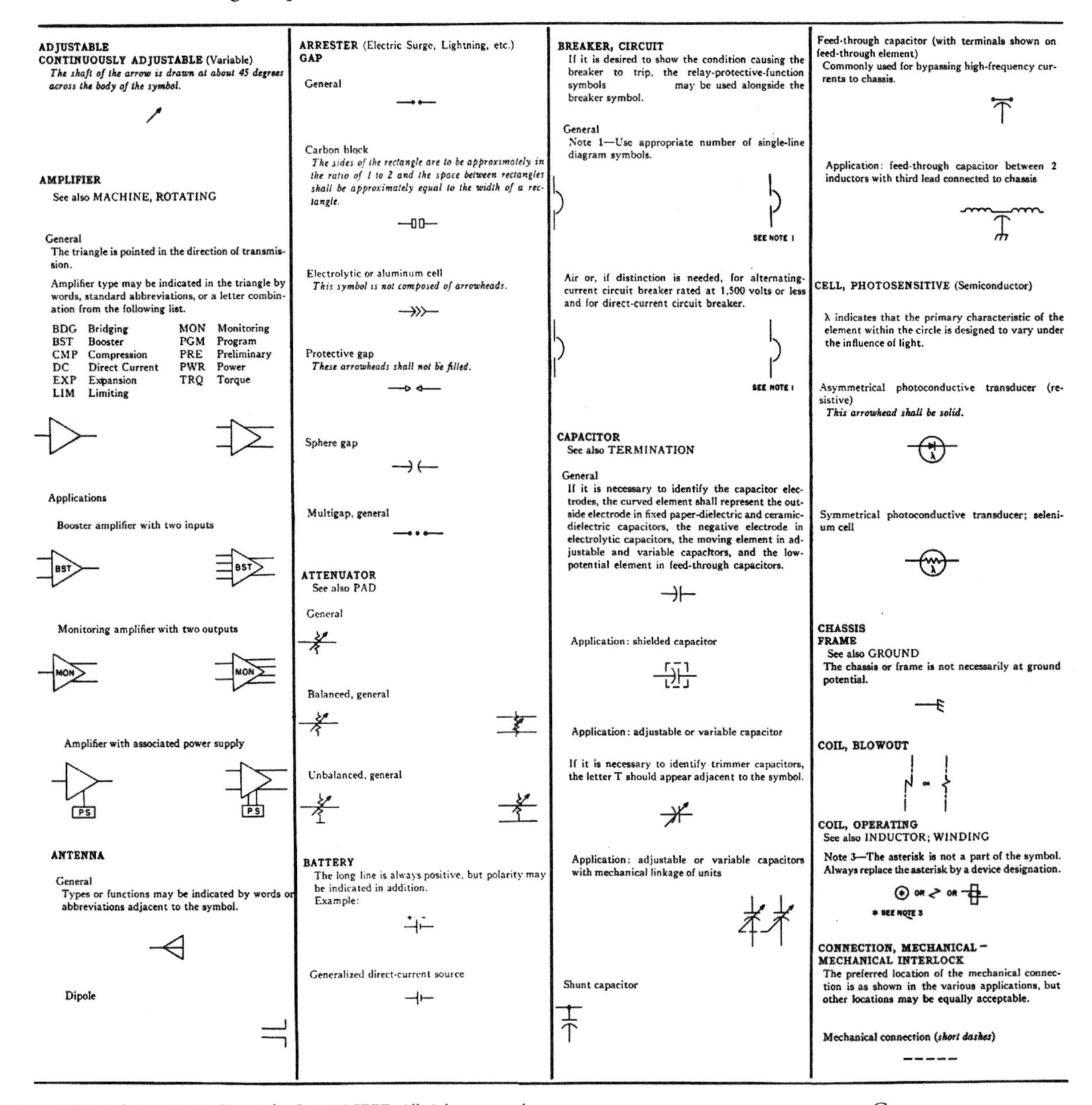

ADJUSTABLE
CONTINUOUSLY ADJUSTABLE (Variable)
The shaft of the arrow is drawn at about 45 degrees across the body of the symbol.

AMPLIFIER
See also MACHINE, ROTATING

General
The triangle is pointed in the direction of transmission.

Amplifier type may be indicated in the triangle by words, standard abbreviations, or a letter combination from the following list.

BDG	Bridging	MON	Monitoring
BST	Booster	PGM	Program
CMP	Compression	PRE	Preliminary
DC	Direct Current	PWR	Power
EXP	Expansion	TRQ	Torque
LIM	Limiting		

Applications

Booster amplifier with two inputs

BST BST

Monitoring amplifier with two outputs

MON MON

Amplifier with associated power supply

PS PS

ANTENNA

General
Types or functions may be indicated by words or abbreviations adjacent to the symbol.

Dipole

ARRESTER (Electric Surge, Lightning, etc.)
GAP

General

Carbon block
The sides of the rectangle are to be approximately in the ratio of 1 to 2 and the space between rectangles shall be approximately equal to the width of a rectangle.

Electrolytic or aluminum cell
This symbol is not composed of arrowheads.

Protective gap
These arrowheads shall not be filled.

Sphere gap

Multigap, general

ATTENUATOR
See also PAD

General

Balanced, general

Unbalanced, general

BATTERY
The long line is always positive, but polarity may be indicated in addition.
Example:

Generalized direct-current source

BREAKER, CIRCUIT
If it is desired to show the condition causing the breaker to trip, the relay-protective-function symbols may be used alongside the breaker symbol.

General
Note 1—Use appropriate number of single-line diagram symbols.

SEE NOTE 1

Air or, if distinction is needed, for alternating-current circuit breaker rated at 1,500 volts or less and for direct-current circuit breaker.

SEE NOTE 1

CAPACITOR
See also TERMINATION

General
If it is necessary to identify the capacitor electrodes, the curved element shall represent the outside electrode in fixed paper-dielectric and ceramic-dielectric capacitors, the negative electrode in electrolytic capacitors, the moving element in adjustable and variable capacitors, and the low-potential element in feed-through capacitors.

Application: shielded capacitor

Application: adjustable or variable capacitor

If it is necessary to identify trimmer capacitors, the letter T should appear adjacent to the symbol.

Application: adjustable or variable capacitors with mechanical linkage of units

Shunt capacitor

Feed-through capacitor (with terminals shown on feed-through element)
Commonly used for bypassing high-frequency currents to chassis.

Application: feed-through capacitor between 2 inductors with third lead connected to chassis

CELL, PHOTOSENSITIVE (Semiconductor)

λ indicates that the primary characteristic of the element within the circle is designed to vary under the influence of light.

Asymmetrical photoconductive transducer (resistive)
This arrowhead shall be solid.

Symmetrical photoconductive transducer; selenium cell

CHASSIS
FRAME
See also GROUND
The chassis or frame is not necessarily at ground potential.

COIL, BLOWOUT

OR

COIL, OPERATING
See also INDUCTOR; WINDING

Note 3—The asterisk is not a part of the symbol. Always replace the asterisk by a device designation.

OR OR

* SEE NOTE 3

CONNECTION, MECHANICAL – MECHANICAL INTERLOCK
The preferred location of the mechanical connection is as shown in the various applications, but other locations may be equally acceptable.

Mechanical connection (*short dashes*)

Cont.

ANSI STANDARD GRAPHIC SYMBOLS FOR ELECTRICAL DIAGRAMS (CONT.)

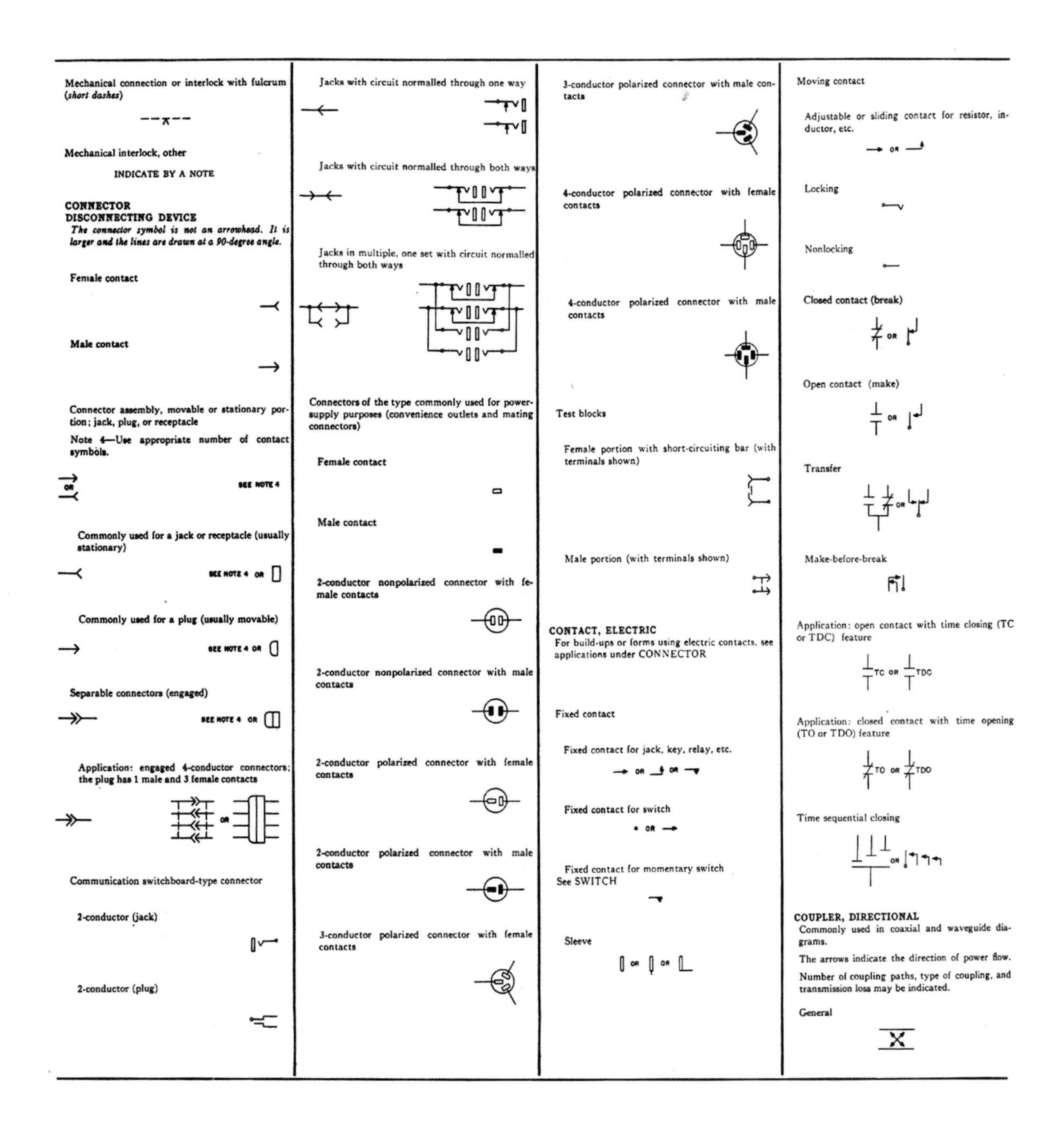
Mechanical connection or interlock with fulcrum (*short dashes*)
Mechanical interlock, other
INDICATE BY A NOTE
CONNECTOR
DISCONNECTING DEVICE
The connector symbol is not an arrowhead. It is larger and the lines are drawn at a 90-degree angle.
Female contact
Male contact
Connector assembly, movable or stationary portion; jack, plug, or receptacle
Note 4—Use appropriate number of contact symbols.
OR
SEE NOTE 4
Commonly used for a jack or receptacle (usually stationary)
SEE NOTE 4 OR
Commonly used for a plug (usually movable)
SEE NOTE 4 OR
Separable connectors (engaged)
SEE NOTE 4 OR
Application: engaged 4-conductor connectors; the plug has 1 male and 3 female contacts
OR
Communication switchboard-type connector
2-conductor (jack)
2-conductor (plug)
Jacks with circuit normalled through one way
Jacks with circuit normalled through both ways
Jacks in multiple, one set with circuit normalled through both ways
Connectors of the type commonly used for power-supply purposes (convenience outlets and mating connectors)
Female contact
Male contact
2-conductor nonpolarized connector with female contacts
2-conductor nonpolarized connector with male contacts
2-conductor polarized connector with female contacts
2-conductor polarized connector with male contacts
3-conductor polarized connector with female contacts
3-conductor polarized connector with male contacts
4-conductor polarized connector with female contacts
4-conductor polarized connector with male contacts
Test blocks
Female portion with short-circuiting bar (with terminals shown)
Male portion (with terminals shown)
CONTACT, ELECTRIC
For build-ups or forms using electric contacts, see applications under CONNECTOR
Fixed contact
Fixed contact for jack, key, relay, etc.
OR
OR
Fixed contact for switch
OR
Fixed contact for momentary switch
See SWITCH
Sleeve
OR
OR
Moving contact
Adjustable or sliding contact for resistor, inductor, etc.
OR
Locking
Nonlocking
Closed contact (break)
OR
Open contact (make)
OR
Transfer
OR
Make-before-break
Application: open contact with time closing (TC or TDC) feature
TC OR TDC
Application: closed contact with time opening (TO or TDO) feature
TO OR TDO
Time sequential closing
OR
COUPLER, DIRECTIONAL
Commonly used in coaxial and waveguide diagrams.
The arrows indicate the direction of power flow.
Number of coupling paths, type of coupling, and transmission loss may be indicated.
General

Cont.

ANSI STANDARD GRAPHIC SYMBOLS FOR ELECTRICAL DIAGRAMS (CONT.)

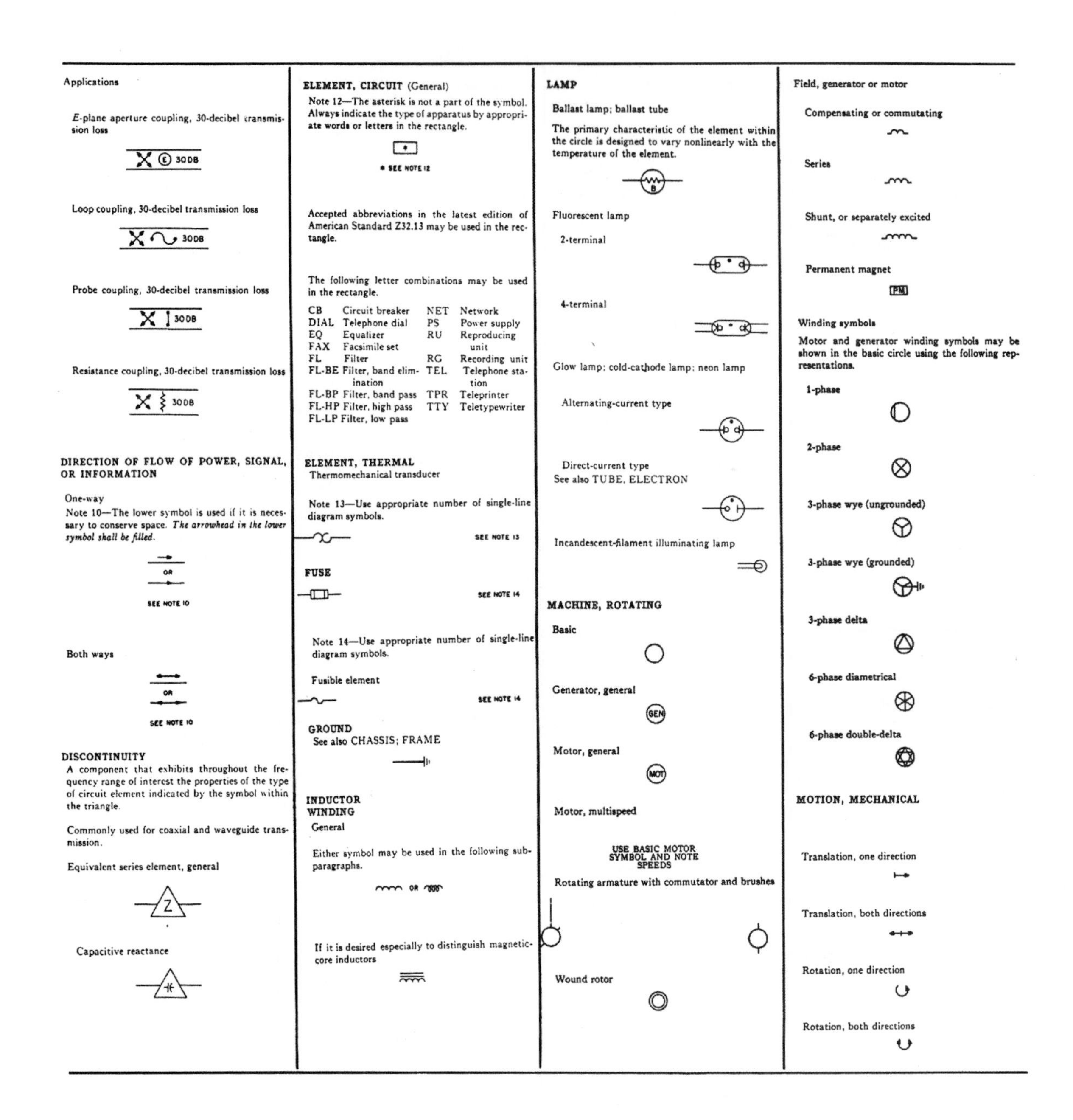
Applications
E-plane aperture coupling, 30-decibel transmission loss
30DB
Loop coupling, 30-decibel transmission loss
30DB
Probe coupling, 30-decibel transmission loss
30DB
Resistance coupling, 30-decibel transmission loss
30DB
DIRECTION OF FLOW OF POWER, SIGNAL, OR INFORMATION
One-way
Note 10—The lower symbol is used if it is necessary to conserve space. The arrowhead in the lower symbol shall be filled.
OR
SEE NOTE 10
Both ways
OR
SEE NOTE 10
DISCONTINUITY
A component that exhibits throughout the frequency range of interest the properties of the type of circuit element indicated by the symbol within the triangle.
Commonly used for coaxial and waveguide transmission.
Equivalent series element, general
Z
Capacitive reactance
ELEMENT, CIRCUIT (General)
Note 12—The asterisk is not a part of the symbol. Always indicate the type of apparatus by appropriate words or letters in the rectangle.
* SEE NOTE 12
Accepted abbreviations in the latest edition of American Standard Z32.13 may be used in the rectangle.
The following letter combinations may be used in the rectangle.
CB Circuit breaker
DIAL Telephone dial
EQ Equalizer
FAX Facsimile set
FL Filter
FL-BE Filter, band elimination
FL-BP Filter, band pass
FL-HP Filter, high pass
FL-LP Filter, low pass
NET Network
PS Power supply
RU Reproducing unit
RG Recording unit
TEL Telephone station
TPR Teleprinter
TTY Teletypewriter
ELEMENT, THERMAL
Thermomechanical transducer
Note 13—Use appropriate number of single-line diagram symbols.
SEE NOTE 13
FUSE
SEE NOTE 14
Note 14—Use appropriate number of single-line diagram symbols.
Fusible element
SEE NOTE 14
GROUND
See also CHASSIS; FRAME
INDUCTOR
WINDING
General
Either symbol may be used in the following subparagraphs.
OR
If it is desired especially to distinguish magnetic-core inductors
LAMP
Ballast lamp; ballast tube
The primary characteristic of the element within the circle is designed to vary nonlinearly with the temperature of the element.
B
Fluorescent lamp
2-terminal
4-terminal
Glow lamp; cold-cathode lamp; neon lamp
Alternating-current type
Direct-current type
See also TUBE, ELECTRON
Incandescent-filament illuminating lamp
MACHINE, ROTATING
Basic
Generator, general
GEN
Motor, general
MOT
Motor, multispeed
USE BASIC MOTOR SYMBOL AND NOTE SPEEDS
Rotating armature with commutator and brushes
Wound rotor
Field, generator or motor
Compensating or commutating
Series
Shunt, or separately excited
Permanent magnet
PM
Winding symbols
Motor and generator winding symbols may be shown in the basic circle using the following representations.
1-phase
2-phase
3-phase wye (ungrounded)
3-phase wye (grounded)
3-phase delta
6-phase diametrical
6-phase double-delta
MOTION, MECHANICAL
Translation, one direction
Translation, both directions
Rotation, one direction
Rotation, both directions

Cont.

ANSI STANDARD GRAPHIC SYMBOLS FOR ELECTRICAL DIAGRAMS (CONT.)

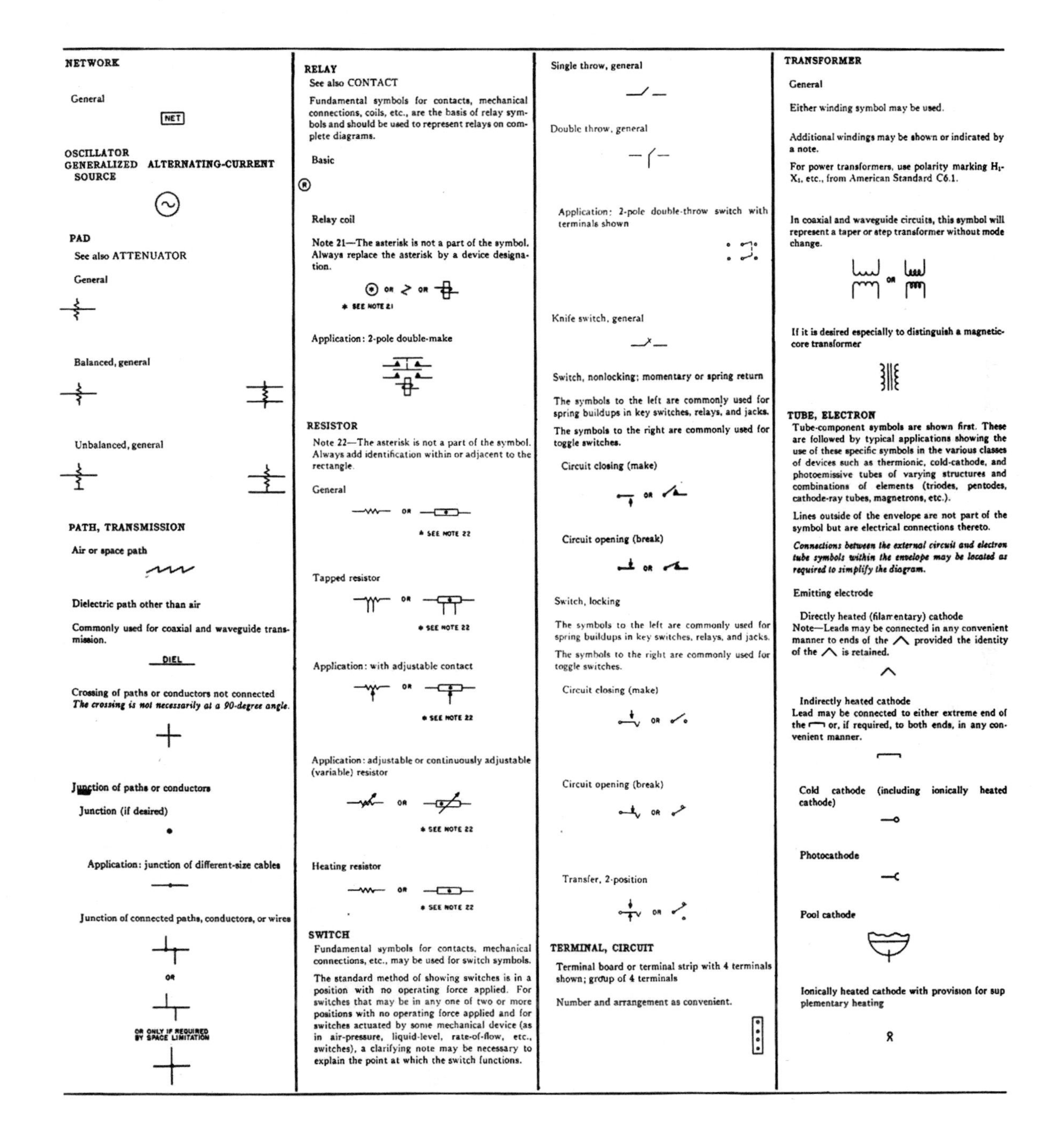

NETWORK

General

NET

OSCILLATOR GENERALIZED ALTERNATING-CURRENT SOURCE

PAD

See also ATTENUATOR

General

Balanced, general

Unbalanced, general

PATH, TRANSMISSION

Air or space path

Dielectric path other than air

Commonly used for coaxial and waveguide transmission.

DIEL

Crossing of paths or conductors not connected

The crossing is not necessarily at a 90-degree angle.

Junction of paths or conductors

Junction (if desired)

Application: junction of different-size cables

Junction of connected paths, conductors, or wires

OR

OR ONLY IF REQUIRED BY SPACE LIMITATION

RELAY

See also CONTACT

Fundamental symbols for contacts, mechanical connections, coils, etc., are the basis of relay symbols and should be used to represent relays on complete diagrams.

Basic

Relay coil

Note 21—The asterisk is not a part of the symbol. Always replace the asterisk by a device designation.

OR OR

* SEE NOTE 21

Application: 2-pole double-make

RESISTOR

Note 22—The asterisk is not a part of the symbol. Always add identification within or adjacent to the rectangle.

General

OR

* SEE NOTE 22

Tapped resistor

OR

* SEE NOTE 22

Application: with adjustable contact

OR

* SEE NOTE 22

Application: adjustable or continuously adjustable (variable) resistor

OR

* SEE NOTE 22

Heating resistor

OR

* SEE NOTE 22

SWITCH

Fundamental symbols for contacts, mechanical connections, etc., may be used for switch symbols.

The standard method of showing switches is in a position with no operating force applied. For switches that may be in any one of two or more positions with no operating force applied and for switches actuated by some mechanical device (as in air-pressure, liquid-level, rate-of-flow, etc., switches), a clarifying note may be necessary to explain the point at which the switch functions.

Single throw, general

Double throw, general

Application: 2-pole double-throw switch with terminals shown

Knife switch, general

Switch, nonlocking; momentary or spring return

The symbols to the left are commonly used for spring buildups in key switches, relays, and jacks.

The symbols to the right are commonly used for toggle switches.

Circuit closing (make)

OR

Circuit opening (break)

OR

Switch, locking

The symbols to the left are commonly used for spring buildups in key switches, relays, and jacks.

The symbols to the right are commonly used for toggle switches.

Circuit closing (make)

OR

Circuit opening (break)

OR

Transfer, 2-position

OR

TERMINAL, CIRCUIT

Terminal board or terminal strip with 4 terminals shown; group of 4 terminals

Number and arrangement as convenient.

TRANSFORMER

General

Either winding symbol may be used.

Additional windings may be shown or indicated by a note.

For power transformers, use polarity marking H_1-X_1, etc., from American Standard C6.1.

In coaxial and waveguide circuits, this symbol will represent a taper or step transformer without mode change.

OR

If it is desired especially to distinguish a magnetic-core transformer

TUBE, ELECTRON

Tube-component symbols are shown first. These are followed by typical applications showing the use of these specific symbols in the various classes of devices such as thermionic, cold-cathode, and photoemissive tubes of varying structures and combinations of elements (triodes, pentodes, cathode-ray tubes, magnetrons, etc.).

Lines outside of the envelope are not part of the symbol but are electrical connections thereto.

Connections between the external circuit and electron tube symbols within the envelope may be located as required to simplify the diagram.

Emitting electrode

Directly heated (filamentary) cathode

Note—Leads may be connected in any convenient manner to ends of the ∧ provided the identity of the ∧ is retained.

Indirectly heated cathode

Lead may be connected to either extreme end of the ⌐¬ or, if required, to both ends, in any convenient manner.

Cold cathode (including ionically heated cathode)

Photocathode

Pool cathode

Ionically heated cathode with provision for supplementary heating

APPENDIX I

SI Units and Conversion Factors

Complete instructions for the use of SI units and conversion factors for other quantities can be found in:

CSA-Z234.1-89	Canadian Metric Practice Guide
ASME S1-1	ASME Orientation Guide for the Use of SI (Metric) Units
ASTM/IEEE SI10-1977	Standard for the Use of the International System of Units (SI)
ISO 1000:1992	SI Units and Recommendations for the Use of Their Multiples and of Certain Other Units plus standards relating to specific industries.

Additional information can be found at:

- www.cssinfo.com
- www.asme.org
- www.ansi.org

BASE UNITS

Unit	Name	Symbol
Length	metre (meter)	m
Mass	kilogram	kg
Time	second	s
Electric current	ampere	A
Thermodynamic temperature	kelvin (not degree kelvin)	K
Amount of substance	mole	mol
Luminous intensity	candela	cd

Derived Units with Special Names

Unit	Name	Symbol	Derived From
Frequency	hertz	Hz	s^{-1}
Force	newton	N	m·kg/s2
Pressure	pascal	Pa	N/m^2
Energy, work, quantity of heat	joule	J	N·m
Quantity of electricity	watt	W	J/s
Celsius temperature	degree Celsius	°C	

SI Prefixes

Prefix	Multiplying Factor	Symbol
exa	10^{18}	E
peta	10^{15}	P
tera	10^{12}	T
giga	10^{9}	G
mega	10^{6}	M
kilo	10^{3}	k
hecto	10^{2}	h
deca	10^{1}	da
deci	10^{-1}	d
centi	10^{-2}	c
milli	10^{-3}	m
micro	10^{-6}	μ
nano	10^{-9}	n
pico	10^{-12}	p
femto	10^{-15}	f
atto	10^{-18}	a

COMMON CONVERSION FACTORS

The values in the following table are rounded.

Item	Conversion Factor	
Mass and Density	1 pound per cubic foot	= 16.018 kg/m^3
	1 pound per gallon (US)	= 99.776 kg/m^3
	1 pound per gallon (UK)	= 119.826 kg/m^3
Energy and Power	1 Btu (mean)	= 1.056 kJ
	1 calorie	= 4.187 J
	1 foot pound force	= 1.356 J
	1 horsepower hour	= 2.685 MJ
	1 horsepower	= 746 W
	1 kilowatt hour	= 3.6 MJ
Force	1 pound force	= 4.448 N
	1 kip (1000 pounds force)	= 4.448 kN
	1 kilogram force	= 9.807 N
Heat		
Conductivity	1 Btu/h.ft.°F	= 1.731 W/m·K
Specific heat	1 Btu/pound mass °F	= 4187 J/kg·K
Length	1 foot	= 0.3048 m
	1 inch	= 25.4 mm (exact)
	1 micron	= 1 μm
	1 mile	= 1.609 km
	1 yard	= 0.9144 m
	1 nautical mile (US)	= 1.852 km
Mass	1 ounce (avoirdupois)	= 28.349 g
	1 pound (avoirdupois)	= 0.454 kg
	1 slug	= 14.594 kg
	1 ton (2000 pounds)	= 0.907 Mg
Pressure (force per unit area)	1 bar	= 100 kPa
	1 inch of water (68°F, 20°C)	= 248.843 Pa
	1 inch of mercury (68°F, 20°C)	= 3.374 kPa
	1 pound force per square inch (psi)	= 6.895 kPa
	1 pound force per square foot	= 47.880 Pa
	1 kip per square inch (ksi)	= 6.895 MPa (exact)

Temperature	Fahrenheit temperature	= 1.8 (Celsius temperature) + 32
Torque	1 foot pound force	= 1.356 N·m
	1 inch pound force	= 0.113 N·m
Velocity	1 foot per minute	= 5.08 mm/s
	1 foot per second	= 304.8 mm/s
	1 mile per hour	= 1.609 km/h
	1 mile per minute	= 26.822 m/s
	1 knot (UK)	= 1.853 km/h
Volume	1 litre (liter)	= 1 dm^3
	1 cubic inch	= 16.387 cm^3
	1 cubic foot	= 28.317 dm^3
	1 gallon (US)	= 3.785 dm^3
	1 gallon (UK)	= 4.546 dm^3
Volume Rate of Flow	1 cubic foot per minute	= 0.472 dm^3/s
	1 cubic foot per second	= 28.317 dm^3/s
	1 gallon (US) per minute	= 63.090 cm^3/s
	1 gallon (UK) per minute	= 75.768 cm^3/s

Index

K

L

M

N

O

P